LIVERPOOL
JOHN MOORES UNIVERSITY
AVRIL ROBARTS LRC
TITHEBARN STREET
LIVERPOOL L2 2ER
TEL. 0151 231 4022

# Rapid Chemical and Biological Techniques for Water Monitoring

# Water Quality Measurements Series

Series Editor

Philippe Quevauviller
*European Commission, Brussels, Belgium*

*Published Titles* in the Water Quality Measurements Series

**Hydrological and Limnological Aspects of Lake Monitoring**
Edited by Pertti Heinonen, Giuliano Ziglio and Andre Van der Beken

**Quality Assurance for Water Analysis**
Edited by Philippe Quevauviller

**Detection Methods for Algae, Protozoa and Helminths in Fresh and Drinking Water**
Edited by Andre Van der Beken, Giuliano Ziglio and Franca Palumbo

**Analytical Methods for Drinking Water: Advances in Sampling and Analysis**
Edited by Philippe Quevauviller

**Biological Monitoring of Rivers: Applications and Perspectives**
Edited by Giuliano Ziglio, Maurizio Siligardi and Giovanna Flaim

**Wastewater Quality Monitoring and Treatment**
Edited by Philippe Quevauviller, Olivier Thomas and Andre Van der Berken

**The Water Framework Directive – Ecological and Chemical Status Monitoring**
Edited by Philippe Quevauviller, Ulrich Borchers, Clive Thompson and Tristan Simonart

**Rapid Chemical and Biological Techniques for Water Monitoring**
Edited by Catherine Gonzalez, Richard Greenwood and Philippe Quevauviller,

*Forthcoming Titles* in the Water Quality Measurements Series

**Groundwater Monitoring**
Philippe Quevauviller, A M Fouillac, D J Grath and R Ward

**Chemical Marine Monitoring: Policy Framework and Analytical Trends**
Philippe Quevauviller, Patrick Roose and Gert Vereet

# Rapid Chemical and Biological Techniques for Water Monitoring

**CATHERINE GONZALEZ**
*Ecole des Mines d'Ales, Ales Cedex, France*

**RICHARD GREENWOOD**
*University of Portsmouth, Portsmouth, UK*

**PHILIPPE QUEVAUVILLER**
*European Commission, Brussels, Belgium*

A John Wiley and Sons, Ltd., Publication

This edition first published 2009
© 2009, John Wiley & Sons Ltd.,

*Registered office*
John Wiley & Sons Ltd, The Atrium, Southern Gate, Chichester, West Sussex, PO19 8SQ, United Kingdom

For details of our global editorial offices, for customer services and for information about how to apply for permission to reuse the copyright material in this book please see our website at www.wiley.com.

The right of the author to be identified as the author of this work has been asserted in accordance with the Copyright, Designs and Patents Act 1988.

All rights reserved. No part of this publication may be reproduced, stored in a retrieval system, or transmitted, in any form or by any means, electronic, mechanical, photocopying, recording or otherwise, except as permitted by the UK Copyright, Designs and Patents Act 1988, without the prior permission of the publisher.

Wiley also publishes its books in a variety of electronic formats. Some content that appears in print may not be available in electronic books.

Designations used by companies to distinguish their products are often claimed as trademarks. All brand names and product names used in this book are trade names, service marks, trademarks or registered trademarks of their respective owners. The publisher is not associated with any product or vendor mentioned in this book. This publication is designed to provide accurate and authoritative information in regard to the subject matter covered. It is sold on the understanding that the publisher is not engaged in rendering professional services. If professional advice or other expert assistance is required, the services of a competent professional should be sought.

The publisher and the author make no representations or warranties with respect to the accuracy or completeness of the contents of this work and specifically disclaim all warranties, including without limitation any implied warranties of fitness for a particular purpose. This work is sold with the understanding that the publisher is not engaged in rendering professional services. The advice and strategies contained herein may not be suitable for every situation. In view of ongoing research, equipment modifications, changes in governmental regulations, and the constant flow of information relating to the use of experimental reagents, equipment, and devices, the reader is urged to review and evaluate the information provided in the package insert or instructions for each chemical, piece of equipment, reagent, or device for, among other things, any changes in the instructions or indication of usage and for added warnings and precautions. The fact that an organization or Website is referred to in this work as a citation and/or a potential source of further information does not mean that the author or the publisher endorses the information the organization or Website may provide or recommendations it may make. Further, readers should be aware that Internet Websites listed in this work may have changed or disappeared between when this work was written and when it is read. No warranty may be created or extended by any promotional statements for this work. Neither the publisher nor the author shall be liable for any damages arising herefrom.

*Library of Congress Cataloging-in-Publication Data*

Gonzalez, Catherine.
  Rapid chemical and biological techniques for water monitoring/Catherine Gonzalez, Richard Greenwood, Philippe Quevauviller.
      p. cm.--(Water quality measurements series)
  Includes bibliographical references and index.
  ISBN 978-0-470-05811-4 (cloth)
 1. Water quality bioassay. 2. Environmental monitoring. 3. Water–Pollution. 4. Water–Microbiology. 5. Water chemistry. 6. Chemistry, Analytic. I. Greenwood, R. (Richard) II. Quevauviller, Ph. III. Title.
 TD367.G66 2009
 628.1'61--dc22
                                                    2008052053

A catalogue record for this book is available from the British Library.

ISBN    978-0-470-05811-4

Typeset in 10/12pt Times by Laserwords Private Limited, Chennai, India.
Printed and bound in Great Britain by TJ International Ltd, Padstow, Cornwall.

# Contents

| | | |
|---|---|---|
| **Series Preface** | | ix |
| **Preface** | | xi |
| **The Series Editor – Philippe Quevauviller** | | xiii |
| **List of Contributors** | | xv |
| **Section 1** | **Screening Methods in the Context of Water Policies** | 1 |
| 1.1 | WFD Monitoring and Metrological Implications<br>*Philippe Quevauviller* | 3 |
| 1.2 | Use of Screening Methods in US Water Regulation<br>*Guillaume Junqua, Estelle Baurès, Emmanuelle Hélias and Olivier Thomas* | 15 |
| 1.3 | Existing and New Methods for Chemical and Ecological Status Monitoring under the WFD<br>*Benoit Roig, Ian Allan, Graham A. Mills, Nathalie Guigues, Richard Greenwood and Catherine Gonzalez* | 39 |
| **Section 2** | **Chemical Methods** | 51 |
| 2.1 | The Potential of Passive Sampling to Support Regulatory Monitoring of the Chemical Quality of Environmental Waters<br>*Graham A. Mills, Branislav Vrana and Richard Greenwood* | 53 |
| 2.2 | Polar Organic Chemical Integrative Sampler and Semi-permeable Membrane Devices<br>*David Alvarez and Audrone Simule* | 71 |
| 2.3 | Main Existing Methods for Chemical Monitoring<br>*Guillaume Junqua, Catherine Gonzalez and Evelyne Touraud* | 79 |

| | | |
|---|---|---|
| 2.4 | UV Spectrophotometry: Environmental Monitoring Solutions<br>*Daniel Constant, Catherine Gonzalez, Evelyne Touraud, Nathalie Guigues and Olivier Thomas* | 91 |

**Section 3   Biological Methods**  107

| | | |
|---|---|---|
| 3.1 | Application of Microbial Assay for Risk Assessment (MARA) to Evaluate Toxicity of Chemicals and Environmental Samples<br>*Kirit Wadhia and K. Clive Thompson* | 109 |
| 3.2 | Bioassays and Biosensors<br>*Marinella Farré and Damia Barceló* | 125 |
| 3.3 | Immunochemical Methods<br>*Petra M. Krämer* | 157 |
| 3.4 | Biomolecular Recognition Systems for Water Monitoring<br>*Benoit Roig, Ingrid Bazin, Sandrine Bayle, Denis Habauzit and Joel Chopineau* | 175 |
| 3.5 | Continuous Monitoring of Waters by Biological Early Warning Systems<br>*Kees J.M. Kramer* | 197 |
| 3.6 | Biological Markers of Exposure and Effect for Water Pollution Monitoring<br>*Josephine A. Hagger and Tamara S. Galloway* | 221 |

**Section 4   Potential Use of Screening Methods and Performance Evaluation**  241

| | | |
|---|---|---|
| 4.1 | Monitoring Heavy Metals Using Passive Sampling Devices<br>*Graham A. Mills, Ian J. Allan, Nathalie Guigues, Jesper Knutsson, A. Holmberg and Richard Greenwood* | 243 |
| 4.2 | On-site Heavy Metal Monitoring Using a Portable Screen-printed Electrode Sensor<br>*Catherine Berho, Nathalie Guigues, Jean-Philippe Ghestem, Catherine Crouzet, Anne Strugeon, Stéphane Roy and Anne-Marie Fouillac* | 263 |
| 4.3 | Field Monitoring of PAHs in River Water by Direct Fluorimetry on $C_{18}$ Solid Sorbent<br>*Guillaume Bernier and Michel Lamotte* | 275 |

Contents vii

4.4 Evaluation of the Field Performance of Emerging
    Water Quality Monitoring Tools 287
    *Catherine Berho, Nathalie Guigues, Anne Togola, Stéphane Roy,
    Anne-Marie Fouillac, Ian Allan, Graham A. Mills, Richard Greenwood,
    Benoît Roig, Charlotte Valat and Nirit Ulitzur*

4.5 Sampling Uncertainty and Environmental Variability for
    Trace Elements on the Meuse River, France 303
    *Anne Strugeon-Dercourt*

**Section 5  Quality Assurance and Validation Method** 333

5.1 Preparation of Reference Materials for Proficiency
    Testing Schemes 335
    *Angels Sahuquillo, Marina Ricci, Ofelia Bercaru, Hakan Emteborg, Franz
    Ulberth, Roberto Morabito, Claudia Brunori, Yolanda Madrid, Erwin
    Rosenberg, Klara Polyak and Herbert Muntau*

5.2 Participation of Screening Methods and Emerging Tools (SMETs) to Proficiency Testing Schemes on the Determination of Priority Substances in Real Water Matrices Organized in Support of the Water Framework Directive Implementation 351
    *Claudia Brunori, Ildi Ipolyi and Roberto Morabito*

5.3 Traceability and Interlaboratory Studies on Yeast-based
    Assays for the Determination of Estrogenicity 371
    *Rikke Brix and Damià Barceló*

**Section 6  Integration of Screening Methods in Water Monitoring
            Strategies** 383

6.1 Assessing the Impacts of Alternative Monitoring Methods and Tools on Costs and Decision Making: Methodology and Experience from Case Studies 385
    *Helen Lückge, Pierre Strosser, Nina Graveline, Thomas Dworak
    and Jean-Daniel Rinaudo*

6.2 Acceptance of Screening Methods by Actors Involved in
    Water Monitoring 397
    *Didier Taverne*

**Index** 405

# Series Preface

Water is a fundamental constituent of life and is essential to a wide range of economic activities. It is also a limited resource, as we are frequently reminded by the tragic effects of drought in certain parts of the world. Even in areas with high precipitation, and in major river basins, overuse and mismanagement of water have created severe constraints on availability. Such problems are widespread and will be made more acute by the accelerating demand on freshwater arising from trends in economic development.

Despite the fact that water-resource management is essentially a local, river-basin based activity, there are a number of areas of action that are relevant to all or significant parts of the European Union and for which it is advisable to pool efforts for the purpose of understanding relevant phenomena (e.g. pollutions, geochemical studies), developing technical solutions and/or defining management procedures. One of the keys for successful cooperation aimed at studying hydrology, water monitoring, biological activities etc. is to achieve and ensure good water quality measurements.

Quality measurements are essential to demonstrate the comparability of data obtained worldwide and they form the basis for correct decisions related to management of water resources, monitoring issues, biological quality etc. Besides the necessary quality control tools developed for various types of physical, chemical and biological measurements, there is a strong need for education and training related to water quality measurements. This need has been recognized by the European Commission which has funded a series of training courses on this topic, covering aspects such as monitoring and measurements of lake recipients, measurements of heavy metals and organic compounds in drinking and surface water, use of biotic indexes, and methods to analyse algae, protozoa and helminths. In addition, a series of research and development projects have been or are being developed.

This book series will ensure a wide coverage of issues related to water quality measurements, including the topics of the above mentioned courses and the outcome of recent scientific advances. In addition, other aspects related to quality control tools (e.g. certified reference materials for the quality control of water analysis) and monitoring of various types of waters (river, wastewater, groundwater) will also be considered.

*Rapid Chemical and Biological Techniques for Water Monitoring* is the eighth book of the series; it has been written by experts in water analyses, including classical and emerging techniques, and offers the reader an overview of existing knowledge and trends in monitoring based on rapid biological and chemical techniques.

**The Series Editor – Philippe Quevauviller**

# Preface

In both the USA and Europe there have been developments of legislation to protect the aquatic environment, including surface waters, ground water, the coastal zone, and wetlands. Effective monitoring of water quality is essential to underpin the legislative frameworks. It is particularly difficult to assess the quality of water bodies where levels of the pollutants can fluctuate in time, and can vary spatially in a water body depending on the nature of pressures present. The successful implementation of strategies to improve the quality of the aquatic environment will depend on the availability and quality of information needed by those charged with managing water quality. There is an urgent need for the development and validation of cost-effective technologies and methodologies that can be adopted widely for the routine monitoring of wastewaters and receiving surface waters. Spot sampling provides only a snapshot of the situation at the instant of sampling, and fluctuations associated with episodic events could be missed, or conclusions could be drawn on the basis of transitory high levels. The cost of incorrect information could be very high and there is, therefore, a need for improved screening methodologies that can provide a complimentary approach to existing quality monitoring systems. However, monitoring tools will be useful only if they are affordable, reliable and produce data that are of comparable quality between times and locations.

The range of promising tools for inclusion in the toolboxes of those charged with managing water quality is expanding, and includes well tried methods such as biomonitoring, online monitoring systems (e.g. the SAMOS system), and biological early warning systems. Commercially available solutions with lower cost implications and greater flexibility than the fixed installations, such as those above, are provided by field test kits for specific pollutants, portable toxicological assay equipment, a wide range of sensors, direct toxicological assays, and passive samplers. Many of these existing methods and ones under development have the potential to be included in the set of useful tools in the toolbox available to those responsible for monitoring and improving water quality under the various legislative frameworks. This volume sets out to examine this range of technologies and methodologies, their properties and their applicability and potential contribution in monitoring programmes. Where possible the utility of the monitoring tools is illustrated by examples of laboratory and field applications.

This textbook brings together a wide range of monitoring tools (both those available, and some under development) and provides an assessment of their potential for underpinning environmental management and legislation. There are major ongoing developments in the latter at a European level and in the USA, and this volume attempts to provide a source of information for those involved in water management at all levels.

*Catherine Gonzalez Richard Greenwood Philippe Quevauviller*

# The Series Editor
# Philippe Quevauviller

Philippe Quevauviller began his research activities in 1983 at the University of Bordeaux I, France, studying lake geochemistry. Between 1984 and 1987, he was Associate Researcher at the Portuguese Environment State Secretary where he performed a multidisciplinary study (sedimentology, geomorphology and geochemistry) of the coastal environment of the Galé coastline and of the Sado Estuary, which was the topic of his PhD degree in oceanography gained in 1987 (at the University of Bordeaux I). In 1988, he became Associate Researcher in the framework of a contract between the University of Bordeaux I and the Dutch Ministry for Public Works (Rijskwaterstaat), in which he investigated organotin contamination levels of Dutch coastal environments and waterways. From this research work, he gained another PhD in chemistry at the University of Bordeaux I in 1990. From 1989 to 2002, he worked at the European Commission (DG Research) in Brussels where he managed various Research and Technological Development (RTD) projects in the field of quality assurance and analytical method development for environmental analyses in the framework of the Standards, Measurements and Testing Programme. In 1999, he obtained an HDR (Diplôme d'Habilitation à Diriger des Recherches) in chemistry at the University of Pau, France, from a study of the quality assurance of chemical species determination in the environment.

In 2002, he left the research world to move to the policy sector at the EC Environment Directorate-General where he developed a new EU Directive on groundwater protection against pollution and chaired European science-policy expert groups on groundwater and chemical monitoring in support of the implementation of the EU Water Framework Directive. Since 2008, he is back to the EC DG Research where he is managing research projects on climate change impacts on the aquatic environment, while ensuring strong links with policy networks.

Philippe Quevauviller has published (as author and co-author) more than 220 scientific and policy publications, 80 reports and 6 books for the European Commission and has acted as an editor and co-editor for 22 special issues of scientific journals and 10 books. Finally, he is Associate Professor at the Free University of Brussels and promoter of Master theses in an international Master on water engineering (IUPWARE programme), and he also teaches integrated water management issues and their links to EU water science and policies to Master students at the Universities of Paris 7, Polytech'Lille and Polytech'Nice (France).

# List of Contributors

| | |
|---|---|
| **Ian Allan** | Norwegian Institute for Water Research (NIVA) Gaustadalleen 21, NO-0349, Oslo, Norway |
| **David Alvarez** | US Geological Survey, Columbia Environmental Research Center, 4200 New Haven Road, Columbia, MO 65201, USA |
| **Damia Barceló** | IIQAB-CSIC, Department of Environmental Chemistry, Jordi Girona 18–26, 08034 Barcelona, Spain |
| | Catalan Institute for Water Research (ICRA), Parc Científic i Tecnològic de la Universitat de Girona, Edifici Jaume Casademont, C/ Pic de Peguera, 15, 17003 Girona, Spain |
| **Estelle Baurès** | Université de Sherbrooke, FLSH, local 1020, Sherbrooke (Québec), J1K 2R1 Canada |
| **Sandrine Bayle** | Ecole des Mines d'Alès, 6 avenue de Clavières, 30319 Alès Cedex, France |
| **Ingrid Bazin** | Ecole des Mines d'Alès, 6 avenue de Clavières, 30319 Alès Cedex, France |
| **Ofelia Bercaru** | European Commission – Joint Research Centre (JRC), Institute for Reference Materials and Measurements (IRMM), Reference Materials Unit, Retieseweg 111–2440 Geel, Belgium |
| **Guillaume Bernier** | Laboratoire de Physico et Toxicochimie de l'Environnement; I.S.M.; Université de Bordeaux 1, 351 cours de la Libération F-33405 Talence, France |
| **Rikke Brix** | IIQAB-CSIC, Department of Environmental Chemistry, Jordi Girona 18–26, 08034 Barcelona, Spain |
| **Claudia Brunori** | ENEA – PROT, Via Anguillarese, 301, 00060 Rome, Italy |
| **Joel Chopineau** | Ecole des Mines d'Alès, 6 avenue de Clavières, 30319 Alès Cedex, France |
| **Daniel Constant** | SECOMAM, 91, Avenue des Pins d'Alep, 30319 Alès cedex, France |

| | |
|---|---|
| **Catherine Crouzet** | BRGM, Service Métrologie, Monitoring, Analyse, 3 avenue Claude Guillemin – BP 36009, 45060 ORLEANS Cedex 2, France |
| **Thomas Dworak** | Ecologic Institute for International and European Environmental Policy, Pfalzburger Strasse 43/44, 10717 Berlin, Germany |
| **Hakan Emteborg** | European Commission – Joint Research Centre (JRC), Institute for Reference Materials and Measurements (IRMM), Reference Materials Unit, Retieseweg 111–2440 Geel, Belgium |
| **Marinella Farré** | IIQAB-CSIC, Department of Environmental Chemistry, Jordi Girona 18–26, 08034 Barcelona, Spain |
| **Anne-Marie Fouillac** | BRGM, Service Métrologie, Monitoring, Analyse, 3 avenue Claude Guillemin – BP 6009, 45060 ORLEANS Cedex 2, France |
| **Tamara Galloway** | University of Exeter, School of Biosciences, Hatherly Laboratories, Prince of Wales Road, Exeter, UK EX4 4PS, UK |
| **Catherine Gonzalez** | Ecole des Mines d'Alès, 6 avenue de Clavières, 30319 Alès Cedex, France |
| **Nina Graveline** | BRGM, Water Division, 1034 rue de Pinville, 34000 Montpellier, France |
| **Richard Greenwood** | University of Portsmouth, Dept. Biological Sciences, King Henry Building, King Henry I Street, Portsmouth PO1 2DY, UK |
| **Nathalie Guigues** | BRGM, Service Métrologie, Monitoring, Analyse, 3 avenue Claude Guillemin – BP 6009, 45060 ORLEANS Cedex 2, France |
| **Denis Habauzit** | Ecole des Mines d'Alès, 6 avenue de Clavières, 30319 Alès Cedex, France |
| **Josephine Hagger** | University of Exeter, Prince of Wales Road, Exeter, Devon, EX4 4PS, UK |
| **Emmanuelle Hélias** | Université de Sherbrooke, FLSH, local 1020, Sherbrooke (Québec), J1K 2R1 Canada |
| **Arne Holmberg** | Alcontrol Sweden, Box 1083, 581 10 Linköping, Sweden |
| **Ildi Ipolyi** | QualityConsult, Via G. Bettolo 4, 00195 Rome, Italy |
| **Guillaume Junqua** | Ecole des Mines d'Alès, 6 avenue de Clavières, 30319 Alès Cedex, France |
| **Jesper Knutsson** | Chalmers University of Technology, Water Environment Transport, SE-412 96 Göteborg, Sweden |
| **Kees Kramer** | Mermayde, P.O. Box 109, NL-1860 AC Bergen, The Netherlands |

# List of Contributors

| | |
|---|---|
| **Petra Krämer** | Institute of Ecological Chemistry, Helmholtz Zentrum München, German Research Center for Environmental Health (GmbH) Ingolstädter Landstr. 1 85764 Neuherberg, Germany |
| **Michel Lamotte** | 3, allée de la Charmille, 33700 Mérignac, France |
| **Helen Lückge** | Ecologic Institute for International and European Environmental Policy, Pfalzburger Strasse 43/44, 10717 Berlin, Germany |
| **Yolanda Madrid** | Departamento de Química Analítica, Universidad Complutense de Madrid, Spain |
| **Graham Mills** | University of Portsmouth, School of Pharmacy and Biomedical Sciences, St Michaels Building, White Swan Road, Portsmouth PO1 2DT, UK |
| **Roberto Morabito** | ENEA – PROT, Via Anguillarese, 301, 00060 Rome, Italy |
| **Herbert Muntau** | Lithos Geoscience. Via Castello 18, 21020, Ranco (Va), Italy |
| **Klara Polyak** | University of Vezsprem, Hungary |
| **Philippe Quevauviller** | European Commission, DG Research (CDMA 03/07), rue de la Loi 200, B-1049 Brussels, Belgium |
| | Vrije Universiteit Brussel (VUB), Dept. Water Engineering, Bd. du Triomphe, B-1060 Brussels, Belgium |
| **Marina Ricci** | European Commission – Joint Research Centre (JRC), Institute for Reference Materials and Measurements (IRMM), Reference Materials Unit, Retieseweg 111, B-2440 Geel, Belgium |
| **Jean-Daniel Rinaudo** | BRGM, Water Division, 1034 rue de Pinville, 34000 Montpellier, France |
| **Benoit Roig** | Ecole des Mines d'Alès, 6 avenue de Clavières, 30319 Alès Cedex, France |
| **Erwin Rosenberg** | Technische Universität Wien, Institut für Chemische Technologien und Analytik, Getreidemarkt 9/164, 1060 Wien, Austria |
| **Stéphane Roy** | BRGM, Service Métrologie, Monitoring, Analyse, 3 avenue Claude Guillemin - BP 6009, 45060 ORLEANS Cedex 2, France |
| **Angels Sahuquillo** | University of Barcelona, Department Quimica Analytica, Av. Diagonal, 647, 3$^a$ pl, Barcelona, Spain |
| **Pierre Strosser** | ACTeon, Innovation, Policy, Environment, Le Chalimont - B.P. Ferme du Pré du Bois, 68370 Orbey, France |
| **Anne Strugeon** | BRGM, Service Métrologie, Monitoring, Analyse, 3 avenue Claude Guillemin - BP 6009, 45060 ORLEANS Cedex 2, France |

| | |
|---|---|
| **Didier Taverne** | Sciences, Territoires et Sociétés, 42 avenue Saint Lazare, n°78, F-34000 Montpellier, France |
| **Olivier Thomas** | Université de Sherbrooke, FLSH, local 1020, Sherbrooke (Québec), J1K 2R1 Canada |
| **Clive K. Thompson** | ALcontrol Laboratories, Templeborough House, Mill Close, Rotherham, South Yorkshire, S60 1BZ, UK |
| **Anne Togola** | BRGM, 3, avenue Claude Guillemin BP 6009, 45060 Orléans, Cédex, France |
| **Evelyne Touraud** | Ecole des Mines d'Alès, 6 avenue de Clavières, 30319 Alès Cedex, France |
| **Franz Ulberth** | European Commission – Joint Research Centre (JRC), Institute for Reference Materials and Measurements (IRMM), Reference Materials Unit, Retieseweg 111–2440 Geel, Belgium |
| **Nirit Ulitzur** | CheckLight, PO Box 72, Qiryat Tivon, 36000, Israel |
| **Charlotte Valat** | AFSSA (Agence Française de Sécurité Sanitaire des Aliments), Zoopôle des Côtes d'Armor, Site de Beaucemaine, BP 53, F-22440 Ploufragan, France |
| **Branislav Vrana** | Vyskumny ustav vodneho hospodarstva, Nabr. arm. gen., L. Svobodu 7, 812 49 Bratislava, Slovakia |
| **Kirit Wadhia** | ALcontrol Laboratories, Templeborough House, Mill Close, Rotherham, South Yorkshire, S60 1BZ, UK |
| **Audrone Simule** | ExposMeter AB, Nygatan 15, SE-702 11 Orebro, Sweden |

# Section I
## Screening Methods in the Context of Water Policies

# 1.1
## WFD Monitoring and Metrological Implications

Philippe Quevauviller*

1.1.1 Introduction
1.1.2 Main Legal Requirements
1.1.3 Non-Legally Binding Recommendations
1.1.4 Further Binding Rules
1.1.5 WFD Monitoring and Its Links with Metrology
    1.1.5.1 Introduction
    1.1.5.2 Basic References: the SI Units
    1.1.5.3 Standardized Methods in the Context of the WFD
    1.1.5.4 Role of Reference Methods
    1.1.5.5 Role of Reference Materials
    1.1.5.6 Proficiency Testing Schemes
    1.1.5.7 Laboratory Accreditation
1.1.6 Conclusions, Perspectives
References

## 1.1.1 INTRODUCTION

The Water Framework Directive (WFD) is certainly the first EU legislative instrument which requires a systematic monitoring of biological, chemical and quantitative parameters in European waters on such a wide geographical scale (European Commission, 2000). The principles are fixed in the legislative text and exchanges of information

---

*The views expressed in this chapter are purely those of the author and may not in any circumstances be regarded as stating an official position of the European Commission.

*Rapid Chemical and Biological Techniques for Water Monitoring* Edited by Catherine Gonzalez, Philippe Quevauviller and Richard Greenwood
© 2009 John Wiley & Sons, Ltd

among experts have enabled them to set out a common understanding of monitoring requirements in the form of guidance documents (European Commission, 2003a, 2006, 2009). While water monitoring is obviously not a new feature, it should be noted that the WFD monitoring programmes are in their infancy in that they had to be designed and reported by the Member States only in March 2007.

Monitoring data produced in 2007–2008 under the WFD will form the basis of the design of programmes of measures to be included in the first River Basin Management Plan (due to be published in 2009), and thereafter used for evaluating the efficiency of these measures. Monitoring data will hence obviously be used as the basis for classifying the water status, and they will also be used to identify possible pollution trends. This is an iterative process in that better monitoring will ensure a better design and follow-up of the impacts of remedial measures, a better status classification and a timely identification of trends (calling for reversal measures) which puts a clear accent on the needs for constant improvements and regular reviews (foreseen under the WFD) and hence on the need to integrate scientific progress in an efficient way.

Monitoring goes hand in hand with metrology. The 'science of measurements' indeed puts a clear accent on issues of data quality and comparability (including traceability aspects) which are covered, albeit in a general way, in the technical specifications of the WFD. This chapter focuses on the chemical monitoring requirements of the WFD (noting that the directive also requires ecological status monitoring for surface waters and quantitative status monitoring for ground waters), and highlights recently discussed metrological features linked to WFD monitoring (Quevauviller, 2007) that are presently under consideration in European expert groups discussing specific implementation issues.

## 1.1.2 MAIN LEGAL REQUIREMENTS

The Water Framework Directive establishes 'good status' objectives to be achieved for all waters by the end of 2015. With regard to surface waters, good status criteria are based on biological parameters (ecological status) and chemistry (chemical status linked to compliance to EU Environmental Quality Standards), while for ground waters, good status refers to quantitative levels (balance between recharge and abstraction) and chemistry (linked to compliance to groundwater quality standards established at EU, national, regional or local levels). Monitoring requirements linked to these status objectives are found in Annex V of the directive.

EU Member States had to design monitoring programmes before the end of 2006 and report them to the European Commission in March 2007. Basic requirements are that monitoring data have to provide a reliable assessment of status of all water bodies or groups of bodies (administrative units defined by Member States). This implies that networks have to consider the representativeness of monitoring points as well as frequency. In addition, monitoring has to be designed in such a way that long-term pollution trends may be detected.

Details about WFD monitoring programmes and examples linked to current practices and research projects will be soon published in a forthcoming volume of the Wiley's Water Quality Measurements Series (Quevauviller *et al*., 2008).

## 1.1.3 NON-LEGALLY BINDING RECOMMENDATIONS

The technical challenges of the WFD have led Member States to request the European Commission to launch a 'Common Implementation Strategy' (CIS), which aims to exchange knowledge and best practices, and to develop guidance documents. in a coordinated fashion. The CIS process is recognized to be a powerful consultative process and an excellent example of governance at EU level (European Commission, 2003b). It comprises expert groups discussing various topics, including a Chemical Monitoring Activity (CMA), which is the object of this paragraph.

The Chemical Monitoring Activity is described elsewhere (Quevauviller, 2006). Along the CIS principles, the expert group (composed of ca. 60 experts from Member States environment agencies or ministries, stakeholder's associations and the scientific community) has discussed various chemical monitoring features concerning surface and ground waters, which resulted in recommendations summarized in two guidance documents (European Commission, 2006, 2009). Issues such as sampling design and representativeness, sampling and analytical methods are discussed in some detail in order to provide a common base for Member States when dealing with chemical monitoring programmes. These recommendations are directly relevant to metrology and provide background rules that should help the development of harmonized approaches in Europe. It is important to note that these guidance documents are not legally binding, i.e. they provide general rules that may be adapted to local or regional situations without requiring a necessary enforcement. In other words, the enforceable obligations are those found in Article 8 and Annex V of the WFD, and these are explained with technical details in guidance documents that are themselves not legally binding. Despite this, the guidance documents will have an obvious impact on the way Member States will perform chemical monitoring programmes in the future.

Guidance recommendations for surface and groundwater monitoring stipulate that, in general, performance-based methods shall be used in surveillance and operational monitoring. They shall be described clearly, properly validated and give the laboratories the flexibility to select from several options when possible and meaningful. It should be noted that, while WFD monitoring recommendations are mostly focused on 'classical' methods (based on sample collection, treatment and laboratory analysis of water, sediment or biota), i.e. currently based on chemical analysis of spot samples taken in a defined frequency, the recognition of the usefulness of other techniques has been highlighted in the Guidance document on surface water monitoring (European Commission, 2009). In particular, they include current advanced methods for environmental assessment such as e.g. in-situ probes for measuring physico-chemical characteristics (e.g. Dissolved Organic Carbon (DOC), pH, temperature, oxygen), biological assessment techniques (e.g. biomarker analysis, bioassays/biosensors and biological early warning systems, immunosensors, etc.), sampling and chemical analytical methods (e.g. sensors, passive sampling devices, test kits, and on-site GC-MS or LC-MS screening methodologies). These techniques are exemplified in several chapters of this book.

Coming back to metrology, the demonstration of data comparability, which should be ensured at EU level, was considered to be the most important aspect of monitoring within the WFD. In this case only, technical specifications were deemed to be necessary in an enforceable way. This is discussed in the following paragraph.

## 1.1.4 FURTHER BINDING RULES

As discussed above, the overall water management and decision-making system of the WFD is closely dependent upon monitoring data. Without clear rules for demonstrating the accuracy (hence the comparability) of data, the system cannot deliver a sound basis for proper decisions. This argument led to open discussions about legally binding rules concerning minimum performance criteria for analytical monitoring methods. The legal background is the paragraph 3 of Article 8 of the WFD, which opens the possibility of developing technical specifications that support monitoring programmes. In this context, a draft Commission Directive 'adopting technical specifications for chemical monitoring in accordance with Directive 2000/60/EC of the European Parliament and the Council' has been developed by the CMA expert group. The proposal includes mandatory requirements for Member State's laboratories regarding the validation of methods, minimum performance criteria (linked to target uncertainty and limit of quantification linked to environmental quality standard values), and participation in quality assurance programmes (including proficiency testing schemes, analyses of reference materials, training). This text has been adopted by Member States under comitology (i.e. by a regulatory committee composed of Member State's representatives, with a consultation of the European Parliament) in Spring 2008 and was adopted by the European Commission in early 2009. With this background, all elements are met to develop a metrological framework that will ensure that chemical monitoring data produced under the WFD are comparable, i.e. traceable to commonly accepted references. The achievement of traceability and related references in the context of the WFD monitoring programmes are discussed below.

## 1.1.5 WFD MONITORING AND ITS LINKS WITH METROLOGY

### 1.1.5.1 Introduction

The above sections have highlighted the importance of data comparability and traceability in the context of WFD chemical monitoring. Let us now examine in detail what references need to be considered for the development of a sound metrological system. Firstly, as a reminder, traceability is defined as 'the property of the result of a measurement or the value of a standard whereby it can be related to stated references, usually national or international standards, through an unbroken chain of comparisons all having stated uncertainties' (ISO, 1993). The ways in which these elements can be applied to chemical measurements were discussed some years ago (Válcarcel and Rios, 1999; Quevauviller, 1999; Walsh, 2000) and those discussions still continue. In this context the basic references are those of the SI (Système International) units, i.e. the kg or mole for chemical measurements. Establishing SI traceability of chemical measurements may, in principle, be achieved in relation to either a reference material or to a reference method (Quevauviller and Donard, 2001). The unbroken chain of comparison implies that no loss of information should occur during the analytical procedure (e.g. incomplete recovery or contamination). Finally, traceability implies, in theory, that the uncertainty of all stated references that contribute to the measurement is duly considered (meaning that the smaller the chain of comparison the smaller the uncertainty of the final result), and it is clear that this is hardly applicable to complex environmental measurements and to sampling and sample pretreatment steps (Quevauviller, 2004a).

As an additional remark, it should be noted that traceability should not be confused with accuracy which covers the terms trueness (the closeness of agreement of the measured value with the 'true value') and precision (the closeness of agreement between results obtained by applying the same experimental procedure several times under prescribed conditions). In other words, a method which is traceable to a stated reference is not necessarily accurate (i.e. the stated reference does not necessarily correspond to the 'true value'), whereas an accurate method is always traceable to what is considered to be the best approximation of the true value (defined as 'a value which would be obtained by measurement if the quantity could be completely defined if all measurement imperfections could be eliminated').

Chemical measurements in the context of the WFD (including water, sediment and biota) are based on a succession of actions, namely:

- sampling, storage and preservation of representative samples;
- pretreatment of a sample portion for quantitation;
- calibration;
- final determination; and,
- calculation and presentation of results.

Starting from this, let us examine how traceability may be understood in the context of the WFD chemical monitoring programme.

### 1.1.5.2 Basic References: the SI Units

The unit that underpins chemical measurements is the unit of amount of substance, the mole. However, in practice, as there is no 'mole' standard, the kg is used, i.e. chemical measurements are actually traceable to the mass unit, the kg. In other words, water-related chemical measurements are based on the determination of amount of substance per mass of matrix. For solid matrices (sediment, suspended matters, biota), these are units corresponding to ultratrace (ng/kg) and trace (µg/kg) concentrations for many organic micropollutants and trace elements, and mg/kg for major elements. For water, results should also in principle be reported in mass/kg of water but the practice is usually that they are reported as mass/volume, e.g. ng/l, µg/l or mg/l, which is already diverging from basic metrological principles.

### 1.1.5.3 Standardized Methods in the Context of the WFD

Routine monitoring measurements often rely on standard methods adopted by official national or international standardization organizations (e.g. ISO, CEN). Written standards aim to establish minimum quality requirements and to improve the comparability of analytical results. Standardized methods are usually developed to be used on a voluntary basis, but they may also be linked to regulations. This is the case of series of standards that appear in paragraph 1.3.6 of Annex V of the WFD, which reads

> methods for the monitoring of the parameters shall conform to the international standards listed below [NB: six standards are presently listed by the annex is opened to any relevant standards] or such other national or international standards which will ensure the provision of data of an equivalent scientific quality and comparability.

As underlined above, the on-going discussions on minimum performance criteria for analytical methods used in chemical monitoring (proposal for a Commission Decision) are embedded in paragraph 3 of Article 8 of the WFD, stipulating that 'technical specifications and standardized methods of analysis and monitoring of water status shall be laid down in accordance with the procedure laid down in Article 21' (NB: this latter article concerns 'comitology', i.e. the possible adoption of legally binding regulations by Member States with consultation of the European Parliament). In this draft decision, the use of alternative methods providing data of equivalent or better scientific quality and comparability than standardized methods is highlighted, providing that these methods are properly validated. This actually enables all kind of methods to be used for chemical monitoring of 'total' amounts of chemical substances. For operationally defined parameters, however, the use of standardized methods is made mandatory.

For operationally defined parameters (e.g. extractable forms of elements using a specific extraction method), results are obviously linked to the methods used, which hence requires that these methods are standardized if comparability of data has to be ensured. This justifies standard methods for these parameters to be made mandatory. In this context, given that analytical results are directly linked to the analytical protocol, standard methods will represent a key element in the traceability chain. In other words, the traceability chain is broken if the protocol is not strictly followed (Quevauviller, 2004b).

In the context of the WFD chemical monitoring programme, recommendations to use standard methods will hence certainly focus on analytical steps that are based on technical operations that may differ from one country to another. Examples are sampling, sample pretreatment (e.g. filtration for water, sieving for sediment), and measurements of 'extractable' forms of pollutants. Efforts are, therefore, on-going to identify existing standard methods that could readily be recommended in the light of WFD chemical monitoring, as well as standards for which research would be needed prior to adoption by the European Standardization Organization (CEN). The principle will be to develop documented guidelines about the methods, their limits of applicability, and other relevant aspects (e.g. sampling representativeness, frequency and techniques).

At the present stage, a range of standardized methods is available to provide references for specific parameters or monitoring steps. However, there are still many cases for which there are no such standards, hence the reference has to be found elsewhere else. This is discussed in the next paragraphs.

## 1.1.5.4 Role of Reference Methods

Complex environmental monitoring measurements are generally based on successive analytical steps, such as extraction, derivatization, separation and detection. This succession multiplies the risk that the traceability chain is broken because of a lack of appropriate references (e.g. pure calibrants, reference materials). Reference methods usually refer to methods with high metrological qualities (they may actually be standardized methods), and examples are provided by primary methods that are directly traceable to SI units without the need to use an external calibration.

If we consider that reference methods correspond to methods exempt from systematic errors (this is the case of primary methods) and only affected by few random errors, we have to admit that these mainly exist for inorganic determinations at the present stage and solely for analysing samples in the laboratory (i.e. there are no 'reference methods' possible for sampling). In the case of determinations of organic or organometallic compounds, the need to include pretreatment steps, e.g. extraction or derivatization, will lead to breaking of the traceability chain and stated references will rely on approximations (recovery estimates). The better these approximations, the closer the traceability of the measurement of the substance in the sample to the true value. In the case of substances requiring an extraction step, primary methods hardly exist since there are no means at present to give proof that extraction or chemical reactions (e.g. derivatization) have yielded a 100% recovery.

Examples of primary methods used in determining multi-isotope trace elements are isotope dilution-based techniques (e.g. isotope dilution mass spectrometry), enabling traceability of the results to SI units. For organometallic compounds, the use of these techniques will hence guarantee traceability to SI units for the compounds in the extracts.

For methods that are based on internal or external calibrations, the link will rely on the availability of calibrants of high purity and verified stoichiometry, which represents the last part of the traceability chain (i.e. calibration of the detector signal). A reference method should provide guarantees that all steps prior to determination are recorded and documented in such a way that the result of the final determination is linked to an unbroken chain of comparisons to appropriate standards. In other words, establishing traceability for reference methods implies that several 'primary' chemical reference materials in the form of ultrapure substances are interlinked by well-known, quantitative, high-precision, high-accuracy chemical reactions (de Bièvre, 1996). This is actually not achieved for the vast majority of chemical monitoring measurements as there are still many 'weak links' in the traceability chain (e.g. extraction recovery estimates, derivatization yield verifications using 'secondary standards').

In conclusion, the denomination of a method as 'reference method' in the context of such a wide-scale monitoring programme as the WFD will require a very careful consideration, and that any weaknesses in the traceability chain are identified and highlighted. As stressed above, methods with analytical steps requiring recovery estimates need to be validated using independent methods (based on different principles) in order to provide data of good scientific comparability. It should be stressed again, however, that comparability is not synonymous with accuracy. Hence, few of the methods may be considered as reference methods unless they are documented with a full description of all the analytical operations and the limits of applicability of the methods. This is the case, for example, of 'official methods' required in the food sector.

Currently there are no technical prescriptions for analytical methods for WFD chemical monitoring,. The principles of validation are now under discussion for agreement at EU level, but these discussions do not go as far as describing methods with specific technical requirements. We have seen that this may envisaged in the context of standardization, but we have also to realize that WFD monitoring relies on Member States' practices and the aim is not to establish an unnecessarily over rigid system, even if it is recognized that operational coordination and common understanding are desirable.

### 1.1.5.5 Role of Reference Materials

The role and use of reference materials are in principle well known, in particular for Certified Reference Materials (CRMs) used as calibration materials or matrix materials representing – as far as possible – 'real matrices' used for the verification of the measurement process, or (not certified) laboratory reference materials (LRMs also known as quality control (QC) materials) used, for example, in interlaboratory studies or in the maintenance of internal quality control (control charts). Examples of reference materials relevant to WFD monitoring (water, sediment and biota) are described in the literature (Quevauviller, 1994; Quevauviller and Maier, 1999).

Control charts used for monitoring the reproducibility of methods (repeated analyses of one or several reference materials) may be considered as long-term references for analytical measurements since they allow the monitoring of analytical variation with respect to an anchorage point, i.e. the reference material(s). This concerns reproducibility checking but not necessarily trueness whose evaluation relies on relevant CRM analysis.

Regarding CRMs, certifying organizations attempt, wherever possible, to estimate true values of parameters in representative matrix materials. This is achieved mainly by employing a variety of methods with different measurement principles in the material certification study. A good agreement between the various methods supports the assumption that no systematic error was left undetected, and that the certified values are the closest estimate to the true value: however, it does not provide an unequivocal demonstration of this. Wherever possible, this approach should include primary methods (see above). In many instances, the use of various independent methods enables consensus values that are accepted as true values reflecting the analytical state-of-the-art and hence ensuring that data comparability is obtained (Quevauviller, 2004a). The wide variety of matrices and substances encountered in water-related measurements, in principle implies that various types of matrix materials (e.g. waters, sediments and biota) should be available. However, a perfect match between a CRM and a natural sample can never be achieved and compromises are in most instances necessary. As discussed previously (Quevauviller, 2004b), reference materials represent 'physical' references to which measurements can be linked, but which call for a cautious evaluation because of matrix differences. It was stressed that because of differences in matrix composition a correct result obtained with a matrix CRM does not give a full assurance that 'correct results' will be achieved when analysing unknown samples (Quevauviller, 2000).

The traceability of water, sediment and biota measurements to SI units on the basis of matrix CRMs (representing complex chemical systems) has been widely debated. This results from the lack of certainty that the certified values correspond to true values, in particular with regard to complex matrix reference materials. This does not diminish the value of CRMs since the 'consensus values' provide the reference necessary to achieve traceability for a given water-related measurement (linking results to a wide analytical community and thus ensuring data comparability) but they do not necessarily allow trueness to be checked. In addition, there are still gaps in availability of CRM and in their representativeness, and there are situations in which they cannot be prepared because of instability. In these cases, other approaches have to be followed to achieve traceability, e.g. interlaboratory studies (see paragraph 1.1.5.6). Where there is good correspondence between the matrix of samples and the matrix of CRMs, this reference is certainly the most appropriate one for checking the accuracy of analytical methods and comparing the performance of one method with that of other methods (and of other laboratories). Similar considerations apply to the representativeness of noncertified reference materials used for internal QC purposes. It is noteworthy that RMs are key QC tools in the context of the monitoring requirements of the WFD. For although they are not specifically referred to in the WFD core text, they are clearly part of the related Commission Decision on minimum performance criteria (see above paragraph on 'further binding rules').

### 1.1.5.6 Proficiency Testing Schemes

Proficiency testing schemes enable laboratories to establish 'external' references for evaluating the performance of their methods. In this framework, one or more reference materials are distributed by a central organization to participating laboratories for the determination of given substances. A comparison of the results obtained by the various laboratories using their different methods makes it possible to detect potential sources of error linked to a specific procedure or to the ways in which a method is applied in a given laboratory. When the testing focuses on a single method, this enables the assessment of performance criteria (such as limits of detection or quantification, and precision). The references establishing the traceability link are again based on reference materials which have to fulfil the usual requirements. However, in contrast with reference materials used for internal QC, it is not important if the samples used in proficiency testing schemes have a limited shelf life, and so they may be used by laboratories to evaluate the analysis of particular parameters for which appropriate stabilized RMs are not available. Examples include water samples containing unstable compounds for which only short-term preservation is possible. Proficiency testing schemes may also be based on reference sites e.g. to evaluate sampling procedures or a bulk common sample to be analysed by participating laboratories at a given time (e.g. a tank full of water for analysing microbiological parameters).

As discussed above for RMs, the measurement values obtained in the framework of interlaboratory studies (using different techniques) may be considered as an anchorage point, representing the analytical state-of-the-art: this offers laboratories a means of achieving comparability (i.e. traceability) of their results with a recognized reference, which in this case is a consensus value (generally the mean of laboratory means).

It must be stressed again that this reference does not provide traceability to the true value of the substance in the medium, but it does represent a very useful method for achieving comparability of environmental measurements using an external QC scheme (Quevauviller, 2004b).

In common with the use of CRM RMs, proficiency testing schemes are not specifically referred to in the WFD core text, but are clearly part of the related Commission Decision on minimum performance criteria (see above paragraph on 'further binding rules').

### 1.1.5.7 Laboratory Accreditation

Accreditation of laboratories following rules established by the ISO 17025 contributes to the overall framework contributing to the achievement of traceability, but it is only an organizational framework and as such does not guarantee the achievement of traceability. Rather it requests laboratories to proceed with internal and external QC of analytical methods as described in the preceding paragraphs.

## 1.1.6 CONCLUSIONS, PERSPECTIVES

This chapter highlights to the requirements for metrology in the context of a large-scale chemical monitoring programme such as that required by the WFD. It hence mixes policy considerations with technical issues, and highlights the complexity of the overall approach that needs to be developed to provide the necessary demonstration of the traceability of chemical monitoring data at EU level. In comparison with physical metrology, the challenges are tremendous as they involve many different chemical specific features (including operational steps such as sampling, sample pretreatment, different matrices, thousands of substances, and needs for reference materials for validating methods) that do not have to be addressed in making physical measurements. A 'simple' translation of physical metrological principles to chemical metrology is, therefore, not possible. This has been extensively discussed by chemical experts over the past ten years (Válcarcel and Rios, 1999; Quevauviller, 1999; Walsh, 2000), and theoretical discussions have not yet led to practical solutions that might be implemented on the scale of an EU-wide monitoring programme. These on-going discussions have the merit that the principles are now well established and that the direction that should be taken to improve the metrological chemical framework is reasonably well understood. However, in order to make this operational, a huge coordination of efforts will be needed, and it would not be realistic to think that what has taken one century for physical metrology will be achieved in few years for chemical metrology. One advantage of the WFD is that it has the potential to provide a very wide testing framework, and the metrological community should take this opportunity to examine how theory may be linked to practice. This is being studied in the context of an EU funded project, 'European Analytical Quality Control in support of Water Information System for Europe' (EAQC-WISE). Actual improvements will emerge progressively as discussions proceed on the soundness of the chemical data produced by Member States in the context of the preparation of river basin management plans. There are, therefore, milestones that could and should be used not only to improve the metrological system but also to enforce technical rules that will underpin the practical implementation of an EU-wide system and increase its chances of being successful. This perspective extends over the next decade, and could be completed by taking the opportunities provided by the reviewing milestones (2015 and every six years thereafter) of the WFD to bring about systematic and ongoing improvements in the monitoring metrological basis of not only on chemical monitoring but also ecological status monitoring.

As a final consideration, one should keep in mind that achieving traceability of water, sediment and biota chemical monitoring measurements in the context of the WFD will have direct implications for the way in which programmes of measures will be designed and made operational to achieve the 'good status' objectives by 2015. This places a strong emphasis on the importance of metrology since any erroneous data could have tremendous (social and economic) consequences.

# REFERENCES

De Bièvre P., 1996. In: H. Günzler (ed.), Accreditation and Quality Assurance in Analytical Chemistry, Springer, Berlin.
European Commission, 2000. Directive 2000/60/EC of the European Parliament and of the Council of 23 October 2000 establishing a framework for Community action in the field of water policy, Official Journal of the European Communities **L 327**, 22.12.2000, p. 1.
European Commission, 2003a. Monitoring under the Water Framework Directive, CIS Guidance Document No. 7, European Commission, Brussels.
European Commission, 2003b. Common Implementation Strategy for the Water Framework Directive, European Communities, ISBN 92-894-2040-5, 2003.
European Commission, 2006. Groundwater Monitoring, CIS Guidance Document No. 15, European Commission, Brussels.
European Commission, 2009. Surface Water Monitoring, CIS Guidance Document No. 19, European Commission, Brussels.
ISO, 1993. International Vocabulary of Basic and General Terms in Metrology, 2nd edition, International Standardisation Organisation, Geneva, Switzerland, 1993.
Quevauviller Ph., 1994. Trends Anal. Chem., **13**(9), 404.
Quevauviller Ph., 2000. J. Environ. Monitor., **2**, 292.
Quevauviller Ph., 2002. Quality Assurance for Water Analysis, Water Quality Measurements Series, John Wiley & Sons, Ltd, Chichester, ISBN: 0-471-89,962-3, pp. 252.
Quevauviller Ph., 2004a. In: Z. Mester, B. Sturgeon (Eds.), Sample Preparation for Trace Analysis, Elsevier, Amsterdam.
Quevauviller Ph., 2004b. Trends Anal. Chem., **23**(3), 171.
Quevauviller Ph., 2006. J. Soil & Sediments., **6**(1), 2.
Quevauviller Ph., 2007. 13th Int. Metrology Conference, Lille, 20–22 September 2007.
Quevauviller Ph., Borchers U., Thompson K.C. and Simonart T. (eds), 2008. The Water Framework Directive – Ecological and Chemical Status Monitoring, Water Quality Measurements Series, John Wiley & Sons Ltd., Chichester.
Quevauviller Ph. and Donard O.F.X., 2001. Trends Anal. Chem., **20**, 600.
Quevauviller Ph. and Maier E.A., 1999. Interlaboratory Studies and Certified Reference Materials for Environmental Analysis, Elsevier, Amsterdam, ISBN: 0-444-82,389-1, pp. 558, 1999.
Válcarcel M. and Ríos A., 1999. Anal. Chem., **65**, 78A.
Walsh M.C., 1999. Trends Anal. Chem., **18**, 616.

# 1.2
# Use of Screening Methods in US Water Regulation

Guillaume Junqua, Estelle Baurès, Emmanuelle Hélias and Olivier Thomas

1.2.1 Introduction
1.2.2 The Clean Water Act
1.2.3 Monitoring and Assessment of Water Quality
1.2.4 Approved Methods and Alternative Test Procedure
    1.2.4.1 Approved Methods
    1.2.4.2 Procedures for a New Alternative Test Procedure or New Method Approval
    1.2.4.3 Homeland Security Issues
1.2.5 Volunteers Monitoring Programs
    1.2.5.1 Volunteers and Cooperative Networks
    1.2.5.2 Resources Available for Volunteers for Monitoring Implementation
    1.2.5.3 Field Kits
    1.2.5.4 River and Lake Parameters
    1.2.5.5 Quality of Data
    1.2.5.6 Quality Assurance Project Plan
1.2.6 Applications
1.2.7 Conclusions
References

## 1.2.1 INTRODUCTION

US water regulation is mainly based on three laws, respectively the Clean Water Act, the Safe Drinking Water Act and the Marine Protection, Research, and Sanctuaries Act. The Clean Water Act which is the most important (CWA, 2002) will be the

---

*Rapid Chemical and Biological Techniques for Water Monitoring*    Edited by Catherine Gonzalez, Philippe Quevauviller and Richard Greenwood
© 2009 John Wiley & Sons, Ltd

only one considered hereafter, along with US Environmental Protection Agency action (USEPA). USEPA is a regulatory agency that proposes complementary regulations and rules that explain the critical technical, operational, and legal details necessary to implement laws. Then, these rules and regulation are published by the Federal Register and codified in the Code of Federal Regulations (CFR), mainly in Title 40. USEPA proposes other recommendations, guidelines and guidance, including those covering monitoring issues.

For these laws and regulations, various tools can be used, including more than 1600 analytical methods recommended by USEPA, but also new alternative methods which can be approved by an alternative test procedure (40 CFR 136). One original innovation in the US system is the encouragement of networks of volunteers (with knowledge, material and financial resources allocated for them) to help governments achieve an efficient and representative water quality monitoring. The dynamism of all these stakeholders allows both the improvement of knowledge exchange and experience and promotion of the use of alternative screening methods to improve US water monitoring.

The CWA implementation, allows the participation, not only of the Environmental Protection Agency (USEPA) and its 10 regional offices, other federal agencies, States and Indian tribes, but also other stakeholder groups (including associations, universities, volunteers networks) more particularly for efficient monitoring of water bodies (USEPA, website). In this context, the use of screening methods for water quality monitoring in the US becomes more and more a reality. Screening methods are often chosen by field operators for rapid and simple measurement of water quality. Actually, screening methods are generally defined as qualitative or semi-quantitative methods but must also include rapid methods or procedures giving quantitative values. They are mainly used for field measurements and are also called alternative methods (Thomas, 2006) or emerging tools (Allan *et al*., 2006). There are different types of alternative method such as ready-to-use methods (test kit method), handheld devices (handheld instrument with generally no reagent needed), online sensors (generally in an industrial context) and methods for biological monitoring (such as bio markers, whole-organism tests).

## 1.2.2 THE CLEAN WATER ACT

The Federal Water Pollution Control Act (FWPCA), usually named Clean Water Act (CWA) since 1977, aims at the protection of surface water quality in the United States. It was written in 1948, but largely reorganized and expanded in 1972, and covers both the chemical aspects for the 'integrity' goal, and the point source pollution issue (municipal and industrial wastewaters mitigation). These priorities have gradually evolved towards physical and biological integrity, and a more holistic strategy, based on an integrated watershed management. Moreover, voluntary programs and a regulatory approach have been used for respectively, 'non-point' runoff and 'wet weather point sources' (USEPA, website). The key CWA elements are summarized in Figure 1.2.1 (USEPA, 2008).

Water Quality Standards (WQS) are defined for a given water body, associated with its main use and water quality criteria: drinking water (treated and untreated), water based-recreation (non, short-term, long-term contact), fishing, aquatic life (warm or cold water species or habitat), agriculture and industrial water supplies. It includes

*The Clean Water Act*  17

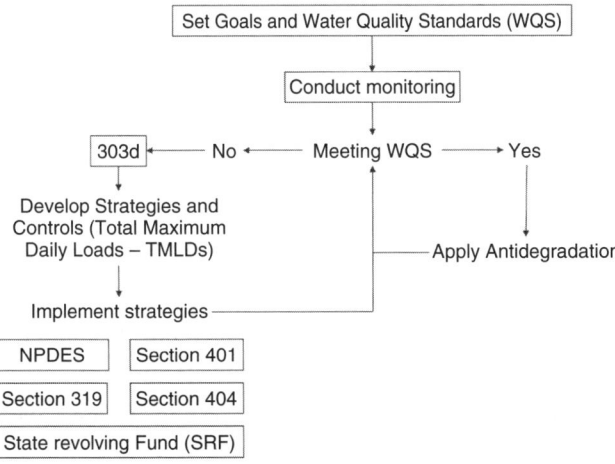

**Figure 1.2.1** Key CWA elements (www.epa.gov/watertrain/cwa/cwa1.htm)

parameters such as individual pollutants concentrations, water quality characteristics (including temperature, pH, oxygen, TSS) or other quantitative indicators and qualitative considerations. Thus, monitoring checks whether WQS are met in each waterbody, and if they are, then antidegradation policies and programs can be defined and implemented by states, tribes or territories in order to maintain a high water quality. If they are not met in a particular water body, then this is included in a threatened and impaired waters list called '303(d) list' by states, tribes or territories and submitted to USEPA for review and approval. Then, it is necessary to develop a strategy based on Total Maximum Daily Loads (TMDLs) or other integrated watershed approaches. These strategies are also implemented with regulatory (Sections 401, 402, 404), voluntary (Section 319) or funding (SRF, Section 319) CWA elements, presented by Table 1.2.1 (USEPA, 2008).

**Table 1.2.1** Key CWA elements to improve quality of threatened and impaired waters (USEPA, 2008)

| Name | Targets |
|---|---|
| Section 402 | Point sources of pollution discharging into a surface waterbody |
| | Also called National Pollutant Discharge Elimination System (NPDES) permit program |
| Section 319 | Non point sources of pollution (farming, forestry operations, ...) |
| Section 404 | Placement of dredged or fill materials into wetlands and others waters |
| Section 401 | Certification from the state, territory or Indian tribes of none violation of WQS for a construction or other activity before federal issuing permits or licences |
| State revolving funds (SRF) | Money for municipal point sources, nonpoint sources and other activities |

Thus, CWA implementation is based on various monitoring methods and procedures which must be adapted to the goals of the different CWA elements. Furthermore, State monitoring program must meet requirements of Section 106 prior to the award of Section 106 grant funds dedicated to pollution control programs (USEPA, 2003). One of these requirements is to submit reports under Section 305(b). Indeed, all states have to provide a state's Biennial Water Quality Report to Congress, also called the 305(b) report (and the 303(d) list if necessary). States must assess their waters in order to know if the quality standards for the designated use of water bodies are met. The information provided is:

- the status of the water bodies in the State, Tribal land or other territory, according to their designated use (healthy, threatened, impaired);
- the presence of pollutants (chemicals, nutrients, sediments, metals, temperature, pH) or other stressors (altered flows, invasive species, ...) which causes impairment to water bodies;
- the sources of pollutants and stressors.

## 1.2.3 MONITORING AND ASSESSMENT OF WATER QUALITY

CWA implementation needs the involvement of multilevel stakeholders for water monitoring (USEPA, website) in order to:

- characterize waters and identify changes or trends in waters over time;
- identify specific or existing or emerging quality problems;
- gather information to design specific remediation for identified pollutants;
- determine whatever programs goals are being met;
- respond to emergencies, such as spills and floods.

At the Federal level, USEPA conducts limited monitoring surveys for specific issues. The regional offices have to comply with these and inspect the monitoring of industrial and municipal discharges. Since 1974 the US Geological Survey (USGS) has been in charge of the National Stream Accounting Network (NASQAN), (http://water.usgs.gov/nasqan) that covers extensive chemical monitoring on large rivers. Since 1991, it has also piloted the National Water Quality Assessment Program (NAWQA) that focuses on status and trends in water, sediment and biota at the regional level (http://water.usgs.gov/nawqa). Others federal agencies conduct water quality monitoring programs such as the US Fish and Wildlife Service, the National Oceanic and atmospheric Administration, the US Army Corps of Engineers, and some water authorities such as the Tennessee Valley Authority.

At the regional level, States and Indian Tribes have the main responsibilities for conducting monitoring programs, and these activities are supported by grants from

USEPA. They have to develop and conduct the State water monitoring and assessment program (presented below). Moreover, local governments (cities, counties environmental offices) can also implement monitoring programs. At the local level, many private organizations (universities, watershed associations groups, permitted dischargers) are becoming increasingly important in water quality monitoring. These local stakeholders are trained and can collect monitoring data for local decision-making and to supplement state water quality data. They have also a great importance in the development and the validation of new screening methods.

Since 2002, States, Tribal, and territories must submit an 'Integrated Water Quality Monitoring and Assessment Report' that satisfies CWA requirements for both Section 305(b) and section 303(d) (www.epa.gov/owow/monitoring). In this context, a Consolidated Assessment and Listing Methodology (CALM) has been proposed by USEPA for improving monitoring and assessment programs of states, territories, and authorized tribes. The approach is an iterative process released with cooperation of all stakeholders (USEPA, 2002). However, the devices, methods, systems and procedures used to monitor and analyse water quality data have not been clearly defined until recently (USEPA, 2003). So, with regard to a large variability between State programs, a list of recommendations has been established in 2003, for harmonizing monitoring programs within the next 10 years (Table 1.2.2). Water quality indicators are particularly important in this depending on the monitoring design and sample site (Table 1.2.3). As part of this monitoring the chemical quality of waters and sediments and the health of organisms can all be analyzed. There are additional core indicators for particular

**Table 1.2.2** Recommended elements of a state water and monitoring program (USEPA 2003)

| Elements | Description |
|---|---|
| Monitoring program strategy | Each State has a comprehensive monitoring program strategy (<10 years) for all waterbody types |
| | Recommended collaborations with all the stakeholders (EPA regions, other federal agencies, other State environmental managers, volunteers monitoring associations, academic institutions) |
| Monitoring objectives | Identification of monitoring objectives, including CWA goals, and able to answer to the following questions |
| | – What is the overall quality of waters in the State? |
| | – To what extent is water quality changing over time? |
| | – What are the problem areas and areas needing protection? |
| | – What level of protection is needed? |
| | – How effective are clean water projects and programs? |
| Monitoring design | Methodology for the selection of monitoring design and sample sites in relation with decision needs |
| | EPA's Environmental Monitoring and Assessment Program (EMAP) and US Geological Survey's National Water Quality Assessment program are recommended |

*(continued overleaf)*

**Table 1.2.2** (*continued*)

| Elements | Description |
|---|---|
| Core and supplemental water quality indicators | The design should be have appropriate levels of precision and confidence (integrating monitoring objective and type of date collected) as well as methods to control and balance possible decision errors |
| | Definition of a core set of indicators for each water resource type which can be used routinely to assess WQS |
| | Definition of supplemental indicators, such as specific pollutants, for example to identify causes and sources of impairments and targeting appropriate source controls |
| Quality assurance | Development maintain and review of Quality Management Plans and Quality Assurance Project Plans according to EPA policy, in relation with an appropriate level of data quality for an efficient decision |
| Data management | Use by the State of a public accessible data system for water quality, fish issue, toxicity, sediment chemistry, habitat and biological data. Various databases can be used, and more particularly STORET which is a national repository, managed by USEPA |
| Data analysis/assessment | Development and implementation of a methodology for assessment of water quality in relation with WQS, with different types of data coming from various sources, based on 4 steps: |
| | – Identification and collection of sources and data, |
| | – Identification and description of requirements for data quality and representativeness |
| | – Identification and referencing of procedures for evaluation of quality of datasets |
| | – Explanation of procedures used to comparing data to WQS |
| Reporting | Production of timely and complete water quality reports and lists required by CWA and others, using wherever possible consolidation and diffusion towards public |
| Programmatic evaluation | Periodic reviews and assessment of State monitoring program and its 10 elements, for example in a continuous improvement approach by the State and consultation with its EPA Region. |
| General support and infrastructure planning | Identification of current and future monitoring resources to implementation of monitoring program strategy, and more particularly staff and training laboratories resources, funding. |

*Monitoring and Assessment of Water Quality*

**Table 1.2.3** Examples of recommended core and supplemental water quality indicators for general designated use categories (USEPA, 2003)

|  | Aquatic life and wildlife | Recreation | Drinking water | Fish/shellfish consumption |
|---|---|---|---|---|
| Core indicators | Condition of biological communities (EPA recommends the use of at least two assemblages) Dissolved oxygen Temperature Conductivity pH Habitat assessment Flow Nutrients Landscape conditions (e.g. % cover of land uses) Additional indicators for lakes: Eutrophic condition Additional indicators for wetlands: Wetland hydrogeomorphic settings and functions | Pathogen indicators (E. coli, enterococci) Nuisance plant Growth Flow Nutrients Chlorophyll Landscape conditions (e.g., % cover of land uses) Additional indicators for lakes: Secchi depth Additional indicators for wetlands: Wetland hydrogeomorphic settings and functions | Trace metals Pathogens Nitrates Salinity Sediments/TDS Flow Landscape conditions (e.g. % cover of land uses) | Pathogens Mercury Chlordane DDT PCBs Landscape conditions (e.g. % cover of land uses) |
| Supplemental indicators | Ambient toxicity Sediment toxicity Other chemicals of concern in water column or sediment Health of organisms | Other chemicals of concern in water column or sediment Hazardous chemicals Aesthetics | VOCs (in reservoirs) Hydrophyllic pesticides Nutrients Other chemicals of concern in water column or sediment Algae | Other chemicals of concern in water column or sediment |

media, such as lakes and wetlands. Finally, data quality is also a priority as well as their management by various local or national databases such as STORET (data STOrage and RETrieval system, www.epa.gov/storet/). These mechanisms allow the timely production of reports as required under the CWA in order to have an approach based on the seeking of continuous improvement. The design of prospective actions has an influence on the development of new monitoring tools and methodologies.

## 1.2.4 APPROVED METHODS AND ALTERNATIVE TEST PROCEDURE

In order to estimate these indicators, it is necessary to use various analytical methods, generally approved by USEPA or standardized by others national or international organizations such as APHA (American Public Health Association) ASTM, AOAC-International or ISO. In addition to reference methods, alternative procedures or new methods can be tested and approved, according to EPA guidelines or APHA, ASTM, AOAC-international, and ISO protocols (AOAC, 1999; APHA, 1998; ASTM, 1999; ISO, 2001). In order to approve and validate new test procedures, confirm laboratory performance and update approved methods (Table 1.2.4) USEPA publishes both approved test methods and procedures and those that have not been promulgated.

In the next paragraphs, only published methods by USEPA and Alternative test procedures will be discussed.

### 1.2.4.1 Approved Methods

In order to ensure quality control, USEPA has listed regulations, rules and guidance on approved analytical methods (test procedures) in the 40 CFR 136. This is especially

Table 1.2.4 Different US methods and procedures relative to analytical test methods

| | |
|---|---|
| Approved methods | Compliance methods to measure pollutants in various media for many CWA applications (Section 304(h), 40 CFR 136). |
| Other approved methods | Compliance methods (40 CFR 136.3, Tables IF and IG) specific to pharmaceutical manufacturing and pesticide chemical point source categories (40 CFR 439 and 455). |
| Other methods | Methods published by USEPA, but not promulgated. |
| Alternative Test Procedures (ATP) | Approval & validation process for submitting new test methods or procedures for USEPA approval (40 CFR 136.4, 136.5, and 141.27). |
| Detection & quantitation | Procedures to confirm laboratory performance, managed by the Federal Advisory Committee on Detection and Quantitation (FACQD). |
| Method updates | Revision of the list of approved analysis and sampling procedures (new methods added, other method withdrew, new sample collection procedures, general analytical requirements for multi-analyte methods, method flexibility requirements). |

**Table 1.2.5** USEPA methods content (USEPA, 1996a)

| | |
|---|---|
| 1.0 Scope and application | 10.0 Calibration and standardization |
| 2.0 Summary of method | 11.0 Procedure |
| 3.0 Definitions | 12.0 Data analysis and calculations |
| 4.0 Interferences | 13.0 Method performance |
| 5.0 Safety | 14.0 Pollution prevention |
| 6.0 Equipments and supplies | 15.0 Waste management |
| 7.0 Reagents and standards | 16.0 References |
| 8.0 Sample collection, preservation and storage | 17.0 Tables, diagrams, flowcharts and validation data |
| 9.0 Quality control | |

designed for use by industries and municipalities for the measurement of chemical and biological components of wastewaters, drinking waters, sediments and other environmental samples. There are about 1600 analytical methods officially described by the USEPA (www.epa.gov/waterscience/methods/). Each method is described following a specified format (Table 1.2.5) in order to allow the comparison of data coming from different laboratories, including their precision and accuracy.

### 1.2.4.2 Procedures for a New Alternative Test Procedure or New Method Approval

The Federal Code of Regulation forecast procedures for the approval of new test methods (40CFR part 136.4 and 136.5 for wastewaters and ambient waters, 141.27 for drinking water) distinguishes between Alternative Test Procedure (ATP) and new (test) methods. 'An ATP is a modification of an approved method or a procedure that uses the same determinative technique and measures the same analyte(s) of interest as the approved method. The use of a different determinative technique to measure the same analyte(s) of interest as an approved method is considered as a new method.' (USEPA, 1999a, 1999b). The function of the ATP Protocol is to demonstrate that an ATP or new method produces better results than, or equal to, those produced by an USEPA appropriate method. The comparison of the performances of an ATP or new method and an USEPA-approved method can be achieved through a side-by-side comparison or a Quality Control (QC) acceptance criteria-based comparison study (USEPA, 2004). The modification to sample preparation techniques (USEPA, 1999a, 2004) in order to have better results or adding new target analytes (USEPA, 1999a) can also be submitted.

The approval of ATP or new methods is linked to the level of use (local or nationwide), for one matrix type or all matrixes. 'A matrix type is defined as a sample medium with common characteristics across a given industrial subcategory' (EPA, 1999a). Depending on level of use, and the applicant (USEPA regional laboratories, states, commercial laboratories, dischargers or anybody), the submission address (USEPA regional administrator, Director of the State Agency, Director of Analytical Methods Staff) and the approval Authority (USEPA regional or national administrator), for microbiological ATP can differ (Table 1.2.6).

**Table 1.2.6** Submission of ATP and new methods (USEPA, 2004)

| Level of use | Applicant | Submit application to | Approval authority |
|---|---|---|---|
| Limited use for wastewater or ambient water | EPA Regional laboratories | EPA Regional Administrator (Regional ATP Coordinator) | EPA Regional Administrator |
| | States, commercial laboratories, individual dischargers, or permittees in States that do not have the authority to administer Clean Water Act and Safe Drinking Water Act monitoring programs | EPA Regional Administrator (Regional ATP Coordinator) | |
| | States, commercial laboratories, individual dischargers, or permittees in States that do not have the authority to administer Clean Water Act and Safe Drinking Water Act monitoring programs | Director of State Agency issuing the NPDES permit | |
| Nationwide use for drinking water, wastewater, ambient water | All applicants | Director, Analytical Methods Staff, EPA Headquarters | EPA Administrator |

Furthermore, any application for an ATP must give some information, and more particularly as shown in Table 1.2.7.

Justification for new method or ATP can be related to, for example, the reduction of cost analysis, limitation of interferences, improvement of safety and waste management compared with the approved method. The method must be presented according to the USEPA format. For ATP, an in-depth comparison between ATP and USEPA approved methods must be produced with a two-column method comparison, providing a detailed discussion of a number of specific aspects each in its own section. The method development information is published in the Federal Register after proposal by USEPA to approve the ATP or new method. Information can be a detailed background and summary of the method, a discussion of the development of QC acceptance criteria, a description and discussion of the interlaboratory validation study, others studies (for instance of method limitations). The study plan is submitted to USEPA for review and comment, prior to conducting all studies. Lastly, the

*Approved Methods and Alternative Test Procedure*

**Table 1.2.7** Application requirements according to type of ATP or new method (from USEPA, 1999a, 1999b and 2004)

| Application information | Microbiological ATP | Organic and inorganic ATP (limited use) | Organic and inorganic ATP (nationwide use) | Organic and inorganic new method (limited use) | Organic and inorganic new method (nationwide use) |
|---|---|---|---|---|---|
| Completed application form | X | X | X | X | X |
| Justification for ATP/new method | X | X | X | X | X |
| Method in USEPA format | X | X | X | X | X |
| Method comparison table | X | X | X | | |
| Method development information | X | | X | | X |
| Study plan (to be approved by USEPA before proceeding with study) | X | | | | |
| Validation study report (final report generally considered part of a complete application) | X | X | X | X | X |

**Table 1.2.8** Elements of the validation study report with the new method/ATP application (from USEPA 1999 a, b and 2004)

| Background | Data Analysis and Discussion |
|---|---|
| Study objectives and design | Conclusions |
| Study implementation | Appendix A - Method |
| Data reporting and validation | Appendix B - Study Plan |
| Results | Appendix C - Supporting Data |
| Development of quality control acceptance Criteria (only for new method) | Appendix D - Supporting References |

applicant must provide a comprehensive study report with the ATP or new method application, with the elements given in Table 1.2.8.

More particularly for new method (but not exclusively), it could be necessary to develop QC acceptance criteria for all of the required QC tests.

### 1.2.4.3 Homeland Security Issues

In response to Homeland Security Presidential Directive 9, USEPA, in collaboration with other agencies, must develop surveillance and monitoring systems in order to provide earlier detection of intentional or accidental water contamination. It is necessary for a preservation of water infrastructures and public health. 'The fundamental challenge to the reliance on a variety of information streams as an indication of a contamination incident is a means of distinguishing anomalous patterns in these data from background signals' (USEPA, 2005). Thus, USEPA has launched the WaterSentinel (WS) program, to design, deploy and evaluate Contamination Warning System (CWS), coupling usual sensors for estimation of water quality, algorithms for data analysis and procedures that aid decision-making. Furthermore, because of previous terrorist attacks against US organizations, standardized analytical methods for environmental restoration following Homely Security events have been proposed by USEPA (USEPA, 2007). These methods could be applied for the monitoring and the control of environmental restoration programs after an intentional contamination of a water body.

## 1.2.5 VOLUNTEERS MONITORING PROGRAMS

One of the key actions of US water monitoring is the use of networks of volunteers. More than 1000 programs are listed in the National Directory of Volunteer Monitoring Programs (http://yosemite.epa.gov/water/volmon.nsf/home?openform). The Volunteers' work is essentially informing communities and citizen of water quality issues, getting data on water bodies that otherwise may go unmonitored, and helping monitoring programs for a better water quality (CSREES, 2002).

### 1.2.5.1 Volunteers and Cooperative Networks

One way for State agencies to face increasing needs for water monitoring is to collaborate with volunteers to encourage all citizens to improve their knowledge of water resources. In turn this supports the implementation of monitoring activities, and the volunteer/state partnerships are provided with access to equipment and with a transfer of expertise from professionals to volunteers who are participating in programs. The State agencies are able to conduct projects that would not otherwise be possible. The volunteers acquire knowledge that helps them to monitor and preserve their own watershed. USEPA has a Volunteer monitoring program that encourages cooperation and exchanges between groups of volunteers through initiatives including newsletters, list servers, development of guidelines for monitoring methods, and many national or international events and conferences. USEPA also provides technical or financial assistance through its regional offices. Other cooperative systems have been implemented. For example, USGS manages the Cooperative Water (Coop) Program with almost 1400 States, cities, counties, Indian tribes and others cooperators (http://water.usgs.gov/coop/).

Volunteers actions and new groups development are encouraged by several means among which the Volunteer Water Quality Monitoring National Facilitation Project (http://www.usawaterquality.org/volunteer/), with funding of US Department of Agriculture and the Cooperative State Research, Education and Extension Service (CSREES). This project offers several guides for growing monitoring programs including a Monitoring Matrix that comprises, for instance, data objectives, examples of activities, equipment and suppliers, education and training, frequency of monitoring, QA/QC level and standards for each monitoring activities. These activities include shoreline activities, assessment of watershed habitats, measurement of physical characteristics, surveying exotic invasive species, biota, water chemistry, and sediment analysis (CSREES, 2008).

This project manages the coordination of almost all volunteers organizations and favours cooperation and knowledge exchange and events such as the Secchi Dip-in, or the world water monitoring day (http://www.worldwatermonitoringday.org/) that aim to increase awareness of the various activities in the area. For example, Secchi Dip-In concept is that individuals in volunteer monitoring programs take a transparency measurement on one day during the weeks surrounding Canada Day and July Fourth, in order to demonstrate the potential of volunteer monitors to gather environmentally important information. There is also the Tribal Colleges and Universities National Facilitation Project. All these water monitoring programs are often coupled with educational goals and may involve citizens as well as Colleges and Universities. Finally, others cooperative projects or networks exist, such as the Environmental Alliance for Senior Involvement (EASI) whose monitoring program constitutes a large part of their activities and programs (SITE EASI) or the NEMO (Nonpoint Education for Municipal Officials) program (http://nemo.uconn.edu/).

### 1.2.5.2 Resources Available for Volunteers for Monitoring Implementation

Volunteer can find information on monitoring programs in many different ways such as events, journals, on line monitoring databases and guides, training programs, award grants and others local resources. The most important event is probably the National Water Quality Monitoring Conference held every two years, which is now a key activity of the Water Quality Monitoring Council (NWQMC, http://acwi.gov/monitoring/). The conference gives a large place to the actions of volunteers and usually includes several presentations concerning many aspects of volunteer monitoring programs, and workshops on how to conduct such projects. For example, one workshop proposed by (and for) the volunteers with the National Monitoring Conference was related to the monitoring of macro-invertebrates. Participants received practical information on field techniques (qualitative and quantitative methods) for monitoring macro-invertebrates, and discussed the importance of a study design for the success of a monitoring program. Some innovative approaches to education and training were presented by Travers (2005). The NWQMC also publishes technical reports on water monitoring. For example, the technical report no. 3 is a user guide of water quality data elements sets that are the minimum elements necessary to facilitate the exchange of chemical, microbiological, population/community and (eco)toxicological assessment data (NWQMC, 2006).

The Volunteer Monitor is the national newsletter of Volunteer Water Quality Monitoring and is published twice a year (www.epa.gov/volunteer/vm_index.html). It is partially funded under a cooperative agreement with the USEPA. All volumes are dedicated to different issues (including technical, knowledge management, organization management, financial, regulation, partnership), and can be read at or downloaded from www.epa.gov/owow/monitoring/volunteer/index_updated.pdf. Online water monitoring databases provide repositories of monitoring results, and the guidelines and procedures used to obtain them. A document summarizes the main guides provided by USEPA or regional and state organizations, for activities including monitoring of lakes, estuaries, streams, wetland, seawaters, low-coast waters and riparian ecosystems or for the management of quality assurance (CSREES, 2003). Another source of information and guidance is provided by the National Environmental Methods Index (NEMI). This is a free database containing chemical, microbiological and radio-chemical summaries of laboratory and field protocols for regulatory or nonregulatory water quality analyses (www.nemi.gov). The Pennsylvania Senior Environment Corps of the Environmental Alliance for Senior Involvement (EASI) maintains its open source on the web (www.easi.org), but unfortunately this is not yet generally available.

### 1.2.5.3 Field Kits

States and regulatory agencies do not accept most data that have been obtained by analysis conducted in the field by technologies such as test kits or portable meters. The main reason is that field equipment has to be meticulously and accurately calibrated and maintained. Another reason is that most portable kits do not have the ability to detect low concentrations of chemicals in surface waters because they are mainly designed for monitoring pollutants in high concentrations such as are found in effluents from wastewater treatment plants or in highly polluted water. However, these sorts of equipment can be used depending on needs and ability of the end user. Field portable test kits and hand held instruments can produce accurate and reliable data but they must be calibrated properly and be used in appropriate conditions. These kits or devices use prepackaged sets of chemicals or not. Some field test kits can be simple tests and provide general results while others are more precise (Picotte and Boudett, 2005). It is possible to increase the confidence in the results by assessing accuracy with known standards and precision with replicate analysis of samples. The USEPA has identified five major criteria for data from measurements of water quality data (Picotte and Boudett, 2005):

- *Precision.* Measured by the difference between samples taken from the same place at the same time.
- *Accuracy.* Given by comparing analysis of a known standard or reference sample with its actual value.
- *Representativeness.* How closely samples represent the true environmental condition or population at the time a sample was collected.
- *Completeness.* Whether enough valid or usable data has been collected (comparing what was originally planned and what has actually been collected).
- *Comparability.* How data compares between sample locations or periods of time within a project or between volunteers.

## 1.2.5.4 River and Lake Parameters

The 5th National Directory of Volunteer Environmental Monitoring Program (http://www.epa.gov/owow/monitoring/dir.html) had shown ten years ago that monitoring the quality of rivers and lakes were the main goals of the 772 networks considered (respectively 76 and 38% of the monitoring networks*), far above wetlands and estuaries (around 20%). This inventory had revealed that chemical monitoring was already implemented along with physical and biological measurements. Table 1.2.9 gives the percentage of water monitoring programs using a given parameter.

Some comments can be made on the contents of Table 1.2.9:

- The three basic physico-chemical parameters are obviously the most commonly recorded, but curiously although it is very easy to measure conductivity, it is measured by only 25% of the networks.

**Table 1.2.9** Ranking of parameters used in volunteer monitoring programs (N = 772) (from http://www.epa.gov/owow/monitoring/dir.html/)

| Rank | Parameter | n° (%) of progr. | Rank | Parameter | n° (%) of progr. |
|---|---|---|---|---|---|
| 1 | Water temperature | 602 (78%) | 21 | Human use surveys | 147 (19%) |
| 2 | Dissolved oxygen | 527 (68%) | 22 | Wildlife | 145 (19%) |
| 3 | pH | 523 (68%) | 23 | BOD | 138 (18%) |
| 4 | Macroinvertebrates | 401 (52%) | 24 | Debris monitoring | 135 (18%) |
| 5 | Phosphorus | 381 (49%) | 25 | Salinity | 132 (17%) |
| 6 | Nitrogen | 381 (49%) | 26 | Birds | 128 (17%) |
| 7 | Flow/water level | 341 (44%) | 27 | Terrestrial vegetation | 123 (16%) |
| 8 | Turbidity | 324 (42%) | 28 | Stream channel morphology | 114 (15%) |
| 9 | Habitat assessment | 291 (38%) | 29 | Hardness | 111 (14%) |
| 10 | Secchi transparency | 282 (37%) | 30 | Chlorophyll | 105 (14%) |
| 11 | Bacteria | 245 (32%) | 31 | Chloride | 88 (11%) |
| 12 | Land use surveys | 208 (27%) | 32 | Metals | 78 (10%) |
| 13 | Rainfall | 197 (26%) | 33 | Pipe surveys | 75 (10%) |
| 14 | Conductivity | 191 (25%) | 34 | Construction site inspections | 73 (9%) |
| 15 | TSS/TDS | 188 (24%) | 35 | Phytoplancton | 55 (7%) |
| 16 | Aquatic vegetation | 186 (24%) | 36 | Shellfish | 46 (6%) |
| 17 | Fish | 169 (22%) | 37 | Pesticides | 36 (5%) |
| 18 | Alkalinity | 159 (21%) | 38 | Hydrocarbons | 28 (4%) |
| 19 | Photographic survey | 157 (20%) | 39 | Toxicity | 24 (3%) |
| 20 | Exotic/invasive species | 156 (20%) | | | |

---

* Only half the networks (356) were working for a single type of water body.

- The chemical quality is much more frequently monitored than biological parameters.
- Nutrients are monitored by half of the networks, but chlorophyll which is an interesting complementary parameter is not.
- Micropollutants are monitored by very few networks of volunteers, probably because no simple devices were available in the 1990s.

The last observation reflects the lack of simple, reliable, low-cost methods. It seems likely that many more volunteer groups would want to test for these pollutants if appropriate methods were available. Since they are not, volunteer programs rather concentrate on monitoring the biological response of organisms such as macroinvertebrates, aquatic vegetation, fish, and other wildlife. The abundance, diversity, and/or condition of these organisms will reflect the overall health of the system and suggest whether toxic levels of pollutants are present (Herron, 2008).

### 1.2.5.5 Quality of Data

Generating reliable data requires the fulfilment of all quality constraints depending on what you need to do before, during and after your monitoring effort (CSREES, 2004, Table 1.2.10).

Three elements combine to form the Quality System: quality assurance, quality control and quality assessment.

- *Quality assurance* is the broad plan for maintaining quality in all aspects of a program. It guides the selection of parameters and methods, how data will be managed, analyzed and reported, and what steps will be used to determine validity of the selected procedures.
- *Quality control* procedures are the mechanisms established to control errors and make analyses more accurate and precise. Quality control procedures help you

**Table 1.2.10** Data Quality System (from CSREES, 2004)

| Before – Plan | During – Implement | After – Assess |
|---|---|---|
| Quality Assurance | Quality Control | Quality Assessment |
| Study design | Training | Data proofing/review |
| Quality Assurance Project Plan | Follow the written monitoring manual | Outside performance evaluation |
| Develop training program and materials | Follow standard operating procedures (SOPs) | Reconcile data with objectives |
| | Document changes | Revise SOPs as needed |
| | Proficiency testing | |

discover a problem quickly, allowing timely action to be taken to remedy problems. They also offer confirmation that you are doing your work correctly.

- *Quality assessment* is the process by which the various phases of data generation are reviewed after data collection. Assessment provides verification that sampling and analytical processes operated within analytical or operational limits and that enough data were collected to permit reasonable interpretation. Together these three components help ensure that the data are reliable.

It is very important to know the reliability of water monitoring data obtained by volunteers, and there have been numerous studies of this. These have generally concluded that the volunteer data is comparable with that obtained by professional operators (Herron *et al.*, 2006).

A study on the validation of data generated from lake samples collected by volunteers in a project carried out by the Lakes of Missouri Volunteer Program (LMVP) concludes that data generated from volunteers are of the same quality as data collected by laboratories. Different methods of evaluation were used. A comparison between volunteer data and university collected samples showed that trophic state classification of lakes were the same for 74% of estimates based on total phosphorus, 84% of those based on total nitrogen and 89% for those based on chlorophyll. There were no significant differences found between estimates of suspended solids, chlorophyll or total nitrogen in a split sample study. Differences were found between analyses of total phosphorus analyses, and these were shown to be due to differences in storage methods (Obrecht *et al.*, 1998). It is important that all operators are aware of errors linked to sample containers, sample preservation, sample conservation and storage, documenting methods and material used. For these reasons, it may be easier to use accepted scientific methods or protocols, but in all such work it is necessary to adopt Standard Operating Procedures (SOPs) for the collection and analysis of environmental samples. SOPs are step-by-step directions for methods including calibration and maintenance procedures for field and laboratory analytical instrumentation (Picotte and Boudette, 2005).

### 1.2.5.6 Quality Assurance Project Plan

A Quality Assurance Project Plan (QAPP) can be designed to support monitoring activities. This is a written document that outlines the procedures (SOPs or best practices) that will be used. The aim is to ensure that the samples collected and analysed, and the data and the reports are of sufficiently high quality to meet the quality goal. It is required for all monitoring efforts funded by the USEPA and for all volunteer monitoring programs supported by the Vermont Water Quality Division. A QAPP must be very detailed with elements prescribed and formatted to meet the needs of reviewers (Picotte, 2005). Software assistance such as SWAMP (Surface Water Ambient Monitoring Program) QAAP Advisor is available. This is a computer-based tool that helps users in producing a QAAP efficiently and accurately (Tadesse and Keith, 2008).

## 1.2.6 APPLICATIONS

Applications of screening methods could be presented according to operators (e.g. environment professionals, other professional, volunteers, students), monitoring needs (e.g. survey, investigation), waterbody (e.g. rivers, lakes, estuaries), methods types (e.g. chemical, biological, physical) or methods principles (e.g. colorimetry, optical methods, Elisa tests). However, the number of studies reported in literature is rather limited, and as discussed above, the main applications have been directed towards evaluating monitoring data from Volunteer networks, and comparing them with those obtained by professionals. On the basis of these comparisons, the authors are able to provide some information on the screening methods used. One of the simplest and most widely used methods (by about 40% of networks) is the Secchi disk for the measurement of transparency in lakes and rivers. This is one of the easiest to use, most affordable and relevant tool available, though there is no standard method for its use. The overall precision (1–3%) achieved by volunteers is excellent, and comparable with that achieved by professional operators using different protocols (Schloss and Craycraft, 2006).

Measurement of turbidity is also a widespread alternative method for both particulates and colloids. It is often measured as Total Suspended Solids (TSS) although this cannot be measured on site. Reid *et al.* (2006) evaluated the effectiveness of using transparency tubes to estimate TSS and turbidity by a volunteer phosphorus monitoring network. The transparency tube is a good predictor of turbidity ($R^2 = 0.78$), but there is a poorer correlation ($R^2 = 0.55$) with TSS. The latter is due to variation between sites in the size, shape and composition of the particles that makes each site unique. On the basis of this study a program of training in the use transparency tubes was proposed. Carlson *et al.* (2006) made a comparison of ocular turbidity instruments (Secchi disk of all the colours, turbidity tube, horizontal black disk and clarity tube) used by volunteer monitoring programs in lakes and streams for shallow waters in Delaware and Chesapeake bays tributaries and main stream. The main conclusion is that the relationships between Secchi disk and turbidity tube appears linear and calibrations seem possible.

Transparency or turbidity is often measured on site with a coupled electrode for the measurement of temperature and dissolved oxygen (more and more measured by optical sensor or optrode). For Rhode Island Volunteers, complementary determinations (pH, alkalinity, TP, chl-a, Ca, Mg, Na and Cl) are carried out in a laboratory and results are statistically similar to those collected by professionals (Herron, 2006).

Wagner (2006) and EPA Region 9 support the Volunteer Network for monitoring bacteria in San Francisco Bay area Creeks, with technical assistance, loans of field sampling equipment, and a limited number of laboratory analyses. Total Coliforms, *E. coli, and Enterococcus sp.* have been analysed (IDEXX methods) on sample collection from four creeks (San Pedro, Cerrito, Temescal and Mont Diablo) during wet and dry season. The results showed:

- sites greatly exceeding the bacterial water quality criterion indicated sewer line leaks;
- absence of bacteria can indicate chlorinated water leaks;

- at the beach site, both indicators (*E. coli and Enterococcus*) exceeded state standards.

The aims of this volunteer monitoring were to form partnerships for increasing the level of stewardship and of public information and especially to reduce bacterial sources. This monitoring program could be improved by checking dissolved oxygen, turbidity, *E. coli* counts, pH, temperature and changes in the macro invertebrate community, because these parameters can indicate an impact of a nearby treatment plant discharge, or leakages of wastewater.

Chemicals can also be measured in the field. For instance nitrates can be measured using a field meter (UV) or colorimetric kit usually used for the determination of other inorganic analytes. Chlorophyll-a can be analysed by a fluorometric device, that can also be used for measuring cyanotoxins (estimated by the phycocyanin concentration) or hydrocarbons. Passive samplers are beginning to be used in the monitoring of organic micropollutants, Multi parameter sensors are frequently used for dissolved oxygen, temperature, pH, conductivity and turbidity. Geary *et al.* (2008) has developed an automated total water quality embedded monitoring device. This device has an optical sensor (optical fibre) that detects faecal matter indicator organisms, and can semi-quantify low levels 50–200 CFU/100 mL of *Escherichia coli*. This is combined with a colorimetric phosphate sensor (range between 0 and 20 mg/L with LOD of 0.3 mg/L). Both sensors are connected to a central monitoring station and could send an alarm signal to warn of the presence of dangerous levels of faecal matter, elevated phosphate levels and deleterious changes in other water quality parameters of interest (such as turbidity, pH and dissolved oxygen). It is still a prototype, but 21 days are scheduled in April 2009 for using 10 sensors. The field test will be carried out on Chicago public beaches.

In another monitoring field, optical brightener sampling devices are used to help identify pollution from faulty septic systems, sewage leaks, storm drain cross connections and to differentiate human/animal waste. Optical brighteners are fluorescent white dyes that are added to almost all laundry soaps and detergent. The method is quite simple: an untreated cotton pad is inserted into a rigid sampling device that is placed into storm drains or pipe outlet. Then the cotton pad is viewed in a dark room under ultraviolet fluorescent light.

For the NJDEP (New Jersey Department of Environmental Protection; NJDEP, 2007; Poretti and Franken, 2008) 40 lakes are sampled 3 times per year (spring, summer and fall). The trophic status of the 40 lakes was determined on the basis of Carlson's Trophic State Index (TSI), linked to TP, transparency and levels of chlorophyll-a:

- TSI from 0 to 40, 15% was oligotrophic (but no lake was oligotrophic at all times)
- 41–50: 33% was mesotrophic
- 51–70: 37% was eutrophic
- >70: 15% was hypereutrophic

The TSI measurement is largely used in the US as included in field protocols and daily operations for crews to use in the Survey of the Nation's Lakes (USEPA, 2007b). This aggregated method allows the lakes to be classified according to their trophic state.

A UV spectrophtometry procedure for the rapid estimation of TSI has been found to provide useful estimates that correlate well ($R^2 = 0.80$) with TSI, and has been proposed for adoption in Canada (Thomas and Pouet, 2005). This screening method takes only one minute against several hours for the measurement of TSI parameters.

A final illustration is provided by developments in the monitoring of cyanotoxins. These are necessary to support the increased requirements for monitoring in response to the huge increase in occurrence of green-blue algal blooms in North America. In order to address the concerns associated with the increases in algal contamination, USEPA is working with states, tribes and others to survey the quality of the nation's lakes, ponds and reservoirs (Tarquinio and Olsen, 2008). This program contains a preliminary assessment of the extent of occurrence of cyanotoxin (Loftin *et al*., 2008). Graham *et al*. (2008) provides guidance for the development of scientific studies of cyanobacteria and associated by-products in lakes and reservoirs. This guidance includes background information on cyanobacteria, toxins, and taste-and-odour compounds; spatial and temporal considerations that are unique to the cyanobacteria in lakes and reservoirs, and information on sample handling, preparation, processing, and shipping. In 2007 lakes were selected randomly for inclusion in the USEPA National Lake Assessment (NLA). 1150 samples have been collected in photic zone in the deepest part of lake for a total microcystin analysis. All samples were analysed by ELISA tests enzyme-linked immunosorbent assays (Kamp *et al*., 2008; Humphries *et al*., 2008) and 2% of samples will be analysed by LC/MS/MS for confirmation. Microcystins were found throughout the United States, especially in the Upper Midwest, with a mean of 3.0 ppb MCLR equivalent and a median value of 0.52 ppb. Nine lakes (0.7%) exceeded the World Health Organization (WHO) recreational guidelines (20 ppb) and 143 lakes (12%) exceeded WHO drinking water guidelines (1.0 ppb) for microcystins.

### 1.2.7 CONCLUSIONS

In the US, as in Europe, the use of screening methods for water quality monitoring is relatively confidential. However, for some years the US regulations have opened the way for the use of screening (alternative) methods by defining comparison and validation procedures. Currently this development is limited by the lack of available innovative solutions.

One of the main differences between Europe and USA is the development of Volunteers Networks for water monitoring, working in a complementary way with water authorities or State agencies. Volunteer actions are numerous and increasing, involve a wide range of citizens, including students and seniors, and are facilitated by a variety of programmes programs some of which are supported by the USEPA. They are beginning to use screening methods, mainly in response to environmental crises such as that caused by cyanobacterial blooms and cyantoxin-related risks. In this case, for example, Elisa tests were developed quickly and efficiently in response to a growing need.

There is now an urgent need to boost research in the design and development of screening tools through partnerships between laboratories and companies, and to achieve this without waiting for another crisis. In the US one stimulus for the industrial development of such products might the large and growing numbers of end users associated with the Volunteers Networks.

# REFERENCES

Allan, I.J., Vrana, B., Greenwood, R., Mills, G.A., Roig, B. and Gonzalez, G., 2006. A 'toolbox' for biological and chemical monitoring requirements for the European Union's Water Framework Directive, *Talanta*, **69**(2): 302–22.

AOAC, 1999. Qualitative and Quantitative Microbiology Guidelines for Methods Validation, *Journal of AOAC International*, **82**(2).

APHA, 1998. *Standard Methods for the Examination of Water and Wastewater*. 20th Edition. American Public Health Association. 1015 15th Street, NW, Washington, DC 20005.

ASTM, 1999. D4855-91: Standard Practice for Comparing Test Methods, ASTM Standards on Precision and Bias for Various Applications. *1999 Annual Book of ASTM Standards: Water and Environmental Technology*, Volume 7.02. 100 Barr Harbor Drive, West Conshohocken, PA 19,428.

Carlson, R., Pasko, S., Mulder, J., Reiter, M. and Shalles J., 2006. A comparison of ocular turbidity instruments for shallow waters, *5th National Monitoring Conference*, San José, California.[†]

CFR, 2008. Title 40, *Code of Federal Regulations*.

CSREES, 2002. K. Addy, L. Green, E. Herron and K. Stepenuck, Why volunteer water quality monitoring makes sense, Factsheet II, available from http://www.usawaterquality.org/volunteer/

CSREES, 2003. E. Herron, K. Stepenuck, K. Addy, Getting Started Finding Resources in the Guide for Growing CSREES Volunteer Monitoring Programs, Factsheet III, available from http://www.usawaterquality.org/volunteer/

CSREES, 2004. E. Herron, L. Green, K. Stepenuck and K. Addy, *Building Credibility: Quality Assurance and Quality Control for Volunteer Monitoring Programs*, Factsheet IV, available from http://www.usawaterquality.org/volunteer/

CSREES, 2008. Matrix of Monitoring Activities, available from http://www.usawaterquality.org/volunteer/

CWA, 2002. Federal Water Pollution Control Act, As Amended Through P.L. 107–303, 27 November 2002.

Geary J.R., Nijak, G.J. and Talley, J.W. 2008. Networked in-situ water monitoring, *6th National Monitoring Conference*, Atlantic City, New Jersey.

Graham, J.L., Loftin, K.A., Ziegler, A.C. and Meyer, M.T. 2008. Guidelines for design and sampling for cyanobacterial toxin and taste-and-odor studies in lakes and reservoirs: *US Geological Survey Scientific Investigations Report 2008*, 5038, 39p.

Herron, E., 2008. Volunteer lake water quality data – an effective tool for lake management, *6th National Monitoring Conference*, Atlantic City, New Jersey.

Herron E., Green L.T., Gold, A.J., 2006. QA/QC assessment of volunteer monitoring in Rhode Island, *5th National Monitoring Conference*, San José, California.

Humphries, E.M., Painter, K. and Pressly, B. 2008. Cyanotoxin ELISA testing in Delaware: Process Development, *6th National Monitoring Conference*, Atlantic City, New Jersey.

---

[†] Considering the importance of this conference (more than 500 presentations since the 5th edition), the reader will find the related information (copy of presentation or abstract) at : http://acwi.gov/monitoring/

ISO, 2001, CD17994, 2001. Water Quality - Criteria for the Establishment of Equivalence Between Microbiological Methods, Final Version, June 15, 2001.

Kamp, L., Church, J. and Rubio, F., 2008. Development of sensitive Immunoessay formats for algal toxin detection, *6th National Monitoring Conference*, Atlantic City, New Jersey.

Loftin, K.A., Graham, J.L., Meyer, M.T., Ziegler A.C. and Dietze, J.E., 2008. Preliminary assessments of cyanotoxin occurrence in the United States, *6th National Monitoring Conference*, Atlantic City, New Jersey.

NJDEP Water monitoring and standards, 2007. Ambient Lake Monitoring Network, Panel 1 Lake Report, Vol. 1 of 2.

NWQMC, 2006. Data Elements for Reporting Water Quality Monitoring Results for Chemical, Biological, Toxicological and Microbiological Analytes, Technical Report no. 3.

Obrecht, D.V., Milanick, M., Perkins, B.D., Ready, D. and Jones, J.R., 1998. Evaluation of data generated from lakes samples collected by volunteers, *Journal of Lakes and Reservoir Management*, **14** (1): 21–7.

Picotte, A. and Boudette, L., 2005. Vermont Volunteer Surface Water Monitoring Guide, Vermont Department of Environmental Conservation.

Poretti, V. and Franken, J., 2008. New Jersey ambient lake monitoring, *6th National Monitoring Conference*, Atlantic City, New Jersey.

Reid, N., Herbert, J. and Baas, D., 2006. Transparency tube as a surrogate for turbidity and suspended solids in rivers and reservoirs, *5th National Monitoring Conference*, San José, California.

Roig, B., Valat, C., Allan, I.J., *et al*., 2007. The use of field studies to establish the performance of a range of tools for monitoring water quality *TrAC Trends in Analytical Chemistry*, **26**(4): 274–82.

Schloss, J. and Craycraft, R., 2006. Gaining clarity on transparency measurements. *5th National Monitoring Conference*, San José, California.

Tadesse, D. and Keith, L., 2008. SWAMP Advisor: a computer-based tool for developing QAPPs. *6th National Monitoring Conference*, Atlantic City, New Jersey.

Tarquinio, E. and Olsen, D., 2008. Survey of the nation's lakes: overview and preliminary results, *6th National Monitoring Conference*, Atlantic City, New Jersey.

Thomas, O., 2006. Alternative methods. In: P. Quevauviller, O. Thomas and A. Van der Beken (eds), *Wastewater Quality Monitoring and Treatment*, John Wiley & Sons, Ltd, Chichester, 53–66.

Thomas, O. and Pouet, M-F., 2005. UV absorption UV spectrum, a fingerprint of organic matter in lake water, *40th Central Symposium*, CAWQ, Burlington.

Travers, K., 2005. The leaf-stream connection, *The Volunteer Monitor*, **17**(2): 16–17.

USEPA, 1996a. Guidelines and Format for Methods to be proposed at 40 CFR Part 136 or Part 141. EPA- 821-B-96-003.

USEPA, 1999a. Protocol for EPA Approval of Alternate Test Procedures for Organic and Inorganic Analytes in Wastewater and Drinking Water. EPA 821-B-98-002.

USEPA, 1999b. Protocol for EPA Approval of New Methods for Organic and Inorganic Analytes in Wastewater and Drinking Water, EPA 821-B-98-003.

USEPA, 2002. *Consolidated Assessment and Listing Methodology – Toward a Compendium of Best Practices*. ( www.epa.gov/owow/monitoring/calm.html)

USEPA, 2003. Elements of a State Water Monitoring and Assessment Program. EPA841-B-03-003.

USEPA, 2004. EPA Microbiological Alternate Test Procedure (ATP) Protocol for Drinking Water, Ambient Water, and Wastewater Monitoring Methods, Guidance. EPA 821-B-03-004.

USEPA, 2005. Overview of Event Detection Systems for WaterSentinel, Draft, Version 1.0. EPA 817-D-05-001.

# References

USEPA, 2007. Standardized Analytical Methods for Environmental Restoration following Homeland Security Events, Revision 3.1. EPA-600-R-07-136.

USEPA, 2007b. *Survey of the Nation's Lakes. Field Operations Manual.* EPA 841-B-07-004. US Environmental Protection Agency, Washington, DC.

USEPA, 2008. EPA's Watershed Academy Web, Introduction to the Clean Water Act, 70 pp. (http://www.epa.gov/watertrain).

Wagner, A., 2006. Volunteer monitoring for bacteria in San Francisco Bay area creeks. 5th National Monitoring Conference, San José, California

# 1.3
# Existing and New Methods for Chemical and Ecological Status Monitoring under the WFD

Benoit Roig, Ian Allan, Graham A. Mills, Nathalie Guigues, Richard Greenwood and Catherine Gonzalez

1.3.1 Introduction
1.3.2 Emerging Techniques
    1.3.2.1 Monitoring of Physico-chemical Characteristics
    1.3.2.2 Monitoring of Chemical Priority Substances
    1.3.2.3 Monitoring of Effect of the Pollutants
1.3.3 New Trends
1.3.4 Conclusions
References

## 1.3.1 INTRODUCTION

The Water Framework Directive (WFD) is a legislative framework to protect and improve the quality of all water sources, including lakes, rivers, transitional and coastal waters, and groundwater in the European Union (EU). This overarching Framework will eventually replace most of the existing water legislation in the EU, and aims to achieve by 2015 'good status' in all these types of water across the member states. Definitions of high, good and moderate ecological status may be found in the WFD 2000/60/EC (Directive, 2000). In addition, it is expected to contribute to the protection, prevention of deterioration and improvement of the quality of aquatic ecosystems, the promotion of sustainable water use, the reduction of pollution, and mitigation of floods and droughts. Monitoring is required to cover a number of quality elements:

---

*Rapid Chemical and Biological Techniques for Water Monitoring*    Edited by Catherine Gonzalez, Philippe Quevauviller and Richard Greenwood
© 2009 John Wiley & Sons, Ltd

hydromorphological, physico-chemical, biological, and levels of specific priority pollutants, both anthropogenic and naturally occurring, according to the type of water body being monitored (specified in WFD Annex V1.2). As the WFD requires management at a river basin level and since many waters cross national boundaries, it is necessary to ensure that monitoring under the WFD is harmonized so that the water quality information collected is comparable, reliable and consistent data across all member states. These aspects are discussed in detail in Chapter 1.1 of this book.

The success of the implementation of the WFD will depend on the information available to those charged with managing water quality being fit for purpose in both quantity and quality. Three modes of monitoring are specified in the Directive:

- surveillance monitoring to assess long-term changes;
- operational monitoring to provide extra data on water bodies at risk or failing to meet the environmental objectives of the WFD;
- investigative monitoring to determine the causes of such failure where they are unknown.

More generally, information obtained through surveillance monitoring will be used to determine requirements for future monitoring, and whether to implement operational or investigative monitoring. Moreover, surveillance and operational monitoring programmes should be established for adequate time periods and reviewed according to monitoring results.

The monitoring network should be designed to provide a coherent and comprehensive overview of ecological and chemical status within each mapped river basin.

For *biological elements*, as for example phytoplankton, the monitoring is mainly based on the taxonomic composition of the community, and abundance in relation to type-specific physico-chemical conditions. Assessments using macrophytes are undertaken by investigating taxonomic composition and average changes in abundance,. The taxonomic composition and abundance, the ratio of disturbance (ratio of sensitive to insensitive taxa), and the level of diversity all contribute to the assessments based on benthic invertebrates. Monitoring based on fish takes into account species composition, and abundance, the presence of disturbance-sensitive species, and the age structure of the fish community. Various models and associated software tools provide an assessment of water quality on the basis of differences between the observed ecological data at the sampling site and those expected under ideal conditions at pristine sites. Importantly, data from pristine sites at different locations across Europe are likely to differ since the species assemblages present in different regions can be markedly different. Hence many of the computer-based models are very similar in the way that they work, but differ in reference conditions that they use. These tools are usually based on benthic invertebrates (AQEM, AusRivAs, ECOPROF and RIVPACS,), fish (CITYFISH and FAME) or a combination of invertebrates, algae, fish and macrophyte (ECOFRAME and PAEQANN). Most, if not all of these tools are for freshwater environments. As for any type of monitoring, the results from biomonitoring are operationally defined, that is they are totally dependent on the sampling methodology and equipment used. For this reason it is important to use standard protocols or methods, and some CEN and ISO standards are available to provide guidance when undertaking biological sampling of fish or invertebrates. The STAR project collates updated protocols to be used to sample biological organisms.

*Hydromorphological monitoring* generally relies on series of elements such as: assessment of quantity and dynamics of flow, connections to groundwater, continuity, channel patterns, width/depth variations, substrate conditions, structure and condition of the riparian zone in the case of rivers, residence time, lake depth variation, tidal flow regime, substrate condition and structure and condition of the intertidal zone, freshwater flow regime and direction and speed of dominant currents for transition/coastal waters.

Monitoring of both biological and hydromorphological elements requires mainly observational and macroscopic counting systems. On the contrary, monitoring of general physico-chemical conditions (e.g. nutrients, salinity, pH, oxygen balance, acid neutralizing capacity, temperature, transparency) and chemical quality in terms of levels of specific synthetic or nonsynthetic pollutants generally involves measurement systems. Currently the most commonly used procedures for measuring these parameters are classical analytical techniques performed under laboratory conditions. These techniques are very sensitive, specific, and cover the totality of the possible parameters to be measured, but are expensive, often time-consuming, and require specialized personnel. In addition, in most cases, spot (bottle) sampling is employed to collect the matrix to be analysed. This has a number of disadvantages, including cost and the fact that it provides only a snapshot of the situation at the instant of sampling. This is an important factor since levels of pollutants can vary with time even at a fixed location, and fluctuations associated with episodic events could be missed, or conclusions could be drawn on the basis of transitory high levels. Taking into account these disadvantages, there is a need for improved monitoring methodologies that can provide the necessary information in simpler ways and at lower cost than is currently possible. There is also a need to provide a complementary approach to the quality monitoring required within the legislative Framework in order to provide representative information in the periods between the infrequent spot samples that form the basis of current practice. This chapter aims to provide an overview of some of the methods that can be used as alternatives to the classical approach, and some of these will be considered in more detail in the later chapters.

In order to provide a more representative picture, sampling and analytical procedures such as automatic sequential sampling to provide composite samples over a period of time (usually 24 hours), continuous on-line monitoring, frequent sampling, or in-field measurement must be used. The first two involve the use of expensive equipment that requires a power supply, and needs to be deployed in a secure site, and the third would be expensive because of transport and labour costs. The last one can be considered as emerging techniques and represents a very useful alternative to current practice. Field techniques (both sampling and measurement methods) can be cost effective, and can provide a rapid representative overview of the parameter of interest.

The different types of field measurement systems are summarized in Figure 1.3.1 (Greenwood *et al.*, 2007a). They are all performed directly in the field. On-line and in-field systems (portable or transportable equipment, chemical tests kits and immunoassays tests kits) require a sampling step (spot sampling, sampling belt). Some

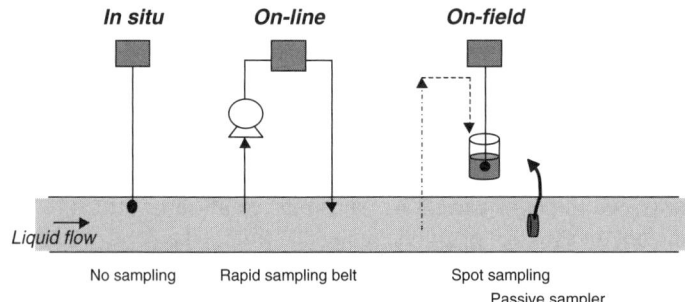

**Figure 1.3.1** Different field measurement techniques (Reproduced, with permission from Elsevier, from Greenwood *et al.*, 2007a)

**Table 1.3.1** Comparison of properties between classical and emerging monitoring techniques

|  | Classical techniques | Emerging techniques | | |
|---|---|---|---|---|
|  | Laboratory methods | *In situ* | On-line | On-field |
| Quantification | +++ | ++ | ++ | ++ |
| Identification | +++ | ++ | + | + |
| Sampling | S | None | SB | S |
| Measurement frequency | -- | +++ | +++ | +++ |
| Speed of response | --- | +++ | +++ | +++ |
| Sensitivity | +++ | − | +/− | +/− |
| Specificity | +++ | ++ | +/− | +/− |

S = spot sampling, SB = sampling belt

other methods allow *in situ* measurements (sensors, probes, on line systems, biological early warning systems (BEWS)). In this case, the pollutants are continuously monitored and data can be stored online. Some of these methods provide qualitative rather than quantitative information.

Figure 1.3.1 summarises the configurations of the various monitoring techniques used to quantify and identify the presence of pollutants in waters.

## 1.3.2 EMERGING TECHNIQUES

This terminology has been used to describe existing methods and new methods that can be employed for rapid and/or representative monitoring of water quality. In the context of the SWIFT-WFD project (European Union's Sixth Framework Project, funded by DG Research), Allan *et al.* (2006) produced a directory of existing and emerging techniques for water quality monitoring. This inventory summarises those emerging methods that could potentially be used in the water monitoring that is necessary within the context of the WFD. This includes the assessment of physico-chemical, biological

*Emerging Techniques*

and chemical quality elements and parameters (but excludes hydromorphological elements). These tools (some commercially available and some in development) include methods/equipment for measuring:

- physico-chemical characteristics (e.g. total organic carbon, pH, temperature, nutrients);
- chemical priority substances (e.g. PAH, pesticides, metals);
- effects of the presence of the pollutants (e.g. mortality, and sublethal effects such as oestrogenicity, reduction in feeding or locomotory activity).

More detailed considerations of some of the various categories of tools are provided in subsequent chapters. Here the principles underlying the various approaches are outlined, and the added value of the various methods and technologies is briefly considered.

### 1.3.2.1 Monitoring of Physico-chemical Characteristics

A number of parameters may be used to indicate the physico-chemical status of a water body. Some other parameters such as pH may be measured directly whilst for others such as salinity indirect measures such as conductivity are used. Nutrients such as ammonium, nitrite, nitrate, phosphate, and more generally total nitrogen and phosphorus may be useful indicators in monitoring programmes as they are involved in eutrophication processes, and some (e.g., nitrates) may contaminate groundwater after fertilizer applications. Other parameters may be used to characterize the degree of oxygenation of a water body, and include dissolved oxygen, the chemical oxygen demand, the biochemical oxygen demand, redox conditions, or respirometry measurements. The amount of dissolved, suspended organic matter can be assessed in a number of ways: by measuring total organic matter, total organic carbon, aromaticity, and the turbidity of the water. In turn the presence and levels of organic matter strongly influence the chemical oxygen demand of a water sample. Different types of devices are available based on a range of specific electrodes, optical sensors, UV, visible spectroscopy, colorimetry, chemiluminescence, titrimetric methods or ion chromatography. Table 1.3.2 summarizes the available techniques:

### 1.3.2.2 Monitoring of Chemical Priority Substances

Monitoring the chemical quality of water involves taking samples of water and analysing those samples for the chemicals of interest or importance. In most legislation governing this area there are lists of priority pollutants that are regarded as being harmful or potentially harmful. These are normally classified under three major categories: nonpolar organic compounds (e.g. some pesticides, and some industrial chemicals such as PAHs), polar organics (some pesticides, and pharmaceuticals), and heavy metals (e.g. mercury and cadmium), and the methods for monitoring these are

**Table 1.3.2** Available technique (commercial or in development) for physico-chemical monitoring

|  | Specific electrode | Optical techniques | Ionic chromatography | Polarography | Titimetry |
|---|---|---|---|---|---|
| Ammonium | ✓ | UV, V, C, Ch | ✓ |  | ✓ |
| BOD |  |  |  |  |  |
| COD | ✓ | UV, V, Ch |  |  | ✓ |
| Conductivity | ✓ |  |  |  |  |
| Dissolved oxygen | ✓ |  |  | ✓ |  |
| Organic matter |  | UV |  |  |  |
| pH | ✓ |  |  |  |  |
| Phosphate |  | V, C, N | ✓ |  | ✓ |
| Redox | ✓ |  |  |  |  |
| TOC | ✓ | IR, UV |  |  |  |
| Total nitrogen |  | UV-V |  |  |  |
| Total phosphorus |  | V, C, N |  |  |  |
| Turbidity |  | N, UV, IR |  |  |  |

Ch = chemiluminescence, C = colorimetry, IR = infrared, N = nephelometry, UV = ultraviolet V = visible

very different in detail. A number of steps is involved, and most of these are common to all classes of chemicals.

1. sampling
2. stabilization
3. transport
4. storage
5. preparation for analysis
6. analysis

Depending on the monitoring methods used, some of these steps may be omitted. For instance on-site methods obviate the need for transport and storage. Sampling usually comprises taking a water sample from a water body and placing it in a bottle, but in some cases samples can be pumped through a pipe for collection and/or analysis (especially for groundwater), and some methods can be carried out *in situ* so that the sampling step is omitted completely. The analytical step can be effected using a wide range of methods. The best established and currently most tightly monitored in terms of quality assurance and control are chromatographic methods (gas chromatography, or liquid chromatography) linked to a sensitive detector (e.g. flame ionization, electron capture, mass spectrometer, fluorescence spectrometer, UV spectrometer), and for metals methods such as inductively coupled plasma mass spectrometry, or graphite oven atomic absorption spectrometry. However, a range of other methods (based on for instance chemical, optical, electrochemical of biological techniques) is available for carrying out the analytical step, depending on the chemical to be measured. All of the methods used for monitoring chemical parameters have one feature in common, a recognition system that is specific for the target molecules/parameter. Whatever the technique used there is, as discussed in the first chapter of this book, an absolute requirement for established quality assurance and quality control procedures.

The chapters in this book will concentrate on the alternatives to the currently used classical methods, and on methods that can complement current practice to provide more representative pictures of the chemical quality of water bodies.

A range of analytical methods that can be used in the laboratory or in some cases in the field is available (see Section 2 of this book (Chemical Methods)). Some methods are based on specific complexation reactions whose products (usually coloured) can be quantified in a colorimeter or a spectrophotometer. Some can be measured directly on the basis of modification of an electromagnetic beam to produce a characteristic spectrum (e.g. UV and IR spectroscopy), or through emission of a characteristic spectrum (e.g. fluorescence and luminescence spectroscopy). Electrochemical techniques are based on the measurement of changes in a potential (or a current) generated between two electrodes placed in a solution to be analysed. These methods are for measuring either directly (e.g. pH, concentrations of some metals) or indirectly by using an intermediary substrate that recognizes the analyte of interest (as used for some organic molecules). Immunoassays comprise a further set of assays or methods for measuring levels of different chemicals in water samples. They rely on the interaction between a biological material (the recognition step) called an antibody, and this is specific to the analyte under study. These assays are available in a range of formats such as bound to magnetic particles in 96-well plates or coated at the bottom of test tubes. An amplification system is incorporated and often provided by a linked-enzyme system that provides a coloured product from a colourless substrate. Other systems for quantifying the bound analyte include the use of fluorescent or radio-labelled tags. These methods are very sensitive, and are applicable to a wide range of compounds e.g. PAHs, pesticides, phenols, surfactant residues, heavy metals, mutagens and other PCBs. Many of the assays described here are commercially available in kit form.

Some of the above methods can be miniaturized, and the instrumentation can be adapted for use in field situations (see Section 4 of this book (Potential Use of Screening Methods and Performance Evaluation)), and some can be used only in the laboratory. In the former case a spot sample can be taken, processed and analysed in the field without the need for sample preservation, transport, or storage. Since many of these methods are rapid, they can provide either quantitative (concentration) or qualitative (above or below a threshold) data on water quality in a time-scale that enables a timely and appropriate response (for instance in the case of an accidental spillage) or the rapid mapping of water quality in a wide area. Table 1.3.3 shows the main classes of chemical priority substances and the different methods that can be used for their analysis.

### 1.3.2.3 Monitoring of Effect of the Pollutants

Measurements of biological responses to the levels of contaminants present in the environment can provide an alternative or complementary approach to measurements

**Table 1.3.3** Main emerging monitoring methods used for chemical priority substances

|  | Example | Optical techniques | Biosensor/ Bioassays | Immuno-assay | Electro-chemistry |
|---|---|---|---|---|---|
| Metals | Cd, Hg, Ni, Pb | ✓ |  |  | ✓ |
| PAH | Benzo(a)pyrene<br>Benzo(b)fluoranthene | ✓ | ✓ | ✓ |  |
| Polychlorinated compounds | Hexachlorobenzene<br>Pentachlorophenol |  | ✓ | ✓ |  |
| Pesticides | Atrazine, Alachlor, Diuron |  | ✓ | ✓ |  |
| Endocrine disruptors | Nonylphenol, phthalates | ✓ | ✓ | ✓ |  |

of the concentrations of individual chemicals for the assessment of water quality and for the identification of trends in quality (see Section 3 of this book (Biological Methods)). In general, there is a proportional relationship between the concentration of an analyte in the water and its impact on a living organism. This biomonitoring can be based on changes measured at the level of the whole organism, tissues, cells or isolated biochemical mechanisms. The data can be qualitative, semi-quantitative or quantitative. These biological monitoring techniques include bioassays, biomarkers, and BEWS.

Bioassays are based on the use of whole organisms (including a range of animals (both vertebrates and invertebrates), yeast, algae and bacteria) or of isolated parts of organisms (including isolated cells and tissues, enzyme systems). These can be used to detect or quantify levels of organic and inorganic pollutants or to measure the general toxicity of water samples. Since there is generally a complex mixture of contaminants in the water, the latter is the most common application of bioassays. Toxicity depends in part on the bioavailability of the toxicants, and this is affected by factors such as the presence of suspended matter, and the concentration of dissolved organic carbon. It is also affected by physicochemical variables such as pH, temperature, water hardness, oxygen tension, and salinity. Information on limits of detection for various chemicals and standard operating procedures are usually supplied by the producers. These assays usually depend on the use of spot sampling and are laboratory-based. However, it is difficult to add preservatives to samples for transport since in some cases they would markedly affect the outcome of the assay. Some assays can be conducted *in situ* or online by deploying organisms in retaining systems directly in the environment, or by placing them in a flow of water pumped from the water body being monitored. Some assays use a range of species of organisms with varying sensitivities to the different types of pollutant in order to provide maximum sensitivity across the range of pollutants encountered. In some cases the systems measure behavioural changes (e.g. *Daphnia* toximeter) or changes in metabolism (e.g. pulse amplitude-modulated (PAM) chlorophyll fluorometer), and these are monitored continuously using electronic systems, and the data can be transmitted or stored. Such systems need to be housed in a secure location. In some cases (BEWs) the changes in the behaviour or metabolism

*Emerging Techniques* 47

**Table 1.3.4** Some of the commercially available bioassays

| System | Organism used | Stimuli |
| --- | --- | --- |
| AquaTox Control, BehavioQuant®, Truitosem™, Truitel™ | Freshwater fish | Swimming capacity behaviour |
| | Freshwater fish | Swimming capacity behaviour |
| Fish toximeter | Zebra fish | Movement rapidity |
| Daphnia Test® | Daphnia | Daphnia activity, swimming behaviour |
| MosselMonitor® | Mussel | Valve movement |
| ToxAlarm®, Toxiguard® | Algae and bacteria | Respirometry, oxygen production and consummation |
| Algae toximeter | Algae | Photosynthetic activity |
| Fluotox | Algae | Fluorescence natural emission |
| Microtox®, Toxscreen-II test, ToxAlert®, Vitotox®, Biotox™, GreenScreen® | bacteria | Natural bioluminescence inhibition |

are linked to an alarm system so that personnel responsible for the management of water quality are aware of any pollution events and can take appropriate action to safeguard resources such as drinking water production plants or aquaculture facilities from pollution. Most of these assays provide a measure of acute toxicity only, and do not provide information on chronic toxicity produced by long-term exposure to toxicants (possibly present in very low concentrations). Table 1.3.4 gives a list of some of these technologies that are either commercially available or still at the prototype stage. Some of these systems may be able to detect microbiological pathogens as well as chemical toxicants.

The use of biomarkers as a measure of water quality has increased markedly over recent years, and a detailed discussion of these is presented along with some recent applications in Chapter 3.6 of this book. A biomarker is defined as a change in a biological response (ranging from molecular through cellular and physiological responses to behavioural changes) which can be related to exposure to or to toxic effects of environmental chemicals. Biomarkers may be classed into three main types:

1. *Biomarkers of exposure:* These include changes that can be detected within a compartment of an organism, and that are the result of the exposure of part of the biological system (a target molecule or cell component) to pollutants and/or their metabolites, or to some other environmental stressor. Such biomarkers are not necessarily directly related to some specific mode of action, and give a general measure of exposure to adverse conditions. They include a number of proteins. One example is heat shock protein (HSP) where an increase in the amount measured at the cellular level indicates that the animal has been exposed to an environmental stressor (e.g. temperature, osmotic pressure, contact with oxidizing, organic or metallic agents). Other biomarkers include the induction of the expression of specific cytochrome P450 linked oxygenases that can indicate exposure to, or contamination with, a range of compounds including PAH, PCB, dioxins and some

pesticides; and levels of metallothioneins that are produced in response to the accumulation of heavy metals.

2. *Biomarkers of effect:* These comprise measurable biochemical, physiological or other alterations within tissues or body fluids of an organism that can be related to specific causal factors, and can provide information about the magnitude of adverse effects related to those specific causal factors. These include the well-documented cases of the impact of estrogenic compounds in fish where the levels of vitellogens in males act as a biomarker.

3. *Biomarkers of susceptibility:* These provide an indication of an increase in tolerance (inherent or acquired) of an organism to exposure to a specific pollutant, and may include genetic factors, or for instance changes in the level and/or sensitivity of specific receptors which determine the susceptibility of an organism to a particular toxicant or family of toxicants whose mode of action involves interaction with those receptors.

Some of these biological methods are true screening methods, and provide an indication of whether further monitoring is necessary or not. The information they provide may be quantitative, all or nothing in nature, or qualitative, and can complement the information from chemical monitoring. Although, because of the nature of these techniques, the uncertainties associated with them can be large this may not be important providing that the size of the uncertainty is defined. It is just as necessary to apply rigorous quality control to these biological methods as to the chemical monitoring methods where such practice is well established. Various chapters of this book provide more detailed discussions of the biological methods, and examples of their application.

## 1.3.3 NEW TRENDS

Some of the methods that are currently available and that are developing rapidly include passive sampling, and rapid measurement techniques based on developments in combinations of biotechnological and biophysical systems to provide increased sensitivity and robustness.

Passive samplers were originally developed to mimic uptake by living organisms (as used in biomonitoring campaigns), and to indicate the bioavailability of nonpolar organic and metallic pollutants. However, their ability to provide time-weighted average (TWA) concentrations to which they have been exposed over deployments of days to weeks has been recognized as being valuable in the context of monitoring water quality. This allows them to provide representative pictures of water quality over time, even where the concentrations of individual contaminants fluctuate markedly. Such representative monitoring will reduce the uncertainty in assessing average conditions using the current practice of infrequent spot samples. The developments in passive sampling have been driven by the need to provide reliable estimates of TWA concentrations of pollutants.

All passive samplers work on the same principles and have the same basic components, a receiving phase with a high capacity and high affinity for the chemicals to be monitored, separated from the ambient environment by a diffusion limiting layer. Pollutants are accumulated in the receiving phase under Fickian diffusion, and from the mass of chemical accumulated over a given exposure time, and device-specific calibration parameters it is possible to calculate the TWA concentration of freely dissolved contaminant to which the sampler was exposed. A range of passive samplers is now available for metals, nonpolar organics, and polar organics (Namiesnik et al., 2005; Greenwood et al., 2007). The, SPMD (Semi Permeable Membrane Device) for hydrophobic organic compounds and DGT (Diffusive Gradient in Thin films) for metallic compounds are longest established, but other devices (e.g. Chemcatcher®, MESCO, POCIS) with a wide range of operational characteristics are becoming available. In addition to providing representative data on water quality, passive samplers have some other advantages. They provide a direct measure of the bioavailable fractions of contaminants, and can provide information on the speciation of pollutants. Further since they accumulate significant quantities of trace chemicals over time they can bring them within the working range of the analytical method even where they would be below the levels of detection in extracts of spot (bottle) samples of water.

An increasing effort is being put into the development of biotechnological methods for use in environmental applications, and particularly in the area of delivering new detection systems. The application of genomics, proteomics (transcriptomics) in the area of ecotoxicology (ecotoxicogenomics) involves using changes in the genetic and protein responses of organisms in a contaminated (polluted) medium (Snape et al., 2004). These techniques are based on the used of biomarkers. They can provide information on the interaction of toxicants with DNA and with proteins. Developments in biomolecular techniques have enabled the isolation of systems that respond to specific pollutants, and their introduction into other organisms along with convenient signalling systems such as in the recombinant receptor-reporter gene assays for estrogenic compounds (see Chapter 5.3 of this book), and the Greenscreen® yeast genotoxicity assay. There have also been major developments in the detection of pathogenic microorganisms, and toxic bloom producing algae by applying real time PCR techniques to detect specific nucleic acids. There is now the potential to use these techniques in the field because of the development of portable PCR systems (http://www.enigmadiagnostics.com).

Developments in surface plasmon resonance (SPR) have facilitated (improvement of rapidity, sensitivity) the detection of pollutants by measuring their interactions with a biological target (e.g. a receptor, an enzyme, a section of nucleic acid) (Habauzit, 2007). Examples concerning the use of SPR for the detection of pesticides, estrogenic compounds or heavy metals have been described (Gobi et al., 2005; Kim et al., 2006; Mauriz et al., 2006; Habauzit et al., 2008). Moreover, portable systems are now available as well as optical fibre based systems for field applications (Zeng and Liang, 2006).

Another emerging domain is the use of remote sensing for the monitoring and the management of water resources. Indeed, the coupling of field measurements (ground verification) with satellite pictures represents an important source of information that when integrated with other data will allows a better integrated management of water bodies. Approaches combining physical and biological data ground data with satellite observations have been applied for some time, and the main challenge for remote sensing (teledetection or telesurveillance) is to develop and implement models that can take into account the different parameters involved in determining water colour (the most commonly used indicator of water quality). Models should enable an understanding of the complex interactions involved when aquatic ecosystems are perturbed.

## 1.3.4 CONCLUSIONS

The methods outlined above show significant promise as tools that could be used to support the monitoring required under environmental legislation. They may be a useful solution to the problem of balancing a requirement for increased monitoring, made worse by the continuing emergence of new pollutants of concern, with increasing constraints on the resources to achieve this. However, for many of the methods described a lot of work will be necessary to ensure and demonstrate that they are robust, and fit for purpose. These methods will not be adopted by regulatory bodies without the confidence that can be gained only by publication of thorough validation studies and the development and application of appropriate quality-assurance and quality control procedures Progress would be further aided by the development of appropriate norms in many areas of environmental monitoring.

## REFERENCES

Allan, I.J., Vrana, B., Greenwood, R., Mills, G. A., Roig, B. and Gonzalez C., 2006. *Talanta*, **69**, 302–22.
European Commission, 2000. Directive 2000/60/EC of the European Parliament and of the Council of 23 October 2000 establishing a framework for Community action in the field of water policy, L327/1, 2000, p. 72.
Gobi, K.V., Kataoka, C. and Miura N., 2005. *Sens Actuators* **B 108**, 784–90.
Greenwood, R., Mills, G.A. and Roig, B., 2007a. *Trends Anal. Chem.*, **26**, 263–7.
Greenwood, R., Mills, G.A. and Vrana, B. (eds), 2007b. Passive sampling techniques in environmental monitoring. Vol. **48**, in *Comprehensive Analytical Chemistry*, Series Editor in Chief D. Barcelo, Elsevier.
Habauzit, D., Chopineau, J. and Roig, B., 2007. *Anal. Bioanal. Chem.*, **387**, 1215–23.
Habauzit, D., Chopineau, J. and Roig, B., 2008. *Anal. Bioanal. Chem.*, **390**, 873–83.
Kim, S.J., Gobi, K.V., Harada, R., Shankaran, D.R. and Miura, N., 2006. *Sens Actuators*, **B 115**, 349–56.
Mauriz E, Calle A, Montoya A, and Lechuga LM., 2006. *Talanta*, **69**, 359–364.
Namiesnik, J., Zabiegala, B., Kot-Wasik, A., Partyka, M. and Wasik, A., 2005. *Anal. Bioanal. Chem.*, **381**, 279–301.
Snape, J.R., Maund, S.J., Pickford, D.B. and Hutchinson, T.H., 2004. *Aquat. Toxicol.*, **67**, 143–154.
Zeng, J. and Liang D., 2006. *J. Intell. Materia Systems and Structures*, **17**, 701–7.

# Section II
## Chemical Methods

# 2.1
# The Potential of Passive Sampling to Support Regulatory Monitoring of the Chemical Quality of Environmental Waters

Graham A. Mills, Branislav Vrana and Richard Greenwood

2.1.1  Introduction
2.1.2  Passive Sampling Techniques
      2.1.2.1  Devices
2.1.3  Applications
      2.1.3.1  Polar Organic Compounds
      2.1.3.2  Nonpolar Organic and Organometalic Compounds
2.1.4  Conclusions
References

## 2.1.1  INTRODUCTION

Long established water monitoring schemes around the world depend on the collection of spot (bottle or grab) samples, which are transported to a laboratory for analysis. The latter is usually carried out in accredited laboratories that routinely handle large numbers of samples using standard, well validated methods (e.g. the US EPA 500 and 600 series of analytical methods) for the analytes of interest. The high quality of the chemical analysis is the result of sustained efforts over the last twenty years to develop quality assurance (QA) and quality control (QC) protocols, and the production of large volumes of certified reference materials for a range of analytes in different environmental matrices for use in inter-laboratory trials. In contrast the sampling step has received far less attention, and yet the reliability of the data obtained from the monitoring process

---

*Rapid Chemical and Biological Techniques for Water Monitoring*  Edited by Catherine Gonzalez, Philippe Quevauviller and Richard Greenwood
© 2009 John Wiley & Sons, Ltd

depends on the way in which the sample is taken, preserved, transported, stored and treated prior to analysis. For some compounds, significant losses or changes in the proportions of species present can occur due to processes that include: adsorption on the walls of sampling and storage containers, volatility of the analyte of interest, e.g. mercury or organic solvents, degradation due to both biological and physical chemical actions. Depending on the properties of the individual pollutants, different proportions will be in free solution or bound to particulate matter or dissolved organic carbon. If the sample is pretreated, for instance by filtration or ultra-filtration, then the total concentration measured after this could be substantially less (orders of magnitude for very nonpolar analytes) than that measured in untreated samples.

Legal regulation of water quality goes back over a century, but in recent years there have been major developments in the management and control of environmental pollution, and water quality. The implementation and enforcement of this legislation depends on the availability of reliable, fit for purpose monitoring data. In the European Union (EU) in 2000 a range of existing regulations was incorporated into an overarching framework, the Water Framework Directive (WFD) (Directive 2000/60/EC). A novel development in the WFD is the setting of ambitious targets and a tight time scales for improving the quality of water (inland, transitional, and coastal surface waters) across Member States and achieving sustainable use by 2015. A daughter directive (the Ground Water Directive: Directive 2006/118/EC) has been developed to extend this approach to cover the quality and sustainable use of all ground water in the EU. These developments are reflected in developing legislation in many other areas of the world (e.g. Australia and USA). In the EU, monitoring is required at river basin level and involves transboundary cooperation between Member States, and for surface waters approved monitoring schemes have been in operation since 2007. The WFD does not mandate any particular method of monitoring or chemical analysis, and allows for continual improvement of monitoring practice. It provides for periodic review of the performance of the monitoring programmes and opportunities for updating the methods and technologies used in light of scientific and technical developments. This encourages the development, evaluation, validation, and adoption of novel methods and technologies.

In most countries water quality legislation is based on comparisons of levels of specific pollutants found in spot samples with defined concentrations of pollutants that are perceived to be of environmental or human health concern. In the WFD this takes the form of a list of priority pollutants that are known to be particularly harmful in these respects. This list is continuously evolving as compounds become of less concern following their reduction and/or removal from use, or new compounds are identified as posing an environmental risk. Currently this list is based on 33 hazardous chemicals (COM (2006) 397 final), of which 12 are recognized as priority hazardous substances of which emissions must end within 20 years. The listed compounds represent a wide range of chemical classes including nonpolar and polar organics, metals, and organometalics. For each of these pollutants two environmental quality standards (EQS) are defined, and these form the basis of compliance monitoring. There is an annual average EQS (AA-EQS) based on chronic toxicity measurements, and a maximum allowable concentration EQS (MAC-EQS) that reflects the acute toxicity of the priority pollutant. Currently infrequent (typically monthly) spot samples taken at a series of fixed sampling stations are analysed by classical laboratory-based methods and the measured concentrations are compared with the MAC-EQS and the average of these with the AA-EQS. This provides the basis of checking for compliance, risk assessments, monitoring trends, and managing water quality. There is a continuing trend to lower the EQS concentrations in light of improving toxicological and environmental information. This often presents challenges in terms of limits of detection for the currently available analytical methods that use small volume (1-5 L) spot samples of water.

The main aims of the WFD are to improve the quality of all water (surface and ground water) in the EU, and to achieve sustainable use. These objectives may not always be best supported by current monitoring practice on its own. One of the problems in some water bodies is that concentrations of pollutants can fluctuate in time, for instance with seasonal applications of pesticides, seasonal or diurnal fluctuations in sewage treatment plant inputs and effluents, sporadic industrial discharges from batch processes, and wash-off following weather events. Under these circumstances infrequent spot sampling can fail to provide an accurate picture of water quality even on annual average basis. Further difficulties can be caused by marked spatial variation that can occur for instance where effluents mix very slowly with the main flow of water, and remain concentrated in a tight plume for long distances below the outfall (monitoring in compliance with the WFD avoids the problems associated by mixing through recommendations that sampling stations should be outside mixing zones). If there is marked spatial and/or temporal variation then infrequent spot sampling at a few selected sites may yield biased information on the chemical quality of a water body. Since decision making is based on this information, it could lead to inappropriate actions or failure to take action where intervention is needed. In some cases this could be very costly.

There are several ways of improving the information provided by monitoring campaigns, and these include frequent spot sampling, continuous online sampling, biological monitoring, and passive sampling. All of these could potentially be used since the WFD does not mandate any particular method of monitoring or chemical analysis, and provides only a range of guidelines reflecting best available practice (CMA, 2007). However, it does impose minimum quality standards, and requires that comparable methods, both of sampling and analysis, are used. Measurements must be sufficiently accurate and precise to allow important differences between water bodies to be detected, and trends to be identified reliably. It is essential that all data are reliable and representative, but it is difficult to judge the accuracy of a measurement unless the objective is clearly defined. For instance, if a spot sample is taken, and the concentrations of defined pollutants are measured and compared with an EQS, then that information could be used within a legal framework as part of enforcing compliance. However, if the variable of real interest is the chemical quality of the water body, and not just the concentrations of defined pollutants at the instant that a bottle of water is taken, then infrequent spot sampling may not provide an accurate or useful measurement. In some cases the concentration measured is operationally defined, and it is important to understand what is being measured by the sampling and analytical methods used. Different methods will measure different fractions or combinations of fractions of a pollutant. Some of these will be more biologically relevant than others. For instance some species of metals are more toxic than others, and in some cases this reflects their relative bioavailability. Fractions of organic compounds that bind to suspended particulate matter and to dissolved organic carbon may not be bioavailable to some species of animals. The method of sampling and sample treatment (e.g. centrifugation or filtration) before analysis may thus modify not only the concentration value measured, but also its biological relevance. When selecting a method that is to be used in support of a framework of water quality regulations it is important to consider the biological relevance, and representativeness, as well as the precision of the data obtained. This is particularly relevant where EQS values are defined on the basis of toxicity data. A further important consideration is the cost of the method.

A range of sampling and analytical technologies and methodologies that could provide useful information within the context of the WFD were investigated within SWIFT-WFD (an EU funded Sixth Framework project: www.swift-wfd.com) and some of these form the subject of other chapters in this volume. All of these have strengths and weaknesses, and provide different types of information. This chapter will consider the utility of passive sampling within the context of regulatory monitoring: however, it will not provide a comprehensive review of this technology since a number of these have been published recently (Stuer-Lauridsen, 2005; Vrana *et al.*, 2005; Kot-Wasik *et al.*, 2007). Rather, we will concentrate on the sampling steps, and in particular on the potential of passive sampling to provide support within a regulatory context, and explore aspects of calibration and validation.

## 2.1.2 PASSIVE SAMPLING TECHNIQUES

Whilst frequent spot sampling, time- or flow-weighted sampling, or continuous online methods can provide useful pictures of water quality their use is limited by cost, and for the latter also by limitations in location due to the need for security and controlled laboratory conditions. Biomonitoring techniques provide information that is of direct biological relevance, but are costly, have ethical implications, and involve difficult sample preparation steps. Biological variation is an important factor that increases the overall uncertainty associated with these methods. Passive samplers can mimic uptake by living organisms (Ke *et al.*, 2007) since they monitor the freely dissolved species of pollutants that are considered to be the more biologically relevant fractions. They have an important advantage over biomonitoring; they can be placed in a wide range of conditions where organisms could not survive (Smedes, 2007). This method also provides time-weighted average (TWA) concentrations of pollutants over periods of days to weeks, and at relatively low cost. These methods have the potential to complement or replace spot sampling. They yield extracts that are similar to those from spot samples, and that can be analysed using standard, validated laboratory methods equivalent to those developed for use with water samples. The most important factor that hinders the acceptance of these methods for use in a regulatory context is a lack of suitable QA and QC procedures for the calibration steps necessary to calculate the TWA concentrations to which the samplers were exposed during field deployments.

Passive sampling technology was developed to measure TWA concentrations of pollutants in air, and in water. Its application in air monitoring for instance for measuring worker exposure to volatile organic compounds, and measuring indoor air quality is well established, and a range of commercially available devices, and standard procedures for their use are in place (Bartkow *et al.*, 2007; Goia *et al.*, 2007). The passive sampling technologies for use in water are lagging behind in the development of standard operating procedures, and accepted QA and QC protocols to allow them to be used in support of regulations. One standard (BSI, 2006) is available; this covers deployment in surface waters.

All passive samplers work on similar principles, and comprise the same main components. There is a receiving phase that accumulates pollutants of interest, and has a high affinity for them so that their concentrations remain close to zero at its surface. This phase is separated from the bulk water of the environment by a diffusion limiting layer. In some cases the latter is provided by a membrane that serves not only to limit the rate of diffusion from the bulk phase to the receiving phase, but also provides protection from physical and biological fouling. The mass of a contaminant accumulated during a fixed deployment period will depend on only its freely dissolved concentration in the bulk phase, and the sampling rate of the sampler ($R_s$) for that analyte. The latter is independent of the concentration of the pollutant in the bulk phase of the water. Material bound to particulate matter or to dissolved organic carbon is in most cases not available to drive diffusion into the sampler.

Accumulation of materials in a passive sampler can be described in terms of an exponential approach to a maximum, and samplers can be used in two modes, equilibrium and kinetic. The former mode depends on the samplers being deployed for sufficient time to allow the accumulated material to approach the maximum (equilibrium) concentration in the receiving phase. For kinetic sampling the deployment time must be restricted so that the samplers are operating in the region before equilibrium is approached. Ideally the samplers should be deployed for no more than the half time to maximum since in this early phase uptake is approximately linear. Equilibrium samplers have been used to measure concentrations of pollutants where concentrations are relatively constant (e.g. in lakes, ground water, and sediment pore water) (Mayer *et al.*, 2003; Vroblesky, 2007). Kinetic samplers have been used more widely, and in situations (e.g. rivers, waste water discharges, tidal waters) where concentrations of pollutants can fluctuate widely over periods of hours to days. Over a long deployment period the samplers may accumulate more analyte than could be extracted from standard spot sample volumes (typically 1–5 L), and thus allow the measurement of concentrations in the water that would be below the limits of detection when using spot sampling in conjunction with standard analytical methods.

### 2.1.2.1 Devices

A wide range of kinetic samplers has been developed and used in recent years, and some are commercially available. Among the most widely used are the well-established semi-permeable membrane devices (SPMDs) for hydrophobic organic pollutants and the diffusive gradients in thin films (DGTs) for metals and inorganic ions. Several novel passive sampling devices (e.g. the membrane-enclosed sorptive coating (MESCO), the nonpolar Chemcatcher®) have been developed to monitor nonpolar organic pollutants, and the inorganic Chemcatcher® to monitor metals. Two samplers (polar Chemcatcher®

and the polar organic integrative sampler (POCIS)) have been developed to monitor a range of polar organic chemicals (e.g. some pesticides, pharmaceuticals and personal care products). Silicon rods (Paschke *et al.*, 2007) and naked polymeric chromatographic phases stabilized in extraction disks (e.g. Empore™ extraction disks) (Tran *et al.*, 2007) have also been used for monitoring polar organics. A version of the Chemcatcher® has been developed to monitor organo-metalics (Aguilar-Martínez *et al.*, 2008a), and SPMDs (Følsvik *et al.*, 2000) have also been used for this purpose. Other samplers for nonpolar organic compounds include low-density polyethylene (LDPE) strips, and silicone rubber sheets. These samplers have been trialled alongside living organisms and have been found to provide robust, sensitive models of bioaccumulation, and show potential for routine use in a range of aggressive environments where it would be difficult to deploy living organisms (Booij *et al.*, 2006a; Smedes, 2007).

Whilst passive samplers have shown promising potential as tools to be deployed in support of water quality legislation they have not been adopted. This is because of, amongst other factors, a lack of robust QA/QC procedures for their calibration, and of long-term field validations alongside the current methods. The former would be facilitated by the provision of reliable reference materials, and the establishment of inter-laboratory trials similar to those used in the field of analytical chemistry. However, because large volumes (thousands of litres) would be required it is not possible in terms of both cost and availability to use existing reference materials. Another approach that could be beneficial would be the establishment of standard reference test sites where the water chemistry can be well characterized. Another factor that would further enhance the acceptability of the method would be the development of samplers that could be manufactured in large numbers to a high standard with tight QC, but at a low cost, and that are robust, and easy to handle and deploy. Even with such samplers it would be necessary to use blanks to check for contamination during manufacture, and handling, and this adds to the total cost of monitoring. However, samplers that lend themselves to automatic sample preparation would help to reduce the analytical costs.

## Calibration methods

Calibration of passive samplers needs to take into account the impact of environmental variables such as temperature and turbulence on the sampling rate ($R_s$). Turbulence is difficult to measure in the field, and for some samplers it can modify $R_s$ through changing the thickness of the water boundary layer at the surface of the diffusion limiting membrane. This boundary layer forms part of the resistance to diffusion, and in some samplers can be the rate limiting layer. The DGT was designed to have a thick diffusion limiting layer (in the form of a hydrogel) so that fluctuations in the thickness of the boundary layer will have a negligible effect of the sampling rate. Another important factor that can affect $R_s$ is biofouling of the sampler surface. This is difficult to incorporate into routine calibration studies since the nature of the biofouling can vary between locations, and between times within locations (Richardson *et al.*, 2002; Booij *et al.*, 2006b). Some workers have attempted to prevent biofouling by incorporating biocides in the system, or using copper meshes (Mills *et al.*, 2008).

Various approaches have been taken to calibration (Stephens and Müller, 2007). For the DGT the calibration is based on diffusion coefficients in the hydrogel layer for a range of inorganic species at a range of temperatures (Warnken *et al.*, 2007). For the Chemcatcher®, MESCO, POCIS and SPMD samplers used in kinetic mode, calibration parameters are obtained from laboratory-based tank experiments. Tank experiments have been carried out in a number of ways; static, semi-static (batch renewal) where concentrations can decrease over time, and continuous (through) flow that uses a fixed constant concentration of test analytes. Ideally these should provide estimates that cover a range of water temperatures and turbulence conditions that are typically found in field exposures. Turbulence is varied by using a range of stirring speeds in a tank, or in some cases by rotating a carousel that houses the samplers at a range of speeds. One potential problem is the bias that can be introduced in the measurement of $R_s$ for highly nonpolar compounds. These can bind to components of the calibration system, to dissolved organic carbon and to particulate organic matter, and this can lead to an overestimation of the freely dissolved fraction in the water phase, and hence an underestimation of $R_s$. For some other samplers including the LDPE strips and silicone rubber sheets when used in equilibrium mode, calibration is based on measured or calculated partition coefficients (Smedes, 2007) between the receiving phase and water. The problems of bias in the measurement of partition coefficients of very hydrophobic compounds are similar to those described above for kinetic samplers. Calibration is further complicated for dissociable compounds where pH of the water can affect the proportions of dissociated (polar for acids, less polar for bases) and undissociated (less polar for acids, and more polar for bases) fractions available for uptake. Different configurations of sampler (receiving phase and diffusion limiting layer) may be needed for sampling the two fractions. Estimates of the degree of dissociation can be obtained from simple physical chemical considerations, but even this becomes more complex where compounds (such as some pharmaceuticals) have more than one $pK_a$ value. Salinity of the water is a further possible factor that should be taken into account for some field deployments, but little work has been done in this area.

Currently both field and tank experiments used to demonstrate the performance of passive samplers use spot samples of water as a benchmark. This is not ideal since as discussed above measurements in unfiltered water samples will overestimate the concentration of the freely dissolved fraction of many pollutants, especially those that bind to dissolved organic carbon and particulate material. If filtered or ultra-filtered spot water samples are used to estimate the freely dissolved fractions of analytes, then this may underestimate the concentrations of very nonpolar compounds that tend to adsorb to glassware and filter components. A further problem with this approach is that even frequent (e.g. daily) spot samples may not give an accurate estimate of the TWA concentration over the trial period if there is marked temporal fluctuation. Work is urgently needed to establish standard test sites not just for passive samplers, but also for other emerging technologies (e.g. sensors) for monitoring water quality, and to establish analytical protocols for estimating the concentrations of freely dissolved analytes in water samples. For organic compounds one approach could be the use of quantitative structural activity relationship or linear solvation energy relationship models to predict freely dissolved fractions on the basis of the physicochemical properties of the analytes and measured dissolved organic carbon levels.

Transfer of laboratory calibrations to the field is associated with some uncertainty since calibrations are usually conducted under constant conditions, and in the field conditions can fluctuate markedly over a deployment period, and can be determined by weather events. However, in practice laboratory calibrations have been found to provide adequate estimates of concentrations of pollutants in water even where concentrations are known to fluctuate. One potential problem is in the measurement of short-term (over a period of hours) fluctuations in concentration such as can occur in small rivers following a storm event. If a sampler has a significant lag-phase such events could go undetected, or be only partially integrated into the TWA concentration. More research is needed in this area for the different types of sampler available as this is an area of practical interest.

For samplers used to measure concentrations of nonpolar organic analytes one method of overcoming some of the problems associated with the impact of fluctuating environmental conditions (temperature and turbulence) on $R_s$ is the use of performance reference compounds (PRCs). These are compounds (typically deuterated analogues of the compounds to be measured) and are loaded onto the receiving phase of the sampler prior to deployment. These PRCs offload from the receiving phase over the calibration or deployment period. Where the kinetics of uptake and offloading are isotropic, that is the rates of offloading of the PRCs are affected by environmental variables in a manner similar to the uptake rates of pollutants, the rates of offloading of PRCs can be used to correct the uptake rates of pollutants in field deployments. This approach effectively provides an *in situ* calibration method. There is also some evidence that the offloading rates of PRCs can be used to compensate for the impact of biofouling on uptake, however, more work is needed in this area (Booij *et al.*, 2006b; Booij *et al.*, 2007). The use of PRCs can increase confidence in field data, and may provide a means of introducing QA/QC into the calibration procedures. Currently it is not possible to use this approach with samplers for polar compounds or metals. This is because the receiving phase accumulates those pollutants by adsorption or chelation respectively rather than partition.

## 2.1.3 APPLICATIONS

In this section some field applications of the various samplers will be discussed in terms of their potential utility in support of regulatory monitoring requirements. This chapter will focus on polar and nonpolar organic pollutants, and organometalics; metals will be dealt with in another chapter (Chapter 4.1).

### 2.1.3.1 Polar Organic Compounds

Samplers for polar organic pollutants have been developed relatively recently compared with those for nonpolar organic compounds. Two main types are available; the POCIS (Alvarez *et al.* 2004, Alvarez *et al.* 2007; and Chapter 2.2 in this book) and the Chemcatcher® (Kingston, 2002). There has been an increased interest in monitoring polar compounds in recent years since the European Community has become aware of the presence of pharmaceuticals, components of personal care products and some pesticides in surface waters and in some cases drinking water. Work to obtain calibration data for these compounds in passive samplers is ongoing, and a number of studies has indicated the potential utility of the technology within the context of regulatory monitoring. Calibration experiments have been carried out in static systems (Hernando et al., 2005), including large volume microcosms (Mazzella et al., 2008); static systems with renewal (Arditsoglou and Voutsa, 2008; MacLeod et al., 2007); and in through flow systems (Gunold et al., 2008; Harman et al., 2008; Kingston et al., 2000; Tran et al., 2007). In other studies calibrations have been effected in an artificial stream system using spiked waste water (Vermeirssen et al., 2008), and in situ in a river (Zhang et al., 2008). In the latter case both flow and temperature, and concentration varied in time, but measurements were made only daily. A further complicating factor is that the concentration of dissolved organic carbon and density of suspended solids would also fluctuate, and this would affect the dissolved concentrations of pollutants that associate with those fractions to a significant extent. This could explain the observed differences in degree of correspondence between concentrations estimated by spot sampling and passive sampling. Under these conditions there will be a high degree of uncertainty associated with the sampling rates obtained. Field applications to assess the performance of passive samplers alongside spot sampling have been effected in rivers with variable flow rates, and multiple inputs (Vermeirssen et al., 2008), and in rivers and waste water discharges (MacLeod et al., 2007; Arditsoglou and Voutsa, 2008), small agricultural streams (Schäfer et al., 2008a), sea water (Hernando et al., 2005) and canals (Arditsoglou and Voutsa, 2008). There is still much work to be done to provide the range of validated laboratory calibration data that can be transferred to the field with confidence.

The potential of passive sampling for monitoring concentrations of polar organic compounds has been demonstrated in a number of studies. There has been interest in the ability of passive samplers to provide information on short-term fluctuations in concentrations of pollutants such as occur in rivers during storm events in regions where pesticides are used, or in agricultural ditches following spray application. A study (Schäfer et al., 2008b) using an artificial stream where pulses of pesticides could be introduced to simulate such episodic events provides a useful example of the sort of study that is needed to demonstrate the utility of the technology. The outdoor artificial stream system used in this work comprises a number of parallel channels (20 m long by 0.25 m wide (at the bed)), each with an approximate volume of 1000 L, and recycling of the water. The bed of the stream is covered with silt and gravel, and well established (one year) aquatic plants are present in the channels to simulate natural conditions. The slope is shallow (2%) and the flow rate used was relatively slow (0.1 m s$^{-1}$). The two concentrations (0.32 and 100 µg L$^{-1}$) of thiacloprid used were achieved by mixing aliquots of stock solution in a receiving tank at the end of the stream. The water from the receiving tank is pumped directly to the head of the stream. This method provided a pulse of insecticide that decayed over the eleven day deployment period (TWA concentration 18 µg L$^{-1}$). There was a rapid decay initially that was explained in terms of adsorption to components of the test system. Both prefouled (deployed for nine days before the introduction of the pesticide) and clean Chemcatcher® samplers with and without a polyethersulphone membrane were deployed. The samplers with the polyethersulphone membrane showed no significant biofouling. One set of samplers (the biofouled) was removed after one day of exposure to the insecticide, and the second set was removed after a ten day exposure to the pesticide. The sampling rate in the fouled samplers with a naked receiving phase was one-quarter of that in the unfouled equivalents, indicating that for this configuration biofouling is potentially a problem. The samplers where the receiving phase was protected by a diffusion limiting membrane showed a much lower sampling rate (one-tenth) compared with the samplers with no membrane over the one day exposure than over the ten day exposure where the relative rate was half that of the samplers without membrane. These results indicate that uptake by the samplers with a membrane has a significant lag-phase, and this configuration is not useful for monitoring short-term events. However, they show potential for monitoring TWA concentrations over longer deployments (in this case ten days). This work provides a number of useful lessons concerning the validation of passive samplers: there is a need to measure any lag-phase since this will impinge on the minimum deployment over which the sampler will be useful; the impact of biofouling indicates the utility of a protective membrane with a low surface energy in longer deployments. Another important outcome of this work is the indication of the potential of using artificial stream systems for the *in situ* calibration and validation of a wide range of passive samplers. The results of this study in an artificial system were consistent with those found in an extensive field trial (Schäfer *et al.*, 2008b) in which polar organic Chemcatcher® samplers were deployed alongside event driven water samplers in agricultural streams subject to run-off from surrounding cultivated fields. Here the samplers were deployed without a diffusion limiting membrane for 14 days, and yielded TWA concentrations of a range of polar pesticides that were significantly correlated ($r = 0.79$, $p < 0.01$, $n = 75$) with those obtained from the event driven water sampler. Other demonstrations of the utility of this approach have involved the POCIS that has been used for monitoring a wide range of polar organic contaminants that have recently become of concern, including polar pesticides, pharmaceuticals and components of personal care products. One such study (Alvarez *et al.*, 2005) compared the performance of POCIS with water-column sampling in a New Jersey (USA) stream that received agricultural input in it head waters, and treated waste water in its lower reaches. The passive samplers were used in a qualitative way to map the pollutants present, and 32 were detected in the POCIS compared with 24 in the water column samples. This reflects the ability of the samplers to detect compounds where the concentration fluctuates in time, and to accumulate quantities of compounds that are higher than the levels of detection in standard analytical procedures. Both methods picked up pollutants from a range of groups including pharmaceuticals, pesticides, flame retardants, components of domestic and personal care products, including fragrances, preservatives, and anti-foaming agents. The samplers have major advantages over repeated water sampling, and these include higher sensitivity, greater representativeness, and lower costs because of reductions in the number of trips to field sites, and in the number of samples to be analysed. Calibration experiments (Mazzella *et al.*, 2007) carried out in a microcosm system have provided some calibration data for the POCIS, but more work is needed to enable all of the polar samplers to be used in a quantitative way for the wide range of compounds that they have been demonstrated to sequester. Until this is available, applications of these types of passive sampler will lag behind those for the longer established samplers for nonpolar organics and metals.

### 2.1.3.2 Nonpolar Organic and Organometalic Compounds

In contrast with the samplers developed for monitoring polar organic compounds, those developed for monitoring nonpolar organics have been available for much longer, and the technology is more mature. PRCs are available for this class of sampler, and calibration data are published for a wide range of commonly occurring industrial chemicals and nonpolar pesticides. These samplers have been demonstrated in a number of studies to provide useful representative information on the levels of a wide range of industrial chemicals and nonpolar pesticides in aquatic environments including marine waters, rivers, lakes, and waste water.

Several studies have compared the concentrations obtained by spot sampling with those estimated using passive samplers. Although it is known that the concentration and nature of the dissolved organic matter in natural waters is an important factor affecting bioavailability, and hence toxicity, of both metals and nonpolar organic pollutants, there have been few detailed studies. The need for such studies has been highlighted (Vrana *et al.*, 2007) as a result of extensive field trials with the Chemcatcher® passive sampler where the difference between estimates of concentrations of hydrophobic organic pollutants obtained from passive sampling and filtered (0.45 µm) differed from by 20% (phenanthrene and fluoranthene) to by a factor of two (pyrene and chrysene). Recently a large-scale trial (Tusseau-Vuillemin *et al.*, 2007) in the River Seine and River Marne (France) investigated the impact of dissolved organic matter on the availability of PAHs to SPMD samplers. This work used six sites, four above the city of Paris and two downstream. Of the latter, one was in an industrial area, and the other was downstream of the discharge point of a major (8 million inhabitants, flow $24\,\text{m s}^{-1}$) domestic waste water treatment plant. These sites provided a range of conditions, patterns of pollution, and levels and types of dissolved organic matter. SPMDs spiked with PRCs were deployed at each site, and spot water samples (2 L) were taken three times during the deployment period. Spot samples were filtered at 0.7 µm, and extracted using solid-phase extraction cartridges in preparation for analysis. In order to increase the sensitivity, extracts from 6 L of water were pooled for analysis. Dissolved organic matter was measured as dissolved organic carbon, and the nature of the dissolved organic carbon was investigated using ultraviolet spectrophotometry, and the specific ultraviolet absorbance was calculated. The latter gives a measure of the aromaticity of the dissolved organic carbon. The concentration of chlorophyll *a* in the water was also measured. The authors found a large variation between the total concentrations of a range of PAHs found in spot samples. This is what might be expected in river that has a lot of boat traffic, and receives intermittent accidental inputs, and multiple industrial and domestic discharges. The variation between the passive samplers was much smaller since the estimates of concentrations of pollutants are TWA values. Dissolved organic matter was shown to be a major factor in determining the fraction of the larger PAHs available for uptake by the passive samplers, and by organisms. More mature dissolved organic matter (as found upstream of the city) showed a greater capacity for interaction with the PAHs than did the newer dissolved organic matter derived from sewage discharges downstream of the city. The use of PRCs allowed for the differences in temperature and turbulence between the sites, and provided a robust basis for the *in situ* calibration of the samplers. A number of important conclusions can be drawn from this study. The results confirm the importance of the dissolved organic matter, and the need to measure this, and to identify its properties when determining the potential bioavailability and hence risk of pollutants in the aquatic environment.

Several studies, using other designs of passive samplers for nonpolar compounds, have also indicated that the uptake of nonpolar pollutants by passive samplers is driven by the freely dissolved fractions of highly hydrophobic compounds, and this has been postulated to approximate to that available to living organisms. A field study (Adams *et al.*, 2007) in Boston Harbour (USA) using low- and high-density polyethylene strips (spiked with PRCs) as passive samplers to measure concentrations of hydrophobic organic compounds (PAHs and PCBs) showed the importance of careful and robust calibration studies, and gave further support for the idea of using passive samplers to predict bioavailability. Trials (Paschke *et al.*, 2006) using the MESCO sampler in a highly polluted creek (the Spittelwasser near Bitterfeld, Saxony-Anhalt, Germany) for monitoring concentrations of priority organic pollutants (chlorobenzenes, HCH, PAHs and PCBs) showed a good agreement between the fraction available to solid phase microextraction fibres in spot samples, and to the receiving phase of the sampler deployed in the creek. Both sample the freely dissolved fraction. A large-scale trial in the Netherlands (Smedes, 2007) examined a number of sites, including areas influenced by the rivers Meuse, Rhine and Scheldt, around the coast. Concentrations of a range of hydrophobic pollutants in the water were measured using spot sampling and passive sampling (silicone rubber polydimethylsiloxane sheets), and compared with uptake by the marine mussel *Mytilus edulis*. There was a close relationship between concentrations found in the molluscs and those in the passive samplers, and even seasonal variations in concentrations, and ratios of the various compounds found in the mussels were mirrored in the samplers. This provides good evidence of the biological relevance of the concentrations measured by passive samplers. A laboratory-based comparison (Gourlay *et al.*, 2005) of uptake by SPMDs and the amphipod crustacean *Daphnia magna* found that uptake of three model PAHs by the samplers underestimated uptake by the organism, but that the differences were relatively small when the biological variation and large differences between dissolved organic matters from different sources were taken into account. A study (Verweij *et al.*, 2004) of the bioavailable fractions of PAHs, PCBs, and some organochlorine pesticides in water compared levels of pollutants and their metabolites found in the bile of exposed fish with the SPMD available fraction. This work also compared concentrations (calculated using SPMDs) of pollutants in the bulk water with those calculated on the basis of concentrations in sediments. The passive samplers were found to provide useful measures of exposure of fish to some PAHs; however, generally the concentrations in water-based SPMD data were higher than those based on the fish data. Another important observation was a significant variation in bio-sediment accumulation factor and the sampler-sediment accumulation factor between the various test sites. This may be attributable to differences in the properties of the sediment structure and sediment components, and is consistent with the impact discussed above of dissolved organic matter on availability of nonpolar organic compounds to both organisms and passive samplers. This is particularly important for highly bioactive substances such as the organotins that were widely used as antifouling agents, are very stable, and bind strongly to components of sediments. Despite their removal from the market, residues of these compounds in sediments continue to pose a problem for marine molluscs. When sediments are disturbed then these compounds can become redistributed in the water column and available for uptake by organisms. The freely dissolved fractions of these compounds form only a very small proportion of the total material present. EQS values for these compounds are based on the filtered concentrations (a measure of dissolved materials), and are being set at very low levels because of the high toxicity of these compounds. This poses an analytical challenge when spot samples are used to compare environmental levels with the EQS values, particularly those based on acute toxicity. Some studies (Følsvik *et al.*, 2002; Harman *et al.*, 2008) have demonstrated that passive samplers provide a good measure of the bioavailability of organotins to marine molluscs, and that they can be used to measure very low concentrations of these compounds (Følsvik *et al.*, 2002; Aguilar-Martínez *et al.*, 2008a; Aguilar-Martínez *et al.*, 2008b; Harman *et al.*, 2008). Passive samplers could provide a valuable surrogate for organisms in the prediction of risk. If standard test sites are to be used in validation studies of passive samplers, then on the basis of the above studies it would be important to find a location with relatively stable properties, and that is well characterized not only in terms of the pollutants present, but also in terms of the properties of the sediments and concentrations and nature of the dissolved organic matter present.

Where information is needed for making an assessment of risk, then it is necessary to combine estimates of exposure with estimates of hazard. Whilst this is relatively straightforward for pure samples of single toxicants in the laboratory, the situation in the field is more complicated because the toxicant of interest is mixed with a large number of other chemicals covering a range of toxicity. Further, the relative concentrations of the various materials present, and the amount of dissolved organic matter present can fluctuate in time. As discussed above, passive samplers can provide reliable measures of exposure of organisms to a wide range of pollutants. This has been used to advantage in a number of recent studies where whole extracts from samplers have been assessed in direct measurements of toxicity in a range of standard bioassay systems. Mixtures of estrogens are present in the aquatic environment at low levels, and passive samplers can sequester these from the equivalent of several litres of water over a deployment period. The potential utility of combining passive sampling with toxicological assays to identify toxic fractions by combining toxicological assays with chemical analysis has been demonstrated in studies using SPMD (Rastall *et al.*, 2006), and POCIS and SPMD (Alvarez *et al.*, 2008). The possibility of using passive samplers in routine screening for the toxicity of environmental waters has been investigated by a number of workers. In one study (Vermeirssen *et al.*, 2005) the total 17β-estradiol equivalents (ng L$^{-1}$) in extracts from POCIS samplers, and from spot samples were obtained using a recombinant yeast estrogen assay. This demonstrated the equivalence of the two sampling methods, and the increased sensitivity possible with passive sampling. In a study (Escher *et al.*, 2008) of methods to measure the ecotoxicological hazard potential of polar contaminants in surface water and waste water, raw water, and extracts obtained using solid-phase extraction were screened using a battery of *in vitro* mode of action based toxicity assays (a bacterial bioluminescence inhibition test, a growth rate inhibition assay, and a specific inhibition of photosynthesis assay based on green algae, an acetylcholine esterase inhibition assay, and a yeast estrogen assay for estrogens). None of the bioassays showed positive results using the raw water samples, but clear responses were obtained using the enriched extracts from Empore™ solid-phase extraction disks. Since these are used as the receiving phase of Chemcatcher® samplers, this indicates the potential for combining extracts from passive samplers that accumulate pollutants over several weeks with batteries of *in vitro* toxicological screens to provide TWA measures of ecotoxicological hazard associated with polar pollutants. Here results can be expressed in terms of toxic equivalent concentrations. This approach combining passive sampling and *in vitro* assays would provide particularly useful diagnostic information to help regulatory organizations to make decisions on the need for intervention.

## 2.1.4 CONCLUSIONS

Whilst being well established, current practice in regulatory monitoring is flawed. Pictures of the chemical quality of water bodies provided by spot sampling alone are associated with a great deal of uncertainty, and may be misleading. It would be very difficult to use the information currently obtained to measure trends in water quality, and hence to assess the impact of any intervention measures taken to make improvements in this. The uncertainty associated with intermittent spot sampling could be reduced by combining it with passive sampling. Added benefits of this approach would be that more representative and biologically relevant information would be available. This would also facilitate the use of batteries of toxicological screens to monitor ecotoxicological hazards. However, before passive sampling can be fully integrated into regulatory practice it is necessary to introduce robust QA and QC protocols for the calibration and validation of the various samplers. Research is needed to develop appropriate reference materials, and standard methods. One approach that might prove effective would be the use of carefully regulated test sites with artificial stream systems and mesocosms. This would facilitate robust inter-laboratory calibration trials and validation experiments using longer term field exposures. Investment in this area would accelerate the use of passive sampling that can provide representative information in a cost effective manner, and can decrease the uncertainty associated with monitoring data. This would reduce the possibility of wrong decisions that could prove very expensive in terms of money, and environmental damage.

## REFERENCES

Adams, R.G., Lohmann, R., Fernandez, L.A., Macfarlane, J.K. and Gschwend, P.M., 2007. *Environ. Sci. Technol.*, **41**, 1317.
Aguilar-Martínez, R., Greenwood, R., Mills, G.A., Vrana, B., Palacios-Corvillo, M.A. and Gómez-Gómez, M.M., 2008a. *Intern. J. Environ. Anal. Chem.*, **88**, 75.

Aguilar-Martínez, R., Palacios-Corvillo, M.A., Greenwood, R., Mills, G.A., Vrana, B. and Gómez-Gómez, M.M., 2008b. *Anal. Chim. Acta*, **618**, 157.

Alvarez, D.A., Petty, J.D., Huckins, J.N., *et al*., 2004. *Environ. Toxicol. Chem.*, **23**, 1640.

Alvarez, D.A., Stackelberg, P.E., Petty, J.D., *et al*., 2005. *Chemosphere*, **61**, 610.

Alvarez, D.A., Huckins, J.N., Petty, J.D. *et al*., 2007. Tool for monitoring hydrophilic contaminants in water: Polar Organic Chemical Integrative Sampler (POCIS). In: Greenwood, R., Mills, G.A. and Vrana, B. (eds), *Passive Sampling Techniques in Environmental Monitoring*. Comprehensive Analytical Chemistry Series, Barcelo, D (series ed.), Elsevier, Amsterdam, 171–97.

Alvarez, D.A., Cranor, W.L., Perkins, S.D., Clark, R.C. and Smith, S.B., 2008. *J. Environ. Qual.*, **37**, 1024.

Arditsoglou, A. and Voutsa, D., 2008. *Environ. Pollut.*, **156**, 316.

Bartkow, M.E., Orazio, C.E., Gouin, T., Huckins, J.N. and Müller, J.F., 2007. Towards quantitative monitoring of semivolatile organic compounds using passive air samplers. In: Greenwood, R., Mills, G.A. and Vrana, B. (eds), *Passive Sampling Techniques in Environmental Monitoring*. Comprehensive Analytical Chemistry Series, Barcelo D (series ed.), Elsevier, Amsterdam, 125–37.

Booij, K., Smedes, F., Van Weerlee, E.M. and Honkoop, P.J.C., 2006a. *Environ. Sci. Technol.*, **40**, 3893.

Booij, K., Van Bommel, R., Mets, A. and Dekker, R., 2006b. *Chemosphere*, **65**, 2485.

Booij, K., Vrana, B. and Huckins, J.N., 2007. Theory, modelling and calibration of passive sampling devices in water monitoring. In: Greenwood, R., Mills, G.A. and Vrana, B. (eds), *Passive Sampling Techniques in Environmental Monitoring*. Comprehensive Analytical Chemistry Series, Barcelo, D (series ed.), Elsevier, Amsterdam, 141–64.

BSI, 2006. British Standards Institute Publicly Available Specification: Determination of Priority Pollutants in Surface Water Using Passive Sampling (PAS-61), May.

CMA, 2007. WFD chemical monitoring guidance for surface water. http://circa.europa.eu/Public/irc/env/wfd/library?l=/framework_directive/chemical_monitoring/technical_2007pdf/_EN_1.0_&a=d

COM (2006) 397 final: Proposal for a Directive of the European Parliament and of the Council on environmental quality standards in the field of water policy and amending Directive 2000/60/EC.

Directive 2000/60/EC of the European Parliament and of the Council of 23 October 2000 establishing a framework for Community action in the field of water policy, *Off. J. Eur. Comm.* **L327** (2000) 1.

Directive 2006/118/EC of the European Parliament and of the Council of 12 December 2006 on the protection of groundwater against pollution and deterioration, *Off. J. Eur. Union* **L 372** (2006) 19.

Escher, B.I., Bramaz, N., Quayle, P., Rutishauser, S. and Vermeirssen, E.L.M., 2008. *J. Environ. Monit.*, **10**, 622.

Følsvik, N., Brevik, E.M. and Berge, J.A., 2000. *J. Environ. Monit.*, **2**, 281.

Følsvik, N., Brevik, E.M. and Berge, J.A., 2002. *J. Environ. Monit.*, **4**, 280.

Goia, R., Jones, K.C. and Harner, T., 2007. The use of different designs of passive samplers for air monitoring of persistent organic pollutants. In: Greenwood, R., Mills, G.A. and Vrana, B. (eds), *Passive Sampling Techniques in Environmental Monitoring*. Comprehensive Analytical Chemistry Series, Barcelo, D (series ed.), Elsevier, Amsterdam, 33–53.

Gourlay, C., Miege, C., Noir, A., Ravelet, C., Garric, J. and Mouchel, J.-M., 2005. *Chemosphere*, **61**, 1734.

Gunold, R., Schäfer, R.B., Paschke, A., Schüürmann, G. and Liess, M., 2008. *Environ. Pollut.*, **155**, 52.

Harman, C., Bøyum, O., Tollefsen, K.E., Thomas, K. and Grung, M., 2008. *J. Environ. Monit.*, **10**, 239.
Hernando, M.D., Martínez-Bueno, M.J. and Fernández-Alba, A.R., 2005. *Bol. Inst. Esp. Oceanogr.*, **21**, 37.
Ke R., Luo J., Sun L., Wang Z. and Spear P.A., 2007. *Environ. Sci. Technol.*, **41**, 6698.
Kingston, J., 2002. The Development of a Passive Sampling System for the Determination of Time-averaged Concentrations of Organic Pollutants in Aqueous Environments. PhD.
Kingston, J.K., Greenwood, R., Mills, G.A., Morrison, G.M. and Persson, B.L., 2000. *J. Environ. Monit.*, **2**, 487.
Kot-Wasik, A., Zabiegata, B., Urbanowicz, M., Dominiak, E., Wasik, A. and Namieśnik, J., 2007. *Anal. Chim. Acta*, **602**, 141.
MacLeod, S.L., McClure, E.L. and Wong, C.S., 2007. *Environ. Toxicol. Chem.*, **26**, 2517.
Mayer, P., Tolls, J., Hermens, J. and Mackay, D., 2003. *Environ. Sci. Technol.*, **37**, 184A.
Mazzella, N., Dubernet, J.-F. and Delmas, F., 2007. *J. Chromatogr. A*, **1154**, 42.
Mazzella, N., Debenest, T. and Delmas, F., 2008. *Chemosphere*, **73**, 545.
Mills, G.A., Greenwood, R., Allan, I.J., et al., 2008. Application of passive sampling techniques in monitoring the aquatic environment. In: Namieśnik, J. and Szefer, P. (eds), *Analytical Measurements in Aquatic Environments*, Taylor & Francis, London.
Paschke, A., Brümmer, J. and Schüürmann, G., 2007. *Anal. Bioanal. Chem.*, **387**, 1417.
Paschke, A., Schwab, K., Brümmer, J., Schüürmann, G., Paschke, H. and Popp, P., 2006. *J. Chromatogr. A*, **1124**, 187.
Rastall, A.C., Getting, D., Goddard, J., Roberts, D.R. and Erdinger L., 2006. *Environ. Sci. Pollut. Res.*, **13**, 256.
Richardson, B.J., Lam, P.K.S., Zheng, G.J., McCellan, K.E. and De Luca-Abbott, S.B., 2002. *Mar. Pollut. Bull.*, **44**, 1372.
Schäfer, R.B., Paschke, A. and Liess, M., 2008a. *J. Chromatogr. A*, **1203**, 1.
Schäfer, R.B., Paschke, A., Vrana, B., Mueller, R. and Liess, M., 2008b. *Water Res.*, **42**, 2707.
Smedes, F., 2007. Monitoring of chlorinated biphenyls and polycyclic aromatic hydrocarbons by passive sampling in concert with deployed mussels. In: Greenwood, R., Mills, G.A. and Vrana, B. (eds), *Passive Sampling Techniques in Environmental Monitoring*. Comprehensive Analytical Chemistry Series, Barcelo D (series ed.), Elsevier, Amsterdam, 407–48.
Stephens, B.S. and Müller, J.F., 2007. Techniques for quantitatively evaluating aquatic passive sampling devices. In: Greenwood, R., Mills, G.A. and Vrana, B. (eds), *Passive Sampling Techniques in Environmental Monitoring*. Comprehensive Analytical Chemistry Series, Barcelo D (series ed.), Elsevier, Amsterdam, 329–346.
Stuer-Lauridsen, F., 2005. *Environ. Pollut.*, **136**, 503.
Tran, A.T.K., Hyne, R.V. and Doble, P., 2007. *Environ. Toxicol. Chem.*, **26**, 435.
Tusseau-Vuillemin, M.H., Gourlay, C., Lorgeoux, C., et al., 2007. *Sci. Total Environ.*, **375**, 244.
Vermeirssen, E.L.M., Asmin, J., Escher, B.I., Kwon, J.-H., Steimen, I. and Hollender, J., 2008. *J. Environ. Monit.*, **10**, 119.
Vermeirssen, E.L.M., Korner, O., Schonenberger, R., Sutter, M.J.F. and Burkhardt-Holm, P., 2005. *Environ. Sci. Technol.*, **39**, 8191.
Verweij, F., Booij, K., Satumalay, K., Van der Molen, N. and Van der Oost, R., 2004. *Chemosphere*, **54**, 1675.
Vrana, B., Allan, I.J., Greenwood, R., et al., 2005. *TrAC: Trends Anal. Chem.*, **24**, 845.
Vrana, B., Mills, G.A., Kotterman, M., Leonards, P., Booij, K. and Greenwood, R., 2007. *Environ. Pollut.*, **145**, 895.
Vroblesky, D.A., 2007. Passive diffusion samplers to monitor volatile organic compounds in ground-water. In: Greenwood, R., Mills, G.A. and Vrana, B. (eds), *Passive Sampling*

*Techniques in Environmental Monitoring*. Comprehensive Analytical Chemistry Series, Barcelo D (series ed.), Elsevier, Amsterdam, 295–307.

Warnken, K.W., Zhang, H. and Davison, W., 2007. In situ monitoring and dynamic speciation measurements in solution using DGT. In: Greenwood, R., Mills, G.A. and Vrana, B. (eds). *Passive Sampling Techniques in Environmental Monitoring*. Comprehensive Analytical Chemistry Series, Barcelo D (series ed.), Elsevier, Amsterdam, 251–278.

Zhang, Z., Hibberd, A. and Zhou, J.L, 2008. *Anal. Chim. Acta*, **607**, 37.

# 2.2
# Polar Organic Chemical Integrative Sampler and Semi-permeable Membrane Devices

David Alvarez and Audrone Simule

2.2.1 Introduction
2.2.2 Semi-permeable Membrane Devices
2.2.3 Polar Organic Integrative Sampler
References

## 2.2.1 INTRODUCTION

The polar organic chemical integrative sampler (POCIS) and semi-permeable membrane device (SPMD) are passive integrative samplers for monitoring of organic pollutants in the aquatic environment. Since the first publications on the development of SPMDs in 1990 (Huckins *et al.*, 1990) and POCIS in 2004 (Alvarez *et al.*, 2004), these passive monitors have received widespread global recognition by environmental researchers and later by practitioners as tool for (Bergqvist *et al.*, 1998; Petty *et al.*, 2000; Sabaliūnas *et al.*, 2000; Booij *et al.*, 2002; Jones-Lepp *et al.*, 2004; Johnson *et al.*, 2004; Petty *et al.*, 2004; Poiger *et al.*, 2004; Söderström *et al.*, 2005; Vermeirssen *et al.*, 2005; Vrana *et al.*, 2005; Huckins *et al.*, 2006; Mathiessen *et al.*, 2006; Alvarez *et al.*, 2007): point pollution source identification; screening of pollutants presence; monitoring of temporal pollution trends; toxicity screening; investigative monitoring; comparative monitoring; and water quality assessment.

---

*Rapid Chemical and Biological Techniques for Water Monitoring*   Edited by Catherine Gonzalez, Philippe Quevauviller and Richard Greenwood
© 2009 John Wiley & Sons, Ltd

**Table 2.2.1** WFD priority pollutants isolated in SPMD and POCIS

| Sampler | Pollutants from WFD priority list |
|---|---|
| SPMD | Anthracene, Benzene, Brominated diphenylether, $C_{10\text{-}13}$-chloroalkanes, Chlorfenvinphos, Chlorpyrifos, Di(2-ethylhexyl)phthalate (DEHP), Endosulfan, Fluoranthene, Hexachlorobenzene, Hexachlorobutadiene, Hexachlorocyclohexane, Naphthalene, Nonylphenols, Octylphenols, Pentachlorobenzene, Pentachlorophenol, Polyaromatic hydrocarbons, Tributyltin compounds, Trichlorobenzenes, Trifluralin |
| POCIS | Alachlor, Atrazine, Chlorfenvinphos, Chlorpyrifos, Diuron, Isoproturon, Nonylphenols, Octylphenols, Pentachlorobenzene, Pentachlorophenol, Simazine, Trichlorobenzenes, Trifluralin |

The SPMD and POCIS are integrative samplers which provide time-weighted average (TWA) concentrations of chemicals over deployment periods ranging from weeks to months. The term passive sampler means that it has no mechanical or moving parts, requires no power nor supervision during use. They sample chemicals from the dissolved phase, mimicking the respiratory exposure of aquatic organisms and are largely unaffected by many environmental stressors that affect biomonitoring organisms. This provides a highly reproducible means for monitoring trace (i.e. <1 part per billion) and ultra trace (i.e. <1 part per trillion) concentrations of contaminants. These samplers also enable in situ concentration of trace organic contaminant mixtures for toxicity assessments and toxicity identification evaluation (TIE) approaches.

Configuration of these samplers has been described in detail by Alvarez *et al.* (2004, 2007) and Huckins *et al.* (2006) as well in the UK standard PAS 61:2006 Determination of priority pollutants in surface water using passive sampling (PAS 61:2006). Both devices consist of a receiving phase (sorbent or lipid) enclosed in a diffusion membrane.

SPMD and POCIS are designed to concentrate nonpolar and polar organic pollutants such as polycyclic aromatic hydrocarbons (PAHs), polychlorinated biphenyls (PCBs), chlorophenols, chlorobenzenes, dioxins, furans, petroleum hydrocarbons, organochlorine pesticides, organotin compounds, polar pesticides, pharmaceuticals, hormones, and wastewater-related chemicals covering a wide range of log octanol-water partition coefficients, log $K_{ow}$ (Huckins *et al.*, 2006). SPMDs generally sample chemicals with log $K_{ow}$ greater than 3 and the POCIS samples chemicals with log $K_{ow}$ less than 3. The combined use of both samplers will sample nearly all of the organic compounds listed as priority hazardous substances by the European Directive 2455/2001/EC to be monitored (European Commission, 2001), Table 2.2.1.

Accumulated chemicals are extracted from the samplers in the laboratory and measured using suitable analytical techniques. Estimation of the concentration of chemicals in the water from amounts accumulated in the devices is described by general relationship:

$$C_w = C_s V_s / R_s t \qquad \text{(Huckins } et\ al., 1990)$$

*Semi-permeable Membrane Devices* 73

**Table 2.2.2** Passive Sampler Study Design Considerations

| Study stages | Consideration | Notes |
|---|---|---|
| Planning | Number of samplers | To provide adequate mass of the target contaminant for analysis. |
|  | Deployment time | Cover time frame of interest while considering factors such as time to same sufficient chemical mass, biofouling, and site access. |
| Field | Sampler transport, storage, and handling | Avoid exposure to air, transport at near-freezing conditions, store in airtight containers frozen $\leq -15^0$C. |
|  | Water deployment | Consider influence of temperature, flow speed, exposure to sun, biofouling and vandalism. |
| Lab | Processing | Determine types of cleanup and fractionation steps that are required. |
|  | Analysis | What type of sample manipulation is required prior to chemical analysis or toxicity testing. |

where $C_w$ is the estimated average concentration in the water, $C_s$ is the concentration of chemical in the sampler, $V_s$ is the volume of the sampler, t is the exposure time, and $R_s$ is the sampling rate for the chemical.

Designing a successful study using passive samplers involves careful planning. Items which need to be considered during the early stages of planning include: the goals and ambitions of the project; environmental conditions influencing chemical uptake; behaviour of the targeted chemicals; and the analytical limitations of the final analysis. Some basic considerations are shown in Table 2.2.2.

Paramount in the study design is to ensure that sufficient numbers of samplers and/or exposure times are used to meet the goals of the study. Knowledge of how much chemical needs to be sampled to meet the detection criteria of the chemical analysis or toxicity tests will influence the study design. The number of samplers needed to satisfy the study requirements can be estimated using the following equation:

$$R_s t n C_c P_r E_t > MQL V_i \qquad \text{(Huckins } et~al., 2006)$$

where $C_c$ is the chemical concentration of concern, t is the deployment time, $R_s$ is sampling rate in liters of water extracted per day, $P_r$ is the overall method recovery for the analyte, n is the number of samplers combined into a single sample, $E_t$ is the fraction of the total sample extract which is injected into the instrument for quantitation, MQL is the method quantitation limit, and $V_i$ is the volume of standard injection (Huckins *et al.*, 1990).

## 2.2.2 SEMI-PERMEABLE MEMBRANE DEVICES

A standard SPMD consists of 91.4 cm long, 2.5 cm wide layflat tube of low density polyethylene (LDPE) membrane and 1 ml of $\geq$ 95% or $\geq$ 99% purity triolein (Huckins *et al.*, 1990). SPMDs constructed with 95% pure triolein is often used when the sample is designated for chemical analysis, however, additional sample cleanup and processing will be required to remove interfering impurities. Whenever possible, SPMDs constructed containing ≥ 99% purity triolein should be used and is essential if the sample is to be subjected to toxicity or bioindicator testing. SPMDs can be made of various lengths provided they maintain the standard 460 cm$^2$ mL$^{-1}$ surface area to volume ratio consisting of approximately 20% triolein.

Theory and modelling of chemical partitioning and uptake from water phase to the SPMD has been described in detail by Huckins *et al.* (1990). Briefly, dissolved, nonionic lipophilic contaminants from water column diffuse across various rate-limiting barriers which consist of the water boundary layer, the biofilm, and the hydrophobic LDPE membrane prior to concentrating in the triolein.

Uptake of contaminants by SPMD follows first-order kinetics (Huckins *et al.*, 1990). The rate of mass transfer of the contaminant from water to the SPMD is linearly proportional to the difference between the concentration of the contaminant in the water phase $C_w$ and that in the SPMD $C_s$.

$$C_s/dt = k_u C_w - k_e C_s \qquad \text{(Johnson } et\ al.,\ 2004)$$

where $k_u$ is the uptake rate constant, $k_e$ is the elimination rate constant.

When equilibrium is attained, the concentration of the compound in the SPMD is equal to the concentration in water multiplied by ratio of compound uptake and release rate constants. The ratio describes partition coefficient between SPMD and water, $K_{sw}$. Equilibrium for some chemicals with high fugacities may occur in less than one month.

Published laboratory or field-derived sampling rates are dependent on environmental conditions such as temperature, water flow and/or turbulence, and biofouling. Behaviour of performance reference compounds (PRC), which are added into the SPMD during manufacture, provides a means of accurately accounting for these site-specific variables (Huckins *et al.*, 1990; Booij *et al.*, 2002). Caution must be used in selecting chemicals to be used as PRCs. The PRC must not be found in the environment, must be stable (i.e. will not degrade during deployment), cannot be also used in the instrumental method for other purposes, and ideally can be analyzed for in the same instrumental analysis run. Deuterated PAHs are a common selection for PRCs, however, many PAHs are sensitive to photolysis so care must be exercised in their use. Commercially available deployment devices can protect SPMDs from mechanical damage and baffle flow/turbulence around the SPMD, but do little to protect against photolysis (Figure 2.2.1).

## 2.2.3 POLAR ORGANIC INTEGRATIVE SAMPLER

The Polar Organic Chemical Integrative Sampler or POCIS is designed to sample water-soluble (polar or hydrophilic) organic chemicals from aqueous environments (Alvarez *et al.*, 2004; Jones-Lepp *et al.*, 2004; Petty *et al.*, 2004; Alvarez *et al.*, 2007). It consists of a solid material (sorbent) contained between two microporous

# Polar Organic Integrative Sampler

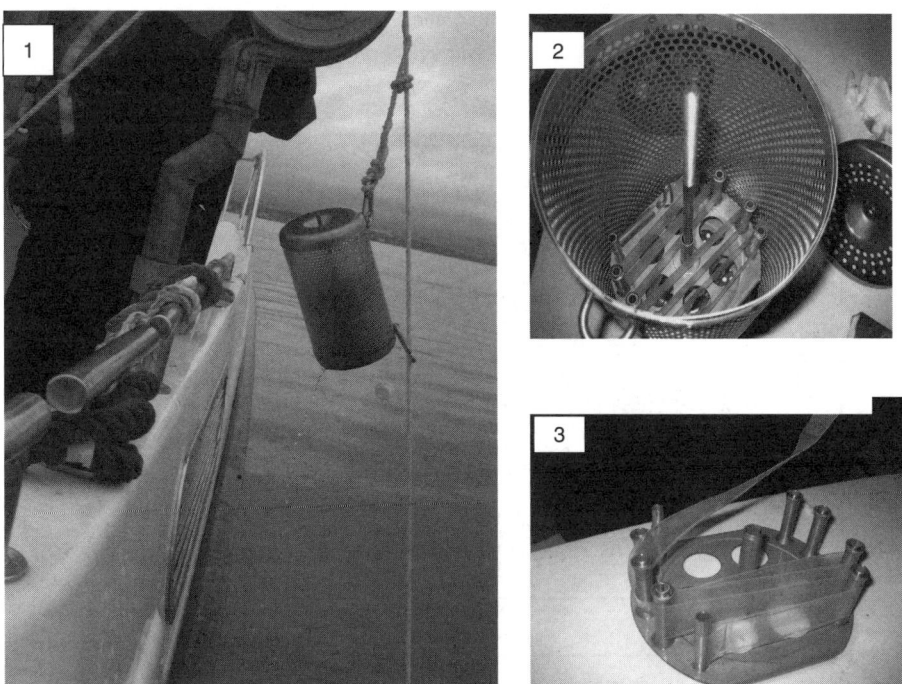

**Figure 2.2.1** Field retrieval of SPMD. Commercially available deployment cage (1), support racks inside the cage (2) and SPMD mounted on the racks (3)

polyethersulfone membranes (Figure 2.2.2). The membranes allow water and dissolved chemicals to pass through to the sorbent where the chemicals are trapped. Larger materials such as sediment and particulate matter are excluded. The inherent properties of the membrane impede the growth of a biofilm which can significantly reduce the uptake rates of some chemicals. The type of sorbent used can be changed to specifically target certain chemicals or chemical classes. A standard POCIS consists of a sampling surface area (surface area of exposed membrane) to sorbent mass ratio of $\cong 180\ cm^2/g$. A typical field deployed POCIS has an effective sampling surface area of 41 $cm^2$.

Although the POCIS can be modified to contain speciality sorbents and/or membranes to suit specific needs, two main configurations of the POCIS are commonly used. A 'generic' configuration contains a mixture of Isolute ENV+ and Ambersorb 1500 dispersed on S-X3 BioBeads and is used for most pesticides, natural and synthetic hormones, many wastewater-related chemicals, and other water-soluble organic chemicals. The 'pharmaceutical' configuration contains Oasis HLB and is designed for sampling most pharmaceutical classes and for toxicity testing. Current research suggests that the 'pharmaceutical' configuration is suitable for the widest range of polar chemicals, therefore it is recommended to use this configuration for most applications. It is common to deploy POCIS of several different configurations together to maximize the types of chemicals sampled.

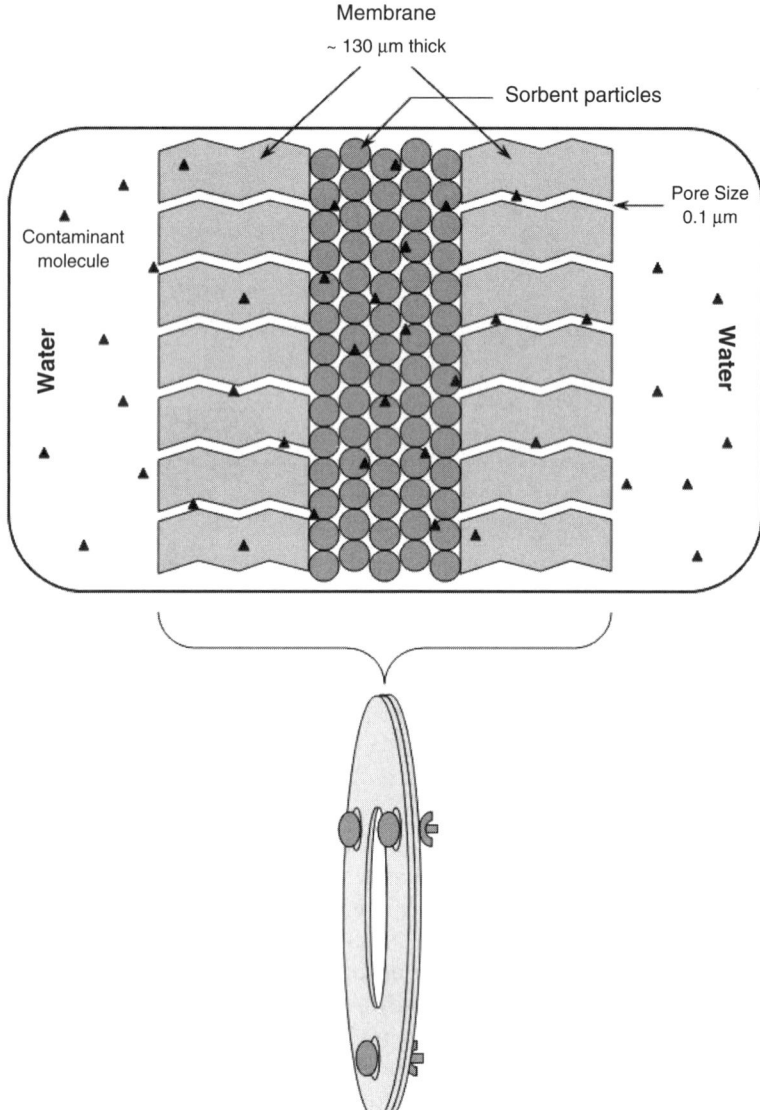

**Figure 2.2.2** Cross-section of a POCIS showing chemical transport through water-filled pores and across the polymer matrix

## REFERENCES

Alvarez D.A., Huckins J.N., Petty J.D., *et al.*, 2007. Tool for monitoring hydrophilic contaminants in water: Polar Organic Chemical Integrative Sampler (POCIS). In R. Greenwood *et al.* (eds), *Passive Sampling Techniques*. Comprehensive Analytical Chemistry, vol **48**, Elsevier.

Alvarez D.A., Petty J.D., Huckins J.N., *et al.*, 2004. *Environ. Toxicol. Chem.*, **23**, 1640–8.

# References

Bergqvist P-A., Strandberg B., Ekelund R. and Rappe C., 1998. *Environ. Sci. Technol.*, **32**, 3887–92.
Booij K., Smedes F. and van Weerle E. M., 2002. *Chemosphere*, **46**, 1157–61.
Booij K., Zegers B.N. and Boon J.P., 2002. *Chemosphere*, **46**, 683–8.
European Commission, 2001. Decision No 2455/2001/EC of the European Parliament and the council of 20 November 2001 establishing the list of priority substances in the field of water policy and amending Directive 2000/60/EC, *Official Journal of the European Communities*, L **331**, 1–5 (15.12.2001).
Huckins J.N., Tubergen M.W., Lebo J.A., Gale R.W. and Schwartz T. R., 1990. *Chemosphere*, **20**, 533–52.
Huckins J.N., Petty J.D. and Booij K., 2006. *Monitors of Organic Chemicals in the Environment. Semipermeable Membrane Devices*, Springer Science+Business Media, LCC, New York.
Johnson B.T., Petty J.D., Huckins J.N. and Lee K., 2004. *Environ. Toxicol.*, **19**, 329–35.
Jones-Lepp T.L., Alvarez D.A., Petty J.D. and Huckins J.N., 2004. *Arch. Environ. Contam. Toxicol.*, **47**, 427–39.
Matthiessen P., Arnold D., Johnson A.C., *et al.*, 2006. *Scie. Total Environm.*, **367**, 616–30.
PAS ratio61:2006, Determination of priority pollutants in surface water using passive sampling, BSI (2006).
Petty J.D., Huckins J.N., Alvarez D.A., *et al.*, 2004. *Chemosphere*, **54**, 695–705.
Petty J.D., Jones S.B., Huckins J.N., *et al.*, 2000. *Chemosphere*, **41**, 311–21.
Poiger T., Buser H-R, Balmer M.E., Bergqvist P-A. and Müller M.D., 2004. *Chemosphere*, **55**, 951–63.
Sabaliūnas D., Lazutka J. R. and Sabaliūnienedot I., 2000. *Environ. Pollut*, **109**, 251–65.
Söderström H., Hajšlová J., Kocourek V., *et al.*, 2005. *Atmospheric Environment*, **39**, 1627–40.
Vermeirssen E.M., Körner O., Schönenberger R., Suter M.J., Burkhardt F. and Holm P., 2005. *Environ. Sci. Technol.*, **39**, 8191–8.
Vrana B., Mills G.A., Allan I.J., *et al.*, 2005. *Trends Anal. Chem.*, **24**, 845–68.

# 2.3
# Main Existing Methods for Chemical Monitoring

Guillaume Junqua, Catherine Gonzalez and Evelyne Touraud

2.3.1 Introduction
2.3.2 Main Parameters Measured for Water Quality Monitoring
    2.3.2.1 Chemical Parameters
    2.3.2.2 Physical Parameters
2.3.3 Design of Sampling
2.3.4 Laboratory-based Methods
    2.3.4.1 Sampling
    2.3.4.2 Sample Preparation
    2.3.4.3 Detection
2.3.5 Conclusion
References

## 2.3.1 INTRODUCTION

A range of laboratory methods has been applied in the area of water monitoring. One set of applications relates to the control of the quality of industrial or urban wastewaters before, during and after their treatment. A further, important application is the estimation of various global or specific parameters defined by the Water Framework Directive (WFD; European Commission, 2000). So, the various methods are required to fulfil different objectives in terms of, for instance, diagnosis of sewage network, discharge impact, treatment efficiency, and regulation compliance. Some of these standardized methods are described by Dupuit (2006). However, they are often based on spot sampling, and in some cases sample processing is required before analysis, and these can be expensive, complex and time consuming. It is necessary to adapt them for a simple, rapid and low cost field monitoring in order to implement the surveillance,

---

*Rapid Chemical and Biological Techniques for Water Monitoring*   Edited by Catherine Gonzalez, Philippe Quevauviller and Richard Greenwood
© 2009 John Wiley & Sons, Ltd

operational and investigative monitoring programs of the WFD (Allan *et al.*, 2006a; Allan *et al.*, 2006b). Some laboratory-based methods have been developed along these lines. They have been identified by a European Union's Sixth Framework Project called 'Screening methods for Water Data Information in Support to the Implementation to the Water Framework Directive' (SWIFT-WFD, www.swift-wfd.com, Allan, 2006a; Greenwood and Roig, 2006). Sampling procedures need to be adapted to the type of sensor used. We can distinguish in-situ, online and on-site field measurements with optical (UV, visible, IR, nephelometry, chemiluminescence), electrochemical, chemical, or chromatographic detection techniques (anion and cation detection), with or without pretreatment.

## 2.3.2 MAIN PARAMETERS MEASURED FOR WATER QUALITY MONITORING

Some parameters must be estimated in order to evaluate the water quality. Nutrients such as ammonium, orthophosphate, total nitrogen and phosphorus may play an important role in the management of water quality since they are involved in eutrophication processes, or may contaminate groundwater, e.g. nitrate contamination resulting from fertilizer applications. A number of parameters may be used to characterize the oxygenation level of a water body. They include the measurement of dissolved oxygen, the chemical oxygen demand, and the biochemical oxygen demand. Redox potential is also used in German standards (DIN 38404-6). The acidification status may be obtained through the measurement of pH, while salinity may be given by measuring conductivity. The amount of dissolved, suspended organic matter can generally be assessed by total organic carbon, and the turbidity of the water. In addition, organic matter or aromatic organic matter can be estimated, usually using methods defined in national standards (DIN 38404-3 in Germany, for example). Indeed, the presence and levels of organic matter strongly influence the chemical oxygen demand of a water sample. Table 2.3.1 shows the main physico-chemical parameters measured for the water quality assessment. Estimation methods may be standardized by international or national organizations, such as the International Standardization Organization (ISO, www.iso.org), the European Committee for standardization (CEN, www.cen.eu), the 'Association Française de NORmalization' (AFNOR, www.afnor.org), or the 'Deutsches Institut für Normung' (DIN, www.din.de). Some of these methods have been automated, facilitating their field transposition. They are briefly presented in the next paragraphs, and are available in detailed form on ISO, CEN, AFNOR and DIN websites.

### 2.3.2.1 Chemical Parameters

Biochemical Oxygen Demand (BOD) estimation is based on a biodegradation. It is defined as the potential for removal of oxygen from water by aerobic heterotrophic bacteria which utilize organic matter for their metabolism and reproduction (Dupuit, 2006). Thus, it can be also considered as a biological parameter. This test requires a period of incubation (several days) at a controlled temperature in a sealed bottle in

*Main Parameters Measured for Water Quality Monitoring* 81

**Table 2.3.1** Main physico-chemical parameters of waters and their possible standardization

| Parameters | Standardization |
| --- | --- |
| Biochemical Oxygen Demand (BOD) | ISO 5815-1 and 2:2003; NF EN 1899-1 and 2; EPA 405.1 |
| Chemical Oxygen Demand (COD) | ISO 15705:2002; ISO 6060:1989; NF T90-101; EPA 410.3 |
| Total Organic Carbon (TOC) | ISO 8245:1999; NF EN 1484; EPA 415.1; 5310B, C and D |
| Total nitrogen (TN) | NF EN ISO 11905-1; ISO/TR 11905-2; ISO/CD 29441; NF EN 12260; EPA 352.1 |
| Total phosphorus (TP) | NF EN ISO 6878; ISO 15681-1 and 2; EPA 365.1, EPA 365.2, EPA 365.3 |
| Ammonium | NF EN ISO 11732; NF EN ISO 14911; ISO 5664:1984; ISO 6778:1984; ISO 7150-1:1984; NF T90-015-1 and 2; EPA 350.1 |
| Orthophosphate | NF EN ISO 6878; NF EN ISO 10304-1; NF EN ISO 15681-1 and 2; EPA 365.1 |
| pH | ISO 10523:1994; ISO/DIS 26149; NF T90-008; DIN 38404-5; EPA 150.1, ASTM D 1293; ASTM D 5464 |
| Total Suspended Solids (TSS) | ISO 11923:1997; NF T90-105; EPA 160.2 |
| Temperature | DIN 38404-4, EPA 170.1 |
| Conductivity | ISO 7888:1985; NF EN 27888; EPA 120.1; ASTM D 1125 |
| Turbidity | NF EN ISO 7027; EPA 180.1; ASTM D 1889; ASTM D 6698; ASTM D 6855; ASTM D 7315 |
| Dissolved oxygen | ISO 5813:1983; ISO 5814:1990; NF EN 25814; NF EN 25813; EPA 360.1 |

the dark. For example, EPA 405.1 method defines 5 days for incubation. The oxygen consumption in $mgO_2.L^{-1}$ is determined, in order to estimate the BOD value.

Chemical Oxygen Demand (COD) is based on the total oxidation of organic compounds presents in the sample to carbon dioxide. For this, strong oxidizing agents, such as potassium dichromate, are used under acid conditions, at high temperature. The amount of oxygen used for this oxidation can be estimated throughout the titration of the excess oxidizing agent (in $mgO_2.L^{-1}$). However, interferences may occur, for example with the presence of chlorides, or other reducer salts (Dupuit, 2006). ISO 15705:2002 uses a sealed tube method, allowing an on-site analysis (2h).

Total Organic Carbon (TOC) is usually released by wet chemical or high temperature catalytic oxidation reactor (WCO or HTCO, respectively), for a complete oxidation of organic carbon into dioxide carbon, coupled with a carbon dioxide gas detector, in order to estimate the carbon content of water samples (Dupuit, 2006).

Total Nitrogen (TN) consists of the estimation of the sum of the different mineral (nitrates, nitrites, ammonium, gaseous ammonia) and organic forms of nitrogen in water. For this analysis all the various forms are mineralized by combustion into nitrogen oxides which are detected and quantified by chemiluminescence (NF EN

12260, ISO/TR 11905-2). Another method oxidizes nitrogen forms into nitrates which can be then estimated (NF EN ISO 11905-1). Lastly, an ISO/CD 29441 method, under development, uses UV digestion, followed by flow injection analysis (FIA), continuous flow analysis (CFA) with spectrophotometric detection, and allows automation of the method.

Total Phosphorus (TP) takes into account orthophosphate, polyphosphates and organic phosphate. Polyphosphates must be hydrolysed and organic phosphate oxidized into orthophosphate form, prior to analysis, usually by oxidizing agents in acid conditions. The orthophosphate content is then quantified by colorimetric method with ammonium molybdate (NF EN ISO 6878). This procedure of detection can be automated by FIA and CFA techniques (NF EN ISO 15681-1 and 2, respectively). The orthophosphate content can be estimated using various techniques including a colorimetric method (NF EN ISO 6878), ionic chromatography (NF EN ISO 10304-1), automated FIA and CFA techniques and colorimetric detection (NF EN ISO 15681-1 and 2, respectively).

Ammonium quantification may be performed by various standardized methods, and more particularly by distillation and titrimetry (ISO 5664:1984, NF T90-015-1), colorimetry (ISO 7150-1:1984; NF T90-015-2), potentiometry (ISO 6778:1984), ionic chromatography (NF EN ISO 14911). An automated colorimetric method by FIA or CFA has also been standardized (NF EN ISO 11732).

Finally, pH can be determined by a potentiometric method using a glass electrode (NF T90-008). This technique can be used for field measurements.

### 2.3.2.2 Physical Parameters

Total Suspended Solids (TSS) estimation is related to particulate forms of organic or mineral matter in water. A separation of TSS by filtration or centrifugation is used, and is followed by drying at 105°C. The remaining mass corresponding to TSS is then determined. Water temperature is a physical parameter which can be determined according to a German standard (DIN 38404-4). This technique may be used for field measurements. Conductivity of water is measured by an electrochemical technique (ISO 7888:1985; NF EN 27888) that may be used in the field as well as in the laboratory. Turbidity may be estimated by four techniques described in an international norm (NF EN ISO 7027). Two semi-quantitative methods are based on determining the transparency of the water (with the Secchi Disk, for example). Others are based on optical analysis, more particularly by nephelometry. The standardized unit is FNU (Formazin Nephelometric Units). These methods are suitable for field applications. Lastly, dissolved oxygen is the $O_2$ concentration contained in a water sample, and is measured by iodometric or electrochemical (with gas membrane) methods (NF EN 25813 or NF EN 25814, respectively). Electrochemical methods using a probe technique can be used in the field.

Whilst most of these standardised methods must be performed within laboratories, some of them (mainly probe techniques such as those for measuring for instance pH, conductivity, temperature, turbidity, and dissolved oxygen) may be readily adapted for use in the field and can be used for making in-situ measurements. Others methods have been automated with standardized systems, based on FIA or CFA systems. This adaptation may allow field online measurements to be made. There are some on-site standardized methods, such as the ISO 15705:2002 for sealed tubes COD estimation, that are available for use in the field. Furthermore, standardization organizations have developed norms for the validation of alternative methods which may be suitable for use in the field. For example, the experimental French norm NF XP T90-210 provides guidance on the validation of alternative methods by comparing them with equivalent standardized methods. Inter laboratories studies can be also implemented according to the NF EN ISO 5725. The SWIFT-WFD project has developed a validation procedure for field-measurement systems, and this is based-on laboratory or alternative methods (Gonzalez, 2007). However, all these methods need an adapted sampling procedure, and this issue is dealt with in the next paragraph.

## 2.3.3 DESIGN OF SAMPLING

Laboratory based analytical methods usually depend on the transport of samples (usually collected by spot, sequential or composite sampling) from the field to the laboratory. The sampling step is one of the main factors affecting the quality of the information obtained, even where the analytical step is reliable, and fit for purpose. Indeed, infrequent spot samples may be inadequate for determining the quality of waters that are heterogeneous or exhibit temporal variation in their composition, as expressed by Thomas (2006). This is particularly the case for wastewaters where the volumes and the nature of inputs can vary on a diurnal basis, depending on the behaviour of the population in a catchment area. Moreover, water and wastewater composition can be unstable during transportation and storage after sampling because of the presence of three factors. Firstly, heterogeneity, where the presence of suspended solids, colloids, volatile organic compounds (VOCs) in a sample may lead to possible adsorption or flocculation phenomena as well as to changes in VOCs composition. Secondly, the chemical nature of compounds may generate reactions (such as reduction, complexation, and acido-basic). This is particularly a problem where surfactants are present in the sample. Thirdly, water or wastewater may contain microorganisms that may modify both the composition and the quality of a water sample (Thomas, 2006). For example, Baurès et al. (2004) showed that some parameters such as TSS and COD can vary by up to 20% in few hours, during their transportation and their storage, even when the samples are refrigerated. The use of field sampling procedures, based on in-situ, online or on-site techniques, as described by Greenwood et al. (2007) may reduce these alteration phenomena that occur during transportation and storage steps. The use of in-situ and online sampling may also provide a more representative picture of water quality where there is marked temporal variability of water and wastewater quality.

Even though these field sampling procedures are efficient, the identification of critical control points remains a real challenge for all sampling procedures (Thomas, 2006). A wrong strategy of sampling points (spatial and temporal deployment) could lead to a nonrelevant monitoring of water quality and thus increase drastically the monitoring costs. In order to avoid these problems it is recommended that assistance is sought in the design of sampling campaigns before their implementation (Thomas, 2006). The NF

EN ISO 5667 series provides some guidance on the design of sampling programmes, and on the choice of adapted sampling techniques. During the design process it is necessary to take into account a range of possible sources of error during the sampling procedure. Madrid and Zayas (2007) have identified them for each stage of the sampling process:

- definition and subdivision of the field: heterogeneity of the sample, spatial and or temporal variation in concentrations of pollutants, and identification of hot spots;
- sampling method: no representative statistics, skewed distribution, contamination or low concentrations of analytes;
- number of samples: few replicates, absence of representativeness;
- sample mass: absence of representativeness;
- timing of sampling: seasonal changes, climatic conditions;
- sampling;
- experimental conditions: matrix effects, lixiviation or irreproducible deposits;
- bottling: contamination or extraction by the equipment or container material, volatilisation;
- storage during sampling: contamination or losses by volatilisation, chemical reactions or microbial action.

## 2.3.4 LABORATORY-BASED METHODS

Laboratory methods have been adapted for field-measurement, taking into account three main steps: sampling, possible pretreatment and detection.

### 2.3.4.1 Sampling

Some of the laboratory methods adapted for use in the field can be used in situ. Those techniques (e.g. for measuring pH, conductivity, temperature, turbidity, and dissolved oxygen) that use a probe can be used directly in the water body to be sampled, thus avoiding the need for removing a sample from environment. Methods for measuring some parameters (e.g. COD, TOC, TN, and TP) can be deployed on-site and use online sampling combined with FIA or CFA techniques. Some new specific electrochemical methods are available for in situ measurement of ammonium and orthophosphate, other methods for these analytes can be used with online sampling linked to modified laboratory-based methods. On-site measurements are less used, because they require spot sampling, and are less convenient for routine surveillance operations. A further consideration is that on-site it is difficult to provide controlled operating conditions (e.g. temperature) such as those found in laboratories. BOD still requires the use of spot sampling followed by laboratory analysis.

*Laboratory-based Methods*

**Table 2.3.2** Main reactions prior to parameter detection

| Parameter | Possible reactions |
|---|---|
| Biological Oxygen Demand (BOD) | Biodegradation |
| Chemical Oxygen Demand (COD) | Catalytical oxidation, microwave oxidation, combustion oxidation, electrochemical oxidation, ozonation oxidation |
| Total Organic Carbon (TOC) | Combustion catalytic oxidation, combustion oxidation, photochemical oxidation, ozonation oxidation |
| Total Nitrogen (TN) | Combustion catalytic oxidation, combustion oxidation, photochemical oxidation |
| Total phosphorus (TP) | Photochemical oxidation, thermo photochemical oxidation, thermochemical oxidation, colorimetric reaction |
| Ammonium | Acido-basic reaction, colorimetric reaction |
| Orthophosphate | Colorimetric reaction |

### 2.3.4.2 Sample Preparation

Some monitoring tools, particularly online and on-site techniques (Table 2.3.2), may require some pretreatment of samples. Most of them are derived from laboratory methods.

The BOD method is based on the processes of degradation by bacteria, and there is a range of operational conditions and reactors can be different depending on the manufacturer (biosensors S.L., Kelma, LAR, STIP ISCO). Since the incubation takes several days is no incentive to convert this into an on-site method. For some methods that can be used online an oxidation step is required before analysis, and this can be achieved in a variety of ways, for instance oxidation processes can be based on thermal, catalytic, chemical, or photochemical reactions, for the estimation of COD, TOC, TN and TP in waters and wastewaters. Photochemical oxidation is determined by persulphate and UV irradiation. These oxidation processes may also be coupled for better efficiency. These processes achieve mineralization of TOC, TN and TP into $CO_2$, NO and $PO_4^{3-}$, respectively, and these can be detected by a range of sensors.

### 2.3.4.3 Detection

Table 2.3.3 (adapted from Greenwood and Roig (2006) – inventory of available technologies for water monitoring) presents an extensive (but not complete, because of a rapid evolution) range of methods and technologies that are commercially available and that may be used for monitoring physico-chemical properties of water bodies. Details are given for each technology and include the parameter measured, method used for the measurement, type of water it may be used for, average sampling time, precision of the measurement and range, possible standardization, and the institution/company responsible for production and/or commercialization.

**Table 2.3.3** Methods and technologies commercially available for the monitoring of physico-chemical properties of water bodies

| Parameter | Type of water | Main sampling procedure | Main reaction | Principle of detection | Sampling time | Precision | Range | Possible standardisation | Examples of companies |
|---|---|---|---|---|---|---|---|---|---|
| BOD | Wastewaters, other waters | On-line | Biodegradation | Measurement of oxygen consumption | Few seconds to few minutes | 3-5% | 1-500 000 $mgO_2.L^{-1}$ | Possible correlation with DIN 38409-H51, APHA-AWWA-WPCF 5210 B | Biosensores SL, Kelma, Isco Stip, LAR |
| COD | All | On-line, on-site | Oxidation processes, oxidation/ colorimetry | Photometry (UV, Vis, IR) Measurement of residual ozone by electrode Measurement of residual $O_2$ by $ZrO_2$ electrode Titrimetry | Few seconds to few minutes | 2-4% | 0-150 000 $mgO_2.L^{-1}$ | AFNOR 90-101 | LAR, ANAEL, AWA Instruments, Bran+Luebbe, EFS LAC Instruments et systèmes, Secomam, Martec, Seres, Kelma |
| TOC | All | On-line, on-site | Oxidation | Photometry (measurement of $CO_2$ by IR) Photometry (UV, Vis) Specific electrode | Few seconds to few minutes | 2-5% | 0-50 000 $mg.L^{-1}$ | ISO 8245, NF EN 1484, EPA 415.1, EPA 5310C | LAR, ANAEL, Shimadzu, Seres, Apollo Instruments, Bran+Luebbe, Datalink, EFS LAC Instruments et systèmes, WTW, Hach, Environment SA, Martec, Sick Maihak |
| TN | Wastewaters | On-line, on-site | Oxidation proceses | Chemiluminescence (detection of NO) Electrochemistry Photometry (measurement of nitrates) | 4-25 minutes | <3% | 0-4 000 $mgN.L^{-1}$ | DIN 38409 part 27, NF EN 12260, ISO/TR11905-2 | Bran+Luebbe, LAR, ANAEL, Shimadzu |

## Laboratory-based Methods

| | | | | | | | | |
|---|---|---|---|---|---|---|---|---|
| TP | Raw waters, wastewaters | On-line, on-site | Oxidation/ Colorimetry | Photometry | 5-28 minutes | 1-5% | 0-60 mgP.L$^{-1}$ | DIN 38405 | Bran+Luebbe, Seres, WTW, Anhydre |
| Ammonium | All | On-line, on-site, in-situ | Colorimetric reaction | Specific electrodes Photometry chemiluminescence Titrimetry Ionic chromatography | Few seconds to few minutes | <1% to 10% | 0-15000 mgN.L$^{-1}$ | NF EN ISO 11732 NF EN ISO 14911 | ABB Instrumentation, ANAEL, Anhydre, Apollo Instruments, Aqualyse, Aquams, AWA Instruments, Bran+Luebbe, Datalink, Endress Hauser, Environnement SA, Equipements scientifiques SA, Brasten Applikon, GFGHydrolab, Isco Stip, Metrohm, OTT, Hach, Seres, ShimadzuSwan, WTW, YSI Hydrodata |
| Phosphate | All | On-line, on-site | Colorimetry, oxidation/ colorimetry | Photometry (UV, Visible) Ionic chromatography | Few minutes | 1-3% | 0-12610 mgP-PO$_4^{3-}$·L$^{-1}$ | NF EN ISO 15681 NF EN ISO 10304 | ABB Instrumentation, Anael, Anhydre, Aquams, Bran+Luebbe, Endress Hauser, GE Panametrics, Martec, Metrohm, Seres, WTW |
| pH | All | On-site, on line, in-situ | No | Glass, polymer electrolyte, LIS electrodes | Few seconds | | 0.01 0-14 pH | NF T90-008 | HEYL, IRIS Instruments, Hydrolab, Kobold, Mettler Toledo, Hach, Hydrodata, Metrohm |

*(continued overleaf)*

Table 2.3.3 (continued)

| Parameter | Type of water | Main sampling procedure | Main reaction | Principle of detection | Sampling time | Precision | Range | Possible standardisation | Examples of companies |
|---|---|---|---|---|---|---|---|---|---|
| Conductivity | All | On-site, on-line, in-situ | No | Electrode techniques | Few seconds | 1% | 0-20 S.cm$^{-1}$ | NF EN 27888 | Aquacontrol, Aqualyse, Hanna Instruments, Heyl, Hydrolab, Kobold, Panametrics, Hach, WTW, YSI Hydrodata, Metrohm |
| Turbidity | All | On-site, on-line, in-situ | No | Nephelometry, photometry (UV, IR, multi-wavelength) Electrodes (specific, glass, polymer, LIS and reference) | Few seconds | 1-10% | 0-4000 NTU | ISO 7027 DIN 38404 USEPA 180.1 | ABB Instrumentation, Anael, Aquacontrol, Aqualyse, Aquams, AWA Instruments, Bamo mesures, Datalink, EFS Lac Instruments et systèmes, Endress Hauser, Equipements scientifiques SA, Hydrolab, Martec, Mettler Toledo, OTT, Hach, Seres, Cole Parmer, WTW, YSI Hydrodata |
| Dissolved oxygen | All | On-site, on line, in-situ | No | Optical probe method Specific electrodes Polarography Rapid pulse oxygen sensor | Few seconds | <0.5 to 15% | 0-90 mgO$_2$.L$^{-1}$ | NF EN 25 814 | Aquacontrol, Kobold, Panametrics, Hach, WtW, YSI Hydrodata, Nereides, Hydrolab, VWR |

BOD is determined using biodegradation processes in a bioreactor, and an activated sludge technique can be used, and this precludes its use in the field. The analytical step involves the estimation of oxygen consumption, and the same methods can be used to define the toxicity and the oxygen concentration of a water body. The latter can be measured on-site because of the lack of sample preparation, and the use of membrane polarography. Some companies have developed instruments for the estimation of COD, and two groups can be defined on the basis of the method used. The first uses analysis by UV/visible spectrophotometry of waters followed by mathematical processing of the UV/Visible spectra. The second is based on an oxidation process, with an estimation of oxidant consumption ($O_2$, $O_3$). This means that COD can be measured in the field using either continuous or spot sampling depending on whether it is based on electrode, photometric, UV absorption/spectrometric or other titrimetric methods (Greenwood and Roig, 2006).

Commercial tools are now available for measuring a number of the variables that are important in assessing water quality, and a directory of these is available (Greenwood and Roig, 2006). Most of the commercially available instruments for measuring TOC are based on oxidation processes by combustion or UV irradiation followed by IR detection of carbon dioxide gas. Other instruments, mostly using photometric methods, are available for measuring the organic matter content of water. Devices that use spot samples and others that are linked to continuous flow are available for the estimation of organic matter aromaticity, and some instruments have been designed for measuring total nitrogen (LAR and Anael, Shimadzu, Bran+Luebbe) and total phosphorus in water. A wide range of devices (based on for instance specific electrodes, colorimetric, UV absorption and spectrophotometric, chemiluminescence, and titrimetric methods) is available for measuring ammonium ions. These form an alternative to ion chromatography. Most of methods developed to measure phosphate in water are based on optical or ion chromatographic methods. Monitoring of pH and conductivity is usually based on electrode measurements, and these can be used in-situ, in-line or in the laboratory. The latter depends on either transport of spot samples, or where the laboratory is on-site (e,g, in a sewage treatment works) on the use of pumped samples. Turbidity monitoring is mainly based on continuous in situ nephelometric measurement. Finally, some of the devices such as multi-parameter probes allow the in situ measurement of multiple parameters simultaneously (for example Shimadzu, Metrohm, LAR).

## 2.3.5 CONCLUSION

Laboratory-based methods have been developed for field-measurement of the main water quality parameters, and their use can be standardized. They are generally based on the same principles as the equivalent laboratory based methods (e.g. oxidation, colorimetry, photometry) but use simplified procedures in order to overcome the constraints of working in the field. Currently there are numerous commercially available devices for online and on-site use, and these provide efficient tools for surveillance, operational and investigative monitoring in the frame of WFD. These techniques are suitable for such applications as incident detection in water treatment plants, detection of accidental pollution, and measurement of spatial and temporal variation in water quality. These additions to the toolbox available to those responsible for monitoring water quality provide an opportunity to use the most cost-effective method that can provide information that is fit for the purpose in hand. In some cases sampling procedures or traditional and emerging monitoring techniques may be required to provide quantitative information, but in some applications it is sufficient to indicate the presence or absence of specific pollutants in the water. Whilst information on levels of some individual pollutants is necessary, this does not always provide a measure of the impact of a single chemical on living organisms where this is just one component of a complex mixture. A validated method that could provide a measure of global toxicity would be the ideal tool for measuring overall water quality, but this is still some way off, and until such a method is available it will be necessary to develop, refine, and validate the range of available tools to provide the robust data needed to support the legislation at a lower cost than is currently possible.

## REFERENCES

Allan I.J., Vrana B., Greenwood R., Mills G.A., Roig B. and Gonzalez C. (2006a) *Talanta*, **69**(2): 302–22.

Allan I.J., Vrana B., Greenwood R., et al. (2006b) *Trends in Analytical Chemistry*, **25**(7): 704–15.

Baurès E., Berho C., Pouët M-F. and Thomas O. (2004) *Water Science and Technology*, **49**(1), 47–52.

Dupuit (2006) Standard methodologies. In: *Wastewater Quality monitoring and Treatment*, Quevauvillier P., Thomas O., Van der Beken A. (eds), John Wiley & Sons, Ltd, Chichester, pp. 35–52.

European Commission (2000) *Water Framework Directive (WFD)*, 2000/60/CE.

Gonzalez C. Prichard E., Spinelli S., Gille J., Touraud E. (2007) *Trends in Analytical Chemistry*, **26**(4): 315–22.

Greenwood R. and Roig B. (2006) Deliverable 5: Directory of screening tools-A toolbox of existing and emerging methods for chemical and ecological status monitoring under the WFD, SWIFT-WFD, Screening methods for water data information in support of the implementation of Water framework Directive, Sixth Framework Programme Contract n°SSPI-CT-2003-502492.

Greenwood R., Mills G.A. and Roig B. (2007) *Trends in Analytical Chemistry*, **26**(4): 263–7.

Madrid Y. and Zayas Z.P. (2007) *Trends in Analytical Chemistry*, **26**(4): 293–9.

Thomas O. (2006) Sampling assistance. In: *Wastewater Quality monitoring and Treatment*, Quevauvillier P., Thomas O., Van der Beken A. (eds), John Wiley & Sons, Ltd, Chichester, pp. 24–34.

# 2.4
# UV Spectrophotometry: Environmental Monitoring Solutions

Daniel Constant, Catherine Gonzalez, Evelyne Touraud, Nathalie Guigues and Olivier Thomas

2.4.1 Introduction
2.4.2 UV Deconvolution Method
2.4.3 Portable and Online Technologies
    2.4.3.1 Portable Instrument
    2.4.3.2 Online Analyser
    2.4.3.3 Sofware Description
2.4.4 Examples
    2.4.4.1 Spatial and Temporal Variability of Water Quality
    2.4.4.2 Trophic State Assessment of Lakes
    2.4.4.3 Water Quality Monitoring
2.4.5 Conclusions
References

## 2.4.1 INTRODUCTION

The adoption of the European Union Water Framework Directive (WFD) has set new requirements for water quality monitoring in Europe. Innovative tools could play a main role in the design and implementation of operational, surveillance and investigative monitoring as described by WFD. Moreover, the monitoring requirements for successfully implementing the WFD will directly depend upon available measurement techniques of demonstrated quality, which will be able to deliver reliable data at an affordable cost.

---

*Rapid Chemical and Biological Techniques for Water Monitoring*    Edited by Catherine Gonzalez, Philippe Quevauviller and Richard Greenwood
© 2009 John Wiley & Sons, Ltd

There is an obvious need to develop, to propose a set of new methods/techniques that could matching the new demands of the WFD with the characteristics of the existing monitoring system, budgetary constraints.

In this context, portable or online UV spectrophotometer systems could comply with new monitoring requirements and could offer a relevant alternative to the classical approach (spot sampling and laboratory analysis) for environmental monitoring:

- spatial and temporal variability of water quality,
- trophic state assessment of lakes,
- accidental pollution and early warning systems,
- monitoring and prevention of water resources.

## 2.4.2 UV DECONVOLUTION METHOD

UV examination has been proved to be a relevant method for the study of water and wastewater quality using deconvolution methods of UV spectra. The absorbency spectrum of water can be decomposed from a few number of characteristic spectra (reference spectra). Therefore, a given spectrum can be reconstructed with a linear combination of reference spectra and all additive parameters can be computed with the same linear combination. Qualitative and quantitative results in terms of classical parameters such as TOC, COD, $BOD_5$, TSS, nitrate, ... can be provided.

The deconvolution methods are multi-wavelength procedures which can be classified with regard to the selection procedure of reference spectra. These spectra can be chosen from specific compounds (Maier, 1981), from independent spectra of real samples (Thomas *et al*., 1993), statistically selected (Gallot and Thomas, 1993) or from a mixed choice of spectra of specific compounds and of real samples. Reference spectra are not universal: recently, according to the complexity of the composition of wastewater, SECOMAM has developed UVPro software based on advanced UV spectral deconvolution (Patent 00402038-4, 17 July 2000) which allows creating dedicated models and determination of reference spectra from a set of studied water and wastewater UV spectra: an automatic calibration step is carried using parameters values obtained by standard or reference method. Deconvolution is used in order to find a linear relation between measured and UV estimated values for any parameter.

In practice, the shape of the UV spectrum can be considered as a linear combination of better defined spectra ($REF_1, \ldots, REF_p$) related to characteristic states or compounds of the water (Thomas *et al*., 1996):

$$Sw = \Sigma a_i . REF_i + / - r$$

where $a_i$ and r are respectively the coefficient of the $i^{th}$ reference spectra and the admitted error. This assumption transforms a nonlinear phenomenon into a simple linear combination, the coefficients of which can be determined with any UV spectrophotometer and multi-component software. The quality of the model can be appreciated by the computation of the quadratic error between measured and reconstructed spectra.

*Portable and Online Technologies* 93

If the error is greater than the admitted value, another basis of reference spectra can be selected to improve the restitution of the sample spectrum.

From the above relation, any additive parameter can be calculated:

$$Pw = \Sigma a_i . PAR_i +/- r_p$$

where Pw and $PAR_i$ are respectively the values of the parameter with the unknown spectrum Sw and the reference spectra $REF_1, \ldots, REF_p$ and rp is the error on the computation of the parameter value. The spectra and parameters to be estimated are measured and for each parameter, a multiple regression is carried out between the measured values and the coefficient values of the reference spectra (calibration step).

Indeed, the optimization of the choice of reference spectra within UVPro allows the improvement of characterisation of water and wastewater. Several applications have been carried in chemistry, petrochemistry, food industry, cosmetics and pharmaceutical industry.

### 2.4.3 PORTABLE AND ONLINE TECHNOLOGIES

On the basis of the deconvolution method and the developed software, SECOMAM Company commercializes two spectrophotometers, a portable instrument (PASTEL UV) and an online analyser (STAC), presented in Figure 2.4.1.

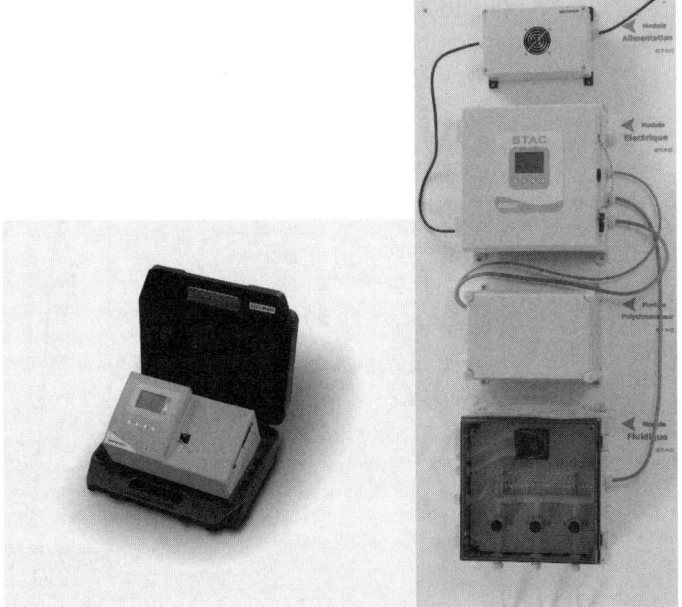

**Figure 2.4.1** Pastel UV (portable instrument) and STAC (on-line analyser) (reproduced by permission of Secomam)

## 2.4.3.1 Portable Instrument

Concerning the portable instrument (PASTEL UV), a Diode Array technology has been developed and the optical system is designed as shown on Figure 2.4.2.

Built on a strong aluminium bench, this optical design avoids any moving parts and insures a perfect stability of the optical specification. The UV source is a pulsed deuterium lamp which has been chosen rather than xenon light sources due to its high light power in the far UV where most of the sample signal is, and due to its smooth light emission which allows optimizing the background noise. The polychromator is based on plane 1200 L/mm holographic grating. This technology strongly optimizes any stray light effect and guarantees the best reading range. The complete optical system is enclosed in a sealed box to keep the performances stable vs. time and humidity.

The embedded microcomputer board applies the ASD (Advanced Spectral Deconvolution) algorithm to the Diode Array signal, stores the spectra, the results together with the traceability data (time, user, sampling site...). All the results are displayed on a bright graphic screen with the traceability data and the multi-parameters concentrations.

PASTEL UV can be used for a wide range of applications such as:

- Water quality assessment by measuring simultaneous, in less than one minute, several parameters (COD, BOD, TOC, $NO_3^-$)
- Due to its ease portability, it is therefore possible to multiply the sampling points along a river improving for example the investigation of WWTP impact, illegal discharges or influent dilution effect.

As the expected water profile could be loaded in the PASTEL UV, it is a powerful tool to detect accidental pollution both in natural waters and sewages.

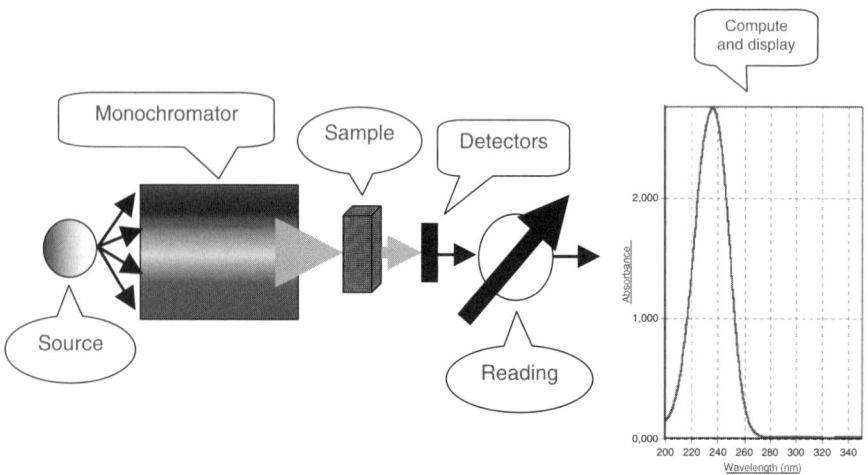

**Figure 2.4.2** Design of the optical system used in case of portable instrument (See Plate 1 for colour representation)

### 2.4.3.2 Online Analyser

Concerning the STAC, this online analyser is composed of three main boxes (sampling system, a Diode Array polychromator, a control box) allowing to isolate the major fluidic circuits from the electronic one.

The general optical design is the same as described for the PASTEL UV (Diode Array technology, Deuterium lamp for high measuring dynamic and reproducible results).

The STAC optical system is equipped with a Flow Though Cell and the reading chamber size is 5, 10 or 50 mm long depending on the measuring range (5 mm cell is mostly used for waste water and 10 mm to 50 mm cells are designed for natural water applications).

The online STAC analyzer can be used for many applications:

- For natural water, STAC can provide water quality monitoring with continuous multi-parametric determination and nitrate measurements. This online analyser could be also very efficient in case of accidental pollution detection in particular for organic contaminants or in case of urban WWTP dysfunction investigation.

- For Waste Water Treatment plans, STAC can monitor both influents and effluents. Same as natural water, STAC can monitor water pollution (COD, BOD, TSS, $NO_3^-$) and quality thanks to the conformity test.

- The SECOMAM STAC finds also its place in industrial processes such as High Level Chloride monitoring or organic chemical.

### 2.4.3.3 Sofware Description

UVPro is software developed by SECOMAM and is intended for the acquisition and the digital processing and graphic of PASTEL UV & STAC measurements made in the ultraviolet light. UVPro software can be generally used for five applications:

- The routine measurements (Module "Open a model"), using an already existing calibration model.

- Measure and spectra processing (Spectrum module), this module applies different models, compares them and choose the most appropriate for the studied sample. Moreover, the module allows different spectrum calculation (spectrum shift, derivative, spectra addition, spectra subtraction),

- Instrument calibration or models creation (Recalibration module), this module is used when the existing models are not fitting well with the sample composition. In this case, it is necessary to create a new model on the base of a set of relevant spectra (minimum of 20 spectra). The parameters values obtained by laboratory methods (COD, TOC, BOD...) are included in the data base in order to found the more suitable model.

- Communication with SECOMAM analyzer and its remote control (peristaltic pump, deuterium lamp, flow through cell, sample measurement...).

- The uninterrupted acquisition and safeguard of the spectra and their associated results (reading module only used for online instrument).

## 2.4.4 EXAMPLES

As described previously, the deconvolution method of UV spectra allows to measure quantitative parameters (nitrates, surfactants for example) as well as to estimate some physico-chemical parameters (TOC, COD, $BOD_5$, TSS). The obtained information (qualitative and quantitative) are enough relevant and robust to be integrated in the decision making process and water resources management. Indeed, the performances of Portable UV spectrophotometer have been evaluated and compared to a reference method and have shown good correlation (Gonzalez *et al.*, 2007). In addition, performance criteria have been verified in field conditions in order to assess the impact of these conditions and to demonstrate the portability of the instrument.

### 2.4.4.1 Spatial and Temporal Variability of Water Quality

*Spatial and temporal variability of water quality along Orlice river (CZ Republic)*

The Czech Republic is a landlocked state lying in the middle of a mild band of the northern hemisphere in the centre of Europe. The Czech Republic has state borders with Poland, Germany, Austria and Slovakia. The total length of the state borders is 2290 km, of which 738 km are defined by rivers – the so-called 'wet border'. The territory of the Czech Republic is on the basins of three important European rivers – the Elbe, the Odra and the Morava (Danube), dividing the basins of the North Sea, the Baltic and the Black Sea. The Orlice River is one of the rivers which composed the Elbe River basin.

For evaluating pollution, data is taken from the state network monitoring the quality of surface water operated by the Czech Hydrometeorological Institute based on several indicators ($BOD_5$, $N-NH_4$, $N-NO_3^-$, Ptotal and the saprobic index of macrozoobenthos).

In order to improve the spatial and temporal variability of water quality along Orlice River, the portable UV instrument and some chemical test kits ($NH_4$ measurement) have been deployed increasing the frequency of physico-chemical parameter measurements.

Several sampling points have been implemented (Table 2.4.1) and correspond to standard monitoring sites (Figure 2.4.3).

Figure 2.4.4 shows the evolution of nitrate concentration at the same sampling point (three times) during a period of 12 days. Even if the raw values of the nitrate concentration level measured by PASTEL UV are slightly overestimated in comparison with those obtained by laboratory method, this device can be used to assess trends in water quality evolution and then could help decision making in water management resources. The results obtained for ammonium concentration by Merck test kit lead to more or less to the same conclusions compare to reference method.

Table 2.4.1 Sampling points (Orlice River basin, CZ Republic)

| Sampling site | River | Location |
| --- | --- | --- |
| SP 1 | Orlice | Hradek Kralove, Na Mlejnku |
| SP 2 | Dědina | Třebechovice, jez |
| SP 3 | Divoká Orlice | Litice, jez |
| SP 4 | Bělá | Kvasiny, jez (Vídeňská) |
| SP 5 | Tichá Orlice | Černovír |
| SP 6 | Tichá Orlice | Ústí nad Orlicí |

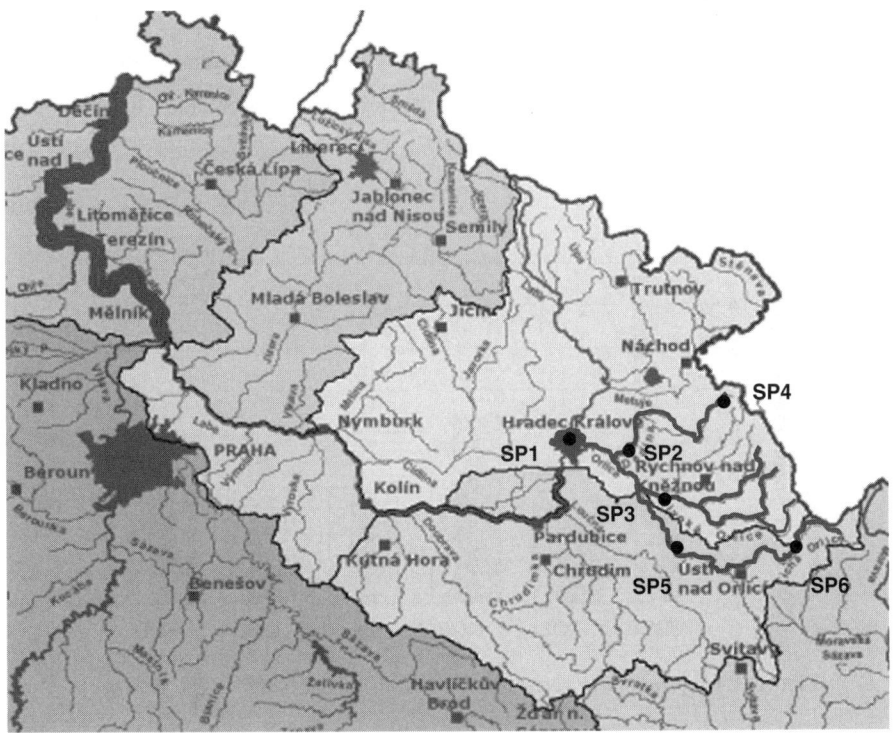

Figure 2.4.3 Sampling sites (Orlice River basin, CZ Republic)

In addition, PASTEL UV was also used to assess the variability of Elbe River in Hradec Kralove (CZ Republic) and the obtained results have shown (Figure 2.4.5) a nonsignificant variability for 6 days, except for TSS concentration level. The increase of TSS concentration (x4) between 8 November and 11 November is probably due to spot sampling procedure and not to the TSS measurement.

The variability of nitrate concentration (Figure 2.4.6) has been also studied depending on the sampling point's location along Orlice River (Table 2.4.1). This approach

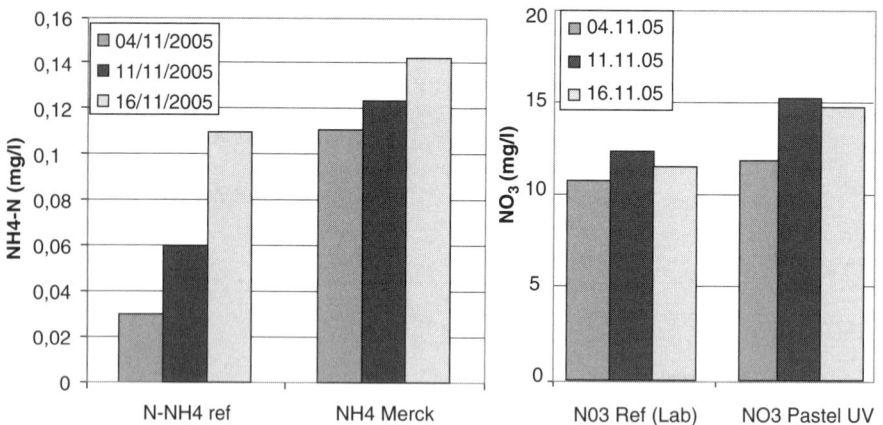

**Figure 2.4.4** Temporal variability of nitrate and ammonium concentration levels on Orlice River

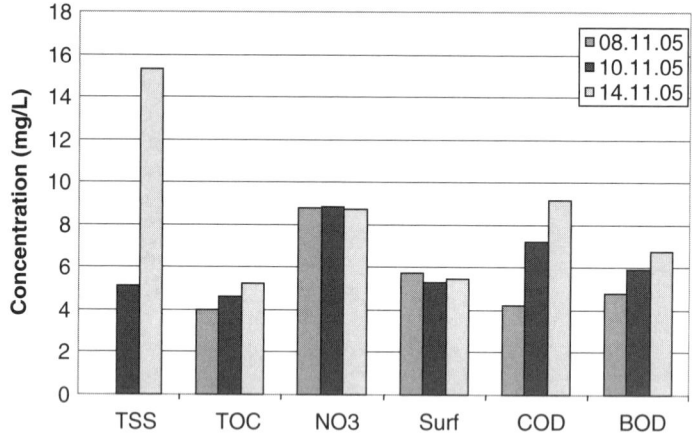

**Figure 2.4.5** Temporal variability data from Elbe River (Hradec Kralove)

could be interesting to identify potential pressure points and assess the impact of a human activity on water quality (urban and agriculture activities for example).

## Transects profile of physico-chemical parameters in lake (city of Daugavpils, Latvia)

The testing activities have been deployed in the city of Daugavpils (Latvia) in order to carry out transects profiles of the Lake to estimate the spatial variability of the physico-chemical parameters and biogens in the lake (Figure 2.4.7). Indeed, no monitoring data were available for this lake and the main objective of the field trials was to achieve a first set of data.

*Examples*

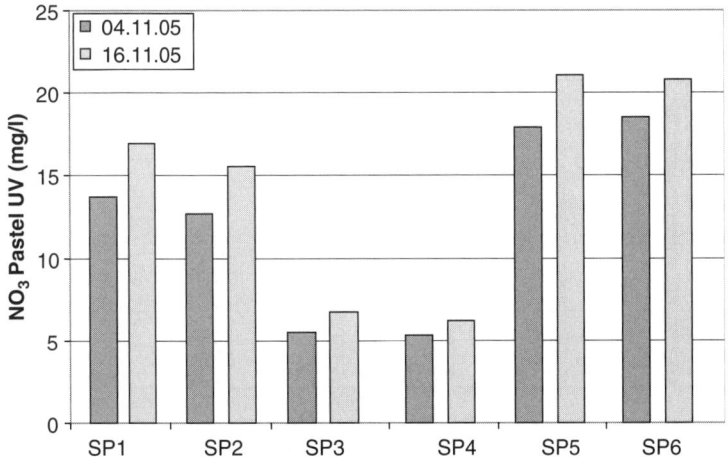

**Figure 2.4.6** Spatial variability

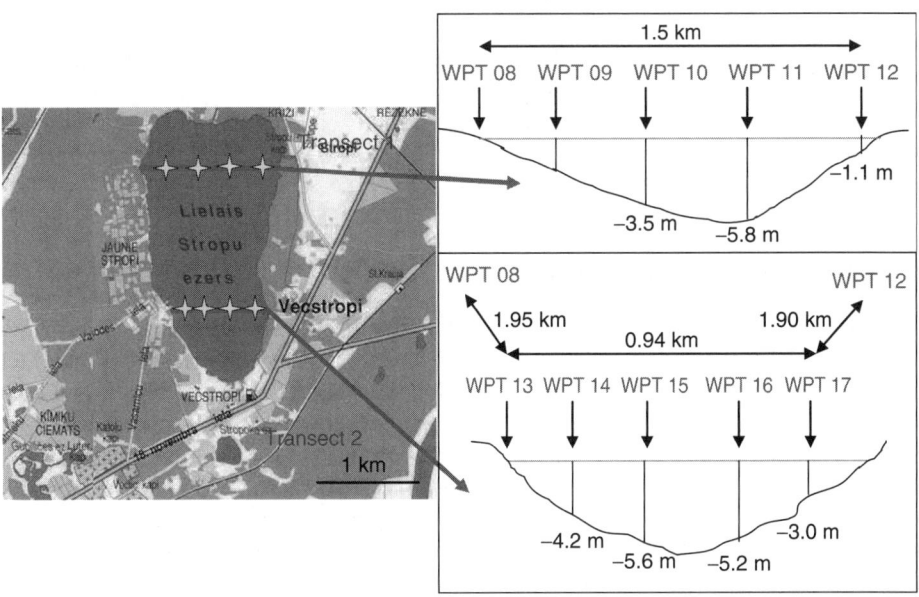

**Figure 2.4.7** Map of the sampling sites – Lake Stropu in Daugavpils (Latvia)

Samples were collected at two depths along the transect when it was possible for rapid on site measurements of $NO_3^-$, COD, BOD, TOC, SPM, DBS.

The parameters obtained by the PASTEL UV instrument demonstrated the water chemistry homogeneity of Lake Stropu (Figure 2.4.8). These results have been confirmed by the profile obtained by an YSI probe (T°, conductivity, pH, dissolved

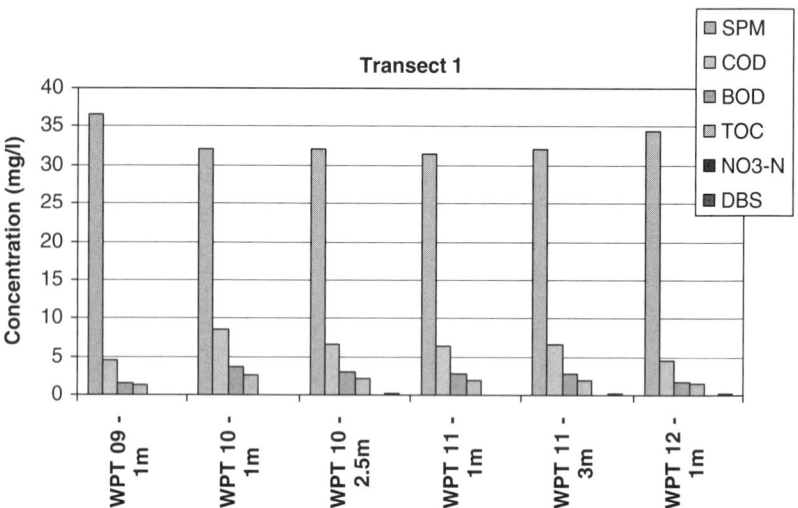

**Figure 2.4.8** On site PASTEL UV measurements – Lake Stropu (See Plate 2 for colour representation)

oxygen). All the results are detailed in a report produced by the consortium of the SWIFT-WFD project (SWIFT-WFD, 2006).

This portable instrument could be easily used to characterize in field conditions the water bodies (not only for lakes but also rivers, and groundwater) on the basis of physico-chemical parameters.

### 2.4.4.2 Trophic State Assessment of Lakes

In a recent book on UV-visible spectrophotometry for water and wastewater quality monitoring, several applications have been presented for the monitoring of lakes water quality (Pouet *et al.*, 2007). The first application was the estimation of nitrate and total organic carbon of some alpine French lakes, using a semi deterministic deconvolution method (Thomas *et al.*, 1993). Another application was a classification attempt, from UV spectra shapes, of trophic state of several Southern Québec lakes. The three main classes of trophic state (oligotrophic, mesotrophic and eutrophic correspond to groups of spectra increasing with eutrophication (Figure 2.4.9). This classification is rather close to the one from water quality parameters values (Table 2.4.2).

A last application of the previous reference (Thomas *et al.*, 2005) was the study of pollution sources of lake Brome (Southern Québec) from its tributaries, giving useful information related to agricultural pressure (manure spreading), wastewater management (urban runoff discharges) or golf management (use of fertilizers and pesticides).

Another recent study (Normand-Marleau, 2007) also included the lake Brome with several other lakes and rivers from Southern Québec (123 samples). For the estimation of nitrate and total organic carbon with the above deconvolution method, the results were quite good with a correlation coefficient of 0.90 for both parameters. For nitrate, the concentration range was rather low (0 to 5 mg N/L) contrary to TOC range (0 to

# Examples

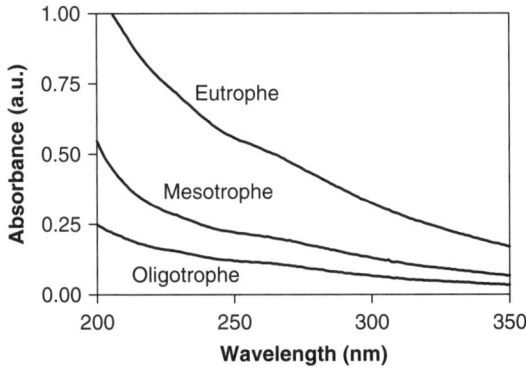

**Figure 2.4.9** Relation between UV spectra (path-length 10 mm) and trophic state of lakes

**Table 2.4.2** Comparison of UV classification of trophic site of lakes (Pouet et al., 2007)

| UV Classification | Number of lakes | Ptotal (μg/l) | Ntotal (mg/l) | Chloro A (μg/l) | Z Secchi (m) | DOC (mg/l) | UV254* (a.u.) |
|---|---|---|---|---|---|---|---|
| Oligotrophic | 6 | 4-12 | 0.2-0.3 | 1-2.5 | 3-7.5 | 2.5-5 | 0.06-0.15 |
| Mesotrophic | 9 | 10-23.5 | 0.31-0.52 | 3-7.5 | 1.8-3 | 5.9-15 | 0.13-0.60 |
| Eutrophic | 6 | 21.5-100 | 0.7-1.3 | 11.5-30 | 0.7-1.5 | 11-22 | 0.35-0.86 |

*For a path-length of 10 mm at 254 nm

30 mg C/L), and the estimation quite good even for low concentration (respectively <1 mg N/L and <2 mg C/L). In this study, the UV estimation of the Trophic State Index or TSI (Carlson, 1977), was also carried out. The TSI is calculated from the phosphorus and chlorophyll concentrations but also from the transparency of water (measured with the Secchi disk). Thus, the trophic state of any lake can be deduced from the TSI: oligotrophic for TSI below 90, mesotrophic between 90 and 150, and eutrophic above. The UV estimation of TSI (by deconvolution) has given interesting results with a correlation coefficient of 0.80 for 31 lakes (Figure 2.4.10).

More generally, UV spectrophotometry can be considered for the monitoring of all compartments of lakes environment. For example, in a recent unpublished work on a eutrophic lake diagnosis, UV spectra of lake and tributaries waters as well as sediment extracts, have been acquired.

Four sampling campaigns have been carried out in summer, and spectra of lake water show that the trophic state is varying from the beginning (June) to the plenty summertime (July and August), changing from a mesotrophic towards a eutrophic state (Figure 2.4.11). This has been confirmed by the appearance of algae blooms of cyanobacteria in July and August.

The spectra of the lake must be related to the ones of its main tributaries (Figure 2.4.12). Only one tributary brings almost 90% of water to the lake, but to upstream main rivers (tributaries 4 and 6) are contributing to the main tributary. The

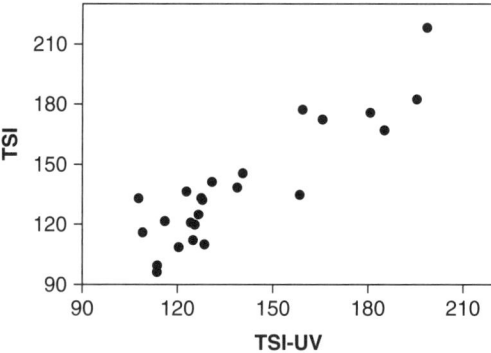

**Figure 2.4.10** Relation between 'measured' and UV estimated TSI

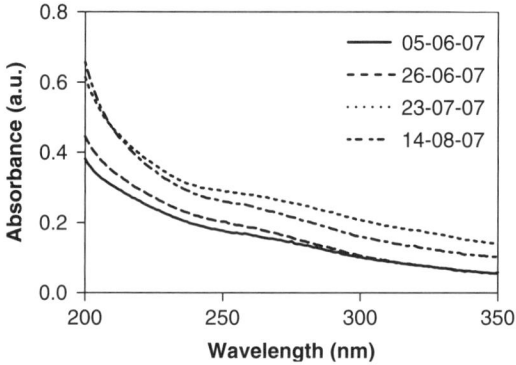

**Figure 2.4.11** UV spectra of the studied lake (path-length 10 mm)

examination of raw spectra (Fig. 2.4.12 a and c) shows that the organic load is increasing in July and August, and that the spectra shape of the lake can be explained by a combination of the ones of the tributaries. More precisely, UV spectra show that tributary 4 is under agricultural influence (presence of nitrate in June) and tributary 6 under more natural influence with spectra close to the ones of humic substances. The normalized spectra show that variability of water quality (Thomas *et al.*, 2005) is more important for tributary 4 (agricultural influence) than for tributary 6 (Figure 2.4.11 b and d).

Finally, lake sediment has been sampled for the study of UV spectra of extracts. In order to characterize the organic matter of sediments, a simple and fast two steps procedure has been designed with an ultrasonic extraction and a final UV analysis (Chabrol, 2007). After sediment drying, an ultrasonic assisted leaching step is carried out with three different solvents: (1) water for soluble substances, (2) acetonitrile for anthropogenic substances and specific polar compounds, and (3) sodium hydroxide (0.1 mol/L) for humic substances. Figure 2.4.13 presents the UV spectra of extracts for sediment of the studied lake. The results of the physico-chemical characterization of sample are the following: Sediment very liquid (8.9% of dry matter, 33.2% of

# Examples

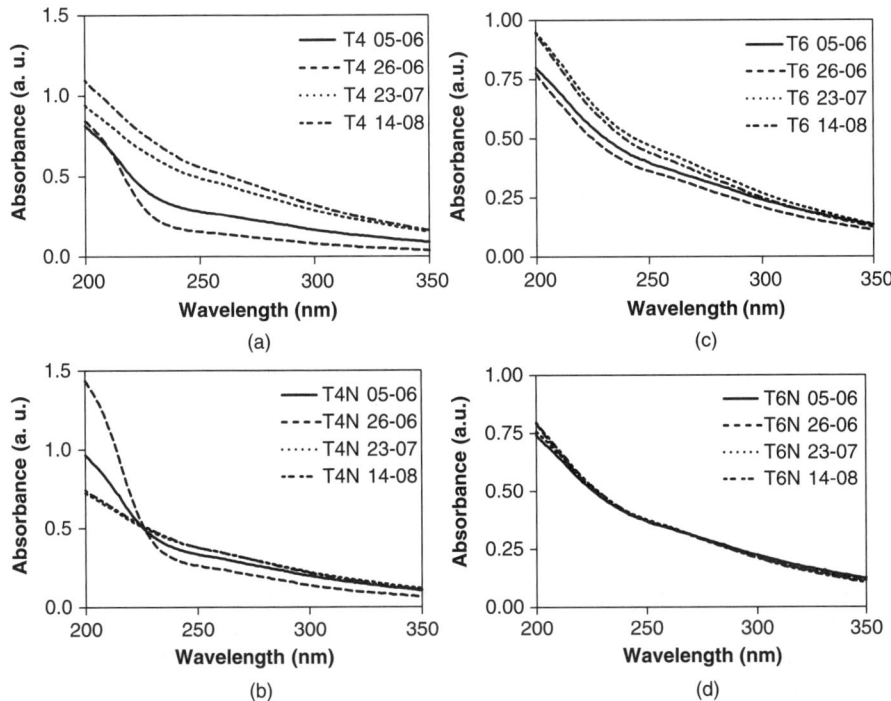

**Figure 2.4.12** UV spectra of tributaries (a and b, raw and normalised spectra of tributary 4, c and d, raw and normalized spectra of tributary 6), (path-length 10 mm)

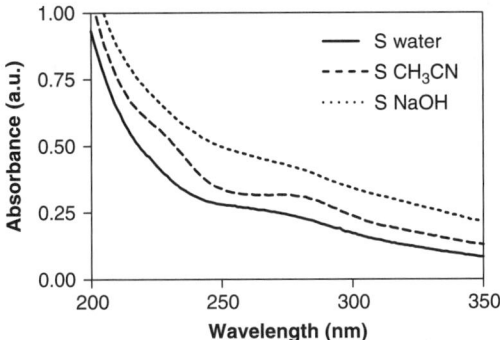

**Figure 2.4.13** UV spectra of water, acetonitrile (dilution 5) and sodium hydroxide extracts of lake sediments (path-length 10 mm)

organic matter) of fine structure (95% w/w $< 500\,\mu m$) with a C/N ratio of 10.8 and a P concentration of 1319 mg/kg. The spectra give complementary information, with a high content of natural organic matter (vegetal degradation products and humic substances related both to water and basic extracts) and a rather high amount of anthropogenic substances (acetonitrile extract), mainly coming from a wastewater discharge (without treatment) of a large camping between confluence of tributaries 4 and 6 and lake.

### 2.4.4.3 Water Quality Monitoring

*Accidental pollution and early warning system*

In case of accidental pollution or process monitoring, there is an obvious need to react immediately if any problems occur in order to take the appropriate decision and to minimize the impact on the aquatic ecosystem. Field measurement systems as portable PASTEL UV could be very efficient and could deliver rapid information on pollution level or the technical hitch. Figure 2.4.14 shows the monitoring of a wastewater treatment plant by measuring the water quality before and after the biological treatment (upstream and downstream).

In a rapid view, a good efficiency of the treatment process can be observed in terms of degradation of organic matter. Nevertheless, a problem in the denitrification step has been highlighted. Indeed, it can be noticed a high increase in the level of nitrate discharge in the environment. After discussion with the WWTP manager, an oxygenation disturbance has occurred. Moreover, the denitrification problem has been also confirmed by nitrate measurement obtained by other chemical test kits. All the results are detailed in a report produced by the consortium of the SWIFT-WFD project (SWIFT-WFD, 2006).

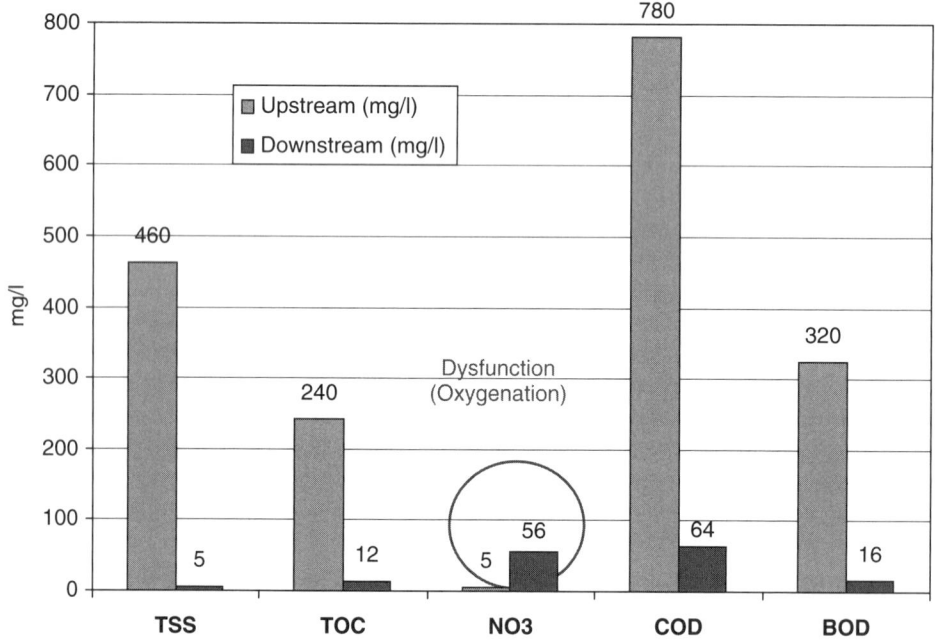

**Figure 2.4.14** Alert system by Pastel UV

## Monitoring and prevention of water resources for drinking water production

In order to avoid contamination of drinking water treatment plant of the Marseille city, the Durance River is monitored by the online system (STAC). This online analyzer measures in continue the quality of the river water and compares to the theoretical spectrum of the river water without any pollution. When some pollution occurs, the shape of the river water spectrum is modified and the difference between the two spectra could be used as an alarm of pollution contamination.

### 2.4.5 CONCLUSIONS

The given examples presented in this section illustrated the potential of UV spectrophotometry for improving information on the quality of water bodies and highlighted the perspectives of their integration on new approaches on water management. The portable instrument is in particular useful for the spatial and temporal water quality survey (river basin, lake…) based on measurements of some well-known physico-chemical parameters (TSS, TOC, COD, BOD, $NO_3^-$, surfactants). Nevertheless, the UV spectrophotometry and the suitable software developed to enhance the deconvolution of UV spectra allows to propose others applications, in particular qualitative interpretation in order to assess, for example, the trophic states of lakes.

More generally, the UV spectrophotometry (portable as well online technologies) could be very efficient tools for water managers in case of decision making (crisis management in case of accidental pollution, corrective actions to comply good water quality requirements) and for water quality survey (prevention of water resources).

## REFERENCES

Carlson, R.E., 1977. *Limnol. Oceanogr.*, **22**, 361–9.
Chabrol O., 2007. *Caractérisation rapide de sédiments lacustres en relation avec le niveau trophique*, MEnv report, University of Sherbrooke, Québec, Canada.
Gallot S. and Thomas O., 1993. *Fresenius J Anal Chem*, **346**, 976–83.
Gonzalez C., Prichard E., Spinelli S., Gille J. and Touraud E., 2007. *Trends Anal. Chem.*, **26**(4), 315–22.
Maier W.J., 1981. Multiwavelength absorbance measurements for monitoring trace organics in water in chemistry. In *Water Reuse*, Vol **1**, W.J. Cooper (ed.), Ann Arbor Science, Michigan.
Normand-Marleau M., 2007. *Application de la spectrophotométrie ultraviolette à la caractérisation d'eaux naturelles du Québec*, MEnv report, University of Sherbrooke, Québec, Canada.
Pouet M-F., Theraulaz F., Mesnage V. and Thomas O., 2007. Natural water. In: *UV-Visible Spectrophotometry for Water and Wastewater*, Thomas O., Burgess C. (eds), Elsevier, Amsterdam, 163–88.
SWIFT-WFD, 2006 (FP 6 project funded by DG Research). Report of performances evaluation of screening methods (Deliverable D43) available on the website ( swift-wfd.com).

Thomas O., Theraulaz F., Domeizel M. and Massiani C., 1993. *Environ. Technol.*, **14**, 1187–92.
Thomas O., Theraulaz F., Agnel C. and Suryani S., 1996. *Environ. Technol.*, **17**, 251–61.
Thomas O., Baurès E. and Pouet M-F., 2005. UV Spectrophotometry as a nonparametric measurement of water and wastewater quality variability. *Water Quality Res., Journal of Canada*, **40**(1), 51–8.
UVPro, 'Procédés et dispositifs pour la manipulation, l'archivage, la caractérisation, l'analyse d'un échantillon liquide', Patent 00402038-4, 17 July 2000.

# Section III
Biological Methods

# 3.1
## Application of Microbial Assay for Risk Assessment (MARA) to Evaluate Toxicity of Chemicals and Environmental Samples

Kirit Wadhia and K. Clive Thompson

3.1.1 Ecotoxicity Testing
3.1.2 Bacterial Tests
3.1.3 Microbial Assay for Risk Assessment (MARA)
3.1.4 MARA Intra-laboratory Trial
    3.1.4.1 Repeatability
    3.1.4.2 Reproducibility
    3.1.4.3 Sensitivity
3.1.5 MARA Inter-laboratory Trial
3.1.6 Conclusions
References

## 3.1.1 ECOTOXICITY TESTING

The conventional approach to assessing toxicity has been by means of chemical analysis. Analyses based on priority substances provided a measure of the toxic loading (Scroggins, 1999). Based on the concentrations of specific chemicals, the potential prediction of toxicity is determined. With this approach, owing to the complexity of matrices of environmental media, a real assessment of the extent of toxic interactions and actual toxicant availability in the ecosystem is subject to misinterpretation (Munawar *et al.*, 1989). In using an ecotoxicity testing approach with the

---

*Rapid Chemical and Biological Techniques for Water Monitoring*  Edited by Catherine Gonzalez, Philippe Quevauviller and Richard Greenwood
© 2009 John Wiley & Sons, Ltd

employment of bioassays, an effective assessment of toxicity of environmental samples has been made possible (Suter *et al*., 2000; Ferguson *et al*., 1998). This approach of *direct toxicity assessment* (DTA) allows evaluation to take into account interactions that may be additive, antagonistic or synergistic. The utilization of bioassays in the regulatory and monitoring frameworks has been realized and implemented worldwide (Power and Boumphrey, 2004; Thompson *et al*., 2005). The need to use a battery of tests in ecotoxicity evaluation has been advocated in view of the fact that tests commonly employed use specific single species (for example *Pseudokirchneriella subcapitata* or *Daphnia magna*) pertaining to a particular trophic level (Kahru *et al*., 2000; Pascoe *et al*., 2000; Girling *et al*., 2000; Mariani *et al*., 2006; Vosylienë, 2007). The use of multi-species tests has obvious financial implications. Tests using bacterial species potentially offer the scope for low-cost routine testing.

## 3.1.2 BACTERIAL TESTS

Utilization of bacteria in ecotoxicity tests has significant benefits:

- the relative size of the microorganisms means that concurrent effects measured pertain to large number (millions) of test organisms;
- the duration of the tests is substantially reduced owing to short generation times;
- the metabolic and physiological activities in bacteria are likely to be impacted by toxicants much more rapidly than those in higher organisms;
- ethical issues, particularly associated with vertebrate species, are not a concern;
- and costs associated with bacterial tests are significantly lower than those of invertebrate and vertebrate ecotoxicity tests (Bitton and Dutka, 1986; Cairns *et al*., 1992).

The potential use of microbial tests has been recognized with the evident proposals submitted by regulatory and standardization organizations (Mayfield, 1993).

## 3.1.3 MICROBIAL ASSAY FOR RISK ASSESSMENT (MARA)

MARA is an innovative bioassay devised for the evaluation of toxicity of chemicals and environmental samples. The assay utilizes a taxonomically diverse array of ten bacterial species (prokaryotes) and a yeast (eukaryote). The assay is performed in a 96 well micro titre plate and involves exposure of the microorganisms provided in a freeze-dried state. The toxicity of the test sample using a concentration gradient is determined with the employment of the redox dye tetrazolium red (TZR). The dye is transformed from a soluble colourless state to a red insoluble form upon reduction. The dye is a growth indicator and detects enzyme systems by acting as an electron acceptor.

The inhibitory effects evident in the formation of the visible pellets are examined electronically by capturing the image of the test plate using a flat bed scanner.

A purpose-built software converts the scan to produce numerical output generating a *microbial toxic concentration* (MTC) for each constituent species and a mean MARA MTC value.

## 3.1.4 MARA INTRA-LABORATORY TRIAL

In order to evaluate the performance of the MARA test, proposed testing was implemented at a reputable accredited contract laboratory. The testing was performed to attain an objective and independent assessment of the capabilities of the test with reference to potential commercial application. To assess the MARA test it was prudent to investigate the test's performance variability in terms of repeatability and reproducibility. To evaluate the MARA test, the investigations considered to be pertinent and drafted in the proposal included assessment of:

- inorganic and organic toxicants
- confounding variables
- environmental samples testing including:
  - waters – raw and final
  - industrial effluents
  - sewages
  - soil leachates.

### 3.1.4.1 Repeatability

In order to determine the inherent variability of the MARA test, testing was implemented using potassium dichromate ($K_2Cr_2O_7$). Ten plates were simultaneously set-up using 100 mg $l^{-1}$ stock solution.

The mean MTC value, 95% confidence limit (CL) and coefficient of variation (CV) for each species are given in Table 3.1.1.

The mean MTC value, 95% confidence limit (CL) and coefficient of variation (CV) for each species are graphically summarized in Figure 3.1.1.

**Table 3.1.1** Mean MTC, 95% CL and CV (%) for testing performed using $K_2Cr_2O_7$

| | \multicolumn{11}{c}{Species} |
|---|---|---|---|---|---|---|---|---|---|---|---|
| | 1 | 2 | 3 | 4 | 5 | 6 | 7 | 8 | 9 | 10 | 11 |
| **Mean MTC** | 10.37 | 40.82 | 13.00 | 78.86 | 12.50 | 30.24 | 23.71 | 29.20 | 22.67 | 20.81 | 55.72 |
| **CL (95.0%)** | 1.02 | 4.80 | 1.06 | 9.11 | 1.39 | 1.17 | 3.41 | 5.15 | 5.58 | 5.13 | 14.15 |
| **CV (%)** | 13.73 | 16.45 | 11.36 | 16.15 | 15.50 | 5.42 | 20.08 | 24.65 | 34.40 | 34.45 | 33.04 |

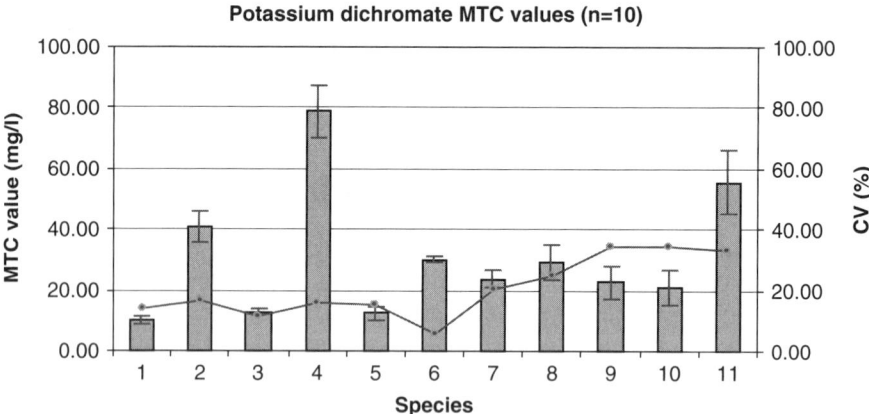

**Figure 3.1.1** Mean MTC, 95% CL and CV (%) for testing performed using $K_2Cr_2O_7$

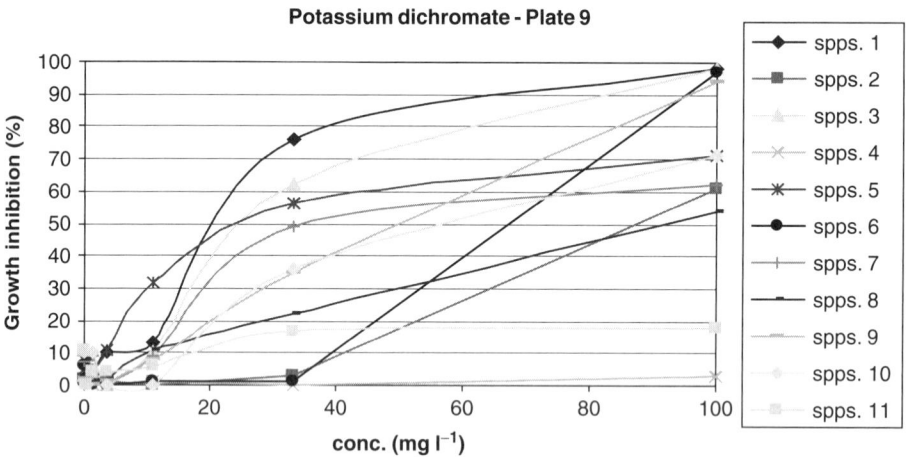

**Figure 3.1.2** MARA species inhibition plots for testing performed using $K_2Cr_2O_7$

The species found to be most sensitive to potassium dichromate was No. 1; Species No. 4 was the least sensitive. This is illustrated for the results obtained with plate 9 presented in Figure 3.1.2.

The MARA test mean (all species) MTC value for $K_2Cr_2O_7$ with replicate (n = 10) testing was found to range from 22.4 to 27.0 mg $l^{-1}$. The species exhibiting the least variation (CV = 5.42%) was No. 6. Five species (Nos. 1 to 5) had CV values ranging from 11.36 to 16.45%. CV values for species 7 to 11 ranged from 20.08 to 34.45%.

ANOVA performed on the data in Table 3.1.1 indicated that the difference between plates was not significant (p>0.05-NS). It was apparent that the response of the species was significantly different (p<0.001-***).

Table 3.1.2 Range of $K_2Cr_2O_7$ $EC_{50}$ values for ecotoxicity tests

| Test method | $EC_{50}$ range (mg $l^{-1}$) |
| --- | --- |
| *Daphnia magna* 48hr immobilisation | 0.5–1.5 |
| Freshwater Algae *Pseudokirchneriella subcapitata* | 0.2–1.7 |
| Marine Algae *Skeletonema costatum* | 1.3–4.6 |

In order to compare the relative sensitivity of the MARA test with some test methods employed for regulatory purposes range of $EC_{50}$ values for three pertinent tests are given in Table 3.1.2.

In comparison with other ecotoxicity tests employed for implementation of environmental legislation (e.g. IPPC), the species utilized for the MARA test are less sensitive.

### 3.1.4.2 Reproducibility

To attain a measure of reproducibility of the MARA test repeated testing on different days using a suitable reference toxicant (3,5 dichlorophenol) was implemented.

The mean MTC value (n = 6), 95% confidence limit (CL) and coefficient of variation (CV) for each species are given in Table 3.1.3.

The mean MTC value, 95% confidence limit (CL) and coefficient of variation (CV) for each species are graphically summarized in Figure 3.1.3. The overall mean MTC value (of all species) obtained for each plate is given in Table 3.1.4.

The species found to be the most sensitive to 3,5 dichlorophenol was No. 6; Species No. 9 was the least sensitive. The MARA species mean MTC value for 3,5 dichlorophenol with replicate (n = 6) testing was found to range from 3.8 to 29.2 mg $l^{-1}$. The species exhibiting the least variation (CV = 5.19%) was No. 7. Six species had CV values ranging from 11.76 to 20.82%. CV value (%) for species No. 2 was 26.03, and species No. 11 and 10 had the highest values of 31.51 and 45.78 respectively.

ANOVA performed on he data showed that the difference between plates was not significant (p>0.05 - NS). And as expected the response of the species was significantly different (p<0.001-***).

In order to compare the relative sensitivity of the MARA test with some ecotoxicity tests, representative plots of AQC data routinely obtained at an accredited laboratory

Table 3.1.3 Mean MTC, 95% CL and CV (%) for testing performed using 3,5 dichlorophenol

| | Species | | | | | | | | | | |
| --- | --- | --- | --- | --- | --- | --- | --- | --- | --- | --- | --- |
| | 1 | 2 | 3 | 4 | 5 | 6 | 7 | 8 | 9 | 10 | 11 |
| **Mean** | 5.73 | 4.59 | 7.70 | 4.51 | 19.83 | 3.80 | 4.41 | 7.35 | 29.17 | 10.53 | 6.21 |
| **CL (95.0%)** | 0.85 | 1.25 | 1.33 | 0.99 | 4.52 | 0.65 | 0.24 | 2.56 | 3.60 | 5.06 | 2.05 |
| **CV (%)** | 15.10 | 26.03 | 16.43 | 20.82 | 21.73 | 16.36 | 5.19 | 33.16 | 11.76 | 45.78 | 31.51 |

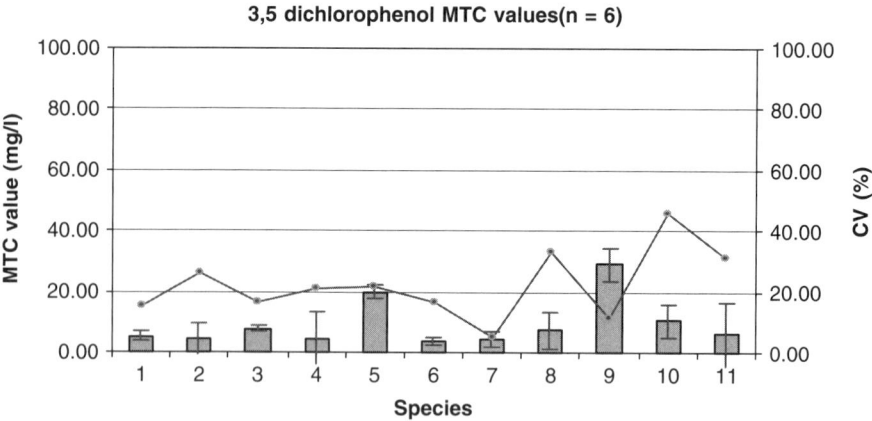

**Figure 3.1.3** Mean MTC, 95% CL and CV (%) for testing performed using 3,5 dichlorophenol

**Table 3.1.4** MARA test plate MTC values

| Plate | 1 | 2 | 3 | 4 | 5 | 6 |
|---|---|---|---|---|---|---|
| MTC (mg l$^{-1}$) | 6.97 | 7.12 | 7.61 | 6.82 | 7.95 | 7.12 |

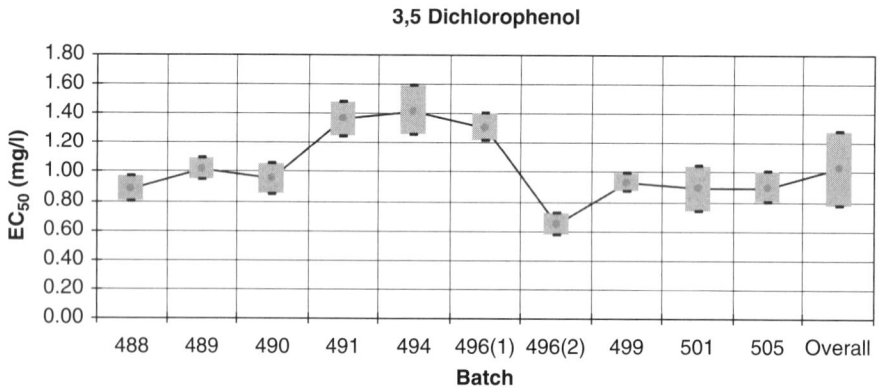

**Figure 3.1.4** Range of EC$_{50}$ values obtained with *Nitrosomonas* using 3,5 dichlorophenol

are presented for different methods to demonstrate the range, variability and relative sensitivity.

The nitrification inhibition test AQC data for different batches of the test organism, *Nitrosomonas*, has given a range of EC$_{50}$ values of 0.76 to 1.28 mg l$^{-1}$. The plot of EC$_{50}$ values for 3,5 dichlorophenol is given in Figure 3.1.4.

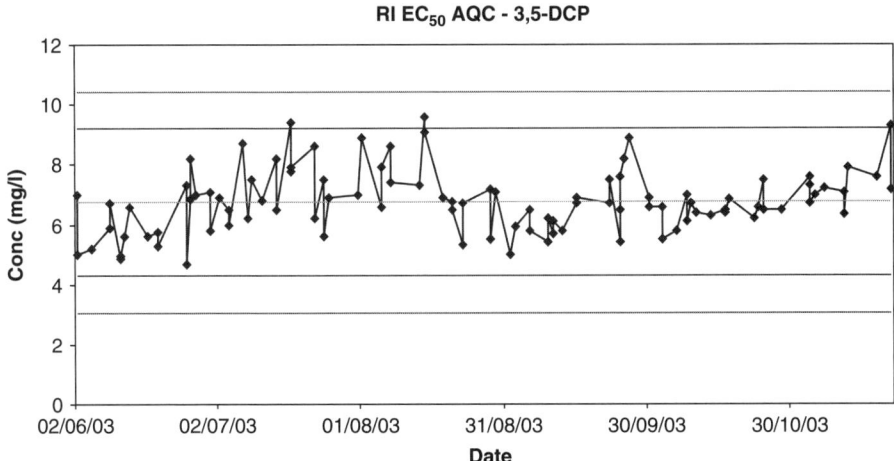

**Figure 3.1.5** Range of $EC_{50}$ values obtained with Respiration inhibition test using 3,5 dichlorophenol

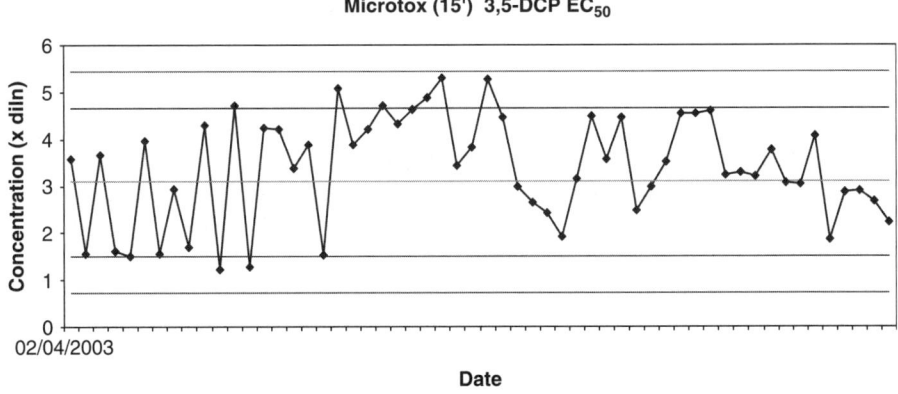

**Figure 3.1.6** Plot of $EC_{50}$ values obtained with Microtox® using 3,5 dichlorophenol

For the Respiration Inhibition (RI) test performed routinely at the aforementioned laboratory the range of $EC_{50}$ values obtained over a period of several months ranged from 4.5 to 10.0 mg $l^{-1}$. The AQC plot of these values is given in Figure 3.1.5.

The AQC 3,5 dichlorophenol data plot for Microtox® is given in Figure 3.1.6. The values are plotted as dilutions of the 10.0 mg $l^{-1}$ stock solution. The range of $EC_{50}$ values obtained equates to 2.0 to 6.7 mg $l^{-1}$.

For phytotoxicity assessment, the range of $EC_{50}$ values that are expected to be obtained using the freshwater and marine algal tests are given in Table 3.1.5 (Environment Agency, 2006).

**Table 3.1.5** Range of acceptable 3,5 dichlorophenol EC$_{50}$ values for algal tests

| Test method | EC$_{50}$ range (mg l$^{-1}$) |
| --- | --- |
| Freshwater Algae *Pseudokirchneriella subcapitata* | 0.9–4.4 |
| Marine Algae *Skeletonema costatum* | 0.6–2.4 |

**Table 3.1.6** Ecotoxicity tests EC$_{50}$ values obtained using 3,5 dichlorophenol

| Test method | EC$_{50}$ range (mg l$^{-1}$) |
| --- | --- |
| MARA test | 12–14 |
| Respiration inhibition | 4.5–10.0 |
| Microtox® | 2.0–6.7 |
| Freshwater Algae *Pseudokirchneriella subcapitata* | 0.9–4.4 |
| Marine Algae *Skeletonema costatum* | 0.6–2.4 |

The overall result of the MARA test to 3,5 dichlorophenol is dependent on the range of sensitivity associated with the eleven constituent species. Approximately seven-fold difference was evident between the most and the least sensitive species.

The species found to be the most sensitive and the least sensitive to 3,5 dichlorophenol were No. 6 and No. 9 respectively. The range of MTC values obtained for 3,5 dichlorophenol was from 6.82 to 7.95 mg l$^{-1}$.

In comparison with the EC$_{50}$ values of the aforementioned ecotoxicity tests, summarized in Table 3.1.6, the MARA test is the least sensitive to 3,5 dichlorophenol. However, this does not necessarily imply that the MARA could not potentially be used to assess the toxicity of environmental samples (Wadhia *et al.*, 2007).

The evaluation of reproducibility showed that there was no significant difference between replicates (plates). A comprehensive inter-laboratory comparison was subsequently implemented to ascertain a definitive assessment of reproducibility (see Section 4.5).

### 3.1.4.3 Sensitivity

Sensitivity of the MARA test was evaluated by testing inorganic and organic chemicals and environmental samples. To examine the effect of inorganic chemicals which are not deemed to be toxic but may have affect on toxicity evaluation, in other words 'confounding variables', were tested using MARA.

The criteria for selection of the chemicals, in the case of metal and organic toxicants, were based on the monitoring and legislative perspective of toxicants posing toxic concern to the environment.

# MARA Intra-laboratory Trial

**Table 3.1.7** Summary of MARA result and other ecotoxicity tests

|  | Daphnia magna | T. platyurus | Microtox® | MARA Lowest MTC | MARA Mean MTC |
|---|---|---|---|---|---|
|  | 24 h $EC_{50}$ | 24 h $LC_{50}$ | 30 min $EC_{50}$ |  |  |
| Metals (mg/l) |  |  |  |  |  |
| $Cd^{2+}$ ($CdCl_2.5H_2O$) | 0.7/0.9 | 0.2 | 8 | 4.4 | 17 |
| $Cr^{6+}$ ($K_2Cr_2O_7$) | 0.35 | 0.018 | 15 | 2.8 | 11 |
| $Cu^{2+}$ ($CuSO_4.5H_2O$) | 0.07/0.536 | 0.079 | 0.2 | 4.6 | 27 |
| $Hg^{2+}$ ($HgCl_2$) | 0.022/0.0081 | 0.04 | 0.05 (15 min) | 1.4 | 2.7 |
| $Ni^{2+}$ ($NiCl_2.2H_2O$) | 10.9 |  | 20 | 36 | 80 |
| $Ni^{2+}$ ($NiSO_4.6H_2O$) | 8.55 | 2.21 |  |  |  |
| $Sb^{3+}$ ($SbCl_3$) |  | 5.27 | 11.2 | 3.3 | 14 |
| $Zn^{2+}$ ($ZnCl_2$) |  | 0.23 | 1.1 |  |  |
| $Zn^{2+}$ ($ZnSO_4$) | 2.1/1.2 | 0.69 | 2 | 19 | 154 |
| $As^{3+}$ ($Na_3AsO_3$) |  | 0.3 | 1.3 | 1.4 | 30 |
| $Ag^{2+}$ ($Ag_2SO4$) |  |  |  | 4.8 | 14 |
| $Al^{2+}$ ($Al(SO_4).18H_2O$) |  |  |  | 9.5 | 28 |
| $Cs^{2+}$ ($CsCl_2$) |  |  |  | 597 | 752 |

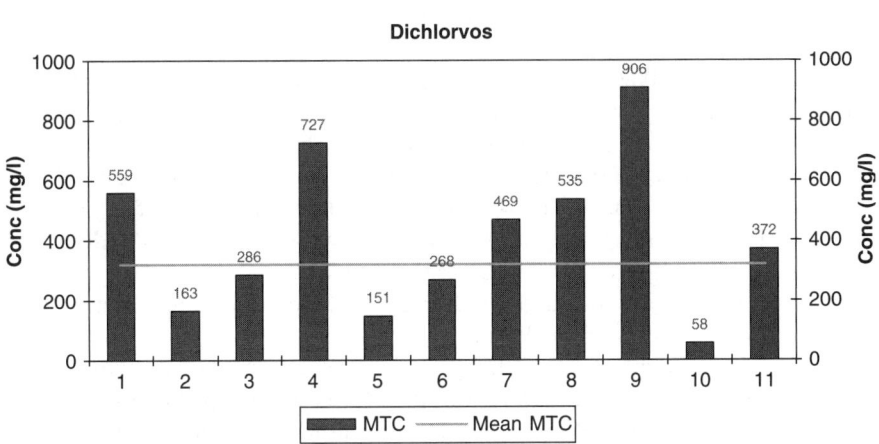

**Figure 3.1.7** MARA species MTC values for testing performed using dichlorvos

The results of the testing performed using MARA to evaluate the sensitivity to metals is summarized in Table 3.1.7.

The predominant difficulty in testing the sensitivity of ecotoxicity tests to organics is the inherent low (aqueous) solubility of many of the test substances. In essence it is prudent to examine the sensitivity to organic chemicals that may be potentially present in the environment, e.g. pesticides. To provide a simple comparison of the performance of MARA, the relative assessment using dichlorvos is presented in Figure 3.1.7.

Table 3.1.8 Comparative sensitivity of ecotoxicity tests to dichlorvos

| | Eclox™ (IC$_{50}$) | MARA# (MTC) | Microtox® (EC$_{50}$) | T. platyurus (LC$_{50}$) |
|---|---|---|---|---|
| | | (mg/l) | | |
| dichlorvos | 354 | 58 | 29 | 19 |

\# most sensitive species
Eclox™ is a test based on chemiluminescence.

Mean MTC value obtained with testing performed using dichlorvos was 319 mg l$^{-1}$.
Results obtained with other ecotoxicity tests in different studies for comparison are presented in Table 3.1.8.
The mean MTC values for solvents obtained with testing performed using MARA were:

- acetone = 1.97%
- DMSO = 2.06%
- methanol = 2.35%

Species 10 was the most sensitive to acetone and methanol, and species 3 was the most sensitive to DMSO. Species 7 was the least sensitive to all three solvents.

Gabrielson et al. (2003) reported MARA MTC values of 0.7 mg l$^{-1}$ (lowest) and 9.0 mg l$^{-1}$ (mean) for pentachlorophenol; and compared these with EC$_{50}$ value of 0.6 mg l$^{-1}$ for Microtox® and *Daphnia magna* from the literature.

To examine the effect of inorganic chemicals likely to be present in the environment but which are not intrinsically classified as toxic but may have affect on toxicity evaluation, in other words 'confounding variables', were tested using MARA.

The results of the testing performed using MARA to evaluate the sensitivity to confounding variables is summarized in Table 3.1.9. Data for comparison with other tests from previous studies are included.

The sensitivity to 'confounding variables' as assessed from the data above indicates that on the whole, invertebrates (*D. magna* and *T. platyurus*) are more sensitive than *Vibrio fischeri* (employed in Microtox®). The overall MARA species sensitivity based on the mean MTC values indicated that *Vibrio fischeri* was more resistant in the case of the above variables.

MARA (data not shown here) exhibited better sensitivity for leachates preparations (soils and sewage ash) than Microtox®.

The MTC values for the various effluents samples tested in the trial ranged from 5 to 48%. Effluents can exhibit a wide range of toxicities. Comparison of the results with nitrification and respiration inhibition tests where applicable showed that the MARA detected toxicity.

Comparative trend of sample toxicity between the assessment made with Microtox® and MARA was not always consistent. Based on the results of the limited number of

Table 3.1.9 Summary of MARA result and other ecotoxicity tests

| | Daphnia magna | T. platyurus | Microtox® | MARA Lowest MTC | MARA Mean MTC |
|---|---|---|---|---|---|
| | 24 h EC$_{50}$ | 24 h LC$_{50}$ | 30 min EC$_{50}$ | | |
| Metals (mg/l) | | | | | |
| Ca$^{2+}$ (CaCl$_2$.2H$_2$O) | 570 | 43 | | 559 | 688 |
| Fe$^{2+}$ (FeSO$_4$.7H$_2$O) | 24.5/30.2 | 41.5 | 71 (5 min) | 1.4 | 84 |
| K$^+$ (KCl) | 980/330 | 410 | | 12 | 501 |
| Mg$^{2+}$ (MgSO$_4$.7H$_2$O) | 400 | 620 | | | 735 |
| Mn$^{2+}$ (MnSO$_4$.H$_2$O) | 10 | 23.3 | 13.7 (15 min) | | |
| Na$^+$ (NaCl) | 400/2200 | 1800 | 20000 (5 min) | 332 | 3930 |
| Anions (mg/l) | | | | | |
| Cl$^-$ (NaCl) | 620 | 2500/2900 | 20000 (5 min) | 708 | 5405 |
| NO$_3^-$ (NaNO$_3$) | | 3800/4400 | | 157 | 2033 |
| SO$_4^{2-}$ (Na$_2$SO$_4$) | 5600 | 3900 | 17000 (5 min) | 5405 | 8388 |

environmental samples tested in the trial, it could be concluded that if toxicity of a sample is detected using Microtox® than it could also be detected using MARA.

The results of the testing performed using the MARA indicated that it is suitable for assessing different types of environmental samples and adequately sensitive to detect toxicity.

## 3.1.5 MARA INTER-LABORATORY TRIAL

The aim of this trial was to validate MARA by evaluating the performance of the assay at different laboratories worldwide. The method of implementation of the testing could have been considered in the same perspective as a collaborative trial used to validate a standard method. The trial was viewed with potential to form the basis of laboratory proficiency testing; defined by ISO as 'determination of laboratory testing performance by means of inter-laboratory comparisons' (Horowitz, 1988).

Twelve different laboratories (and additionally one reference laboratory) from eight countries participated in this trial. These included regulatory agencies, national testing organizations, academia, and contract laboratories (Table 3.1.10).

The inter-laboratory comparison for MARA, unlike in a certification trial or proficiency testing scheme, was organized on the basis of a collaborative trial.

The first phase in the trial involved the implementation of a training workshop for the participating laboratories. Participants from seven laboratories taking part in the trial attended the workshop (Table 3.1.10).

Participants (from Germany, France, Ireland and Japan) unable to attend the workshop were subsequently trained to perform MARA with instructions and all necessary materials (including samples) were sent to the participant's laboratory.

**Table 3.1.10** Laboratories participating in the trial

| Laboratory | Country | Training workshop attendance |
|---|---|---|
| 1 | UK | ✓ |
| 2 | UK | ✓ |
| 3 | UK | ✓ |
| 4 | UK | ✓ |
| 5 | UK | ✓ |
| 6 | Canada | ✓ |
| 7 | Ireland | × |
| 8 | Japan | × |
| 9 | Germany | × |
| 10 | Spain | ✓ |
| 11 | Sweden | × |
| 12 | France | × |

The design of the trial incorporated 3 essential integral elements:

| Laboratories | Samples | Replicates |
|---|---|---|
| 12 + Ref Lab | x  4  A, B, C & D  x | 3 |

The representative samples provided for testing in the trial were:

- *Reference toxicants:*

  Organic – 3,5 dichlorophenol [$125\,mg\ l^{-1}$] (A)

  Inorganic – potassium dichromate {$K_2Cr_2O_7$} [$150\,mg\ l^{-1}$] (C)

- *Environmental samples:*

  Industrial effluent (B)

  Soil leachate (D)

The duration allocated for testing of all samples was a period of two weeks. The laboratories were divided into 2 groups as follows: Group 1 Laboratories 10, 9, 3, 1, 4 and 7 – performed testing commencing the first week; Group 2 Laboratories 2, 6, 12, 5, 8 and 11 – performed testing commencing the second week. Concurrent testing at the reference laboratory was implemented with each group.

In order to assess the relative performance of the participating laboratories z-scores, sometimes referred to as 'standard scores', were utilized.

The z score for an item, indicates how far and in what direction, that item deviates from its distribution's mean, expressed in units of its distribution's standard deviation.

In a given distribution the conversion of all constituent unit values to z-scores results in a transformation that has a mean equal to zero and a standard deviation of one. For each replicate MARA mean MTC result, the z-score for an individual laboratory (i) was calculated as follows:

$z_i = (x_i - X)/s$

$x_i$ = results for laboratory i

X = an assigned value – deemed to be an estimate of the 'true value'

s = standard deviation

*Note – the assigned value (X) employed for the purpose of the assessment in this trial was the overall average MTC value determined using all the laboratories results*

The inference attained from the z-scores was as follows:

- close to zero meant that a result agreed well with the other laboratories;
- greater than 3 i.e. >3 or < −3, identified a result which demonstrated significant variation from the other laboratory results. These results were identified as *'outliers'*;
- a z-score that was an *'outlier'* (as determined above) fell outside ± 3 standard deviations, or 99% confidence.

The distribution criteria applicable for z-scores are:

- $|z| < 2$ will occur with normally distributed results in about 95% of all cases
- $2 < |z| < 3$ will occur with normally distributed results in about 5% of all cases
- $|z| > 3$ will occur with normally distributed results in about 0.3% of all cases

The z-scores for all laboratories participating in the trial are given in Figures 3.1.8a–d.
*Note – asterisk usage in Figures 3.1.8a–d: ** = 2 replicates; * = 1 replicate; none = 3 replicates*

The difference between replicates for all samples (A, B, C & D) tested was found to be not significant (p>0.05 NS). ANOVA outcome indicated that the difference between groups for all samples (A, B, C & D) tested was not significant (p>0.05 NS). The difference between laboratories when assessed with ANOVA performed on the full dataset (using mean replicate MTC values) of all 4 samples was found to be not significant (p>0.05 NS).

The microbial toxic concentration (MTC) pattern of the MARA array for the samples tested in the trial showed that a unique toxicity *'fingerprint'* was evident for the reference chemicals and the environmental samples as shown in Figures 3.1.9a–d.

122  *Application of Microbial Assay for Risk Assessment*

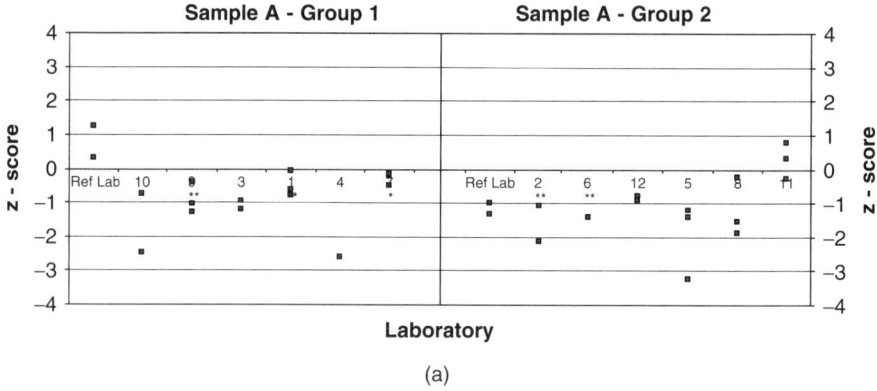

**Figure 3.1.8(a)**  Sample A (3,5 dichlorophenol) z-scores for participating laboratories

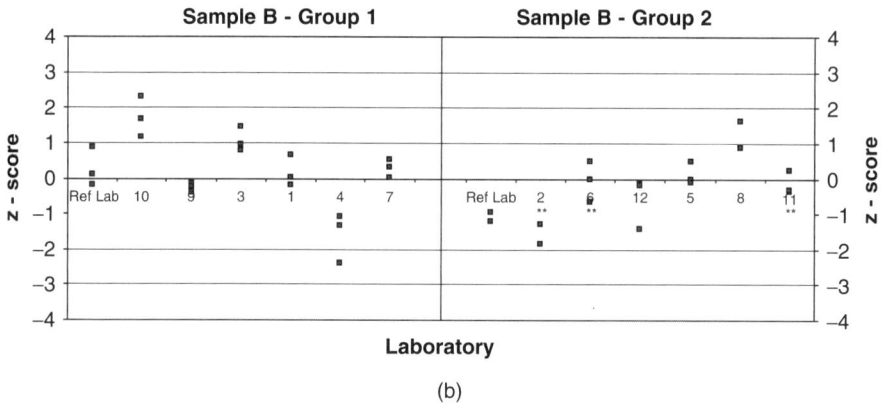

**Figure 3.1.8(b)**  Sample B (industrial effluent) z-scores for participating laboratories

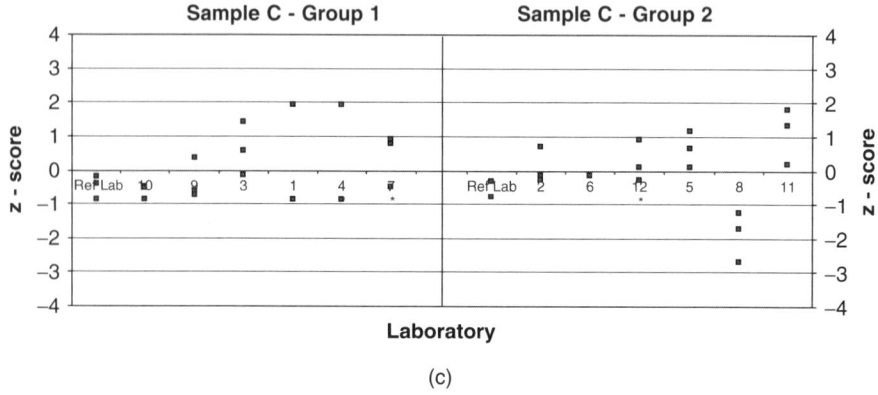

**Figure 3.1.8(c)**  Sample C (potassium dichromate) z-scores for participating laboratories

Conclusions                                                                                                      123

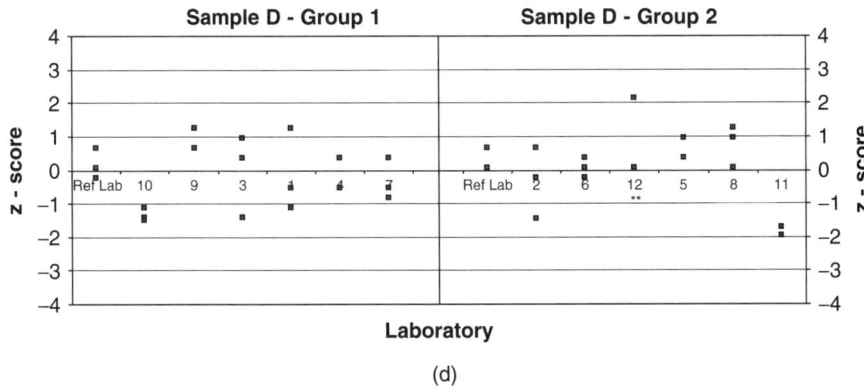

**Figure 3.1.8(d)**   Sample D (soil leachate) z-scores for participating laboratories

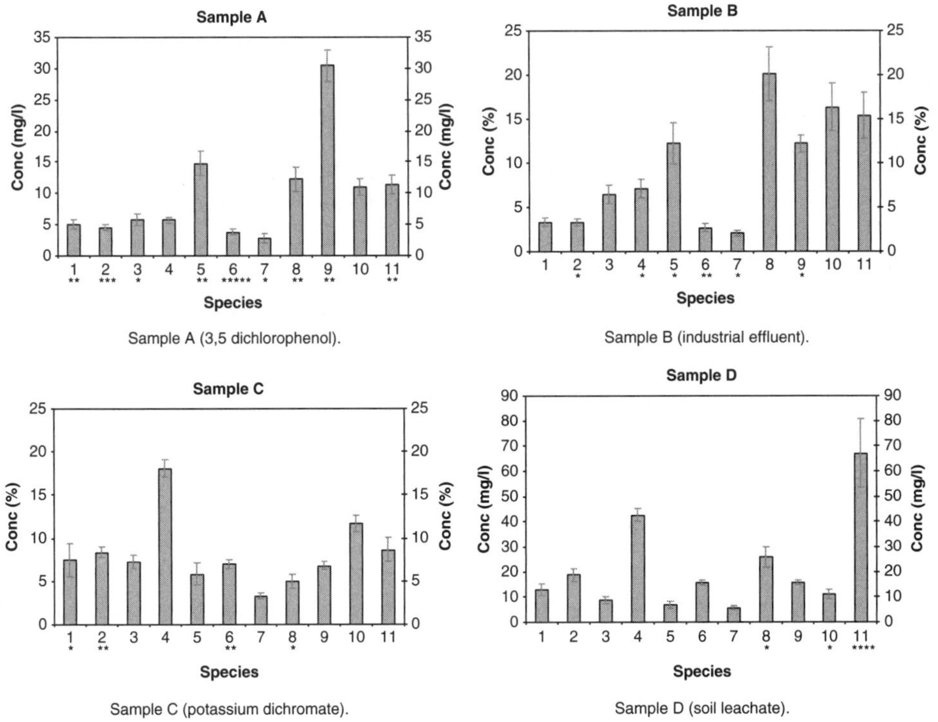

Sample A (3,5 dichlorophenol).

Sample B (industrial effluent).

Sample C (potassium dichromate).

Sample D (soil leachate).

**Figure 3.1.9(a-d)**   Mean MTC values (±95% CL) generated for MARA array

## 3.1.6   CONCLUSIONS

In implementing the intra- and inter- laboratory trials an extensive assessment of specific components of the assay, methodology and pertinent issues of operation, application and production have been examined.

Key aspects of evaluation have focused on examination of variability impacting performance associated with assay technique and potential application considerations.

The outcome of the research investigations and developments showed that MARA has significant potential for diverse application and utilization in laboratories worldwide involved in toxicity assessments.

## REFERENCES

Bitton G. and Dutka B.J., 1986. *Toxicity Testing Using Microorganisms.* CRC Press Inc, Boca Raton, USA.

Cairns J., McCormick P.V. and Niederlehner B.R., 1992. *Hydrobiologia*, **237**, 131–45.

Environment Agency (UK), 2006. *Monitoring Certification Scheme (MCERTS): Performance Standard for Laboratories Undertaking Direct Toxicity Assessment of Effluents.*

Ferguson C., Darmenfraid D., Freir K., *et al.* (eds), 1998. *Risk Assessment of Contaminated Sites in Europe*, Vol. **2** Policy Frameworks, LQM Press, Nottingham, UK.

Gabrielson J., Kuhn I., Colque-Navarro P., *et al.*, 2003, *Anal. Chim. Acta*, **485**, 121–30.

Girling A.E., Pascoe D., Janssen C.R., *et al.*, 2000. *Ecotoxicol. Environ. Safety*, **45**, 148–76.

Horowitz W., 1988. *Pure & Appl Chem*, **60**(6), 855–67.

Kahru A., Põllumaa L., Reiman R., Rätsep A., Liiders M. and Maloveryan A., 2000. *Environ. Toxicol.*, **15**, 431–42.

Mariani L., De Pascale D., Faraponova O., *et al.*, 2006. *Environ. Toxicol.* **4**, 373–9.

Mayfield C.I., 1993. Microbial systems. In: *Handbook of Ecotoxicology*, Calow, P. (ed.), Blackwell Scientific Publications, Oxford, UK.

Munawar M., Munawar I.F., Mayfield C.I. and McCarthy L.H., 1989. *Hydrobiologia*, **188/189**, 93–116.

Pascoe D., Wenzel A., Janssen C., *et al.*, 2000. *Water Res.*, **34**, 2323–9.

Power E.A. and Boumphrey R.S., 2004. *Ecotoxicology*, **13**(5), 377–98.

Scroggins R.P., 1999. *Guidance Document on Application and Interpretation of Single-species Tests in Environmental Toxicology*, Report EPS 1/RM/34. Environmental Technology Centre, Environment Canada, Ottawa, Ontario.

Suter G.W. II, Efroymson R.A., Sample B.E. and Jones D.S., 2000. *Ecological Risk Assessment for Contaminated Sites*, Lewis Publishers, CRC Press, Boca Raton, USA.

Thompson K.C., Wadhia K. and Loibner A.P. (eds), 2005. *Environmental Toxicity Testing*. Blackwell Publishing, Oxford, UK. ISBN 1-4051-1819-9.

Vosylienë M.Z., 2007. *Acta Zoologica Lituanica*, **17**, 1

Wadhia K., Dando T. and Thompson K.C., 2007. *J. Environ. Monitoring*, **9**, 953–8.

# 3.2
## Bioassays and Biosensors

**Marinella Farré and Damia Barceló**

3.2.1　Introduction
3.2.2　Main Legal Requirements
　　3.2.2.1　Toxicity Bioassays
　　3.2.2.2　Genotoxicity Assays
　　3.2.2.3　Estrogenicity Assays
3.2.3　Biosensors
　　3.2.3.1　Basis and Fundamentals
　　3.2.3.2　Recognition Element Classification and Biosensors for Environmental Analysis
3.2.4　Future Trends
3.2.5　Conclusions
References

## 3.2.1　INTRODUCTION

During the last three decades, numerous biological techniques have been developed as qualitative analytical methods to assess the effects of chemical pollutants on the environment. The increasing number of potentially harmful pollutants in the environment necessitates cost-effective analytical techniques to be used in extensive monitoring programmes. As a result, the use of biological data to complement chemical analysis, and the development of bio-detectors, such as immunoassays, biosensors and bioassays, have grown steadily in recent years. This chapter describes the principles, advantages and limitations of several biological methods for the screening and diagnosis of organic compounds in environmental samples, and the valuable additional biological information that can be obtained using these methods.

Due to the difficulty of predicting the collective effects of the increasing number of pollutants in receiving ecosystems, there is a need for screening methods for use in environmental monitoring. The implementation of safety programmes calls for environmental analysis comprising two parts:

- screening methods capable of predicting the possible dangerous biological effects, such as the toxicity, genotoxicity or estrogenicity of the cocktail of pollutants, and capable of measuring a large number of samples in a short time;
- re-analysis of the positive samples with chemical methods.

Bioresponse-linked instrumental analysis combines two processes, biological recognition initiating a biological effect, and chemical analysis. The biomolecular components are used as targets for active substances. Although it is possible to apply, for example, binding assays that provide effects-linked equivalents, information on the responsible contaminants is only accessible by chemical analysis. Therefore a subsequent step is provided by the chemical analysis of the substances that are bound by the biorecognition components and which are therefore bioeffective.

Figure 3.2.1 illustrates this principle using a simplified scheme of the hierarchy of the complexity levels in living systems. It takes into account the fact that the fundamental living processes such as metabolism, sensory qualities, growth, development and reproduction require the interaction of genetically determined factors and the processing of environmental influences. It is obvious that bioresponse-linked instrumental analysis is especially effective if communication channels can be used as targets since they connect environmental inputs with changes at the level of the phenotype.

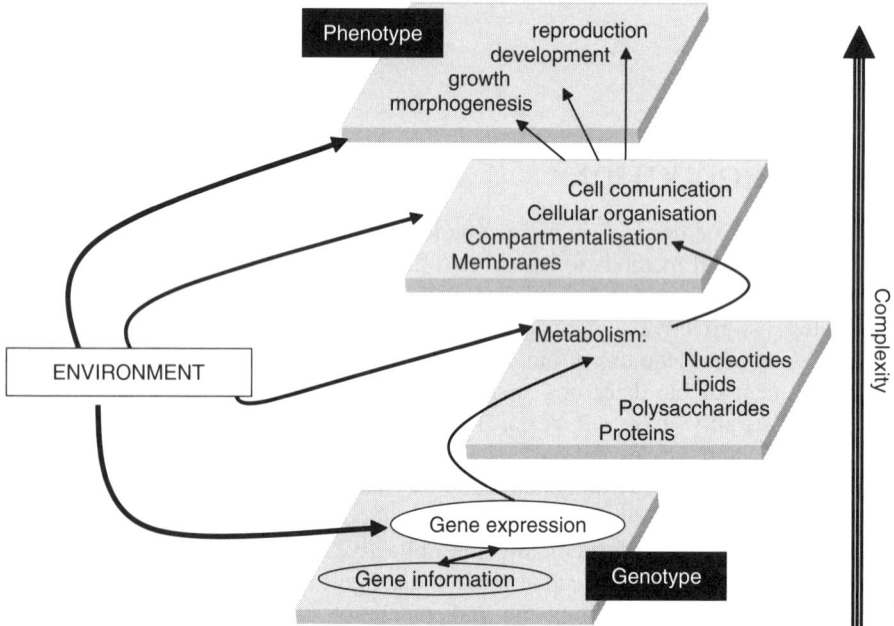

**Figure 3.2.1** Hierarchy of complexity levels of living systems

# Introduction

Biological techniques in environmental analysis can be classified according to:

- the technical principles (e.g. bioassays; biosensors; and immunoassays);
- the type of measurement (e.g. screening method and measurement of specific parameters).

The first biological tools to be applied to the environmental field were acute toxicity assays using mammals, birds, or fish as test species (Tothill and Turner, 1996; Farré and Barceló, 2003). However, these methods suffer some standardization problems, are expensive, time-consuming, and moreover, are associated with ethical difficulties. Due to these reasons, new biological methods using invertebrates, plants, algae or bacteria as test species were developed. A wide variety of bioassays based on microorganisms has been Developer. These methods give a rapid, response, and in many cases allow a large number of samples to be processed at the same time, and at a relatively low cost. Some of these methods are still prototypes for which only preliminary results are available, but they provide a good alternative to the classical methods, especially at the screening level. These tests are presented in easily managed formats that are adaptable to new technologies such as biosensors. The aim of applying bioassays in environmental analysis is to measure global biological effects, such as toxicity, genotoxicity or estrogenicity, or to assess global parameters, such as biological oxygen demand (BOD).

The newest biological approach is the development of biosensors (analytichal devices composed of a biological recognition element immobilized in intimate contact with a physical transducer). These are usually classified either according to:

- the transduction element (for example, electrochemical, optical, piezoelectrical, or thermal);
- the biorecognition principle (for example, enzymatic, immunoaffinity recognition, whole-cell sensors, or DNA).

The main advantages of biosensors are short times of analysis, low cost of assays, they can be incorporated in portable equipment, can make real-time measurements, and can be used as remote devices. These new technologies have been applied not only in the quantitative analysis of target analytes, but also in biological measurements of toxicity, genotoxicity, estrogenicity, or global parameters such as BOD. Currently biological technologies are playing a more prominent role in environmental analysis, and their incorporation into pollution-control programmes is growing. Nevertheless, many of these new technologies are still in the development stage and as with biosensors rely on combined efforts from diverse scientific fields,. In Europe the European Commission (EC) has been the flagship for the development of these new technologies in environmental analysis.

Under the 3th, 4th and 5th Framework Programmes a great number of funded research projects were dedicated to the development of new analytical tools, and more than 300 projects on sensors, biosensors and biological techniques have been carried out (Rodriguez-Mozaz *et al.*, 2005).

Nevertheless, in spite of the intensive development carried out in Europe, USA, and Japan most of these biological approaches are still at the prototype stage, and one of the main objectives for the future must be the validation of these technologies through interlaboratory exercises, comparisons with classical methods, and their progressive marketing and integration into routine analysis. During the 6th Framework Programme one of the thematic priorities has been nanotechnology, and in consequence related technologies such as nano-biosensors have been developed. Other projects have targeted their efforts on the validation, development of verification protocols for, and implementation of new biological tools for environmental monitoring.

## 3.2.2 MAIN LEGAL REQUIREMENTS

The goal of ecotoxicity is to understand how chemicals cause damage in some organisms, which species will be affected, and what effects this will have at the population and ecosystem levels. Bioassays are biological tools for the determination of the biological effects (positive or adverse) of a substance or a mixture of substances by quantifying that effect on living organisms or their component parts. Toxicity can be defined as the degree to which a chemical substance elicits a deleterious or adverse effect upon the biological system of an organism exposed to the substance over a designated time period. Aquatic toxicity, genotoxicity and estrogenicity are all different categories of toxicity.

### 3.2.2.1 Toxicity Bioassays

Most bioassays have been developed for the evaluation of the impact of xenobiotic substances on biota, and they can provide an indication of what overall impact on the environment might be produced by an effluent, sediment, or soil from a contaminated site.

The biological responses in different living organisms to a toxic substance or toxic mixture are diverse, and depend on the sensitivity of the particular species. A batch of different test species each for a different trophic level is highly recommended in order to study the toxicity of a substance or mixture. For instance in a recent study the utility and validity of toxicity tests based on *Vibrio fischeri, Selenastrum capricornotum*, and *Daphnia magna* for monitoring of wastewater treatment were assessed by Hernando *et al.* (2005).

Toxicity bioassays can be classified according to the test species into:

- bioassays on individual species;
- bioassays involving a set of different species;
- bioassays based on infraorganisms (e.g. cells or tissues) or biochemical responses.

According to the exposure time, their toxicity can be grouped into: acute (short exposure time, maximum 96 h); chronic (the test examining the effects of prolonged exposure and); sub-acute (with an exposure between acute and chronic – here the test organism survives the exposure, and measurement is based on growth, or reproduction); and reproductive toxicity tests (may be directed at a single generation or especially for fish in environmental toxicological studies at a complete life cycle). A range of endpoints can be used for all categories of exposure, and include mortality, and sublethal symptoms such as changes in feeding or locomotory activity, changes in growth rate, reproductive capacity, or rate of specific biochemical processes such as photosynthesis. Toxicological assays can also be classified according to the type of the test species involved. Various aquatic vertebrates and invertebrates, earthworms, protozoans, algae and seeds are used for toxicity bioassays, as described in previous reviews (Tothill and Turner, 1996; Farré and Barceló, 2003).

In the past, the majority of fish bioassays have been based on lethal measurements, by exposing a sample(s) of fish to a toxic substance under controlled conditions, during a maximum time exposure of 96 h. Using some form of statistical analysis of the mortality counts over a range of concentrations, the results are expressed as the concentration that would kill 50% of the population ($LC_{50}$). These fish assays can be carried out in systems that are: static (without renewal of the toxicant during the exposure); semi-static (in which the medium containing the toxicant is renewed periodically during the test); and continuous flow-through (in which the concentration of the toxicants is constant). In general, in vivo assays for aquatic toxicity that use higher animals such as fish show good sensitivity and are the nearest approaches to measuring the effects in natural environments. Nevertheless, they are costly, time-consuming, and need specialized equipment and operators with adequate skills. They also suffer some lack of standardization, and require the sacrifice of a great number of animals. Due to these drawbacks, these assays are not suitable for routine analysis of large numbers of samples.

Some alternative ecotoxicological bioassays are available for environmental monitoring, and amongst these the tests based on invertebrates such us *Daphnia magna* (ISO, 1996), microalgae, such as *Skeletonema costatum* (ISO, 1995) and *Selenastrum capricornotum* (EPA, 1982), the marine bacteria *Vibrio fischeri* and *Photobacterium phosphoreum* (ISO, 1998) are well established. These tests use standardized organisms, and are available from a number of commercial companies.

The use of macro invertebrates, such *as D. magna* or *Cerodaphnia*, presents many advantages, such as high sensitivity and a short reproductive cycle. Chronic tests with members of the genus *Daphnia* were initiated in the early 1970s by Biesinger and Christensen (1972) and are currently broadly accepted. Despite the diversity of test species available, in many regulatory schemes the only or preferred, invertebrate species recommended for acute or chronic toxicity testing is the cladoceran, *Daphnia magna*. One reason for the focus on *D. magna* could be its sensitivity to environmental contaminants relative to other invertebrate species (Kimerle *et al.*, 1985). This focus has resulted in a large number of acute and chronic toxicity tests with this species. However, the apparent advantages of *D. magna* over other species such as *Ceriodaphnia* have been a widely debated, and some authors have reported that the acute and chronic toxicity of a broad range of materials, including metal, organic compounds and effluents, are approximately similar for members of the genera *Ceriodaphnia* and *Daphnia* (Versteeg et al., 1997). These toxicity tests can be carried out to determine acute or chronic toxicity using a wide range of toxicological endpoints including lethality, growth, reproduction, mobility, and the population growth rate by exposing the organisms to the tested substances under appropriate controlled conditions.

Experiments using *D. magna* are usually carried out using genetically identical populations and using 3rd to 5th brood offsprings. While the standardized acute lethality toxicity test based on *D. magna* is conducted over 21 days, other alternative tests at screening level are carried out in short exposure times, such as the 24–48 h-screening test with *Daphnia* (Toussaint et al., 1995). The effects produced by pollutants on *D. magna* at suborganismal and organismal levels, and the implications for population dynamics are of great interest. A large number of studies has established the responses of daphnids to different pollutants using different conditions and end points. Some recent examples are the studies of pesticides (Barata et al., 2006; Duquesne, 2006), polyaromatic compounds (Ikenaka et al., 2006), metals (Duquesne, 2006; Gillis et al., 2006), anionic surfactants (Hodges et al., 2006) in *Daphnia*.

In spite of their limited use, a good number of plant based tests have been developed. The main advantages of these assays are that they offer a great variety of end-points, such as germination, enzymatic activity, and root elongation; in addition, they have low maintenance costs, good sensitivities, Further they are good indicators of the ecological risk from contaminated soils (ISO, 1995). The main disadvantage is very long response times. However, the use of algae as bio-indicators of toxicity is growing because they represent an important component of any aquatic community, and form the lowest link in the food chain. For this reason any substance that is toxic to algae can have a strong impact on the ecosystem as a whole. Microalgae, such as *Chlorella fusca, Selenastrum capricornotum, Dunaliella salina* or *Dunaliella tertiolecta* are used as toxicity indicators because of their sensitivity to pesticides (Löschau and Krätke, 2005) and metals (EPA, 1982; Nikookar et al., 2005). Recent research compared the sensitivity of selected freshwater and marine microalgae to some widely used surfactants (six anionic, two amphoteric and one nonionic), that are commonly found as pollutants in surface waters. Marine diatoms were significantly more sensitive to the tested surfactants than freshwater green algae (Diatoms $EC_{50}$ 0.14–1.7 mg L$^{-1}$ for surfactants). All of the surfactants tested can be classified as having toxic effects on the freshwater green alga *Pseudokirchneriella subcapitata*, and some were very toxic to *Scenedesmus subspicatus*, and to the marine diatoms *Skeletonema costatum* and *Phaeodactylum tricornutum* (Pavlić et al., 2005).

Despite the utility of the assay systems described above the most widely used bioassays for evaluating water toxicity in routine laboratories are based on inhibition of the bioluminescence of marine bacteria. The best known species of luminescent marine bacteria are *V. fischeri* and *Photobacterium phosphoreum*, which naturally emit light due to the enzyme bacterial luciferase. Any substance that has a deleterious effect on bacterial metabolism produces a proportional inhibition of the luminescence. The use of luminescent organisms to assess toxicity has been known for more than 40 years (Serat et al., 1965). In 1979 a toxicity bioassay using luminescent bacteria was developed by Bulich 1979 to assess toxicity of wastewater effluents and industrial discharges. This technique facilitates the screening of large numbers of aqueous samples in a rapid, reliable, and inexpensive way, and was commercialized for first time by Microtox and is described in Beckman's Operating Manual (Beckam Instruments, 1978). In addition, the procedure described in Beckman's Operating Manual is the Standard Procedure.

During the last decade, interest has increased in the ecological characterization of real samples by means of combined protocols, involving chemical analysis and toxicological evaluation. These methods combine the advantages of the diagnostic methods for which previous information about the sample is not necessary and which provide indications of potential ecological global effects, with those of targeted quantitative analysis. Table 3.2.1 summarizes some examples. Over the last decade, recombinant technology has created new luminescent bacteria, such as *Pseudomonas fluorescens* P-17, for application in diverse toxicity and genotoxicity tests (Zhang *et al.*, 2001). Organisms widely used in toxicity bioassays are listed in Table 3.2.2.

### 3.2.2.2 Genotoxicity Assays

Genotoxicity is associated with a range of chemical classes, such as phenols, chlorophenols, polychlorinated biphenyls (PCBs), and polyaromatic hydrocarbons (PAHs), and constitutes an early screening for possible carcinogenic activity of pollution. Among those based on microorganisms, we would like to highlight the assays based on the bacterium *Salmonella typhimurium*. The most widespread assay is the Ames test (Ames *et al.*, 1983) that was established as a routine method of analysis. It is based on the retromutation of *S. typhimurium* TA98 (histidine dependent). The umu test is also based on *S. typhimurium*, and the genotoxicity is indicated by the activation of the bacterial SOS repair response to genetic damage. This is measured indirectly through a reporter system based on b-galactosidase activity (Kenyon, 1983). This is a standardized method that is validated (ISO, 2000) for use in the control of water quality. Other genotoxicity bioassays that should be mentioned are those based on *Saccharomyces cerevisiae*, *Bacillus subtilis*, or *Escherichia coli* (Moreau *et al.*, 1976). The last is widely used because of its high sensitivity to certain groups of pollutants (Baun *et al.*, 2000) such as organochlorine pesticides (Houk and DeMarini, 1987) and chlorophenols (DeMarini *et al.*, 1990).

### 3.2.2.3 Estrogenicity Assays

In recent years, many natural and synthesized substances have been associated with endocrine disruption in wildlife, especially in aquatic fauna (Colborn *et al.*, 1993). The presence of numerous endocrine-disrupting compounds (EDCs) in natural waters and sediments has been attributed to the incomplete removal of these substances during wastewater treatment. Several bioassays have been developed to assess the activity of alleged estrogenic substances. In vivo assays based on a wide variety of end points, including cell differentiation and enzyme activities are widely used. However, they are unsuitable for large-scale screening since they are based on complex responses that may be modulated through mechanisms that do not directly involve the estrogen receptor. In order to carry out environmental monitoring of EDCs, a great number of in vitro assays based on the strength of binding of a substance to the estrogen

**Table 3.2.1** Different examples of organic pollution assessment by combined approaches

| Type of sample | Effect | Biological receptor | Test | Chemical analysis | Reference |
|---|---|---|---|---|---|
| Urban wastewater | Toxicidad | Vibrio fischeri | Microtox | SPE-GC-MS, | 1 |
| Wastewater | Toxicidad | Vibrio fischeri | ToxAlert Microtox | SPE-LC-MS | 2 |
| Wastewater | Toxicidad | *Escherechia coli* | Cellsense | SPE-LC-MS | 3 |
| Wastewater | Toxicidad | Pseudomona putida | Cellsense | SPE-LC-MS | 4 |
| Textile wastewater | Toxicidad | *Vibrio fischeri* | LUMIStox | COD. TOC | 5 |
| Sediment | Toxicidad | *Vibrio fischeri* | Microtox | GC-FID, GC-MS | 6 |
| Sewage sludge | Toxicidad | *Vibrio fischeri* | Microtox | GC-MS | 7 |
| Sewage sludge | Toxicidad | *Vibrio fischeri* | Microtox | GC-MS | 8 |
| Wastewater | Toxicity | *Photobacterium phosphoreum* | Microtox | LC-MS | 9 |
| Wastewater | Genotoxicity Toxicity | | HPLC-Toxprint[@] | HPLC | 10 |
| Wastewater, surfacewater | Genotoxicity | | UMU-test | LC-MS/MS GC-MS LC-Q-TOF | 11 |
| Wastewater | Toxicity | *Vibrio fischeri Selenastrum Capricornotum Daphnia magna* | BIOTOX Daphtoxkit FTM magna AlgalToxKit | TOC | 19 |
| Wastewater | Genotoxicity | *DNA* | Biocatalyst/DNA Films | LC-MS/MS | 12 |
| Soil | Genotoxicity | *Human Lymphocyte* | Cultures | GC-MS | 13 |
| Wastewater, surfacewater | Estrogenicity | *Polyclonal antibodies* | VTG | LC-MS | 14 |

[1] N. Paxeus, H.F. Schroder, Water Sci. Technol. 33 (1996) 9.
[2] M. Farré, M.-J. Garcia, J. Riu, D. Barceló, J. Environ. Monit. 3 (2001) 232.
[3] M. Farré, O. Pasini, M.-C. Alonso, M. Castillo, D. Barceló, Anal. Chim. Acta 426 (2001) 155.
[4] M. Farré, D. Barceló, Fresen. J. Anal. Chem. 371 (2001) 467.
[5] C. Wang, A. Yediler, D. Lienert, Z. Wang, A. Kettrup, Chemosphere 46 (2002) 339.
[6] M. Salizzato, V. Bertato, B. Pavón,V. Ghirardini, P.F. Ghetti, Environ. Toxicol. Chem. 17 (1998) 655.
[7] G.A. Harkey, T.M. Young, Environ. Toxicol. Chem. 19 (2000) 276.
[8] F.S. Mowat, K.J. Bundy, Environ. Int. 27 (2001) 479.
[9] L. Lunar, S. Rubio, D. Pérez-Bendito, Environ. Technol. 25 (2004) 173–184.
[10] TH.M. Noij, I. Bobeldijk, N. Fleischmann, G. Langergraber, R. Habert, Water sci. technol 47 (2003) 181–188.
[11] I. Bobeldijk, PG. Stoks, JP. Vissers, E. Emke, JA van Leerdam, B. Muilwijk, R. Berbee, TH. Noij, J Chromatogr A. 970 (2002) 167-81.
[12] M. Tarun, B. Bajrami, J.F. Rusling, Anal. Chem., 78 (2006) 624 – 627.
[13] S.D. Sivanesan, K, Krishnamurthi, S.D. Wachasunder, T. Chakrabarti, Biomed Environ Sci. 17 (2004) 257-65.
[14] M. Solé, MJ. López de Alda, M. Castillo, C. Porte, K. Ladegaard-Pedersen, D. Barceló, Environ. Sci. Technol., 34 (2000), 5076–5083.

Table 3.2.2   Organisms widely used for toxicity assessment

**Plant and algae**

| Species | Parameter measured | References |
|---|---|---|
| Scenedesmus quadricauda | Growth inhibition | 1 |
| Selenastrum capricornutum | Growth inhibition | 2 |
| Dunaliella tertiolecta | Growth inhibition | 20, 21, 3, 4 |
| Skeletonema costatum | Growth inhibition | 19, 5 |
| Chinese Cabbage | Germination | 6, 7 |
| Avena sativa | Germination | 8, 9 |
| Latuca sativa | Root elongation | 10 |

**Microorganism**

| Species | Parameter measured | References |
|---|---|---|
| Vibrio fischeri | Bioluminescence | 23, 12, 13, 14 |
| Escherechia coli | Metabolic status | 15, 11, 12 |
| Pseudomona putida | Metabolic status | 15 |
| Pseudomona fluorescens | Bioluminescence | 16, 17 |
| Activated sludges | Growth | 18, 19, 20 |
| Lipomyces starkeyi | Growth | 21 |
| Bacillus subtilis | Growth | 22 |
| Pseudomonas sp. strain TTO1 | Growth | 22 |

[1] A. Fargašová, M. Drtil, Bull. Environ. Cont. Toxicol. 56 (1996) 993–999.
[2] Y. Kamaya, Y. Kurogi, K. Suzuki, Environ. Toxicol. 18 (2003) 289–294.
[3] J.P. Emblidge, M.E. DeLorenzo, Environ. Res. 100 (2006) 216–226.
[4] K. C. Cheung, M. H. Wong, Y. K. Yung Toxic. Lett. 137 (2003) 121–131.
[5] A. Gélabert, O.S. Pokrovsky, J. Viers, J. Schott, A. Boudou and A. Feurtet-Mazel Geochim. Cosmochim. Acta, 70 (2006) 839–857.
[6] L. Yang, I. Stulen, L. J. De Kok Environ. Exper. Botany, 57 (2006) 236–245.
[7] R-Q Huang, S-F Gao, W-L Wang, S. Staunton, G. Wang, Sci. Tot. Environ., 368 (2006) 531–541.
[8] R. Calvelo Pereira, M. Camps-Arbestain, B. Rodríguez Garrido, F. Macías, C. Monterroso Environmen. Poll., 144 (2006) 210–217.
[9] J.L. Everhart, D. McNear, Jr., E. Peltier, D. van der Lelie, R. L. Chaney, D. L. Sparks Sci. Tot. Environ. 367 (2006) 732–744.
[10] L.A.D. Williams, E. Vasquez, I. Klaiber, W. Kraus and H. Rosner Chemosphere, 51 (2003) 701–706.
[11] M.F. Desimone, M.C. De Marzi, G.J. Copello, M.M. Fernández, F.L. Pieckenstain, E.L. Malchiodi, L.E. Diaz Enzyme and Microbial Technology, 40 (2006) 168–171.
[12] H. Wex, D.M. Rawson, T. Zhang Electrochimica Acta, 51 (2006) 5157–5162.
[13] P. Gikas, P. Romanos J. Hazard. Mat. 133 (2006) 212–217.
[14] R.J. Carr, R.F. Bilton, and T. Atkinson Appl. Environ. Microbiol. 52 (1986) 1112–1116.
[15] R. Duponnois, M. Kisa, K. Assigbetse, Y. Prin, J. Thioulouse, M. Issartel, P. Moulin, M. Lepage Sci. Tot. Environ. Available online 20 September 2006.
[16] C. Barata, D.J. Baird, A.J.A. Nogueira, A.M.V.M. Soares, M.C. Riva Aquatic Toxicol. 78 (2006) 1–14.
[17] Y. Ikenaka, H. Eun, M. Ishizaka, Y. Miyabara, Aquatic Toxicology 80 (2006) 158–165.
[18] P.L. Gillis, C.M. Wood, J.F. Ranville, P. Chow-Fraser Aquatic Toxicology 77 (2006) 402–411.
[19] G. Hodges, D.W. Roberts, S.J Marshall, J.C. Dearden Chemosphere 63 (2006) 1443–1450.
[20] M. Löschau, R. Krätke Environ. Poll. 138 (2005) 260–267.
[21] K. Nikookar, A. Moradshahi and L. Hosseini, Biomol. Engin., 22 (2005) 141–146.
[22] J. Trögl, S. Ripp, G. Kuncová, G.S. Sayler, A. Churavá, P. Pařík, K. Demnerová, J. Hálová and L. Kubicová, Sens. Actuators B: 107 (2005) 98–103.
[23] M. Farré, C. Gonçalves, S. Lacorte, D. Barceló, M. F. Alpendourada Anal. Bional. Chem 373 (2002) 696–703

receptor have been developed, and are well established (White et al., 1994). Their main drawback is the fact that the binding activity does not necessarily imply the estrogenicity of a substance, because that also depends on the ability of the ligand to elicit an estrogen-receptor response.

Yeast reporter gene assays and MCF-7 cell-based proliferation assays (E-screen) are particularly popular. The E-screen, in which proliferation of human breast cancer cells (MCF-7) is measured as a response to estrogen, has also been used to determine the estrogenicity of sewage effluent and surface water (Kórner et al., 2001).

Several assays, such as the ultra-sensitive luminescent ELRA developed by Seifert (2004) with a detection limit of 20 ng $L^{-1}$ for 17β-estradiol, have been reported in the literature. Although these assays are simple to use, many factors, such as differences in culture conditions, cell density or cell-line clones, affect the potency of estrogenic substances, and that makes the standardization of these methods difficult. The induction of several proteins or enzyme activities (e.g. the increasing levels of alkaline phosphatase, cathepsin D, prolactin and vitellogenin as a consequence of progestogens) has also been used to study estrogenicity. However, expression of these proteins or enzyme activities is restricted to specific cell lines and cannot be extrapolated to other tissues or species.

Over the past decade, a great number of recombinant receptor-reporter gene assays have been developed. Basically, these assays are based on simple cell models that express a gene under the control of defined promoters responding to specific substances and that produce an easily quantified signal (Snyder et al., 2001). These assays are of great interest for rapid screening of the estrogenicity of artificial or natural compounds. These cell models are very useful in integrated studies of the synergy or antagonism of different substances and, in the field of environmental research, they are excellent tools for identifying EDCs. Some of them are based on the bioluminescent reporter gene assay. The YES assay using yeast cells (Garcia-Reyero et al., 2001) is a particularly useful example of this technology. Other assays use human breast cancer cells that have been transfected (e.g. with luciferase in the ER-CALUX assay (Murk et al., 2002)).

Concern that the reproductive health of humans is being affected by exposure to xenoestrogens has led to the development of various in vitro and in vivo screening assays for the identification of compounds suspected of falling into this category. However, the estrogenic activity of a chemical determined in vitro may not necessarily predict its activity in the whole organism if the chemical is metabolized during the assay and/or in vivo. Therefore some approaches to investigating the role of metabolism in modulating the estrogenic activity of suspected xenoestrogens have been explored.. In this context Elsby et al. (2001) proposed a two-stage approach involving coupling incubations with either human or rat hepatic microsomes with the yeast estrogenicity (transcription) assay.

### 3.2.3 BIOSENSORS

Biosensors can be defined as an integrated receptor-transducer device that is capable of providing selective quantitative or semi-quantitative analytical information using a biological recognition element. A biosensor converts a biological event into a detectable

# Biosensors

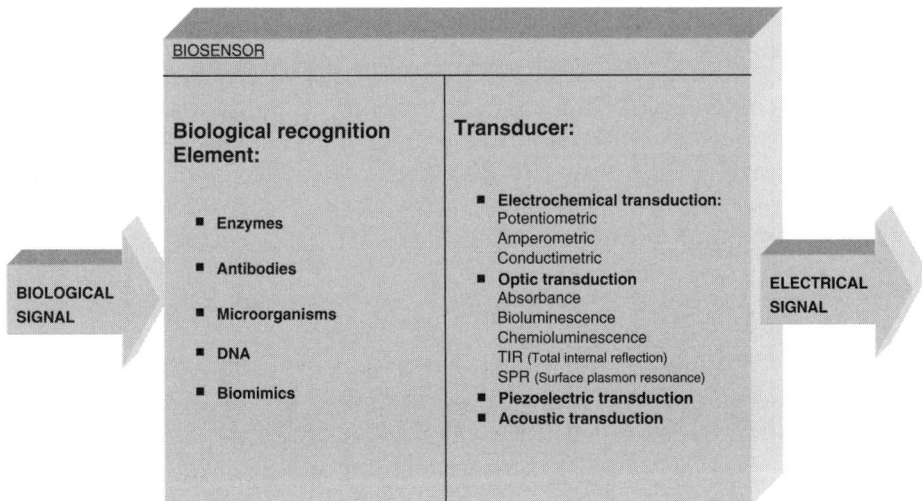

**Figure 3.2.2** A general scheme of a biosensor device

signal by the action of a transducer and a processor. The automation and sometimes the miniaturization of biological analytical techniques, and the development of online and remote measurement equipment, can be achieved through biosensor technology. A general scheme of a biosensor device is shown in figure 3.2.2. In recent years a growing number of biosensor applications have been developed for environmental analysis (Rodriguez-Mozaz *et al.*, 2006) for targeted analysis and for monitoring whole biological effects such as toxicity, or estrogenicity (Rodriguez-Mozaz *et al.*, 2004). Examples of biosensors for environmental applications are summarized in Table 3.2.3.

### 3.2.3.1 Basis and Fundamentals

Biosensors can be classified according to the physicochemical transduction elements used:

1. *Optical transducers* are based on various technologies of optical phenomena, which are the result of an interaction of an analyte with the receptor part. This group may be further subdivided according to the type of optical properties which have been applied in sensing:

   - absorbance, measured in a transparent medium;
   - reflectance is measured in nontransparent media, usually using an immobilized indicator;
   - luminescence, based on the measurement of the intensity of light emitted by a chemical reaction in the receptor system;
   - fluorescence, measured as the positive emission effect caused by irradiation; also, selective quenching of fluorescence may be the basis of such devices;

Table 3.2.3 Examples of enzyme based biosensors reported for environmental analysis

| Analite | Biosensor: Type an basis | Matrix | Sensitivity | Reference |
|---|---|---|---|---|
| Carbamates | Choline oxidase, AChE and acetylcholine. Amperometric detection | Fruit and vegetables | $1.10^{-8} - 4.10^{-7}$ M | 1 |
| Carbamates | AchE, BChE. Potentiometric detection | Synthetic samples | $1.5.10^{-5} - 2.5.10^{-3}$ mol L$^{-1}$ | 2 |
| Carbaryl | AChE and BchE. Amperometric transduction | Fruit and vegetables juices | $0.5-2500$ mg L$^{-1}$ (carbaryl) | 3 |
| Diazinon | Tyrosinase-mediated inhibition with amperometric detection | Aqueous solution | 5 µM | 4 |
| Diazinon | Tyrosinase-mediated inhibition with amperometric detection | Aqueous solution | 5 µM | 52 |
| Dichlovos | Tyrosinase-mediated inhibition with amperometric detection | Aqueous solution | 75 nM | 157 |
| Dichlovos | AchE inhibition amperometric detection | River water | $7.10^{-10}$ M | 5 |
| Dimethyl- and diethyldithiocarbamates | Tyrosinase Amperometric detection | Spiked apple samples | $0.2-2.2$ mmol L$^{-1}$ | 6 |
| Methyl paraxon | Alkaline phoshatase inhibition with chemiluminescence detection | Aqueous solution | 80 ppb | 7 |
| Organophosphate Pesticide | AchE inhibition on a piezoelectric QCM with acoustic detection | Aqueous solution | <100 µg L$^{-1}$ | 8 |
| Organophosphorus pesticides | AChE inhibition optical transduction | | 2 mg L$^{-1}$ | 9 |

| Parathion | Parathion hidrolase enzyme action with amperometric detection | River water | 10 ng mL$^{-1}$ | 10 |
| Paraxon | AChE inhibition optical transduction | Aqueous solution | 152 µg L$^{-1}$ | 56 |
| Paraxon | BChE inhibition with potentiometric detection | Aqueous solution | <10$^{-8}$ | 11 |
| Paraxon | Alkaline phosohatase inhibition with chemiluminescence detection | Aqueous solution | 50 ppb | 159 |
| Propoxur | AChE inhibition and pH detection | Letuce Onion | Propoxur: 0.4 ng [RG1] | 57 |

QCM: quartz crystal microbalances; Acetylcholinesterase (AChE) and butyrylcholinesterase (BChE).

[1] I. Palchetti, A. Cagnini, M. Del Carlo, C. Coppi, M. Mascini, A.P.F. Turner, *Anal. Chim. Acta*, 337 (1997) 315–21.
[2] A.N. Ivanov, G.A. Evtugyn, R.E. Gyurcsanyi, K. Toth, H.C. Budnikov, *Anal. Chim. Acta*, 404 (2000), 55–65.
[3] G.S. Nunes, D. Barceló, B.S. Grabaric, J.M. Diaz-Cruz, M.L. Ribeiro, *Anal. Chim. Acta*, 399 (1999) 37–49.
[4] J.Q. Xu, X.Z. He, Y.X. Zhou, L.T. Liu, J. Cheng, *Chin. Sci. Bull.* 45 (2000) 101.
[5] J.J. Rippeth, T.D. Gibson, J.P. Hart, I.C. Hartley, G. Nelson, *Analyst*, 122 (1997) 1425.
[6] M.T.P. Pita, A.J. Reviejo, F.J.M. Villena, J.M. Pingarrón, *Analytica Chimica Acta*, 340 (1997) 89–97.
[7] M.S. Ayyagari, S. Kamtekar, R. Pande, *et al.*, *Biotechnol. Progr.* 11 (1999) 699.
[8] P. Sritongkham, N. Taravanit, T. Suwannakum, M. Tantitcharoen, K. Kirtikara, in A.P.F. Turner (ed.), Biosensors 94: the Third World Congress on Biosensors, New Orleans, June 1994: Abstracts, Elsevier Advanced Technology, Oxford, 1994.
[9] W. Choi, Y-K Kim, I-H. Lee, J. Min, W.H. Lee, *Biosen. Bioelectron.* 16 (2001) 937–43
[10] V. Sacks, I. Eshkenazi, T. Neufeld, C. Dosoretz, J. Rishpon, *Anal. Chem.* 72 (2000) 2055.
[11] Y.A. Cho, H.S. Lee, G.S. Cha, *et al.*, *Biosensors Bioelectron.* 14 (1999) 435.
[52] I.T. Show, M.B. Show, and L.R. Williams, "Modeling the Contributions of Cross-Reacting Chemicals in Immunoassays", in Aquatic Toxicology and Hazard Assessment: 12th vol., ASTMSTP 1026, ed. by U.M. Cowgill and L.R. Williams, American Society for Testing and Materials, Philadelphia, 1989, pp. 21–33.
[56] K. Streffer, E. Vijgenboom, A.WJW. Tepper, A. Makower, F.W. Seller, G.W. Canters, U. Wollenberger. *Anal. Chim. Acta*, 427 (2001) 201–210.
[57] C. Nistor, J. Emnéus, L. Gorton, A. Ciucu, *Anal. Chim. Acta*, 387 (1999) 309–326.
[157] S. Kröger, A.P. Turner, K. Mosbach, K. Haupt, *Anal. Chem.* 71 (1999) 3698–3702.
[159] R. Suedee, T. Srichana, C. Sangpagai, C. Tunthana, P. Vanichapichat, *Anal. Chim. Acta*, 504 (2004) 89–100.

- refractive index, measured as the result of a change in solution composition. this may include also a surface plasmon resonance effect (SPR);
- optothermal effect, based on a measurement of the thermal effect caused by light absorption;
- light scattering, based on effects caused by particles of definite size present in the sample.

The application of many of these phenomena in sensors became possible because of the use of optical fibres in various configurations. Such devices have also been called optodes.

A high number of optical transduction techniques can be used for biosensor development. These may employ linear optical phenomenon, including adsorption, fluorescence, phosphorescence, polarization, rotation, interference, or nonlinear phenomena, such as second harmonic generation. Total internal reflection fluorescence (TIRF) has been used with planar and fibre optic waveguides as signal transducers in a number of reported biosensors. In these transducers, light is propagated down a waveguide which generates an electromagnetic wave (evanescent wave) at the surface of the optically denser medium of the waveguide and the adjacent less optically dense medium. The amplitude of the standing wave decreases exponentially with distance into the lower refractive index material. The fluorescence of a fluorophore excited within the evanescent field can be collected.

Surface plasmon resonance (SPR) biosensors exploit special electromagnetic waves-surface plasmon-polaritons-to probe interactions between an analyte in solution and a biomolecular recognition element immobilized on the SPR sensor surface. A surface plasmon wave can be described as a light-induced collective oscillation in electron density at the interface between a metal and a dielectric. At SPR, most incident photons are either absorbed or scattered at the metal/dielectric interface and, consequently, reflected light is greatly attenuated. The resonance wavelength and angle of incidence depend upon the permittivity of the metal and dielectric.

The general advantages of optical techniques involve the speed and reproducibility of the measurement, and the main drawback is the high cost of the apparatus.

2. *Electrochemical transductors* transform the effect of the electrochemical interaction between an analyte and the electrode into a primary signal. Such effects may be stimulated electrically or may result from a spontaneous interaction at the zero-current condition. The following subgroups may be distinguished:

- voltammetric sensors, including amperometric devices, in which current is measured in the d/c or a/c mode; this subgroup may include sensors based on chemically inert electrodes, chemically active electrodes and modified electrodes; in this group are included sensors with and without (galvanic sensors) an external current source;
- potentiometric sensors, in which the potential of the indicator electrode (ion-selective electrode, redox electrode, metaVmetal oxide electrode) is measured against a reference electrode;

- chemically sensitized field effect transistor (CHEMFET) in which the effect of the interaction between the analyte and the active coating is transformed into a change in the source-drain current;
- potentiometric solid electrolyte gas sensors, differing from potentiometric sensors because they work in high temperature solid electrolytes and are usually applied for gas sensing measurements.

Electrochemical biosensors are based on monitoring electro active species that are produced or consumed by the action of the biological elements (e.g. enzymes), and can be performed using potentiometric and amperometric measurements. The operation of amperometric biosensors depends on a constant potential applied between a working and a reference electrode. The imposed potential promotes a redox reaction that produces a current whose magnitude is proportional to the concentration of electro active species present in solution. Oxidase enzymes have been the most frequently investigated and applied. A number of amperometric biosensors is based on the monitoring of oxygen consumption, or the generation of hydrogen peroxide. Both are electrochemically active; oxygen can be electrochemically reduced, and hydrogen peroxide can be oxidized. The current generated is proportional to the concentration of the enzyme substrate present in a sample. The use of mediators should permit the replacement of oxygen as an electron acceptor, and operation at a much lower potentials, reducing interference from other electrochemically active species that are found in complex matrices. Potentiometric biosensors are based on the monitoring of the potential at a working electrode with respect to a reference electrode. A potentiometric enzyme biosensor for the direct measurement of organophosphate pesticides was described by Mulchandani *et al.* (1998), and is based on a pH electrode modified with an immobilized organophosphorus hydrolase.

3. *Electrical sensors* are based on measurements where no electrochemical processes take place, but the signal arises from a change of electrical properties caused by the interaction of the analyte.

   - metal oxide semiconductor (MOS) sensors, used principally as gas phase detectors, are based on reversible redox processes of analyte gas components;
   - organic semiconductor sensors are based on the formation of charge transfer complexes, which modify the charge carrier density;
   - electrolytic conductivity sensors;
   - electric permittivity sensors.

4. *Mass sensitive sensors* transform the mass change at a specially modified surface into a change of a property of the support material. The mass change is caused by accumulation of the analyte.

   - Piezoelectric devices used mainly in gaseous phase, but also in solutions, are based on the measurement the frequency change of the quartz oscillator plate caused by adsorption of a mass of the analyte at the oscillator.

- Surface acoustic wave devices depend on the modification of the propagation velocity of a generated acoustical wave by the deposition of a definite mass of the analyte.

The vibration of piezoelectric crystals produces an oscillating electric field in which the resonant frequency of the crystal depends on its chemical nature, size, shape and mass. By placing the crystal in an oscillating circuit, the frequency can be measured as a function of the mass. When the change in mass (m) is very small compared with the total mass of the crystal, the change in frequency (f) relates to m as follows:

$$\text{delta } f = Cf^2 \text{ delta } m/A$$

where f is the vibration frequency of the crystal in the circuit, A is the area of the electrode and C is a constant determined in part by the crystal material and thickness. Piezoelectric crystals, sometimes referred to as a quartz crystal microbalances (QCM), are typically made of quartz and operate at frequencies between 1 and 10 MHz. These devices can operate in liquids with a frequency determination limit of 0.1 Hz, the detection limit of mass bound to the electrode surface is about $10^{-10}$ to $10^{-11}$ g.

Limitations for this transduction method involve format and calibration requirements.

Biosensors based on acoustic transduction have been used mainly for the detection of pathogenic microorganisms such as *Escherechia coli* (Pyun et al., 1998), other examples of applications involve the use of an acoustic sensor to detect genetically modified organisms (GMO) (Mannelli et al., 2003).

5. *Magnetic devices* based on the change of paramagnetic properties of a gas being analyzed. These are represented by certain types of oxygen monitors.

6. *Thermometric sensors* are based on the measurement of the heat effects of a specific chemical reaction or an adsorption process that involves the analyte. In this group of sensors the heat effects may be measured in various ways, for example in catalytic sensors the heat of a combustion reaction or an enzymatic reaction is measured by use of a thermistor. Calorimetric biosensors detect variations of heat during a biological reaction.

### 3.2.3.2 Recognition Element Classification and Biosensors for Environmental Analysis

The main classes of biological elements used in recognition for the development of environmental analysis devices are:

- enzymes
- antibodies
- microorganisms

- dna
- biomimetics

*Enzyme biosensors*

Enzyme biosensors have been described using a range of transduction elements (amperometry, potentiometry, optical and photo-thermal). The first biosensor was described in the literature by Clarck and Lyons (1962a) and was based on the use of glucose oxidase combined with electrochemical detection. Since then, this principle has been widely applied in biosensor development, and the enzyme systems used have been mainly oxido-reductases (e.g. tyrosinase, peroxidase and lactase) (Cass *et al*., 1984; Kulis and Vidziunaite, 2003), and hydrolases (choline esterases) (Andreescu *et al*., 2002; Nunes *et al*., 1998).

Enzyme biosensors can be classified as:

1. those that measure the inhibition of a specific enzyme due to the presence of target analytes;
2. those that measure the catalytic transformation of target analytes by a specific enzyme.

Enzymatic based devices have been widely reported for the detection of organic pollutants (see Table 3.2.4). The selective inhibition of choline esterases, principally acetylcholine esterase (AchE) (Guibault and Das, 1970), and butylcholine esterase (BchE) (Skládal, 1992), have been used for the detection of organophosphorous, and carbamate insecticides. However, there are some limitations.

Sometimes it is necessary to use different substrates, cofactors and mediators in order to increase the sensitivity. Due to the irreversible nature of many analyte-enzyme interactions, sensing elements must either be reactivated or should be disposable elements. An example is the determination of diazinon and dichlorvos that uses tyrosinase immobilized on screen printed electrodes, and a redox mediator (1,2-naphtaquinone-4-sulfonate) (Everett and Rechnitz, 1998). For other systems a further drawback is the lack of selectivity of some of the enzymes involved.

The catalytic transformation of target analytes is the second mechanism used for enzyme-based biosensors. Such sensors are simple in design and operation, and can be configured to operate continuously and reversibly. They can also be configured such that the only required reagent is the analyte of interest.

Most of transduction elements used in enzyme-based biosensors are electrochemical: amperometric or potentiometric. Typically the enzymes used in amperometric biosensors are oxidases. The main advantages of this class of transducer are the low cost; a high degree of reproducibility, and the suitability of many of them for incorporation into disposable electrodes. This type of instrumentation is widely available and can be inexpensive and compact; this allows this makes it possible to use them for making on-site measurements. Limitations of amperometric measurements include potential interferences to the response from any electroactive compounds that are present in

**Table 3.2.4** Examples of biosensors for environmental monitoring

| Analite | Type and basis | Sensitivity | Reference |
|---|---|---|---|
| *Pesticides and Herbicides* | | | |
| Atrazine | Inhibition assay electrochemical detection | 0.03 µg L$^{-1}$ | 1 |
| Atrazine | Inhibition assay Optical detection | 1–10 µg L$^{-1}$ | 2 |
| Chlorsulfuron | Competitive assay amperometric detection | 0.01 ng mL$^{-1}$ | 3 |
| Isoproturon | Indirect assay and TIR-fluorescence detection | 0.01–0.14 µg L$^{-1}$ | 4 |
| Simazine | Competitive immunoreaction potentiometric detection | 3 ng/mL | 5 |
| Triazines | Inhibition assay SPR detection | 0.16 µg L$^{-1}$ | 6 |
| Triazines | Inhibition assay surface refractive index change detection | 15 µg mL$^{-1}$ | 7 |
| Triazines | Flow-through amperometric immunosensor based on peroxidase chip and enzyme channeling system | 1 ng L$^{-1}$ | 8 |
| Pesticides | Competitive immunoassays fluorescence detection | | 9 |
| Organophosphates | Competitive immunoassay SPR detection | 54–64 ng L$^{-1}$ | 72 |
| Phenols | Amperometric tyrosinase biosensor based on a composite graphite–Teflon electrode modified with gold nanoparticles | 3-20 nM | 10 |
| Phenol | Electrochemical nano-bio-chip array with whole-cells | | 96 |
| Phenols | Indirect assay and TIR-fluorescence detection | 0.9, 1.0, 1.0, 3.0 µM and 0.038, 0.037, 0.056, 0.201 mM for 4-chlorophenol, phenol, m-cresol and p-cresol, respectively | 11 |
| Phenols | immobilizing tyrosinase to modified core-shell magnetic nanoparticles supported at a carbon paste electrode | $6.0 \times 10^{-7}$ M | 12 |

| | | | |
|---|---|---|---|
| Phenols, organic acids | whole cell optical biosensor with immobilized bioreporter Pseudomonas fluorescens HK44 | 1.2 and 0.5 mg L$^{-1}$ for naphthalene and salicylate | 13 |
| | PAHs | | |
| PAHs | Electrochemical immunosensor | | 63 |
| PAHs | Sol–gel-derived array DNA | Naphthalene and phenanthrene in the concentration range of 0–10 mg L$^{-1}$ | 80 |
| | EDCs | | |
| Estrogens | A bacterial biosensor (based on Escherechia coli) of endocrine modulators | | 14 |
| Endocrine disruptors | A simple amperometric tyrosinase-based biosensor (Tyr-CPE) has been developed for the detection of phenolic EDCs | | 15 |
| | EDCs | | |
| Hormons | Optical immunosensor based on fluorescence | Sub ng L$^{-1}$ | 16 |
| Bisphenol A | Optical immunosensor based on fluorescence | | 17 |
| Bisphenol A | SPR immunosensor | | 18 |
| | Surfactants | | |
| Alkylphenols and their ethoxylates | Immunosensor using electrochemical transduction | mg L$^{-1}$ range | 19 |
| Nonylphenol | Immunosensor using electrochemical transduction | 10 µg L$^{-1}$ | 20 |
| Surfactants | Amperometric whole cell biosensor using *Pseudomonas* and *Achromobacter* | 25 mg L$^{-1}$ | 21 |
| LAS | Amperometric whole cell biosensor using *Trichosporun cutaneum* | 0.2 mg L$^{-1}$ | 22 |

*(continued overleaf)*

**Table 3.2.4** (Continued)

[1] R.J.M. Fernandez, M. Stiene, R. Kast, M.D. Luque de Castro, U. Bilitewski, *Biosensors Bioelectron.* 13 (1998) 1107.
[2] P. Skladal, Biosensors Bioelectron. 14 (1999) 257.
[3] B.B. Dzantiev, E.V. Yazynena, A.V. Zherdev, Y.V. Plekhanova, A.N. Reshetilov, S.-C. Chang, C.J. MxNeil, *Sens. Actuators*, 98 (2004) 254–261
[4] E. Mallat, C. Barzen, R. Abuknesha, G. Gauglitz, D. Barceló, *Anal. Chim. Acta*, 426 (2001) 209–216.
[5] M.F. Yulaev, R.A. Sitdikov, N.M. Dmitrieva, E.V. Yazynina, V. Zherdev, B.B. Dzantiev, *Sens. Actuators*, 75 (2001) 129–135.
[6] R.D. Harris, B.J. Luff, J.S. Wilkinson, J. Piehler, A. Brecht, G. Gauglitz, R.A. Abuknesha, *Biosensors Bioelectron.* 14 (1999) 377.
[7] F.F. Bier, R. Jockers, R.D. Schmid, Analyst 119 (1994) 437.
[8] J. Zeravik, T. Ruzgas, M. Franek, Biosen. Bioelectron. 18 (2003) 1321–1327.
[9] Tschmelak J, Proll G, Riedt J, Kaiser J, Kraemmer P, Barzaga L, Wilkinson JS, Hua P, Hole JP, Nudd R, Jackson M, Abuknesha R, Barcelo D, Rodriguez-Mozaz S, de Alda MJ, Sacher F, Stien J, Slobodnik J, Oswald P, Kozmenko H, Korenkova E, Tothova L, Krascsenits Z, Gauglitz G, Biosen. Bioelectron. 20 (2004) 1509–1519.
[10] V. Carralero, M.L. Mena, A. Gonzalez-Cortés, P. Yá nez-Sede no and J.M. Pingarr on Biosen. Bioelectron, 22 (2006) 730–736.
[11] J. Abdullah, M. Ahmad, N. Karuppiah, L. Yook Heng, H. Sidek Sens. Actuators 114 (2006) 604–609.
[12] Z. Liu, Y. Liu, H. Yang, Y. Yang, G. Shen, R. Yu Anal. Chim. Acta, 533 (2005) 3–9.
[13] J. Trögl, S. Ripp, G. Kuncová, G.S. Sayler, A. Churavá, P. Paofk, K. Demnerová, J. Hálová and L. Kubicová, Sens. Actuators B: 107 (2005) 98–103.
[14] G. Skretas, D. W. Wood, J. Mol. Biol. 349 (2005) 464–474.
[15] S. Andreescu, O.A. Sadik, Anal. Chem. 76 (2004) 552–560.
[16] J. Tschmelak, M. Kumpf, N. Kappel, G. Proll, G. Gauglitz Talanta 69 (2006) 343–350.
[17] S. Rodriguez-Mozaz, M.J. Lopez de Alda, D. Barceló, Water Res 39 (2005) 20.
[18] Marchesini GR, Meulenberg E, Haasnoot W, Irth H (2005) Anal Chim Acta 528: 37–45.
[19] A. Rose, C. Nistor, J. Emneus, D. Pfeifer, U. Wollenberger , Biosens Bioelectron 17 (2002) 1033–1043.
[20] G.A. Evtugyn, S.A. Eremin, R.P. Shaljamova, A.R. Ismagilova, H.C. Budnikov, Biosen. Bioelectron 22 Issue (2006) 56–62.
[21] L. Taranova, I. Semenchuk, T. Manolov, P. Iliasov, A. Reshetilov Biosens Bioelectron 17 (2002) 635–640.
[22] Y. Nomura, K. Ikebukuro, K. Yokoyama, T. Takeuchi, Y. Arikawa, S. Ohno, I. Karube Biosens Bioelectron 13 (1998) 1047–1053.
[63] S. Kurosawa, Meas. Sci. Technol. 14 (2003) 1882.
[72] P. Lopez-Ortal, V. Souza, L. Bucio, Mutat. Res. 439 (1999) 301–306.
[80] S.S. Babkina, N.A. Ulakhovich, Y.I. Zyavkina, Anal. Chim. Acta 502 (2004) 23–30.
[96] N. Paxeus, H.F. Schroder, Water Sci. Technol. 33 (1996) 9.

the sample, since they can generate false current values. These are exemplified by for instance an amperometric biosensor for hydrogen peroxide can also be used to measure organophosphate pesticides at concentrations as low as $10^{-9}$ M. Organophosphate and carbamate pesticides were quantified using potentiometric transduction in a modified pH electrode that measures the activity of acetyl cholinesterase (Budnikov and Evtugyn, 1998). In a further example potentiometry was used in a system coupling urease in ammonium ion sensors for the determination of heavy metals (Show et al., 1989).

Several examples of enzyme biosensors using optical transduction have been reported (Choi et al., 2001; Doong and Tsai, 2001). During recent years optodes, such as fibre optic biosensors, have been of great interest because they have advantages such as the lack of a need for direct electric connection, ease of miniaturization, possibility of remote sensing, and in-situ monitoring. In consequence a range of different examples has been reported and includes a sol-gel acetyl cholinesterase fibre optic biosensor (Xavier et al., 2000) and an acetyl cholinesterase viologen hetero Langmuir-Blodgett film both of which measure concentrations of organophosphate neurotoxicants (Xavier et al., 2000). Various types of immobilization systems and electrodes (including carbon paste electrodes (CPEs), solid graphite electrodes (Streffer et al., 2001), surface modified electrodes (Nistor et al., 1999), ion-sensitive field effect transistors (ISFETs) and pH-sensitive field-effect transistors (pH-FETs)) (Mai Ahn et al., 2002) have been investigated with the aim of improving the storage stability of enzyme based biosensors.

More recently, other types of enzymatic biosensors have developed based on thermistor or opto-electronic sensors. Pogacnik and Franko (2003) presented a photothermal biosensor for the determination of low concentrations of organophosphate and carbamates pesticides in vegetables without sample treatment.

*Immunosensors*

In these sensors the recognition elements are immunochemical interactions between an antibody (Ab) and an antigen (Ag). This type of devices combines the principles of solid-phase immunoassay with physicochemical transduction elements (electrochemical, optical, piezoelectric, evanescent wave and surface plasmon resonance). Since most small molecular weight organic pollutants in the environment have few distinguishing optical or electrochemical characteristics, the detection of stoichiometric binding of these compounds to antibodies is typically accomplished using competitive binding-assay formats. The main limitation of these techniques is the electrochemical detection of the immunoreaction, because it is necessary to use enzymes that will generate electrochemical active compounds.

Electrochemical immunosensors have been widely used for environmental analysis in amperometric, potentiometric, and conductimetric configurations. Amperometric immunosensors measure the current generated by oxidation or reduction of redox substances at the electrode surface, which is held at an appropriate electrical potential. Wilmer et al. measured concentrations of 2,4-dichlorophenoxyacetic acid (2,4-D) in water by using an amperometric immunosensor with a limit of detection of 0.1 μg $L^{-1}$ (Wilmer et al., 1997). Some examples of new developments are the disposable screen-printed electrodes for the detection of polycyclic aromatic hydrocarbons (PAHs)

(Fahnrich et al., 2003), and the use of recombinant single-chain antibody (scAb) fragments (Grennan et al., 2003) for atrazine determinations.

Acoustic transducers have also been used in immunosensors for water analysis (Guilbault, 1992). The resonant frequency of an oscillating piezoelectric crystal can be affected by a change in mass at the crystal surface. Piezoelectric immunosensors are able to measure a small change in mass. Recent publications have been based on immunosensors using a quartz crystal microbalance (QCM) for the detection of trace amounts of chemical compounds, such as dioxins (Kurosawa, 2003).

In general, a vast number of optical transduction techniques can be used for biosensor development. These may employ linear optical phenomenon (e.g. adsorption, fluorescence, phosphorescence, and polarization) or nonlinear phenomena (e.g. second harmonic generation). The choice of a particular optical method depends on the analyte and the sensitivity needed. Total internal reflection fluorescence (TIRF) has been used with planar and fibre-optic wave-guides as signal transducers in a number of biosensors.

Based on the evanescent wave transducing principle, atrazine was detected at concentrations around $0.1 \mu g\ L^{-1}$ (Schipper et al., 1997; Schipper et al., 1998) and cyclodiene insecticides in the $\mu g\ L^{-1}$ range (Brummel et al., 1997).

A significant high number of surface plasmon resonance (SPR) immunoassays has recently been used for rapid analysis of pollutants in different water matrices. In these assays the incident light is reflected from the internal face of a prism in which the external face has been coated with a thin metal film. At a critical angle, the intensity of the reflected light is lost to a resonant oscillation in the electrons created at the surface of the metal film. This method has been used to measure the binding of Abs to Ags immobilized at the sensor surface because the critical angle depends on the refractive index of the materials at the metal surface. An immune assay approach was exploited by Nabok et al. (2005) for in situ detection of some above low molecular weight toxicants, and used specific antibodies immobilized onto the gold surface by a (poly)allylamine hydrochloride layer, using an electrostatic self-assembly (ESA) technique. A simple and rapid detection method using surface plasmon resonance for dioxins, polychlorinated biphenyls and atrazine was described by Shimomura et al. (2001). This had limits of detection ranging between 0.1 and $5\ ng\ mL^{-1}$, and required 15 min for a single sample measurement. Mauriz et al. (2006), presented an immunoassay based on a binding inhibition test using a portable surface plasmon resonance (SPR) immunosensor for the continuous monitoring of the organophosphate pesticide chlorpyrifos in real water samples. This had limits of detection ranging from 45 to $64\ ng\ L^{-1}$. Sensor reusability was achieved through the formation of alkanethiol self-assembled monolayers (SAMs). Using the same principle Mauriz et al. (2006) recently developed single and multi-analyte surface plasmon resonance assays for the simultaneous detection of cholinesterase inhibiting pesticides.

Among this variety of immunosensor systems, two optical immunosensor devices for environmental applications; RIANA and AWACSS are especially noteworthy. These immunosensors are based on solidphase fluoro-immunoassays combined with an optical transducer chip chemically modified with an analyte derivate. The RIANA immunosensor was applied successfully in a variety of environmental applications (Mallat et al., 1999; Rodriguez-Mozaz et al., 2004b). The other new analytical system AWACSS has been developed and evaluated, it has been shown to be capable of measure concentrations of several organic pollutants at low nanogram per litre levels in a few minutes of analysis, and without the need for any pretreatment steps (Tschmelak *et al.*, 2005).

## DNA biosensors

There is a demand for analytical methods that can diagnose the consequences of the impact of biologically active contaminants. For these reasons bioaffinity methods of analysis, where a high specificity of a ligand-receptor binding is used, appear to be very promising. The structure of DNA is very sensitive to the influence of environmental pollutants, such as heavy metals (Lopez-Ortal *et al.*, 1999; Drevensek *et al.*, 2005), polychlorinated biphenyls (PCBs) (Marrazza *et al.*, 1999), or polyaromatic compounds (PAHs) (Doong *et al.*, 2005). These substances are characterized by a great affinity for DNA, and causing mutagenesis and carcinogenesis. It is therefore very attractive to use DNA-containing systems, e.g. DNA-based biosensors (Bakker, 2004; Chiorcea and Oliveira-Brett, 2004), to perform genotoxic assays, and for rapid testing of pollutants for mutagenic and carcinogenic activity. Lazarides *et al.* developed gold colloidal-nanoparticle aggregates that were linked by short pieces of DNA (Lazarides *et al.*, 2000) and that exhibited a colour change due to electromagnetic coupling between the gold nonspherical nanoparticles after DNA hybridization.

Electrochemical biosensors based on immobilized DNA integrate detection sensitivity with a high specificity for biomolecules. This reduces the consumption of DNA and gives rise to the development of modern methods for the analysis of effectors (including toxicants in both environmental samples and in the biological system) that regulate the functioning of DNA (Lucarelli *et al.*, 2002; Babkina *et al.*, 2004; Mugweru *et al.*, 2004). Mecklenburg *et al.* (1997) reported a fibre-optical device, and Pandey and Weetall (1994) coupled an FIA system with an evanescent wave biosensor for the detection of common intercalating compounds. Disposable electrochemical DNA-based biosensors have been used for the determination of low-molecular weight compounds with a high affinity for nucleic acids and for the detection of the hybridization reaction. The rapid screening of genotoxic compounds using the molecular interaction between surface-linked DNA and the target pollutants or drugs have been applied in a range of different configurations.

In one application the genotoxic compounds were measured by their effect on the oxidation signal of the guanine peak of calf thymus DNA immobilized on the electrode surface and linked with chronopotentiometric analysis by Mascini *et al.* (2001). This type of DNA biosensor is able to detect known intercalating compounds, such as daunomycin, polychlorinated biphenyls (PCBs), aflatoxin B1, and aromatic amines. This technology has been demonstrated to be applicable to river and waste water samples (Lucarelli *et al.*, 2003). Disposable electrochemical sensors for the detection of a specific sequence of DNA were realized by immobilizing synthetic single-stranded oligonucleotides onto a graphite screen-printed electrode. The probes became hybridized with different concentrations of complementary sequences present in the sample. The hybrids formed on the electrode surface are evaluated by chronopotentiometric analysis using daunomycin as indicator of the hybridization reaction.

## Microbial biosensors

The major application of microbial biosensors is in the environmental field. The use of immobilized whole cells (usually bacteria) as the recognition element for biosensor applications has been described for chemical residues such as phenols, pesticides, benzene, toluene, xylene, endocrine disruptors and the water biological oxygen demand (BOD). A large number of publications has appeared describing BOD microbial biosensors, and a range of different devices is commercially available. One of the most relevant applications for whole-cell microbial biosensors is toxicity screening, and several systems are available to support this including the Cellsense system, a commercially available biosensor developed by Rawson (1987).

Biosensors have also been developed using genetically engineered microorganisms (GEMs) that recognize and report the presence of specific environmental pollutants. The basis of GEMs is the use of constructed plasmid in which genes that code for a reporting system such as luciferase or galactosidase are placed under the control of a promoter that recognizes the analyte of interest. Recombinant organisms provide a novel system that exhibits a number of important traits, such as the expression of cellular degradative enzymes, specific binding proteins against a target analyte, and reporter enzymes, for example bacterial luciferase that produces light in response to the presence of a target analyte in the GEM. Recombinant microorganisms that respond sensitively to a broad variety of chemicals have been developed to serve as microbial toxicity biosensors (Bechor *et al.*, 2002). In a recent work, Lee *et al.* presented a cell-based array technology that uses 20 recombinant bioluminescent bacteria to detect and classify environmental toxicity (Lee *et al.*, 2005).

A green fluorescent protein-based *Pseudomonas fluorescens* strain biosensor was constructed and characterized for its potential to measure benzene, toluene, ethylbenzene, and related compounds in aqueous solutions. The biosensor is based on a plasmid carrying the toluene-benzene transcriptional activator (Stiner and Halverson, 2002). Another microbial whole-cell biosensor, using *Escherichia coli* with the promoter luciferase luxAB gene, was developed for the determination of water-dissolved linear alkanes by luminescence (Sticher *et al.*, 1997). The biosensor has been used to detect the bioavailable concentration of alkanes in heating oil-contaminated groundwater samples.

An innovative electrochemical '*lab on a chip*' system has been developed by Popovtzer *et al.* (2006) for the measurement of microbial responses to toxic chemicals. The miniaturized device was designed in two parts to enable multiple measurements: a disposable silicon chip containing an array of nano-volume electrochemical cells that house the biological material, and a reusable unit that includes a multiplexer and a potentiostat connected to a pocket PC for sensing and data analysis.

## MIP-based sensors

Due to the frequently poor stability (thermal, and pH) and short lifetimes of biological components, synthetic molecules with high affinity properties similar to those of biological components are being introduced. One of the most promising groups of biomimetic materials are comprises molecularly imprinted polymers (MIPs). These are becoming an important class of synthetic materials that can mimic the molecular recognition exhibited by natural receptors such as antibodies. In particular noncovalent imprinting, has a great range of applications because of the theoretical lack of restrictions on size, shape or chemical character of the imprinted molecule. The possibility of tailor-made, highly selective receptors at low cost, and with good mechanical, thermal and chemical properties makes these polymers appear ideal chemo receptors. There are great hopes for development of a new generation of chemical sensors using these novel synthetic materials as recognition elements (Haupt, 2001).

In the production of MIPs the selected ligand or *print molecule* is first allowed to establish bond formations with polymerizable functional groups, and the resulting complexes or adducts are then copolymerized with cross-linkers into a rigid polymer. Following the extraction of the print molecule, specific recognition sites are left in the polymer, where the spatial arrangement of the complementary functional entities of the polymer network together with the shape image correspond to the imprinted molecule. MIPs can be prepared by self-assembly, where the preliminary interaction between the ligand and the functional monomers is formed by noncovalent bonding, or metal coordination interactions; or by a preorganized approach, where the aggregates in solution prior to polymerization are maintained by reversible covalent bonds. Substantial rigidity and complete insolubility of the polymers are obtained by using a high percentage of cross linker. The main characteristics of these materials is their resistance in a range of physical and chemical environments to mechanical stress, high temperatures and pressures, treatment with acid, base, or metal ions, and a wide range of solvents. Further these systems can be used repeatedly.

During the last decade the number of application of MIP-based sensors has increased dramatically. The high selectivity and affinity of MIPs for target analytes make them ideal recognition elements in the development of sensors. Capacitive (Panasyuk *et al.*, 2001), conductimetric (Piletsky *et al.*, 1995), field effect (Lahav *et al.*, 2004), amperometric (Kritz and Mosbach, 1995), and voltammetric (Pizzariello *et al.*, 2001), electrochemical transduction systems have been used. Sensors based on conductimetric transduction have been developed by Piletsky *et al.* (1995) for the analysis of herbicides. A system using a $TiO_2$ sol-gel system, and with a linear range of 0.01–0.50 mg $L^{-1}$ for atrazine, without interference of simazine, and chloroaromatic acids has been described by Lahav *et al.* (2004).

MIP-based sensors coupled to piezoelectric transducers is one of the most promising areas, and a variety of devices has been developed for use in the food industry, and in environmental analysis. An example in the latter area is the instrument for the detection of toxic compound such as the polycyclic aromatic hydrocarbons (PAHs) (Dickert *et al.*, 1999).

Applications of MIP based sensors for the analysis of contaminants are summarized in Table 3.2.5.

## 3.2.4 FUTURE TRENDS

Biosensor technology constitutes a rapidly expanding field of research that has been transformed over the past two decades through the discovery of novel material technologies, and means of signal transduction, coupled with the development of powerful

Table 3.2.5  MIP-based sensors for toxicant analysis

| Analyte | Functional monomer | Transduction | Reference |
|---|---|---|---|
| o-Xilene | NS | QCM | 1 |
| PAHs | Phloroglucinol and tiisocyanate | QCM, SAW and optical transduction | 2 |
| 2,4-dichlorophenoxy-acetic-acid | 4-VP | Electrochemical detection using screen printed electrodes | 3 |
| Atrazine | MAA | Conductimetric sensor | 100 |
| Atrazine | MAA+EDMA | Amperometric detection | 4 |
| Trichloroacetic acid and haloacetic acids | 4-VP+EDMA | Conductimetric sensor | 5 |
| Domoic acid | 2-(diethylamino) ethyl methacrylate | SPR-QCM | 6 |
| Sulphamethazine | MAA | Voltammetric | 7 |

Quartz crystal microbalance (QCM); Surface plasmon resonance (SPR); Piezoelectric quartz crystal (PQC); Bulk acoustic wave (BAW); surface acoustic wave (SAW); 4-vinylpyridine (4-VP); Ethylene glycol dimethacrylate (EDMA); Methacrylic acid (MAA); Ethylene glycol dimethacrylate (EDMA); N-phenylacrylamide (PAM); Diethylamino ethyl methacrylate (DEAEM).

[1] F.L. Dickert, O. Hayden, Trends Anal Chem, 18 (1999) 192–199.
[2] H. Zhou, Z. Zhang, D. He, Y, Hu, Y, Huang, D. Chen Anal. Chim. Acta 523 (2004) 237–242.
[3] S. Kröger, A.P. Turner, K. Mosbach, K. Haupt, Anal Chem, 71 (1999) 3698–3702.
[4] R. Shoji, T. Takeuchi, I. Kubo, Anal Chem 75 (2003): 4882–4886.
[5] R. Suedee, T. Srichana, C. Sangpagai, C. Tunthana, P. Vanichapichat, Anal. Chim. Acta. 504 (2004) 89–100.
[6] M. Lotierzo, O. Y. F. Henry, S. Piletsky, I. Tothill, D. Cullen, M. Kania, B. Hock, A. P. F. Turner, Biosen. Bioelectronics, 20 (2004) 75–81.
[7] A. Guzmán-Vázquez de Prada, P. Martínez-Ruiz, A.J. Reviejo, J.M. Pingarrón, Anal. Chim. Acta, 539 (2005) 125–132.
[100] C. Wang, A. Yediler, D. Lienert, Z. Wang, A. Kettrup, Chemosphere 46 (2002) 339.

computer software to control the sensor devices. However, in order to obtain improved and more reliable devices for routine use in the area of environmental monitoring, it will be necessary to focus future research activities on the development of miniaturized devices with robust transduction elements that can provide continuous monitoring, and multi-analyte determinations. Some systems that can provide easy, rapid, on-site measurements have been developed for use in continuous monitoring (Han et al., 2002; Bhattacharyya et al., 2005). They may also be useful for mapping the distribution of contamination when it is important to obtain rapid results in the field, such as after accidental spills or pollution events (Allan et al., 2006). One of the main achievements in this area is the AWACSS system, based on an optical immunosensor that was established as an early-warning system by means of a network of measurement and control stations (Tschmelak et al., 2005). This emphasizes the utility of optical transduction that can provide robustness, versatility, ease of miniaturization, and compatibility with the new technologies such as nanotechnology. Examples, include SPR devices (such as the portable SPR flow-through immunosensor for the analysis of carbaryl in natural water samples (Mauriz et al., 2006) that are amongst the most promising candidate technologies for incorporation in environmental early warning programmes.

Further research is necessary to achieve the important objective of producing sensors capable of determining several analytes simultaneously. Large-scale biosensor arrays, composed of highly miniaturized signal transducer elements, that can achieve real-time parallel monitoring of multiple species of contaminants are an important target and provide an important driving force in biosensor research (Haes and Van Duyne, 2002). In recent years, several examples of multi-analyte determinations have appeared in the literature, and include a portable SPR immunosensor designed for simultaneous on-site analysis of benzopyrene and 2-hydroxybiphenyl (Kawazumi et al., 2005), and an SPR biosensor that uses an SPR prism element to divide wavelengths used on serial sensing channels (Dostálek et al., 2005). A planar array immunosensor, equipped with a charge-coupled device (CCD) as a detector and a diode laser as light source has been applied to either the determination of multiple components, such as viruses, toxins and bacterial spores, in a single sample analysis or a single analyte in multiple samples simultaneously (Rowe-Taitt et al., 2000). These examples provide an indication of the potential of these technologies for use in the development of the multianalyte systems that are needed to underpin the monitoring necessary for the sound management of water quality at a reasonable cost.

Nanotechnology comprises a group of emerging techniques from physics, chemistry, biology, engineering and microelectronics that are capable of manipulating matter at the nanoscale, and it is playing an increasingly important role in the development of biosensors. Self-assembly of biomaterials, such as proteins, lipids, or nucleic acids (Wink et al., 1997) has been proposed as a candidate for the synthesis of nanostructures capable of performing unique functions, and has already inspired the development of novel biosensors. Several research groups have begun to explore alternative strategies for the development of optical SPR biosensors based on the optical properties of metal nanoparticles (Malinsky et al., 2001; Taton et al., 2001). Other nanostuctures that are being considered for development of biosensors include nanotubes, nanofibres, nanorods, nanoparticles and thin films. Boron-doped silicon nanowires (SiNWs) were reported by Cui et al. to create highly sensitive, real-time, electrically based sensors for biological and chemical species (Cui *et al.*, 2001). Microcantilever and nanocantilever biosensors have been developed for high-resolution and label-free molecular recognition, Alvarez *et al.* (2003) reported a system for measuring DDT concentrations as low as 10 nM.

One of the new materials that has emerged in recent years is the quantum dot. This is a nanometer-scale semiconductor particle that may contain a charge in the form of electrons. In solution, they display optical properties that are tunable by regulating their size. There are many possible future applications of these materials as biological labels or as probes by conjugating them with biological compounds. Goldman *et al.* have developed a biosystem for multianalyte determination using antibodies labeled with four colour quantum dots (Goldman *et al.*, 2005).

The goal of producing fit for purpose multianalyte systems for use in environmental monitoring has been made more realistic by the advances in microelectronics and microfluidics that have permitted the miniaturization of analytical systems, by allowing the handling of low volume samples, a reduction in reagent consumption and waste generation, and an increase in sample throughput (Sequeira *et al.*, 2002). This combined with the advent of a range of novel sensing elements with improved affinity, specificity, and for which the molecular recognition components be mass produced holds the key to the future. It may be the latter that ultimately dictates the success or failure of detection technologies. In this context the contributions of molecularly imprinted polymers (MIPs), and of genetic engineering focused on genetically transformed cells and genetically engineered receptor molecules could be very important in the future development of biosensors.

## 3.2.5 CONCLUSIONS

Biological techniques have great potential for delivering small, cost effective devices for real-time and on-site determinations of a wide range of pollutants, and for giving rapid, cost-effective results. They can form the basis of new analytical tools for use in aquatic environmental applications, especially at the screening level. A variety of commercially available systems has been developed, validated, and implemented for the assessment of both general whole organism toxicity effects (e.g. Microtox, or ToxAlert), and of specific toxicity effects such as genotoxicity (e.g. Calux). Biological approaches applied in conjunction with chemical analysis give more complete information about the possible biological impact of complex, polluted effluents than can be obtained from the current regulatory approach that depends on only chemical analysis. However, these newly developed biological detector systems, including biosensors and biological assays, must overcome a number of obstacles before they can be integrated in pollution control programmes. Some are technical issues such as the need for these systems to be incorporated into robust, real-time, miniaturized instruments. Many of these technical obstacles have been overcome thanks to progress in optics, microelectronic, and nanotechnology, and this should permit further improvements in detection systems, such as the miniaturization of diode lasers and a decrease in component costs. Other technical problems to be solved concern the limited lifetimes of some biorecognition components. Stabilization of the biorecognition elements (e.g. enzymes and Abs) continues to be a crucial factor for the future development of biological techniques; though genetic engineering offers possible solutions through the development of genetically engineered receptor molecules and genetically transformed cells. However, even with progressive technical improvements, a further obstacle remains; a lack of sound validation and correlation studies, the need for which is emphasized in the second chapter of this volume. There is an urgent need to demonstrate the validity and equivalence, both in the laboratory and in the field, of any new methods that are to be offered for use in support of regulatory monitoring and the associated risk assessments.

# REFERENCES

Allan I.J, Vrana B., Greenwood R., Mills DW., Roig B. and Gonzalez C., 2006. *Talanta* **69**, 302–2.
Alvarez M., Calle A., Tamayo J., Lechuga LM., Abad A. and Montoya A., 2003. *Biosens Bioelectron* **18**, 649–53.
Ames B.N., Lee F.D. and Durston W.E., 1983. *Proc. Natl. Acad. Sci. USA* **70**, 782.
Andreescu S., Barthelmebs L. and Marty J-L., 2002. *Anal. Chim. Acta.* **464**, 171–80.
Babkina S.S., Ulakhovich N.A. and Zyavkina Y.I., 2004. *Anal. Chim. Acta* **502**, 23–30.
Bakker E., 2004. *Anal. Chem.* **76**, 3285–98.
Barata C., Baird D.J., Nogueira A.J.A., Soares A.M.V.M. and Riva M.C., 2006. *Aquatic Toxicol.* **78**, 1–14.
Baun A., Jensen SD, Bjerg PL, Christensen TH, Nyholm N., 2000. *Environ. Sci. Technol.* **34**, 1647–52.
Bhattacharyya J., Read D., Amos S., Dooley S., Killham K. and Paton G.I., 2005. *Environ Pollut* **134**, 485–92.
Bechor O., Smulski DR., Van Dyk TK, LaRossa RA and Belkin S., 2002. *J Biotech* **94**, 125–32.
Beckman Instruments, 1982. *Microtox System Operating Manual*, Beckman Instruments, Inc., Carlsbad, CA, USA.
Biesinger K.D. and Christensen G.M., 1972. *J Fish Res. Bd. Can.* **29**, 1691–1700.
Brummel K.E., Wright J. and Eldefrawi M.E., 1997. *J. Agr Food Chem* **45**, 3292–8.
Budnikov M.C. and Evtugyn G.A., 1998. Sensitivity and selectivity of electrochemical biosensors for inhibitor determination. In: *Biosensors for Direct Monitoring of Environmental Pollutants in Field*, D. Nikolelis, U. Krull, J. Wang and M. Mascini (eds), Kluwer Academic Publishers, London, pp. 239–53.
Bulich A.A., 1979. Use of luminescent bacteria for determining toxicity in aquatic environments, P. 98–106. In L. L. Markings and R. A. Kimerle (eds), *Aquatic Toxicology*, ASTM 667. American Society for Testing and Materials, Philadelphia, PA.
Cass A.E.G, Francis D.G., Hill H.A.O., *et al.*, 1984. *Anal. Chem.* **56**, 667.
Chiorcea A.M. and Oliveira-Brett A. M., 2004. *Bioelectrochemistry*, **63**, 229–32.
Choi J.W., Kim Y-K, Lee I-H., Min J. and Lee W.H., 2001. *Biosen. Bioelectron.* **16**, 937–43.
Clarck L.C. and Lyons C., 1962. *Ann. N.Y. Acad. Sci.*, **102**, 29.
Colborn T., Vom Saal F. and Soto A.M., 1993. *Environ. Health Perspect.* **101**, 378.
Cui Y., Q. Wei, Park H. and Lieber CM., 2001. *Science* 293.
DeMarini D.M., Brooks H.G. and Parkes D.G., 1990. *Environ. Mol. Mutagen.* **15**, 1.
Dickert F.L., Tortshanoff M., Bulst W.E. and Ficherauer G., 1999. *Anal Chem*, **71**, 4559–63.
Doong R.-A, Shih H-M and Lee S-H, 2005. *Sens Actuators* **B 111**, 323–30.
Doong R.-A and Tsai, H-C., 2001. *Anal. Chim. Acta.*, **434**, 239–46.
Dostálek J., Vaisocherová H. and Homola J., 2005. *Sens Actuators B* **18**, 758–64.

Drevensek P., Zupancic T., Pihlar B., J., *et al.*, 2005. *J. Inorg. Biochem.* **99**, 432–42.
Duquesne S., 2006. *Ecotox. Environ. Saf.* **65**, 145–50.
Elsby R., Maggs J., Ashby J., Paton D., Sumpter J., and Park B.K, 2001. *J Pharmacol Exp Ther* **296**, 329–37.
EPA, 1978. The Selenastrum capricornutum Prinzt algal assay bottle test. Experimental design, application and data interpretation protocol, EPA-600/9-78-018, Environmental Protection Agency, Corvallis, Oregon, USA, 1978. Second US/USSR Symposium: Biological aspects of pollutants elects on marine organisms, EPA-600/3-82-034, Environmental Protection Agency, Corvallis, Oregon, USA, 1982, pp. 112–22.
Everett W.R., Rechnitz G.A., 1998. *Anal. Chem.*, **70**, 807.
Fahnrich K.A., Pravda M. and Guilbault G.G., 2003. *Biosens. Bioelectron.*, **18**, 73.
Farré M. and Barceló D., 2003. *Trends Anal. Chem.* **22**, 299.
García-Reyero N., Grau E., Castillo M., López de Alda M.-J., Barceló D. and Pi na B., 2001. *Environ. Toxicol. Chem.* **20**, 1152.
Gillis P.L., Wood C.M., Ranville J.F. and Chow-Fraser P., 2006. *Aquatic Toxicology* **77**, 402–11.
Goldman E.R., Clapp AR., Anderson GP., *et al.*, 2005. *Anal Chem* **76**, 684–8.
Grennan K., Strachan G., Porter A.J., Killard A.J. and Smyth M.R., 2003. *Anal. Chim. Acta*, **500**, 287.
Guilbault G.G., 1992. *Biosensors Bioelectron.* **7**, 411.
Guilbault G.G. and Das J., 1970. *Anal. Biochem*, **33**, 341.
Haes A.J. and Van Duyne RP., 2002. *J Am Chem Soc* **124**, 10589–95.
Han T-S., Sasaki S., Yano K., *et al.*, 2002. *Talanta* **57**, 271–76.
Haupt K., 2001. *Analyst* **126**, 747.
Haupt K. and Mosbach K., 2000. *Chem. Rev.* **100**, 2495
Hernando M.D., Fernandez-Alba A.R., Tauler R. and Barceló D., 2005. *Talanta* **65**, 358–66.
Hodges G., Roberts D.W., Marshall S.J and Dearden J.C., 2006. *Chemosphere* **63**, 1443–50.
Houk V.S. and DeMarini D.M., 1987. *Mutat. Res.* **182**, 193.
Ikenaka Y., Eun H., Ishizaka M. and Miyabara Y., 2006. *Aquatic Toxicology* **80**(2): 158–65.
ISO, 1995. Water Quality: Marine algal growth inhibition tests with Skeletonema costatum and phaeodactylum tricornutum, ISO 10,253, International Standardization Organization, Geneva, Switzerland.
ISO, 1996. Water Quality: Determination of the inhibition of the Mobility of Daphnia magna Straus (Cladocera, Crustacea), ISO 6341, International Standardization Organization, Geneva, Switzerland.
ISO, 1998. Water Quality: Determination of the inhibitory elect of water samples on the light emission of Vibrio fischeri. (Luminescent bacteria test), ISO 11348-1, 2 and 3, International Standardization Organization, Geneva, Switzerland.
ISO, 2000. Water quality – determination of genotoxicity of water and wastewater using the umu-test, ISO/DIS 13,829, ISO, Geneva, Switzerland.
Kawazumi H., Gobi V., Ogino K., Maeda H. and Miura N., 2005. *Sens Actuators B* **108**, 791–6.
Kenyon C.T., 1983. *Trends Biochem. Sci.* **8**, 8.
Kimerle R.M., Werner A. F. and Adams W. J., 1985. Aquatic hazard evaluation principles applied to the development of water quality criteria. In: *Aquatic Toxicology and Hazard Assessment: Seventh Symposium*, R.D. Cardwell, R. Purdy, and R.C. Bahner (eds), pp. 538–47, ASTM STP 854, American Society for Testing and Materials, Philadelphia.
Körner W., Spengler P., Bolz U., Schuller W., Hanf V. and Metzger J.W., 2001. *Environ. Toxicol. Chem.* **20**, 2142.
Kritz D. and Mosbach K., 1995. *Chen Anal. Chim. Acta* **300**, 71–5.
Kulis J. and Vidziunaite R., 2003. *Biosens. Bioelectron.*, **18**, 319–25.

Kurosawa S., 2003. *Meas. Sci. Technol.* **14**, 1882.
Lahav M., Kharitonov A.B., Katz O, Kunitake T. and Willner I., 2004. *Anal. Chem.* **73**, 1578.
Lazarides A.A., Lance Kelly K., Jensen T.R. and Schatz G.C., 2000. *Mol Struct Theochem* **529**, 59–63.
Lee J.H., Mitchell R.J., Kim B.C., Cullen D.C., Gu M.B., 2005. *Biosens Bioelectron* **21**, 500–7.
Lopez-Ortal P., Souza V. and Bucio L., 1999. *Mutat. Res.* **439**, 01–306.
Löschau M. and Krätke R., 2005. *Environ. Poll.* **138**, 260–7.
Lucarelli F., Authier L., Bagni G., *et al*., 2003. DNA biosensor investigations in fish bile for use as a biomonitoring tool *Analytical Letters* **36**, 1887–1901.
Lucarelli F., Palchetti I., Marrazza G. and Mascini M., 2002. *Talanta*, **56**, 949–57.
Mai Anh T., Dzyadevych S. V., Soldatkin A. P., Duc Chien N., Jaffrezic-Renault N. and Chovelon J.-M., 2002. *Talanta*, **56**, 627–34.
Malinsky M.D., Kelly KL., Schatz GCVD. and Richard P., 2001. *J. Am Chem Soc* **123**, 1471–82.
Mallat E., Barzen C., Klotz A., Brecha A., Gauglitz G. and Barceló D., 1999. *Environ. Sci. Technol.* **33**, 965.
Mannelli I., Minunni M., Tombelli S., Mascini M., 2003. *Biosens. Bioelectron.*, **18**, 129–40.
Marrazza G, Chianella I and Mascini M, 1999. *Anal Chim Acta* **387**, 297–307.
Mascini M., Palchetti I. and Marrazza G., 2001. *Fresenius J. Anal. Chem.* **369**, 15–22.
Mauriz E., Calle A., Abad A., *et al*., 2006. *Biosens Bioelectron* **21**, 2129–36.
Mauriz E., Calle A., Lechuga L.M., Quintana J., Montoya A., Manclús J.J., 2006. *Anal. Chim. Acta* **561**, 40–7.
Mauriz E., Calle A., Manclús J.J., *et al*., 2006. *Sensors and Actuators B* **118**, 2–10.
Mecklenburg M., Grauers A., Jönsson B.R., Weber A. and Danielsson B., 1997. *Anal. Chim. Acta* **347**, 79.
Moreau P., Bailone A. and Devoret R., 1976. *Proc. Natl. Acad. Sci. USA* **73**, 3700.
Mugweru A., Wang B. and Rusling J., 2004. *Anal. Chem.* **76**, 5557–63.
Mulchandani A., Mulchandani P. and ChenField W., 1998. *Anal. Chem.Technol.* **2**, 363–9.
Murk A.J., Legler J., van Lipzig M.M.H, *et al*., 2002. *Environ. Toxicol. Chem.* **21**, 16.
Nabok A.V., Tsargorodskaya A., Hassan A.K. and Starodub N.F., 2005. *Applied Surface Science* **246**, 381–6.
Nikookar K., Moradshahi A. and Hosseini L., 2005. *Biomol. Engin.*, **22**, 141–6,
Nistor C., Emnéus J., Gorton L. and Ciucu A., 1999. *Anal. Chim. Acta.*, **387**, 309–26.
Nunes G., Skládal P., Yamanaka H. and Barceló D., 1998. *Anal. Chim. Acta*, **362**, 59–68.
Panasyuk T.L., Mirsky V.M., Piletsky S.A. and Wolfbeis O.S., 2001. *Anal. Chem.*, **71**, 720–3.
Pandey P. C. and Weetall H. H., 1994. *Anal. Chem.* **66**, 1236.
Pavlić Ž., Vidaković-Cifrek Ž. and Puntarić D., 2005. *Chemosphere*, **61**, 1061–8.
Piletsky S.A., Piletskaya E.V., Elgersma A.V., Yano K. and Karube I., 1995. *Biosens Bioelectro*, **10**, 959–64.
Pizzariello A., Stredansky M., Stredanska S. and Miertus S., 2001. *Sens. Actuators B*, **76**, 286–94.
Pogacnik L. and Franko M., 2003. *Biosensors Bioelectron.*,. **18**, 1–9
Popovtzer R., Neufeld T., Ron E.Z., Rishpon J. and Shacham-Diamand Y., 2006. *Sens. Actuators B*, **119**, 664–72.
Pyun J.C., Beutel H., Meyer J.U. and Ruf H.H., 1998. *Biosens. Bioelectron.* **13**, 839–45.
Rawson D.M., 1987. UK Patent No. 8703119 and European Patent No. 87303411.0, lodged by WRc, Medmenham, United Kingdom.
Rodriguez-Mozaz S., Farre M. and Barcelo D., 2005. *Trends Anal. Chem.* **24**, 165.
Rodriguez-Mozaz S., Lopez de Alda M.J. and Barceló D., 2006. *Anal Bioanal Chem* **386**, 1025–41.

Rodriguez-Mozaz S., Marco M-P, Lopez de Alda M.J. and Barceló D., 2004a. *Anal Bioanal Chem* **378**, 588–98.
Rodriguez-Mozaz S., Reder S., Lopez de Alda M., Gauglitz G. and Barceló D., 2004b. *Biosens. Bioelectron.* **19**, 633.
Rowe-Taitt C.A., Golden JP., Feldstein MJ., Cras JJ., Hoffman KE. and Ligler FS., 2000. *Biosens Bioelectron* **14**, 785–94.
Schipper E.F., Bergevoet A.J.H., Kooyman R.P.H. and Greve J., 1997. *Anal Chim Acta*, **341**, 171–6.
Schipper E.F., Rauchalles S., Kooyman R.P.H., Hock B. and Greve J., 1998. *Anal Chem* **70**, 1192–7.
Seifert M., 2004. *Anal. Bioanal. Chem.* **378**, 684.
Sequeira M., Bowden M., Minogue E. and Diamond D., 2002. *Talanta* **56**, 355–63.
Serat W.F., Budinger F. E. and Mueller P. K., 1965. *J. Bacteriol.* **90**, 832–3.
Shimomura M., Nomura Y., Zhang W., *et al*., 2001. *Anal. Chim. Acta* **434**, 223–30.
Show I.T., Show M.B. and Williams L. R., 1989. Modeling the contributions of cross-reacting chemicals in immunoassays. In *Aquatic Toxicology and Hazard Assessment*: 12th vol., ASTM-STP 1026, U.M. Cowgill and L.R. Williams (eds), American Society for Testing and Materials, Philadelphia, pp. 21–33.
Skládal P., 1992. *Analytica Chimica Acta*, **269**, 281–7.
Snyder S.A., Villeneuve D.L., Snyder E.M. and Giesy J.P., 2001. *Environ.Sci. Technol.* **35**, 3620.
Sticher P., Jaspers M.C., Stemmler K., *et al*., 1997. *Appl Environ Microbiol* **63**, 4053–60.
Stiner L. and Halverson L.J., 2002. *Appl Environ Microbiol* **68**, 1962–71.
Streffer K., Vijgenboom E, Tepper A.WJW, *et al*., 2001. *Anal. Chim. Acta*, **427**, 201–10.
Taton T.A., Lu G. and Mirkin C.A., 2001. *J Am Chem Soc* **123**, 5164–5.
Tothill I.E. and Turner A.P.F., 1996. *Trends Anal. Chem.* **15**, 178.
Toussaint M.W., Shedd T.R., Van der Schaile W.H. and Leather G.R., 1995. *Environ. Toxicol. Chem.* **14**, 907.
Tschmelak J., Proll G., Riedt J., *et al*., 2005. *Biosens. Bioelectron.* **20**, 1499.
Versteeg D.J., Stalmanst M., Dyer S.D. and Janssen C., 1997. *Chemosphere*, **34**, 869–92.
White R., Jobling S., Hoare S.A., Sumpter J.P. and Parker M.G., 1994. *Endocrinology* **135**, 175.
Wilmer M, Trau D, Renneberg R and Spener F, 1997. *Anal Lett* **30**, 515–25.
Wink T.H., van Zuilen SJ., Bult A. and van Bennekom WP., 1997. *Analyst* **122**, 43–50.
Xavier M.P., Vallejo B., Marazuela M.D., Moreno-Bondi M.C., Baldini F. and Falai A., 2000. *Biosen. Bioelectron.*, **14**, 94.
Zhang S., Wei W., Zhang J., Mao Y. and Deng L., 2001. *Int. J. Environ. Anal. Chem.* **82**, 113.

# 3.3

# Immunochemical Methods

Petra M. Krämer

3.3.1 Introduction
3.3.2 Development of Antibodies as Selective Recognition Elements
3.3.3 Different Formats of Competitive Immunoassays
3.3.4 Quality Control
3.3.5 Test-kits
3.3.6 Automation and Multi-Analyte Systems
3.3.7 Contributions of Immunochemical Methods to the Water Framework Directive
3.3.8 Advantages, Drawbacks and Future of Immunochemical Methods
Acknowledgements
References

## 3.3.1 INTRODUCTION

Immunochemical methods for low molecular weight (<1000 D) environmental compounds, such as pesticides, nitro aromatics, and PAHs (polycyclic aromatic hydrocarbons) have been developed since more than 25 years (e.g. Hammock and Mumma, 1980; Hammock *et al.*, 1990; Dankwardt, 2000). The most common method is the immunoassay, which is either carried out with (heterogeneous) or without (homogenous) separation steps. Heterogeneous formats that use microtitre plates, plastic tubes and/or membranes as solid supports are more commonly used.

The antibody-analyte (antigen) (Ab-Ag) bond occurs through multiple noncovalent bonds – hydrophobic, electrostatic, hydrogen bonds, and van der Waals. The Ab-Ag reaction follows the law of mass action (see Equation (3.3.1)):

$$Ag + Ab \underset{k_{d,1}}{\overset{k_{a,1}}{\rightleftarrows}} Ab - Ag \qquad (3.3.1)$$

*Rapid Chemical and Biological Techniques for Water Monitoring*   Edited by Catherine Gonzalez, Philippe Quevauviller and Richard Greenwood
© 2009 John Wiley & Sons, Ltd

where Ag is the antigen (here this means: low molecular weight environmental compound/analyte, and also hapten), Ab the antibody, and Ab-Ag the antibody-antigen complex; $k_{a,1}$ is the association rate constant, and $k_{d,1}$ the dissociation rate constant.

The reaction is not as simple as it appears though, because a labelled substance (tracer) is needed to facilitate the detection of the Ab-Ag binding event. These labels can be radioisotopes (radio immunoassay (RIA), not used in environmental analysis), fluorophores (fluoro immunoassay (FIA)) or enzymes (enzyme immunoassay (EIA)). Therefore a competition of the Ag with the label (L) is involved, and is defined by the association and dissociation rate constants $k_{a,2}$ and $k_{d,2}$ (Equation (3.3.2)):

$$\begin{array}{c} Ag + Ab \xrightleftharpoons[k_{d,1}]{k_{a,1}} Ab - Ag \\ k_{d,2} \updownarrow k_{a,2} \\ L + Ab \end{array}$$

(3.3.2)

Affinity constants $K_{Ag}$ and $K_L$ ($K_{Ag} = k_{a,1}/k_{d,1}$; $K_L = k_{a,2}/k_{d,2}$) will differ between the free analyte and the labelled analyte (Tijssen, 1985), especially when the label is an enzyme. In addition, the analyte (of the standard or sample) competes with a derivative, not with the identical structure, and the difference in recognition between the two will have an effect. And it will also differ when the (derivative of the) analyte is bound to the surface via a protein. All these facts have to be taken into consideration when interpreting the results.

The antibody-antigen reaction is a complex interaction and despite the fact that this binding reaction can be influenced by pH, ionic strength, temperature, and organic solvents, immunoassays are not bioassays. No biological effects can be detected with antibodies; therefore no indication of the toxicity of a compound is given. Immunochemical methods use a biological reagent, namely the antibody, for a very selective recognition of a chemical structure. This is the strength, but maybe also the limitation of this technology. In addition, there is a misconception about the simplicity of immunoassays. Immunoassays are easy to perform, but it should not be concluded that even lay persons can get reliable quantitative results when they run assays to measure analytes in the low ppb to ppt range ($\mu$g/L to ng/L. The quality of the data will depend upon the integrity of the samples and the skill of the analyst. The data interpretation is also very important, therefore the characterization of the assays should be well documented, all relevant cross-reactivities should be known, and effects of different matrices should be taken into consideration. It is extremely important that the matrix in which the standards are set up is comparable to that in which the analyte will be analyzed.

After antibodies (and corresponding reagents such as coating antigens, enzyme-tracers, fluorophore-labelled antibodies, etc.) have been developed, they can be used in a wide range of techniques. These include conventional ELISAs (enzyme-linked immunosorbent assays) on microtitre plates (e.g. Krämer *et al.*, 2004a/b, 2005), dipstick (lateral flow) assays (e.g. Cuong *et al.*, 1999; Shim *et al.*, 2006; Zhang *et al.*, 2006a), immunosensors (e.g. Marco *et al.*, 1995; Meusel *et al.*, 1998; Gonzalez-Martinez *et al.*, 2001; Mallat *et al.*, 2001; Krämer and Rauber, 2007), automated flow injection methods (e.g. Krämer and Schmid, 1991; Krämer *et al.*, 1997;

Barzen *et al.*, 2002; Tschmelak *et al.*, 2005a–c), and techniques based on immunochromatography (e.g. Zhang *et al.*, 2006b/c). As seen from this list, there is enormous potential for the adaptation of these methods to answer a wide range of environmental questions. Despite this fact, the acceptance of this technology is still not as wide as one would expect. Many obstacles have caused this reluctance to accept this technology: (1) Many of the early developments of antibodies and commercial test-kits were carried out in the United States, where the maximum allowable concentrations of pesticide residues in drinking water are not the same for each residue and are in general higher than in Europe. Test-kits were developed in compliance with the legislation in the US. Nevertheless, they were used for European situations and then often failed. (2) The influence of environmental matrices, for example humic substances in water samples, caused many false positive results in immunoassays; therefore the method was considered to be unreliable. (3) Although antibodies can be developed against nearly all compounds of interest, based on the nature of antibodies, the recognition is especially good for large, hydrophilic compounds. Environmental contaminants are on the other hand small and sometimes hydrophobic. (4) Mixtures, such as dioxins, PAHs, PCBs, etc. are not particularly good targets for immunochemical methods. The selective recognition, together with potential cross reactivities (recognition of structurally related and similar compounds to the target compound) might not be appropriate for the selected compounds of one group (e.g. PAHs, dioxins) controlled by regulatory agencies. This also leads to difficulties when immunoassay results have to be compared with data by conventional analysis. (5) Antibodies, which are the heart of these techniques, are in most cases not commercially available, making it difficult for developers of new immunochemical methods to have access to reliable sources and unlimited quantities of the necessary antibodies. When the antibodies are available, the next limitation is often a lack of the other necessary reagents such as enzyme-tracers and coating antigens. (6) A further problem is that these methods were often wrongly categorized as 'bioassays', and were therefore not considered as reliable, quantitative analytical methods by chemists.

Despite all of these obstacles, the development of immunochemical methods has continued over the last two decades, and in many cases they have proved to be fast, cost-effective, easy to use and fit for purpose technologies, that can make an important contribution in the field of environmental analysis. This chapter provides a brief description of the basics and principles of immunoassay, together with a selection of examples of environmental applications, especially with respect to the European Union Water Framework Directive (WFD).

### 3.3.2 DEVELOPMENT OF ANTIBODIES AS SELECTIVE RECOGNITION ELEMENTS

Selective antibodies are the basis of all immunochemical methods and therefore the development of these antibodies is fundamental to this technology. In the early days of immunochemistry, only polyclonal antisera had been developed, and they have been used for more than 25 years both in academia and in commercial test-kits. Although the sensitivity and selectivity of polyclonal antisera are excellent, these reagents are a mixture of antibodies, and even with the same immunogen different animals will produce

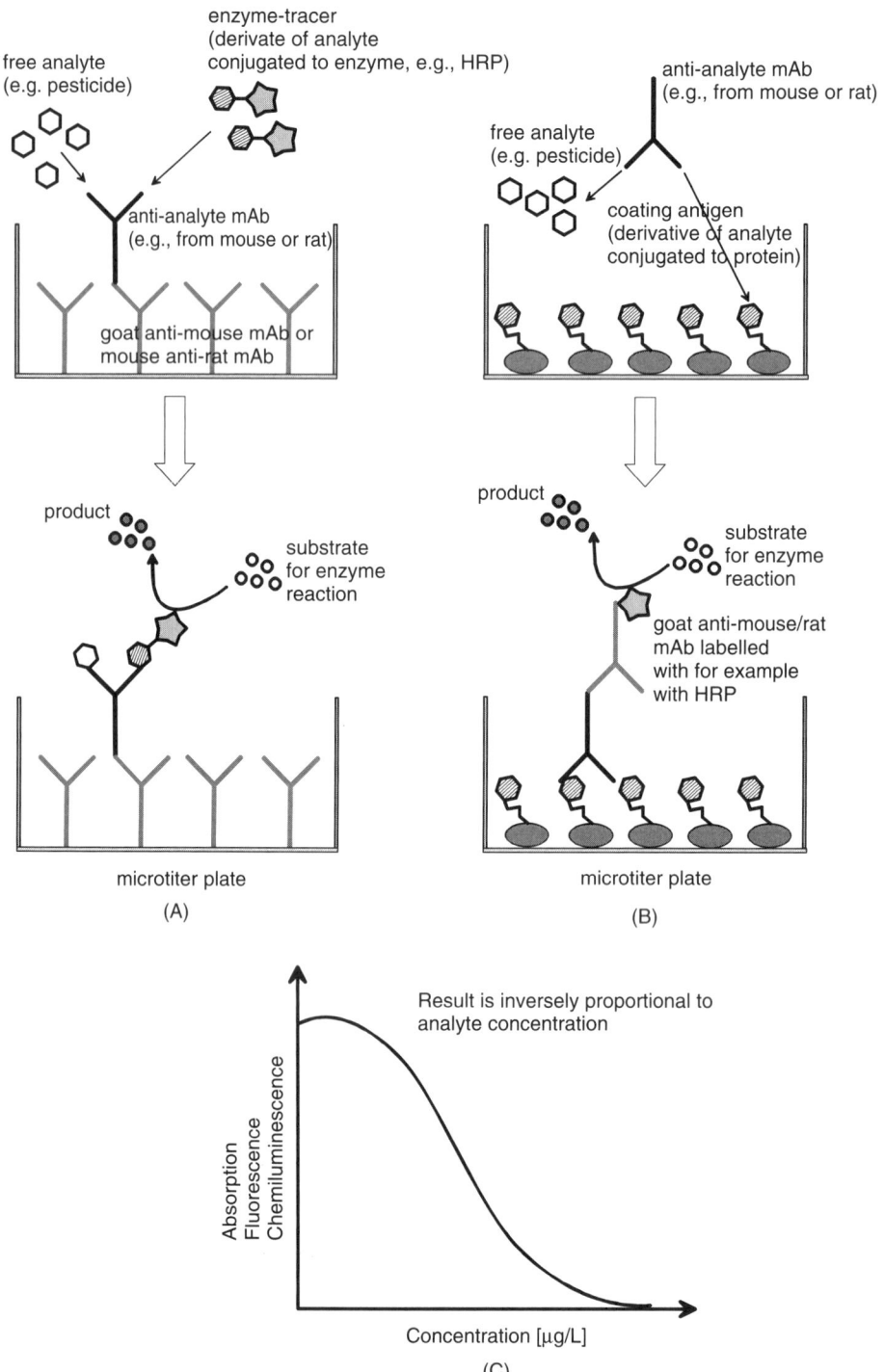

different sera. Thus a further limitation for polyclonal antisera is the volume of serum that can be obtained from one individual animal. Although the development of monoclonal antibodies (Köhler and Milstein, 1975) is more time consuming and expensive, there was a boost in developments in this area in the nineties of the last century (Dankwardt, 2000). The main advantages of monoclonal antibodies are that they are homogenous reagents with defined characteristics, and can be produced in unlimited amounts; they can be purified and labelled and used for different immunosensor techniques.

For the development of antibodies against small molecules, it is necessary to synthesize a derivative of the target molecule. This step is very crucial, because it determines the overall selectivity of an antibody (Goodrow *et al.*, 1990). It is necessary for these so-called haptens to be conjugated to proteins in order to produce the immunogens that can be used for immunization of the animals. In addition, these haptens are needed as an essential component of the assay. There has been some debate in the literature as to whether the haptens for those two purposes should be identical or not. Generally, it is recommended to have structurally different haptens, because antibodies will recognize with a very high affinity the hapten in the immunogen, which will lead to a less sensitive assay for the target analyte. Using a different structure of the hapten for the assay set up is likely to provide a more sensitive assay for the target analyte.

## 3.3.3 DIFFERENT FORMATS OF COMPETITIVE IMMUNOASSAYS

In most cases, competitive immunoassays for the analysis of small molecules are carried out in microtitre plates. The majority of these assays use an enzyme as label, thus leading to the term enzyme immunoassay, and the most commonly used is the ELISA (enzyme-linked immunosorbent assay) that has a heterogeneous format (separation of bound and unbound). This format can be set up either in the *enzyme-tracer format* (Figure 3.3.1A) or in the *coating antigen format* (Figure 3.3.1B). The result shows a

---

**Figure 3.3.1** Formats of competitive EIAs (enzyme immunoassays) for small molecules. (A) Enzyme-tracer format. (A) Upper part: first the microtitre plates are coated with an anti-animal antibody, here either goat anti-mouse or mouse anti-rat antibodies. In the next step the anti-analyte antibody (usually monoclonal, either from mouse or rat) is added and incubated. Free analyte (e.g. isoproturon) and enzyme-tracer (derivative of analyte conjugated to an enzyme, e.g. horseradish peroxidase (HRP)) compete for the limited binding sites of the anti-analyte antibodies. After the addition of substrate for the enzyme reaction (A) lower part), the absorbance of the product is measured. (B) Coating antigen format. (B) Upper part: first the microtitre plates are coated with the coating antigen (derivative of analyte conjugated to a protein, e.g. BSA, OVA). In the next step, free analyte (e.g. isoproturon) and anti-analyte antibody are added. Free analyte and coating antigen are competing for the limited binding sites of the anti-analyte antibodies. All unbound material is washed off, and anti-animal antibody, labeled with HRP, is added (B) lower part). After the addition of substrate for the enzyme reaction, the absorbance of the product is measured. In both assay formats, the result is inversely proportional to the analyte concentration (C).

standard curve with sigmoid shape, when the signal (e.g. absorbance) is plotted versus the logarithm of the analyte concentration. The signal is inversely proportional to the analyte concentration (Figure 3.3.1C). Fitting of standard curves is usually performed via a software program (e.g. SOFTmax Pro™, Molecular Devices, Palo Alto, CA, USA) using the 4-parameter curve fit according to Equation (3.3.3):

$$y = \frac{A - D}{1 + \left(\frac{x}{C}\right)^B} + D \qquad (3.3.3)$$

Absorbance values of standard curves can be normalized either to % control values according to the formula (3.3.4):

$$\text{Control}(\%) = \frac{A}{A_0} \times 100 \qquad (3.3.4)$$

or to $\%B/B_0$ values according to the formula (3.3.5):

$$\frac{\%B}{B_0} = \frac{A - A_{\text{Excess}}}{A_0 - A_{\text{Excess}}} \times 100 \qquad (3.3.5)$$

Where A is the value of absorbance for each standard or sample, and $A_0$ the value of absorbance for the zero standard (40 mM PBS or Milli-Q; the higher asymptote of the curve), and $A_{\text{Excess}}$ the value of the highest analyte concentration (excess concentration; the lower asymptote).

A linear range is obtained around the test midpoint (IC50%, concentration of analyte that causes 50% inhibition). This linear dose-response region (working range) has to be used for determinations, because here the change in absorbance correlates with the analyte concentration. Depending upon the standard deviations and the steepness of the curve, this working range is usually one or two orders of magnitude. It ranges usually from 90% to 10% control (or $\%B/B_0$).

There is another format (the *sandwich format* illustrated in Figure 3.3.2), but this can be used only for large molecules, because it needs two different binding sites. In this method two different antibodies that recognize different parts of the same molecule have to be developed. The so-called detection antibody (mAb2) is either directly labelled with an enzyme (Figure 3.3.2) or a fluorophore, or it can be detected with a labelled anti-animal antibody. This format also yields a sigmoid shaped standard curve, but here the signal (e.g. absorbance) is proportional to the analyte concentration (Figure 3.3.2). Currently this format does not play a role in environmental immunoanalysis, but it might become important for other environmental applications (e.g. detection of microorganisms, viruses, toxins, and 'biopesticides').

## 3.3.4 QUALITY CONTROL

As in all analytical methods, accuracy and precision are very important properties. Several papers and norms that deal with this aspect have been published (Gee *et al.*,

# Quality Control

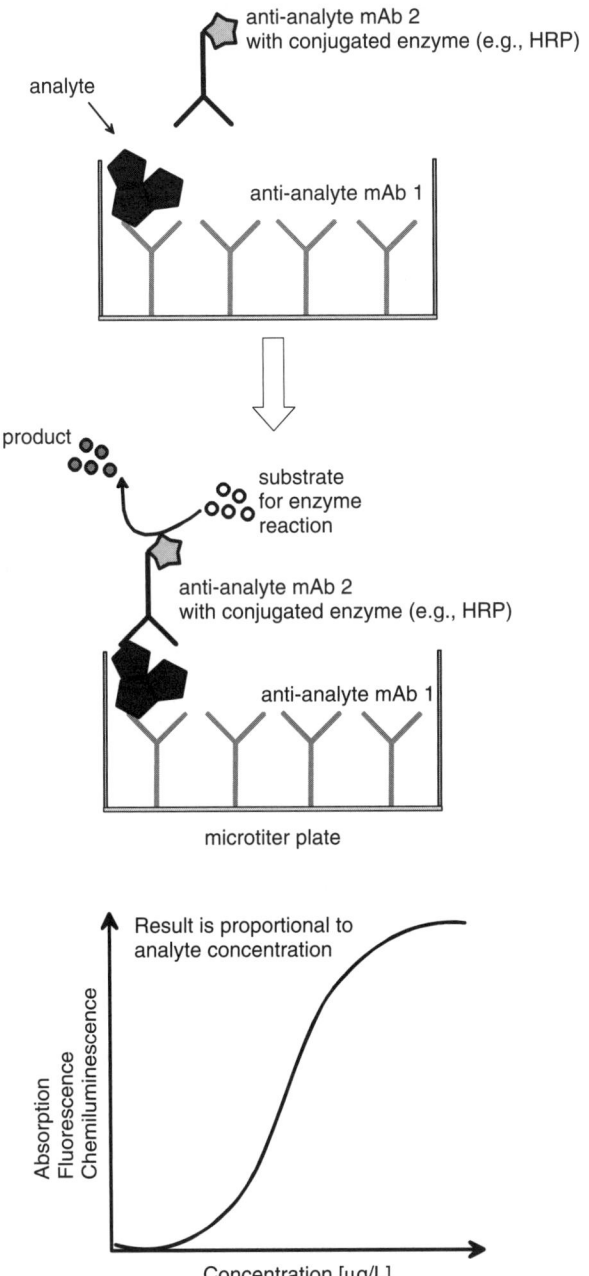

**Figure 3.3.2** Format of sandwich ELISA for antigens (high molecular weight compounds). This format needs a capture antibody (on the microtitre plate surface) and a detector antibody, to which an enzyme (e.g. HRP) can be conjugated. Other possibilities would be that the detector antibody is labeled with a fluorophore or is biotinylated. The latter would use streptavidin-HRP in the detection step

1994; DIN V 38,415-2, 1995; ISO 15,089: 2000, 2001; Brady, 1995; Watts and Hegarty, 1995; Krotzky and Zeeh, 1995). In order for analyses to be fit for purpose, stability of immunoreagents (antibodies, enzyme-tracers, coating antigens) has to be ensured through proper storage and handling.

In brief, the following parameters have to be considered during the development and establishment of an immunoassay in the laboratory:

- type of microtitre plate;
- availability of immunoreagents (antibodies, coating antigens, enzyme-tracers);
- stability of coating antigen or enzyme-tracer (stability on plate or in solution, stability of covalent linkage between hapten and protein/enzyme);
- stability of antibodies;
- blocking steps, buffers;
- washing step (how many times, volume, additives (e.g. Tween 20));
- time of each step in immunoassay;
- number of steps in immunoassay;
- assay temperature;
- readout;
- instrument linearity;
- software program for calculation.

This list gives a first orientation. It might have to be adjusted and extended depending upon the specific properties of an assay and the specific purpose for which it is intended.

## 3.3.5 TEST-KITS

For water and/or soil analysis, different test-kits are commercially available for a variety of pesticide residues, industrial contaminants and hormones (Table 3.3.1). To the author's knowledge, today the main provider of these techniques is Strategic Diagnostics Inc. (SDI), Newark, DE, USA, which has also distributors worldwide. With some exceptions, most test-kits use polyclonal antisera; they use the heterogeneous, competitive ELISA, and in most cases the enzyme-tracer format (Figure 3.3.1A), because the procedure is faster and easier. Usually, test-kits for environmental applications are still very expensive, because the development costs are high and the market is, even after 15 years of applications, still small. Generally, it takes some practice and knowledge to apply these test-kits successfully and reliably. The manufacturer's instructions have to be followed carefully to assure good results. In some cases, companies claim that 96 samples (determinations) can be analysed on one microtitre plate. However, this is not possible, because a complete standard curve should always be run on the same

**Table 3.3.1** Suppliers of immunoassay test-kits for environmentally relevant compounds

| Company | Internet Address | Analytes[1] |
|---|---|---|
| **Abraxis LLC** Warminster, PA, USA | http://www. abraxiskits.com | *Pesticide residues:* Acetochlor; Alachlor; Atrazine; Carbendazin/Benomyl; Cyclodienes, 2,4-D; DDE; Diuron; Glyphosate; Metolachlor; Organophosphate/Carbamate; Pyrethroids; Spynosyn *Surfactants:* Alkyl ethoxylate (AE); Alkylphenol ethoxylate (APE); Linear alkylbenzene sulfonate (LAS) *Estrogens:* 17-ß-Estradiol (E2); Estrogens (E1+E2+E3); Estrone (E1); Ethynylestradiol (EE2) *Industrial chemicals:* Bisphenol A; Coplanar PCBs; PCBs higher chlorinated; PCBs lower chlorinated; PCBs broad reactivity; Polycrominated diethyl ether (PBDE); Trichlosan |
| **Beacon Analytical Systems** Portland, ME, USA | http://www. beaconkits.com | *Pesticide residues:* Alachlor; Atrazine; 2,4-D; r-Metolachlor; s-Metolachlor *Industrial contaminants:* Cyclodienes; Petroleum Fuels; Toxaphene |
| **CAPE Technologies** Portland, ME, USA | http://www. cape-tech.com | *Industrial chemicals:* Dioxin/Furan |
| **Coring System Diagnostix GmbH** Gernsheim, Germany | http://www.coring.de (Distributor for Strategic Diagnostics Inc., USA) | *Pesticide residues:* 2,4-D; Alachlor; Aldicarb; Atrazine; Benomyl/Carbendazim; Carbofuran; Chlordan; Chlorothalonil; Chlorpyrifos; Cyanazine; Diazinon; Hydroflour; Isoproturon; Metolachlor; Simazine; Spinosad; Triazines; Triclopyr *Hormones:* Boldenon; beta-Agonisten; Clenbuterol/Brombuterol; Corticosteroide; 17-ß-Estradiol; Nortestosteron/Trenbolon; Progesteron; Ractopamin; Stilben; Testosteron; Trenbolon; Zeranol *Industrial contaminants:* Inquest OP/Carbamat; MKW (TPH); PAHs; PCBs; PCP; RDX; TNT |

*(continued overleaf)*

**Table 3.3.1**  (Continued)

| Company | Internet Address | Analytes[1] |
|---|---|---|
| **EnviroLogix** Portland, ME, USA | http://www.envirologix.com | *Pesticide residues:* Imidacloprid; Isoproturon; Paraquat; Synthetic Pyrethroid, Organochlorines (Endosulfan) |
| **Sension GmbH** Augsburg, Germany | http://www.sension-gmbh.de | *Pesticide residues:* Atrazine; Diuron |
| **Silver Lake Research Corp.** Monrovia, CA, USA | http://www.watersafetestkits.com | *Pesticide residues:* Atrazine; Simazine |
| **Strategic Diagnostics Inc.** Newark, DE, USA | http://www.sdix.com | *Pesticide residues:* 2,4-D; Alachor; Aldicarb; Atrazine; HS (High Sensitivity) Atrazine; Benomyl/Carbendazim; Carbofuran; Chlorothalonil; Chlorpyrifos; Cyanazine; Cyclodienes; Methomyl; Metolachlor; Procymidone; Simazine; Spinosad; Triclopyr DDT; Lindane; Toxaphene; Triazines *Industrial contaminants/remediation:* BTEX; PAHs; cPAHs; PCBs; PCP; TPH; TNT; RDX |

[1] Information was taken from the homepage of the corresponding company

plate as the samples, and all standards and samples should be run at least in triplicate, or preferably in quadruplicate. This means that 20 samples per plate is a more realistic expectation, and this means that the cost of analyzing one sample may be up to 25 € (in most cases for only one analyte).

It is essential that test-kits are used for the intended purpose in a specific matrix. Thus test-kits can limit the applications in which the technology can be applied. In some cases, it will be necessary to set up an in-house assay format for which it will be necessary to buy the separate individual immunoreagents, such as antibodies, enzyme-tracer or analyte-derivative protein conjugates. In some cases this is difficult because the quality and supply are not always guaranteed. A recent list of antibody manufacturers, suppliers and services is given by Blow, 2007.

## 3.3.6 AUTOMATION AND MULTI-ANALYTE SYSTEMS

Automated instruments based on immunochemical principles have been developed over the past several years (e.g. Krämer and Schmid, 1991; Oroszlan *et al.*, 1993; Krämer *et al.*, 1997; Gonzáles-Martinez *et al.*, 1999; Barzen *et al.*, 2002). Despite the fact that some of these instruments would be very valuable for analysis, screening, and monitoring within the context of the WFD, they are unfortunately not commercially available. The routine usage of such instruments would save, in Europe alone, thousands of litres of organic solvents during analysis and would therefore have a much lower negative environmental impact than methods based on chromatography.

During recent years there has been a clear trend towards the use of multi-analyte systems for environmental contaminants (e.g. Klotz *et al.*, 1998; Weller *et al.*, 1999; Kido *et al.*, 2000; Mauriz *et al.*, 2006; Krämer and Räuber, 2007). This issue is especially interesting for automation for on-site and/or for online monitoring. The most promising analytical system with online multi-analyte analysis is the so-called AWACSS (Automated Water Analyser Computer Supported System) that was developed within an European project as a fully automated online monitoring tool. This system combines immunochemical technology with remote and surveillance facilities (Willard *et al.*, 2003; Tschmelak *et al.*, 2004; 2005a/b/c). It employs fluorescence-based detection of the binding of fluorophore-tagged biomolecules to the surface of an optical waveguide. A fibre-coupled detection array is used to monitor 32 separate fluorescence signals, thus allowing the simultaneous measurement of up to 32 different analytes. Unfortunately, again this system is available only as a prototype and is used in only one laboratory. However, it is still under development, and it is hoped that it will be commercially available in the future.

## 3.3.7 CONTRIBUTIONS OF IMMUNOCHEMICAL METHODS TO THE WATER FRAMEWORK DIRECTIVE

Standardization of immunochemical methods will be needed for their widespread usage in different countries and laboratories. This will include SOPs (standard operating procedures), GLP (good laboratory practice), and participation in proficiency tests (PTs) that will be comparable to those carried out routinely for conventional analysis, but with adaptations to tailor them for immunochemical analysis.

Although there are already several superior formats available, currently the conventional ELISA on microtitre plates is still the most common format used in environmental applications. Within the European project SWIFT-WFD (*Screening methods for water data information in support of the implementation of the Water Framework Directive*) our laboratory participated with conventional microtitre plate ELISAs for isoproturon (Krämer *et al.*, 2004a/b), diuron (Karu *et al.*, 1994), and atrazine (Karu *et al.*, 1991) (Figure 3.3.3 a–c), in two Europe-wide proficiency tests (Brunori *et al.*, 2007). Table 3.3.2 gives an overview of the applied assay set-ups, including volumes, concentrations, and times. Although these PTs were not structured specifically for immunoassays, it was shown that immunoassays performed, in the case of diuron and isoproturon, just as well as conventional reference analysis methods (LC, GC with corresponding sample preparation and detection) (Krämer *et al.*, 2007b), both for spiked water samples and for blind solutions (containing the same pesticide residue pattern in ethyl acetate). In the case of atrazine, the result gained by immunoassay was about six times higher than the consensus value (Krämer *et al.*, 2007b). This result needed more interpretation,

**Table 3.3.2** Standard operating procedures for ELISAs – enzyme-tracer format

| Analyte | Isoproturon | Diuron | Atrazine |
|---|---|---|---|
| **Procedure** | | | |
| Coating of microtiter plates (NUNC MaxiSorp) 200 µL/well, over night, 4°C | Mouse anti-rat TIB 172 [2 µg/mL] in 50 mM carbonate buffer, pH 9.6 | Goat anti-mouse [2.4 µg/mL] in 50 mM carbonate buffer, pH 9.6 | Goat anti-mouse [2.4 µg/mL] in 50 mM carbonate buffer, pH 9.6 |
| Washing (3 times with 300 µL/well) 4 mM PBST, pH 7.2 | | | |
| mAb 150 µL/well in 40 mM PBS, pH 7.6 2h, RT (21–23°C) | Rat IOC7E1 75 or 37.5 ng/mL | Mouse 481.3 61 ng/mL | Mouse AM5D1-3 23 ng/mL |
| Washing (3 times with 300 µL/well) 4 mM PBST, pH 7.2 | | | |
| Analyte 100 µL/well 1 h, RT (21–23°C) | Isoproturon Standards: 0.001–1,000 or 10,000 µg/L in Milli-Q water or in 40 mM PBS, pH 7.6 | Diuron Standards: 0.001–10,000 µg/L in 40 mM PBS, pH 7.6 | Atrazine Standards: 0.001–10,000 µg/L in 40 mM PBS, pH 7.6 |
| Enzyme-Tracer 50 µL/well in 40 mM PBS, pH 7.6 30 min, RT (21–23°C) with shaking of the plates | Isoproturon Standards: 0.01–100,000 µg/L in Milli-Q water or in 40 mM PBS, pH 7.6 III-HRP[1] 1:2000 | Isoproturon Standards: 0.001–10,000 µg/L in Milli-Q water or in 40 mM PBS, pH 7.6 | |
| | Rat IOC7E1/IOC10G7 37.5/20 ng/mL | Diuron I-HRP[1] 1:15,000 | Atr-HRP[1] 1:4,000 |
| | Rat IOC10G7 20 ng/mL | | |
| Washing (3 times with 300 µL/well) 4 mM PBST, pH 7.2 | | | |
| Substrate 150 µL/well 10–20 min, RT (21–23°C) | Substrate for horseradish peroxidase reaction: 0.4 mM TMB (tetramethylbenzidine), 1.3 mM H$_2$O$_2$ in 100 mM sodium acetate buffer, pH 5.5 | | |
| Stop solution 50 µL/well | 2 M H$_2$SO$_4$. Stop enzyme reaction after 10–20 min, depending upon the color development. Note the temperature! | | |
| Absorbance measurement | 450 nm (minus reference: 650 nm) | | |

[1] horseradish peroxidase (HRP)

because the sample material contained not only atrazine, but also other triazines (e.g. simazine, terbutylazine, and deethylatrazine), that showed cross-reactivity with this antibody. When the blind solution was analyzed by immunoassay, two problems occurred: (1) in order to be able to analyze in the proper range of the assay, the blind solution had to be diluted 1:25000, and (2) and although the z-score of the immunoassay result was satisfactory (0.3), the result was 'correct' only by chance. The blind solution contained cross-reacting triazines in addition to atrazine (total of 198.7 mg/L). The concentration (53.1 mg/L of atrazine estimated by the ELISA was some 27% of the total triazines present. By chance the atrazine concentration was 25% of the total concentration of triazines present, and so the result of the immunoassay seemed to be correct.

One outcome of this PT was the conclusion that in order to achieve a proper comparison of different methods it is important to evaluate the results carefully. The matrix of the water samples, and the matrix of pill material, that was used for spiking, might both have had an influence on the performance of the immunoassays.

In another demonstration of the potential utility of immunoassays in the monitoring of water quality necessary to underpin the WFD, samples were taken from tank experiments using spiked river water (Guigues *et al.*, 2007). These tank trials were conducted alongside a field trial in the River Meuse at the RIZA (Rijksinstituut voor Integraal Zoetwaterbeheer en Afvalwaterbehandeling (Engl.: Institute for Inland Water Management and Waste Water Treatment)) monitoring station at Eijsden, NL. The matrix was river water pumped from the River Meuse, spiked with stock solutions of pesticides and circulated through a tank where it was agitated by a rotating carousel containing a range of in situ sampling technologies. The concentrations of the test analytes in the tank were increased and decreased twice over the 14 day experiment to simulate pollution events. Samples from the tanks were transported to the laboratory in glass bottles, and analysed by immunoassays (*enzyme-tracer format*) for isoproturon, diuron, and atrazine. The results were compared with those obtained from LC-MS/MS (Martens *et al.*, 2004). Similar results were obtained from both immunoassay and LC-MS/MS for diuron, and the two cycles of concentration changes were closely mapped. For isoproturon whilst the fluctuations in concentration with time were well defined by the methods, the concentrations determined by but ELISA were consistently 2–3 times higher than those derived from LC-MS/MS. With atrazine, ELISA results were 5–6 times higher than those obtained by LC-MS/MS, and this can be interpreted as a matrix effect and/or the presence of other cross-reacting triazines or metabolites (Krämer *et al.*, 2007b).

It can be concluded that immunoassays can serve very well in the surveillance and operational monitoring mandated within the WFD. Nevertheless, the possibility of interference from cross-reacting compounds and matrix effects must be taken into account.

## 3.3.8 ADVANTAGES, DRAWBACKS AND FUTURE OF IMMUNOCHEMICAL METHODS

Immunochemical methods are very versatile, and a range of antibodies for small environmentally relevant compounds has been developed over many years, but

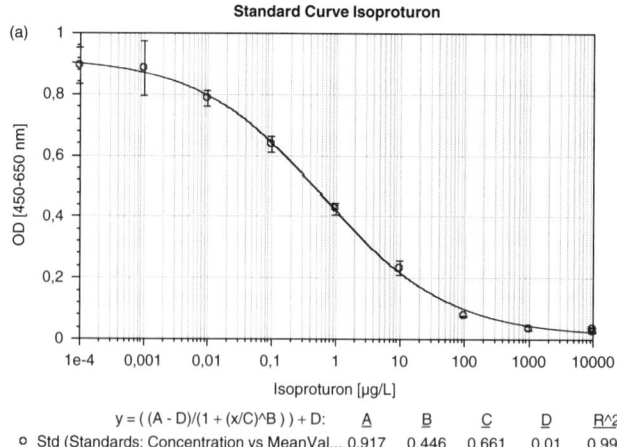

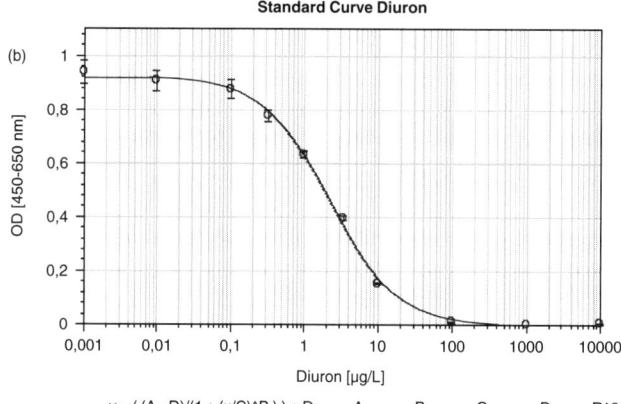

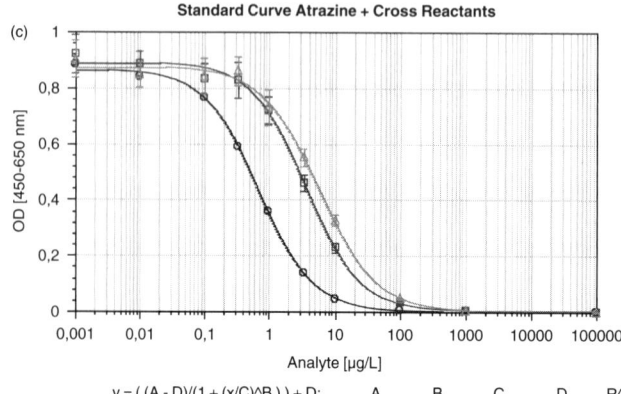

chromatographic methods have a long track record and proven utility in this area, and are generally accepted and validated, and therefore most analytical chemists still prefer to use these methods rather than immunochemical analysis. However, this picture might change when the analyst is faced with more hydrophilic and higher molecular weight compounds, for which GC or HPLC will not be so easy to apply and for which selective antibodies will be very suitable.

Immunoassays are ideally applied for screening large numbers of samples for a rather small number of compounds. They are not very efficient, at least not in the usual microtitre plate format, if one sample has to be analyzed for more than 10 analytes. In principle this is possible, but it will need an automated instrument with a multi-analyte configuration (e.g, AWACSS). However, even then the effort will be worth while only if there is a large sample load; ideally in monitoring situations where a rapid, high throughput of samples would be advantageous. A further important advantage of the immunochemical methods is that sample preconcentration is usually not necessary, and this allows the use of small sample volumes (usually 50 mL is more than sufficient to analyze the sample many times), and this reduces the problems and costs associated with transportation and distribution.

The bottleneck for a widespread application of these technologies is still the lack of commercial or common availability of immunoreagents. So even if an analytical laboratory has all the necessary equipment, such as a microtitre plate reader, microtitre plate washer, and pipettes, it is still not in the position to use immunoanalysis for the routine analysis of pesticides (or other environmentally relevant compounds) of interest. As a general recommendation, cooperation with laboratories experienced in assay development is recommended, and also the training of personnel in such an organization. This would reduce the likelihood of the frustration and failure encountered by some researchers during the establishment of this technology in their own laboratory.

The future of immunochemical analysis will depend upon the development and availability of new immunoreagents that can be used in different applications. These range from conventional ELISA on microtitre plates to immunosensors and immunoaffinity columns. It will further depend upon the success in efforts to stabilize immunoreagents. The achievement of the necessary selectivity and sensitivity of the antibodies will determine the success of the applications. Here, the development of recombinant antibodies might play an important role (e.g. Hock *et al.*, 2002; Churchill *et al.*, 2002; Sheedy *et al.*, 2007). Generally, we will need more applications and usage of these technologies in order to achieve a much broader and cost-effective monitoring of our water bodies.

---

**Figure 3.3.3** (a) Standard curve for isoproturon using a mixture of mAbs IOC 7E1 and IOC 10 G7. Assay was carried out in the enzyme-tracer format; detailed assay conditions, see Table 3.3.2. Working range of this standard curve is from 0.02–25 μg/L. (b) Standard curve for diuron using mAb 481.3. Assay was carried out in the enzyme-tracer format; detailed assay conditions, see Table 3.3.2. Working range of this standard curve is from 0.2–20 μg/L. (c) Standard curve for atrazine and cross reactants terbutylazin (1) and propazine (2) using mAb AM5D1-3. Assay was carried out in the enzyme-tracer format; detailed assay conditions, see Table 3.3.2. Working range of this standard curve is from 0.1–7 μg/L.

## ACKNOWLEDGEMENTS

Mr Stephan Forster and Ms Christina Räuber (formerly HMGU) are acknowledged for immunoanalysis of water samples during proficiency tests and tank experiments. Many thanks to Dr Natalie Guigues (BRGM, Orleans, France) and to Dr Ian Allan (University of Portsmouth, United Kingdom) for providing us samples from the tank experiments. In part, this research was funded by EU project SWIFT-WFD (SSPI-CT-2003-502492).

## REFERENCES

Barzen C., Brecht A. and Gauglitz G., 2002. *Biosens. Bioelectron.*, **17**, 289.
Blow N., 2007. *Nature*, **447**, 741.
Brady J.F., 1995. In: J.O. Nelson, A.E. Karu, R.B. Wong (eds), *Immunoanalysis of Agrochemicals: Emerging Technologies*, ACS Symposium Series 586, American Chemical Society, Washington, DC.
Brunori C., Morabito R., Ipolyi I., et al., Madrid Y., 2007. *Trends Anal. Chem.*, **26**, 993.
Churchill R.L.T., Sheedy C., Yau K.Y.F. and Hall J.C., 2002. *Anal. Chim. Acta*, **468**, 185.
Cuong, N.V., Bachmann, T.T. and Schmid, R.D., 1999. *Fresenius J Anal Chem.*, **364**, 584.
Dankwardt A., 2000. In: R.A. Meyers (ed.), *Encyclopedia of Analytical Chemistry*, John Wiley & Sons, Ltd, Chichester, UK.
DIN V 38,415-2, 1995. German Standard Methods for the Examination of Water, Wastewater and Sludge – Sub-animal Testing (group T) – Part 2: Guideline for Selective Immunotest Methods (Immunoassays) for the Determination of Plant Treatment and Pesticide Agents (T2), Beuth Verlag GmbH, Berlin.
Gee S.J., Hammock B.D., van Emon J.M., 1994. *A User's Guide to Environmental Immunochemical Analysis*, EPA/540/R-94/509, United States Environmental Protection Agency, Office of Research and Development, Washington, DC.
Gonzáles-Martinez M.A., Puchades R., Maquieira A., Manclús J.J. and Montoya A., 1999. *Anal. Chim. Acta*, **392**, 113.
González-Martinez M.A., Puchades R. and Maquieira A., 2001. *Anal. Chem.*, **73**, 4326.
Goodrow M.H., Harrison R.O. and Hammock B.D., 1990. *J. Agric. Food Chem.*, **38**, 990.
Guigues N., Behro C., Roy S., Foucher J.-C. and Fouillac A.-M., 2007. *Trends Anal. Chem.*, **26**, 268.
Hammock B.D. and Mumma R.O., 1980. In: J. Harvey Jr. and G. Zweig (eds), *Pesticide Analytical Methodology*, ACS Symposium Series 136, American Chemical Society, Washington, DC.
Hammock B.D., Gee S.J., Harrison R.O., in: J.M. van Emon, R.O. Mumma (eds), 1990. *Immunochemical Methods for Environmental Analysis*, ACS Symposium Series 442, American Chemical Society, Washington, DC.
Hock B., Seifert M. and Kramer K., 2002. *Biosens. Bioelectron.*, **17**, 239.
ISO 15,089, 2000. *Guidelines for Selective Immunoassays for the Determination of Plant Treatment and Pesticide Agents*, Beuth Verlag GmbH, Berlin, 2001.
Karu A.E., Goodrow M.H., Schmidt D.J., Hammock B.D. and Bigelow M.W., 1994. *J. Agric. Food Chem.*, **42**, 301.
Karu A.E., Harrison R.O., Schmidt D.J., in: M. Vanderlaan, L.H. Stanker, B.E. Watkins, D.W. Roberts (eds) 1991. *Immunoassays for Trace Chemical Analysis: Monitoring Toxic Chemicals in Humans, Food, and the Environment*, ACS Symposium Series 451, American Chemical Society, Washington, DC.

# References

Kido H., Maquieira A., Hammock B.D., 2000. *Anal. Chim. Acta*, **411**, 1.
Klotz A., Brecht A., Barzen C., et al., 1998. *Sens. Actuators B*, **51**, 181.
Köhler G. and Milstein C., 1975. *Nature*, **256**, 495.
Krämer P.M. and Schmid R.D., 1991. *Pest. Sci.*, **32**, 451.
Krämer P.M., Baumann B.A. and Stoks P.G., 1997. *Anal. Chim. Acta*, **347**, 187.
Krämer P.M., Goodrow M.H. and Kremmer E., 2004a. *J. Agric. Food Chem.*, **52**, 2462.
Krämer P.M., Kremmer E., Forster S. and Goodrow M.H., 2004b. *J. Agric. Food Chem.*, **52**, 6394.
Krämer P.M., Kremmer E., Weber C.M., Ciumasu I.M., Forster S. and Kettrup A.A., 2005. *Anal. Bioanal. Chem.*, **382**, 1919.
Krämer P.M. and Räuber C., 2007. The Pittsburgh Conference for Pittcon 2007, Chicago, USA, 27 February–1 March 2007.
Krämer P.M., Weber C.M., Kremmer E., in: I.R. Kennedy, K. Solomon, S. Gee, A. Crossan, S. Wang, F. Sanchez-Bayo (eds), 2007a. *Rational Environmental Management of Agrochemicals: Risk Assessment, Monitoring and Remedial Action*. ACS Symposium Series 966, American Chemical Society, Washington, DC.
Krämer P.M., Martens D., Forster S., Ipolyi I., Brunori C., Morabito R., 2007b. *Anal. Bioanal. Chem.*, **387**, 1435.
Krotzky A.J. and Zeeh B., 1995. *Pure & Appl. Chem.*, **67**, 2065.
Mallat E., Barcelo D., Barzen C., Gauglitz G. and Abuknesha R., 2001. *Trends Anal. Chem.*, **20**, 124.
Marco M.-P., Gee S. and Hammock B.D., 1995. *Trends Anal. Chem.*, **14**, 341.
Martens D., Pantiru M., Forster S. and Kettrup A., 2004. *2nd Asian International Conference on Ecotoxicology and Environmental Safety*, Songkla, Thailand, 26-29 September 2004.
Mauriz E., Calle A., Manclus J.J., Montoya A. and Lechuga L.M., 2006. *Anal. Bioanal. Chem.*, **387**, 1449.
Meusel M., Trau D., Katerkamp A., Meier F., Polzius R., Cammann K., 1998. *Sens. Actuators B*, **51**, 249.
Oroszlan P., Thommen Ch., Wehrli M., Duveneck G. and Ehrat M., 1993. *Anal. Methods Instrum.*, **1**, 43.
Sheedy C., MacKenzie C.R. and Hall J.C., 2007. *Biotechnol. Adv.*, **25**, 333.
Shim W.-B., Yang Z.-Y., Kim J.-Y., et al. 2006. *J. Agric. Food Chem.*, **54**, 9728.
Tijssen, P., 1985. In: R.H. Burdon and P.H. Knippenberg (general eds), *Laboratory Techniques in Biochemistry and Molecular Biology*. Elsevier, Amsterdam, New York, Oxford.
Tschmelak J., Proll G. and Gauglitz G., 2004. *Anal. Chim. Acta*, **519**, 143.
Tschmelak J., Proll G. and Gauglitz G., 2005a. *Talanta*, **65**, 313.
Tschmelak J., Proll G., Riedt J., et al. 2005b. *Biosens. Bioelectron.*, **20**, 1499.
Tschmelak J., Proll G., Riedt J., et al. 2005c. *Biosens. Bioelectron.*, **20**, 1509.
Watts C.D. and Hegarty B., 1995. *Pure & Appl. Chem.*, **67**, 1533.
Weller M.G., Schuetz A.J., Winklmair M. and Niessner R., 1999. *Anal. Chim. Acta*, **393**, 29.
Willard D., Proll G., Reder S. and Gauglitz, G., 2003. *Environ. Sci. Poll. Res.*, **10**, 188.
Zhang C., Zhang Y. and Wang S., 2006a. *J. Agric. Food Chem.*, **54**, 2502.
Zhang X., Martens D., Krämer P.M., Kettrup A.A. and Liang X., 2006b. *J. Chromatogr. A*, **1133**, 112.
Zhang X., Martens D., Krämer P.M., Kettrup A.A. and Liang X., 2006c. *J. Chromatogr. A*, **1102**, 84.

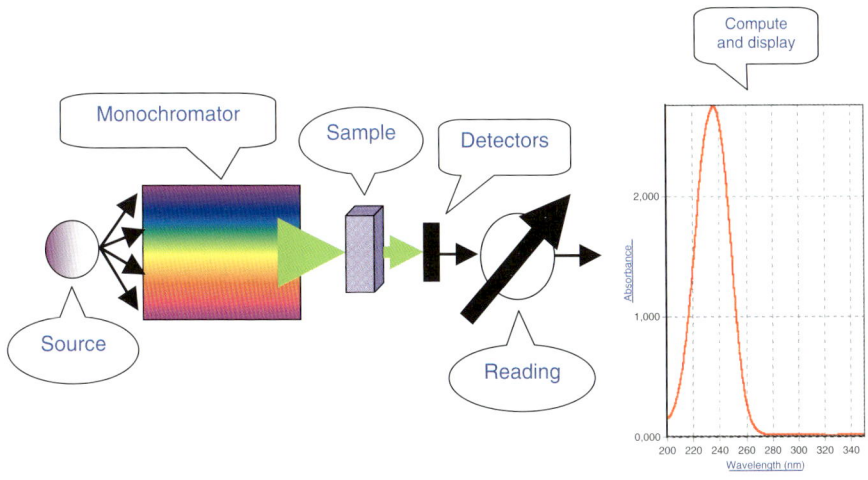

**Plate 1** Design of the optical system used in case of portable instrument

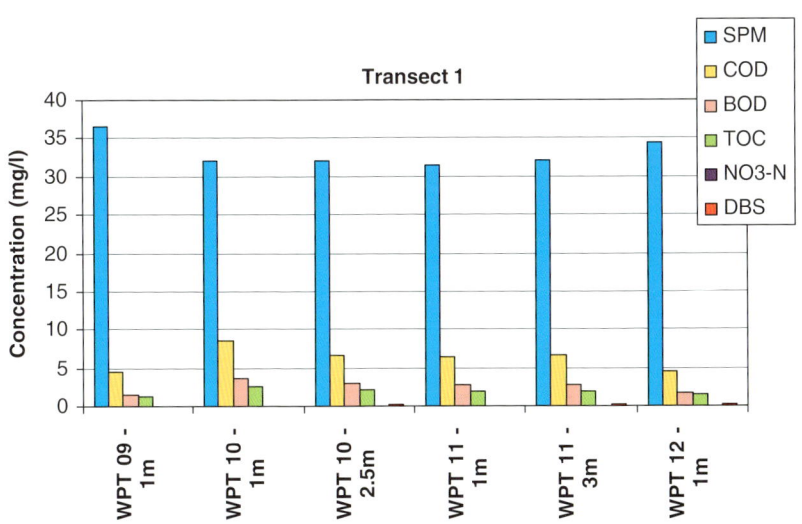

**Plate 2** On site PASTEL UV measurements – Lake Stropu

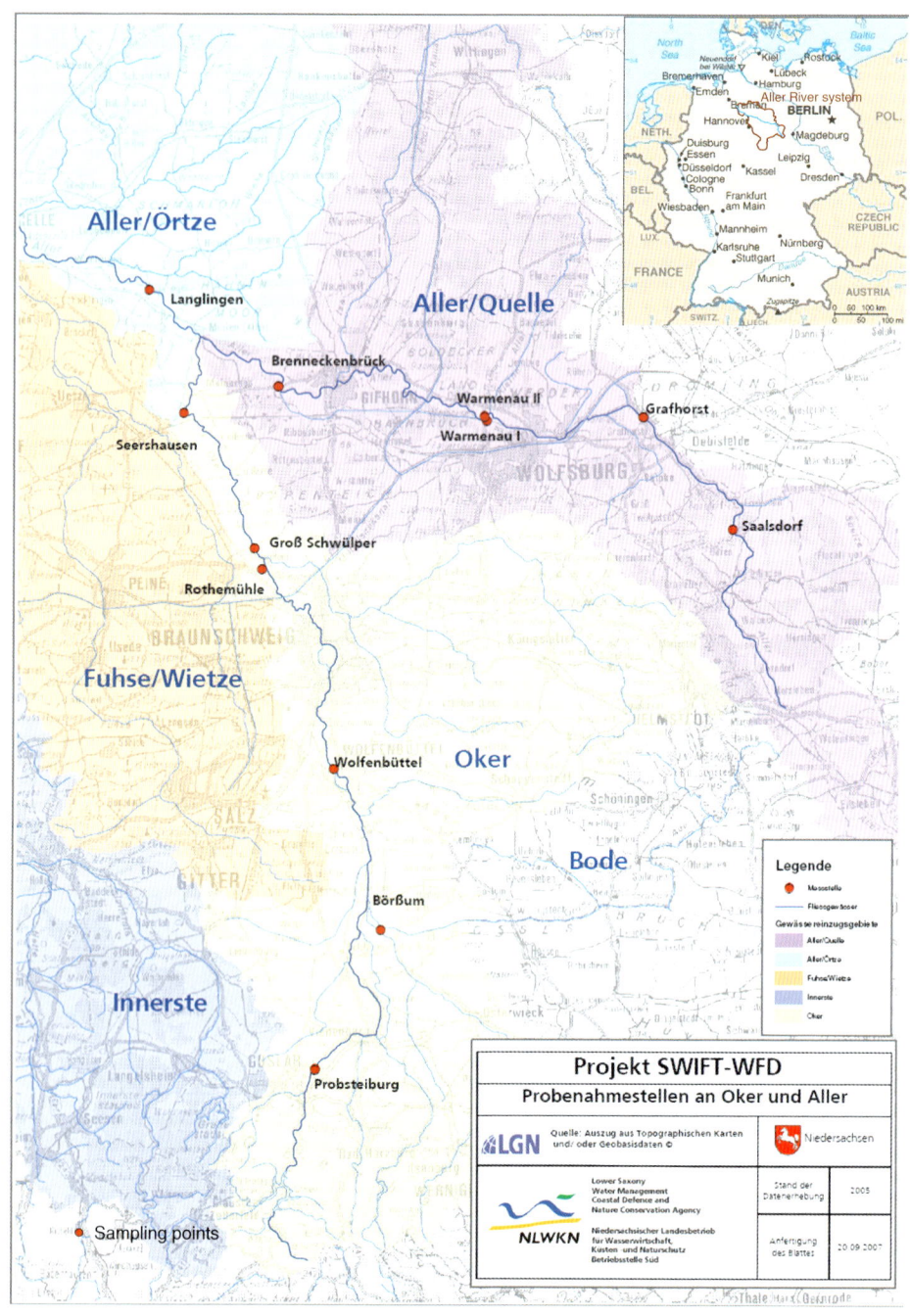

**Plate 3** Sampling point's location

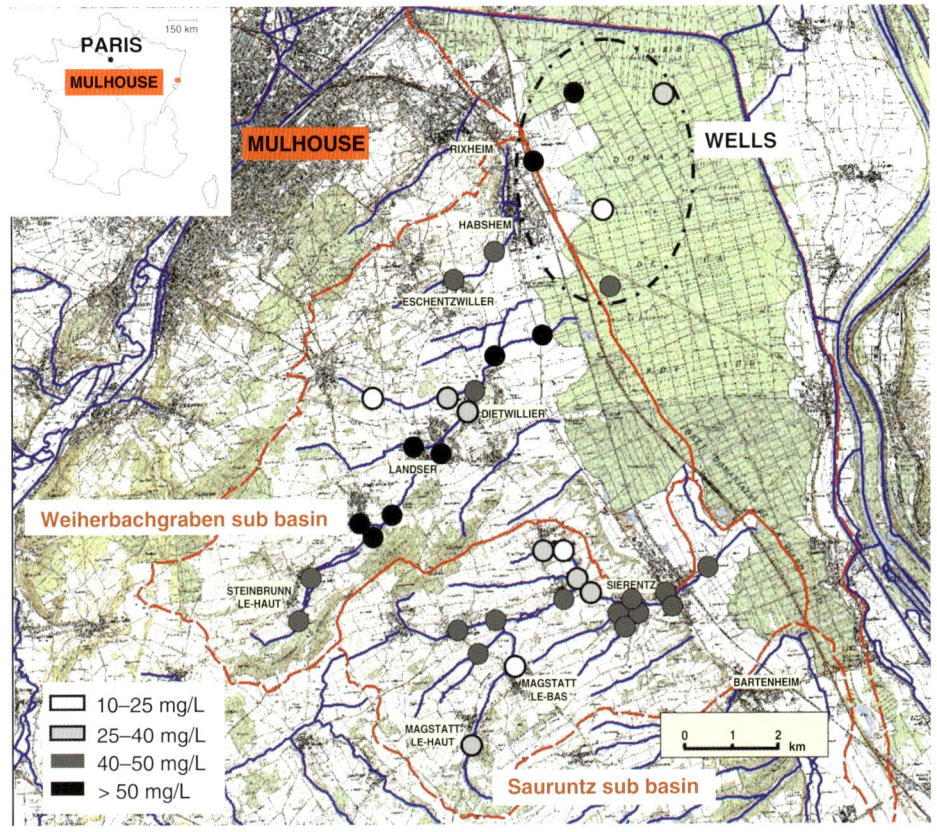

**Plate 4** Nitrate concentration estimated by Pastel UV in the Hardt catchment (Roig *et al.*, 2007)

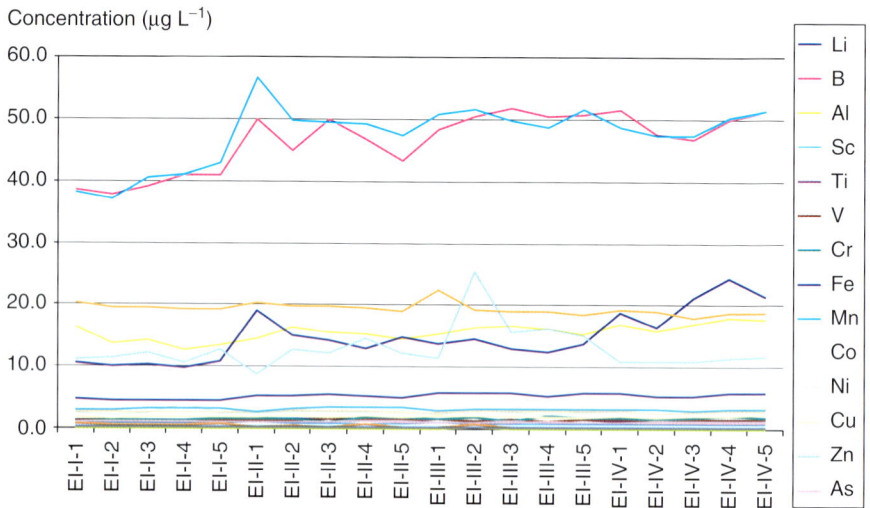

**Plate 5** Variations in concentrations (µg L$^{-1}$) of trace elements along the 4 transects in the Meuse River

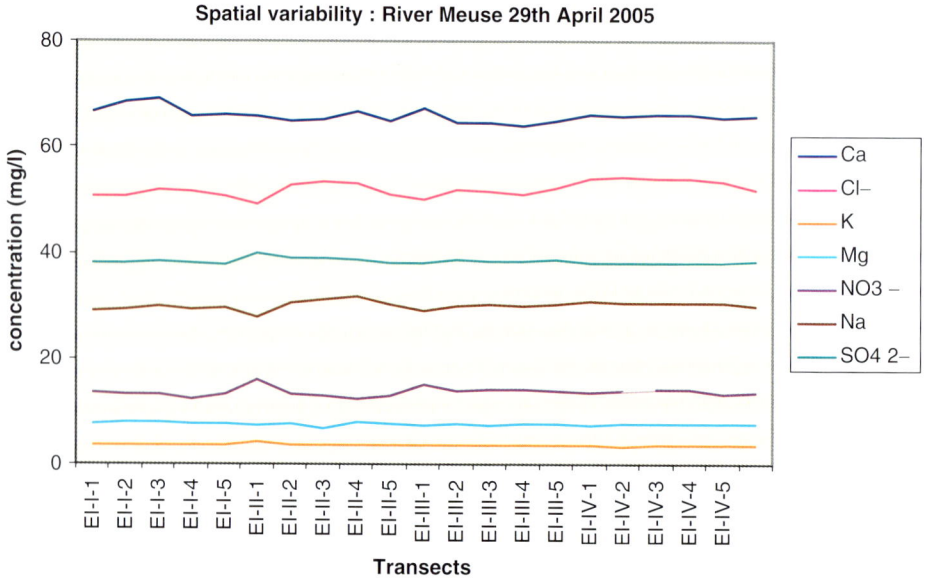

**Plate 6** Spatial variability in the concentrations (mg L$^{-1}$) of the major components

# 3.4
## Biomolecular Recognition Systems for Water Monitoring

Benoit Roig, Ingrid Bazin, Sandrine Bayle, Denis Habauzit and Joel Chopineau

3.4.1 Introduction
3.4.2 Recognition Phase
    3.4.2.1 Receptor Proteins
    3.4.2.2 Promoter Interaction
3.4.3 Transduction Phase
    3.4.3.1 Bioreporter Organisms
    3.4.3.2 Reporter Systems
    3.4.3.3 Bioreporter Technology
3.4.4 Conclusion
References

## 3.4.1 INTRODUCTION

For many years, the analysis of substances in water samples has been very challenging because of the increasing number of substances found in different water bodies at very low concentrations (trace levels). The identification and monitoring of such organic microcontaminants is usually performed by high performance liquid chromatography coupled to a mass spectroscopy detector and often tandem mass spectrometry (LC–MS/MS), which provides additional selectivity and sensitivity. Although there have been improvements in the performance of these analytical techniques, there is still a need to preconcentrate the samples in order to achieve the sensitivity necessary to detect pollutants in the ultra trace (within the ng $L^{-1}$ or low µg $L^{-1}$ range) in which they are found in the environment (Coestier *et al*., 2007). Consequently, the routine application of these techniques is limited by the time required for an individual determination, operational difficulties associated with the quantification of very low trace

*Rapid Chemical and Biological Techniques for Water Monitoring*    Edited by Catherine Gonzalez, Philippe Quevauviller and Richard Greenwood
© 2009 John Wiley & Sons, Ltd

Table 3.4.1  Biological techniques for water monitoring

| | | | |
|---|---|---|---|
| Bioindicator BEWS Bioassays, Biosensors | Biological recognition element (BRE) | Organisms, tissues, cells, organelles, membranes, nucleic acids, cofactors Protein (enzyme, antibody, receptor) | Byfield and Abuknesha, 1994; Hock et al., 2002; |
| | Detection technique | Macroscopic observation (growth, speed, movement ...), Optical method (fluorescence, evanescence SPR, absorbance), Electrochemical methods (amperometric, conductimetric, potentiometric), piezoelectric acoustic, colorimetric, mechanical, thermal methods | Kramer and Hock, 2003; Subrahmanyam et al., 2002; Baeumner, 2003; Nakamura and Karube, 2003 |

levels of compounds and the costs of the analysis. Alternative methods are thus needed to overcome these difficulties.

For several years, there has been an increasing interest in applying biological sciences in environmental monitoring due to the potential of biological systems to react to a particular stress generated by of a xenobiotic present in an environmental medium. Indeed, biological systems have been found to be the most sensitive and promising basis for dealing with the presence of trace of components or the continuous emergence of new substances that are of concern. Several biological systems corresponding to different levels of organization have been used as biological recognition elements (BRE) in water monitoring, and examples include whole organisms (fish; mouse; *Daphnia*), cells (immortalized mammalian cell lines; bacteria; yeasts), and biomolecules (nucleic acids; proteins, including antibodies). Bioindicators, including biological early warning systems (BEWS), bioassays and biosensors, are based on the recognition of a contaminant (by one of these BREs) and the transduction of the interaction into an observable (with a detection system) signal have emerged as useful aids in monitoring water quality, and continue to be developed for environmental applications, and work in this area has been well reviewed (Table 3.4.1).

In contrast with chemical analytical techniques (e.g. chromatography linked to a sensitive detector) that generally produce quantitative data, biological tools can be used for both quantitative and qualitative purposes in environmental monitoring, and can be used to detect the concentration of a contaminant, or just its presence or absence. They can also provide information on the toxicological and/or ecological effects of the contaminant. In this context, different BREs have been employed in a range of different applications (Table 3.4.2):

- Observations, normally over periods of days, of macroscopic symptoms corresponding to phenotypic changes, such as growth, mobility, or reproduction, in whole organisms can, depending on the assay, reveal the consequences of short-term (acute) or long-term (chronic) exposure to a pollutant.

# Introduction

**Table 3.4.2** Response characteristics of biological recognition elements

| Biological recognition elements | Highlighted biological responses | Exposure/ response time | Sensitivity/ specificity |
|---|---|---|---|
| Biomolecules (proteins, nucleic acids, enzymes, antibodies...) | Molecular interaction | Several min | + |
| Cells (eukaryotic, prokaryotic) | Molecular target expression | Several hours | |
| Whole organisms (fish, *Daphnia*) | Phenotypic changes (growth, mobility damage...) | Several days | − |

- More rapid responses to a stressor are observed in assays based on isolated tissues or cells, or microorganisms where contact with a pollutant is identified through an interaction with a molecular target and the expression of biomarkers.

- Assays using isolated biomolecules provide a very rapid measurement since they depend on the direct exposure of a biomolecular recognition site to the pollutant and the production of a measurable response to that interaction between the pollutant and the biomolecule. The delays associated with uptake by an organism, or part of an organism (cell or tissue), and distribution to a site of action, and the development of a measurable symptom of exposure are absent from these biomolecular assays.

Systems that use biomolecules as a recognition element have proved to be the most promising biological tools for use in monitoring water quality. Indeed, these biosensors can detect the presence of a contaminant by specific interaction between the chemical and a recognition site on the biological molecule in a very short time after contact. It can be:

- a specific interaction between an enzyme and its substrate,
- a recognition between an antibody and its antigen,
- a hybridization of a nucleic acid and its complementary sequence,
- a binding of a ligand to a receptor.

Biological recognition elements are chosen because they are able to identify a category of toxic effect such as cellular toxicity, genotoxicity, immunotoxicity, neurotoxicity or endocrine disruption. Table 3.4.3 shows some examples of biomolecular recognition systems used in biosensors and bioassays for environmental monitoring (based on Bilitewski *et al.*, 2000; Brenner-Weiß and Obst, 2003).

The detection of a pollutant using biosensors that use receptor proteins as recognition elements involves two steps (Figure 3.4.1):

- interaction between the protein and its target = RECOGNITION,

**Table 3.4.3** Biomolecular recognition components and their applications

| Biomolecule | Example | Targeted pollutant | Observed effect |
|---|---|---|---|
| Enzymes | Acetylcholine Esterase | Organophosphorus and carbamic compounds | Neurotoxicity |
| Ion channels | $Ca^{2+}$ channel | Algal toxins (maitotoxin) | Neurotoxicity |
| | $Na^+$ channel | Algal toxins | Neurotoxicity |
| Transport proteins | SHBG, CBG, TBG* | Endocrine disrupters | Growth, development, reproduction |
| Receptors | Estrogen receptor | Phytoestrogens, endocrine disrupters (e.g. DDT, nonylphenol) | Growth, development, Reproduction |
| | Androgen receptor | Endocrine disrupters (e.g. tributyltin) | |
| Electron carriers | $Q_B$ protein (protein of plant photosystem II) | Photosynthesis II herbicides (e.g. *s*-triazines, phenylureas), phototoxins | Photosynthesis (cyano-bacteria, algae, plants) |
| DNA | DNA double strands | Intercalating polycyclic aromatics (e.g. ethidium, acridine, caffeine) | Genotoxicity |
| Cytoskeleton | Tubulin | Colchicine, taxol, antitubulin herbicides (e.g. trifluralin, oryzalin) | Cytotoxicity (plants) |
| Ribosomes | rRNA (ricin) | Ribotoxins (ricin, abrin, Shiga toxin) | Cytotoxicity |

*SHBG, sex hormone binding globulin; CBG, corticosteroid binding globulin; TBG, thyroxine binding globulin

- characterization of this interaction and production of a specific signal = TRANSDUCTION + EMISSION.

Whatever the mode of transduction in-vivo/in cellulo (with bioreporter organisms) or in vitro (bioreporter technology), the recognition phase between the target and a receptor protein remain the same (for a given application).

In this chapter, we describe methods using receptor proteins as recognition elements that can be used in water monitoring. We address different aspects from the properties of the receptors themselves to their linkage with engineering components in several kinds of biosensors.

## 3.4.2 RECOGNITION PHASE

Several protein systems have been used as recognition elements for the detection of environmental contaminants, and some examples are:

- a positive regulatory protein (cadC) that is the regulatory protein of the cadmium resistance operon (Endo and Silver, 1995),

*Recognition Phase* 179

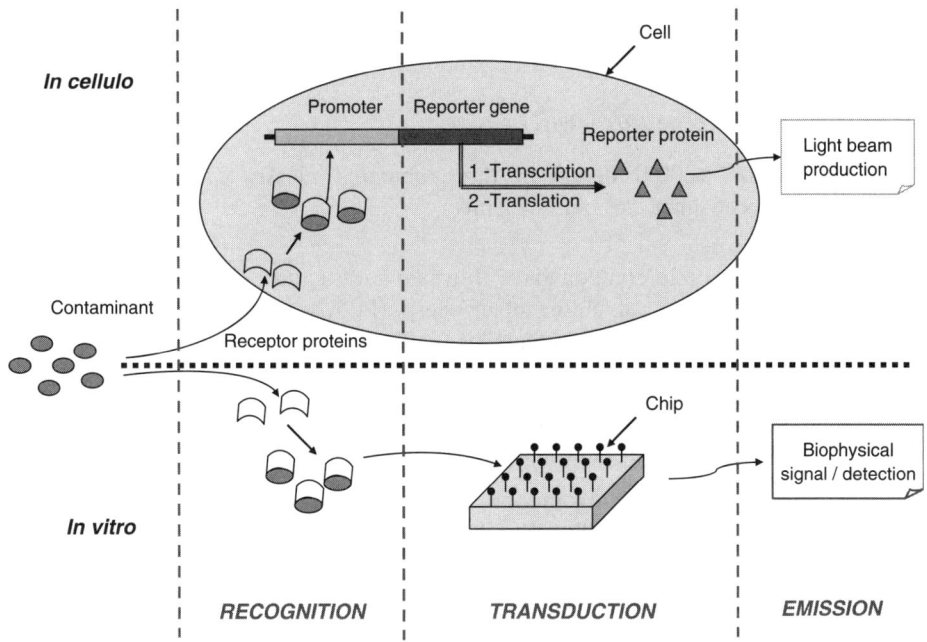

**Figure 3.4.1** Principle of biosensors using protein receptor as recognition element

- a negative regulatory protein (AsrR) which mediates the induction of resistance by arsenic in *Staphylococcus aureus* (Rosenstein *et al.*, 1994),
- a nuclear receptor (Estrogen Receptor (ER)) used in the Yeast Estrogen Screen (Beck *et al.*, 2006) or in human cell lines (MCF7 ERE-CAT) (Matsumura *et al.*, 2005).

### 3.4.2.1 Receptor Proteins

Numerous interactions of environmental pollutants with protein receptors have been described and include either direct binding of xenobiotics to receptor or indirect effects mediated via modulation of associated signalling pathways. Most of identified receptors used in this context are nuclear receptors involved in regulation of a wide range of physiological functions in eukaryotic organisms including cell growth and proliferation, differentiation or maintenance of homeostasis. These receptors normally function by binding a specific ligand that modifies their structural conformation, after which the modified system transfers to the nucleus, where it binds to a corresponding responsive element of DNA and triggers gene expression. Nuclear receptors can be activated by low molecular weight environmental contaminants such as dioxins, polychlorinated biphenyls (PCBs), polycyclic aromatic hydrocarbons (PAHs), phthalates, and xenoestrogens, (Hilscherova *et al.*, 2000; Gray, 1998). This interaction between these receptors and environmental contaminants can be used to design biosensors for use in monitoring applications. Aryl hydrocarbon Receptor (AhR) and Estrogen Receptor (ER) have been the most intensely investigated receptor systems (Janosek et al., 2006).

## Aryl hydrocarbon receptor (AhR)

Hilscherova et al. (2000) described three potential classes of compounds with dioxin-like properties that can bind to AhR:

- hydrophobic aromatic compounds with a planar structure such as PCB congeners and PCDD/PCDFs; polychlorazobenzenes, (PCABs), polychlorazoxybenzenes (PCAOBs), polychlorinated naphthalenes (PCNs), and several high molecular weight polyaromatic hydrocarbons (PAHs);
- compounds with a specific stereochemical configuration; such as polyhalogenated (chlorinated, brominated, fluorinated), anthracenes (PCANs), fluorenes (PCFL);
- weak AhR ligands including indoles, heterocyclic amines, some pesticides, and drugs (these tend to be rapidly degraded by the induced detoxification enzymes).

The determination of dioxin related compounds is based on their action on the AhR signal transduction pathway. The mechanism comprises a number of steps: (1) the contaminant binds to the AhR, (2) the complex is translocated to the nucleus of the cell, (3) where it induces the transcription of a number of genes, and subsequently, (4) the production of proteins including cytochrome P-450 (CYP), aldehyde dehydrogenase, NADPH-quinone-oxidoreductase and glutathione-S-transferase. The replacement of the genes coding for these proteins by a reporter gene allows the interaction between the contaminant and the receptor protein to be detected, and this is a basis for the development of biosensors.

*Several transformed cells have been produced using AhR coupled to various reporter genes, the most frequently employed reporter system is firefly luciferase (Murk et al., 1996) due to in particular its sensitivity. These cellular systems have been employed for the determination of the dioxin-like potential of pure substances, including PCDD/Fs, PCBs (Murk et al., 1996), PAHs (Machala et al., 2001a), PCNs (Blankenship et al., 1999), as well as for detecting dioxin-like effects in a range of environmental samples: sediments (Murk et al., 1996; Machala et al., 2001b), air particle matters (Hamers et al., 2000) and biota (Murk et al., 1998). Detailed discussions of the advantages and limitations of in vitro assays for AhR-mediated effects can be found in some specialized reviews (Hilscherova et al., 2000; Behnisch et al., 2001a, 2001b).*

## Estrogen receptor (ER)

ER is a transcription factor belonging to the steroid receptor family. At least two structurally different subtypes (ERα and ERβ) have been described in mammals, and these monomers form homo or heterodimers in cells. Another subtype (ERγ) may possibly exist in fish (Drummond et al., 2002). ER has been used as recognition element for endocrine disruptors (exogenous chemicals, or mixtures of compounds, that are able to modify the functioning of the female sex steroid modulated endocrine system, and consequently cause adverse health effects in an intact organism or its progeny or in (sub)populations). They are characterized by their ability to interact directly or indirectly with ER by a mechanism similar to that involved in the action of the natural hormone.

Endocrine disruptors can impact on the normal functioning of the system by three modes of action.

- Agonist action: mimicking the action of endogenous hormone by binding to the receptor, activating it, and subsequently activating of the ERE (Estrogen Responsive Element) in nuclear DNA;
- Antagonist action: by binding the receptor but without activating it. In this case, they block the receptor and stop its functioning;
- Nonreceptor based mechanisms: by causing modulations of tissue levels or activities of enzymes participating in the biosynthesis or catabolism of hormones.

A wide range of compounds including natural products, pharmaceuticals and industrial chemicals has been shown to have estrogenic or anti-estrogenic activity (Coldham *et al*., 1997; Mantovani *et al*., 1999; Fang *et al*., 2000; Combes, 2000; Legler *et al*., 1999, 2002; Gutendorf and Westendorf, 2001; Vondracek *et al*., 2002; Machala *et al*., 2004). Due to the complexity of the biochemical toxicity mechanisms of contaminant-induced nuclear receptor modulations, it is possible for contaminants to impact upon this system in a variety of ways:

- nonphysiological binding to ER that can activate its function (e.g. phthalates, Nakai *et al*., 1999; Balaguer *et al*., 1999);
- competitive binding to active site without activation of ER, thus inhibiting the function of natural estrogens and acting as an antihormone (e.g. PCBs, Moore *et al*., 1997);
- modulation of upstream signalling without binding to ER (e.g. ethanol, Combes, 2000).

A further mode of action of endocrine disruptors that has been reported is an ER-independent 'anti-estrogenity' mediated by AhR ligands such as TCDD and/or PAHs. This indicates substantial cross-talk between signalling pathways of different nuclear receptors (Zacharewski *et al*., 1991). Table 3.4.4 reports mode of action of the main ER sensitive environmental contaminants.

Several bioassays have been developed for the assessment of ER/ligand interaction. Most of these are cellular assays and are based on the induction of specific transcripts and proteins controlled by ER-activities (especially those focussing on estrogenity). Human cell lines, have been used and in particular, MCF-7 (Human breast carcinoma (ERα)), (Lebail *et al*., 1998; Legler *et al*., 1999; Hilscherova *et al*., 2000; Kuruto-Niwa *et al*., 2002). These have been transfected with a reporter gene under the control of ERE. HeLa (breast carcinoma (Erα, ERβ)) (Balaguer *et al*., 1999) and BG-1 (ovarian adenocarcinoma (ERα)) (Rogers and Denison, 2000) cell lines have also been proposed (as reviewed by Giesy *et al*. (2002)) as suitable for use in the determination of

**Table 3.4.4** Examples of endocrine disrupting compounds that interact with ER, and their mode of action (adapted from Hilscherova et al., 2000)

| Compounds | Mode of action |
| --- | --- |
| **Phytoestrogens:** | |
| Coumestrol, Genistein | ER agonist |
| **Pharmaceuticals:** | |
| Tamoxifen | ER antagonist or agonist |
| Hydroxytamoxifen | antiestrogenic activity |
| Ethynylestradiol | ER agonist |
| **Additives:** | |
| Parabens | ER agonist |
| **Insecticides:** | |
| DDT, DDD, Endosulfan, Dieldrin, Lindane, Chlordane, | |
| Methoxychlor | ER agonist |
| Carbamate insecticides | ER agonist – after metabolism endocrine modulators, non-ligand binding |
| **Herbicides:** | |
| Atrazine | estrogen, antiestrogen |
| Simazine | antiestrogen |
| Alachlor | ER agonists |
| **Phtalates** | ER agonists |
| **Alkyl phenols:** | |
| Octyl and nonyl phenol | ER agonists |
| Bisphenol A | |
| **PCBs** | ER agonists or antagonists or other mechanism – depending on the degree of substitution |
| **PAHs** | ER agonists – estrogenic, antiestrogenic – various mechanisms |

xenoestrogens in environmental samples. Transfected yeast cells have also been used for the detection of estrogenic activities (Lascombe et al., 2000; Jungbauer and Beck, 2002). Whilst the yeast-based model has some advantages including ease of culture, and genetic manipulations, and the use of steroid free media, it also has some drawbacks. A number of factors can affect the reproducibility of this assay, and in particular variability between strains (Combes, 2000; Zacharewski, 1997).

*Other receptor proteins*

Other receptor proteins such as androgen receptor (AR), retinoid receptor (RR), thyroid receptor (TR), pregnane X receptor (PXR) can interact with environmental xenobiotics.

However, for environmental applications, little information is available concerning these receptors as reviewed by Janosek *et al*. (2006).

Organic contaminants are known to interact directly with AR in a range of different pathways: for example DDT, its metabolites and some fungicides bind with AR and act as competitive inhibitors (Kelce *et al*., 1997; Gray *et al*., 1997), whereas bisphenol, hydroxyphenol (Paris *et al*., 2002), PAHs (Kizu *et al*., 2000; Vinggaard *et al*., 2000) and PCBs (Schrader and Cooke, 2003) seem to be involved in anti-androgenic activity through different mechanisms. There are also examples of mechanisms that activate endocrine modulation by more complex physiological pathways involving interactions between the receptors involved, for instance the action of PXR and estrogen receptors that was recently investigated by Minf *et al*. (2007). Here different classes of estrogenic compounds (including natural, and synthetic estrogens, alkylphenols, and phthalates), known for their affinity for ER have been investigated to determine whether they interact with PXR. A significant proportion (54%) of compounds with estrogenic activity or able to bind with ER was found to comprise PXR activators: in particular, antiestrogens, mycoestrogens and phthalates. It is important to realize in this context that the recognition of an analyte by a receptor protein does not always produce a cellular response.

### 3.4.2.2 Promoter Interaction

In many cases, the hormone receptor proteins stimulate the transcription by binding to DNA sequences called promoter. Promoters are specific DNA sequences that are recognized by specific proteins (transcription factors) that control the transcription of genes that code for proteins that are for instance enzymes involved in cellular metabolism or in detoxification processes (degradation of toxic compounds). Activation mechanisms of promoters are complex and multiple, and often respond to a range of different kinds of stress or chemical compounds. Indeed, activation or inhibition of promoters generally depends on regulatory proteins that modulate transcription by interaction at specific sites on these DNA sequences. These transcription factors can be activated or not by the presence of targeted compound that modify for instance their conformation. Regulatory proteins and their promoters are key elements for the specificity of a monitoring system, and an adapted promoter should be chosen according to toxic effect to be assayed.

Some molecular recognition systems can contain *nonspecific promoters* that are not inducible by a target compound but that respond to a cellular stress. An example is provided by the detection of damages caused to proteins by the presence of a pollutant, and here a range of different promoters (e.g. *gprE, dnaK, Ion, ibp*) can be used. Damages related to DNA are identified by the promoters *recA, uvrA, umuC* (Vollmer *et al*., 1997; Min *et al*., 1999). Oxidative stress (hydrogen peroxide or superoxide) can be detected by using *katG*, and *micF* promoters (Belkin *et al*., 1996; Lee and Gu, 2003). Control of variables like growth limitation, or deficiencies in amino acids, phosphate, nitrogen and carbon is achieved through the *uspA, his, phoA, glnA*, and *lac* promoters (Dollard and Billard, 2003).

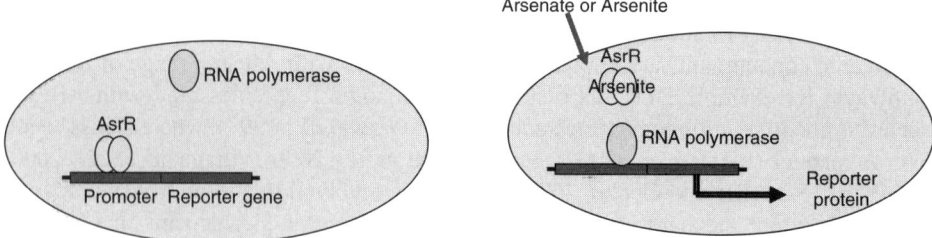

**Figure 3.4.2** Scheme of AsrR control DNA element coupled to a reporter gene. Adapted from Stocker *et al.*, 2003

Biosensors can be also designed from a variety of *specific promoters* that are particularly sensitive to particular chemicals, including:

- inorganic chemicals such as aluminium *(fliC)*, arsenic *(ars)*, cadmium *(cad)*, chromium *(chr)*, mercury *(mer)*, zinc *(smtA)* (Tauriainen *et al.*, 1998; 2000; Corbisier *et al.*, 1999)

- organic compounds such as alkanes *(alkB)*, toluene and derivatives (xylene) *(tol, xyl)* (Paitan *et al.*, 2004) chlorobenzene *(fcbA)*, chlorocatechols *(clc)*, PCBs *(orf)*, naphtalene *(nah)*, (Köhler *et al.*, 2000; Diaz and Prieto, 2000), estrogenic compounds (ERE).

The design of sensors such as those cited above is generally based on knowledge of the operon that controls gene expression. For example, in a luminescent bacterial sensor for cadmium, a positively controlled operon, was constructed by inserting the regulation unit of the cadmium resistance operon from the *Staphylococcus aureus* plasmid pI258 *cadA*. The plasmid carries the firefly luciferase reporter gene under the control of the *cad* promoter and the *cadC* gene of the *cadA* operon (Tauriainen *et al.*, 1998). Other biosensors are based on negative regulation of an operon by a transcriptional repressor, for example the bacterial defence mechanism against arsenic. The arsenic sensing protein, called ArsR, in the absence of the pollutant, binds to a specific element of DNA and prevents transcription of the arsenic defence genes (Daunert *et al.*, 2000).

In presence of arsenic, ArsR binds to the compound and transcription of arsenic defence genes is possible. The gene coding for ArsR protein, the DNA binding site for ArsR and the reporter gene are linked together (Stocker *et al.*, 2003) (Figure 3.4.2).

When receptor systems are used as molecular recognition elements, the operating sequence of the device is similar to those described for AhR or ER, and generally functions within a transfected cell (mainly human cells or yeast) with a reporter gene as the expression system (bioreporter organisms). Janosek *et al.*, (2006) reported a table of the main human cell lines used in bioassays using AR, TR and RR recognition systems. Other transcription systems using bioreporter technology are now emerging and overcome the disadvantages associated with using live higher organisms.

## 3.4.3 TRANSDUCTION PHASE

### 3.4.3.1 Bioreporter Organisms

Living microbial cells can be used as sensitive sensors of environmental stress providing that they contain a unit that recognizes the analyte of interest, and regulates the expression of a specific cell response. In this area, cell based biosensors have been developed using genetically engineered microorganisms (GEMs) that can recognize and report the presence of specific environmental pollutants. Recombinant DNA techniques have been used to generate various specific microbial or mammalian cell sensors. Various biological systems have been used for this purpose, but particularly cells such as bacteria, and mammalian cell lines, because of their large population, rapid growth rate and low cost. Several reviews have addressed different aspects of the use of such genetically engineered microorganisms as environmental bioreporters as reported by Belkin (2003).

Biosensors that incorporate bioreporters are based on the same concept as bioreporter organisms (Figure 3.4.1) and incorporate a plasmid in which genes coding for a reporter protein are placed under the control of a promoter that recognizes the analyte of interest. The gene is turned on (transcripted) when the target agent is present in the cell's environment. The promoter used is a normal bacterial cell promoter that is linked to genes that code for proteins that help the cell in either combating or adapting to the agent of interest. In the case of a bioreporter, these genes, or portions thereof, have been removed and replaced with a reporter gene. It is important that the biosensor is free of any gene expression or activity similar to the desired gene expression or activity that is being measured in order to ensure that the response is specific for the chemical that is being monitored, and that there is no chance of misinterpretation (Daunert *et al.*, 2000).

### 3.4.3.2 Reporter Systems

Several stress responses to exposure to environmental contaminants have been studied and are exemplified by the SOS response (Quillardet *et al.*, 1989; Sassanfar *et al.*, 1990; Kenyon *et al.*, 1998), the heat-shock response (Demple and Harrison, 1994; Lee *et al.*, 2003), and the oxidative damage response (Hidalgo *et al.*, 1994; Asad *et al.*, 1998). These responses can be detected in transfected cells by using reporter genes that encode for mechanisms producing detectable cellular responses (reporter proteins). It is the latter that determine the sensitivity and the detection limits of biosensors. To generate an optimal response in a biosensor, the reporter gene must have an activity that is reliable and that reflects the amount the analyte of interest or the extent of the chemical or physical change that is caused by its presence.

The most common reporter proteins (coded by reporter genes) employed in biosensor systems include firefly luciferase (*lucFF* gene) (Justus and Thomas, 1999), bacterial luciferase (*luxCDABE* gene) (Belkin *et al.*, 1996; Vollmer *et al.*, 1997; Min *et al.*, 1999), Green Fluorescent Protein (*GFP* gene) (Justus and Thomas, 1999, Kostrzynska *et al.*, 2002), β-galactosidase (Kozubek *et al.*, 1990; Nunoshiba *et al.*, 1991; Quillardet *et al.*, 1993). Other ones such as: β-glucuronidase (Saidi *et al.*, 2007) and alkaline

**Table 3.4.5** Advantages and drawbacks of reporter proteins used in biosensor systems (adapted from Daunert et al., 2000)

| Reporter protein | Advantages | Drawbacks |
|---|---|---|
| Firefly luciferase | High sensitivity. High linearity. No endogenous activity in mammalian cells. | Need addition of substrate and ATP in an aerobic medium. |
| Bacterial luciferase | High sensitivity. Needs no addition of substrate. No endogenous activity in mammalian cells. | Heat sensitivity (>30°C). weak linearity |
| GFP | Auto fluorescent. No homologue endogenous activity in most systems. Stable at biological pH. | Moderate sensitivity. Need post-transductional modification. Auto fluorescent for some compounds (aromatic amino acids). Potentially cytotoxic for some cellular types. |
| β-galactosidase | Sensitive and stable. Correct linearity. Applicable in anaerobic medium. | Endogenous activity. Need addition of substrate |

phosphatase (Fahnrich et al., 2003; Paitan et al., 2004) have been also described. Choice of the reporter depends on:

- the endogenous activity of the cell lines employed,
- the gene expression level,
- the transfection efficiency,
- the detection method.

Table 3.4.5 summarizes the advantages and drawbacks of the main reporter systems.

## Firefly luciferase (from Photinus pyralis; lucFF gene)

The reporter gene is the *luc* operon from the firefly *Photinus pyralis*. This operon codes for an enzyme, luciferase, that catalyses the reaction between D-luciferin (the substrate), and oxygen, and requires ATP as a cofactor. This reaction produces the emission light that provides the measure of enzyme activity. The oxidation of one molecule of luciferin involves one ATP molecule and releases one photon (a stoichiometric reaction) (DeLuca and McElroy, 1974; Wood et al. 1989). The reaction can be summarized as follows:

$$\text{Luciferin} + \text{ATP} + O_2 \xrightleftharpoons{Mg^{2+}} \text{AMP} + CO_2 + \text{oxyluciferin} + \text{PPi} + h\nu \ (565 \ nm)$$

A high concentration of AMP can inhibit the luminescent reaction, but the yield of this reaction mainly relies on the concentrations of the substrate (luciferin) and the cofactor (ATP). The visible light produced in this reaction can be measured with a luminometer and optical fibres (Brian et al., 2003). The reaction has a high level of sensitivity. A further advantage of this reporter system is that there are mutant variants available that produce enzymes that produce a range of colours of light from green to red (Daunert et al., 2000). Since firefly luciferase is of eukaryote origin it was used particularly in the development of eukaryote biosensors and for the measurement of ATP. However, it has also become widely used as reporter in bacterial biosensor (Tsein, 1998).

## Bacterial luciferase (from Vibrio fischeri; luxCDABE operon)

Most of bacterial biosensors are based on the operon *luxCDABE* that codes for the bacterial luciferase founded in the marine bacteria *V. fischeri* and *V. harveyi*, and for an essential aldehyde substrate that would otherwise have to be supplied exogenously. The cluster *luxAB* cassette codes for the luciferase whereas *luxCDE* encodes a fatty acid reductase complex. The latter enzymes are responsible for the synthesis of the long-chain aldehyde that is required as substrate in the bioluminescence reaction (Meighen and Dunlap, 1993; Hakkila et al., 2002). Luciferase catalyses the oxidation reaction of flavin mononucleotide (FMNH2). A long-chain (7 to 16 carbons) aldehyde is reduced in presence of oxygen by the aldehyde reductase. The outcome of the bioluminescent reaction can be expressed as follows:

$$FMNH_2 + R-CHO + O_2 \xrightarrow{Luciferase} FMN + H_2O + RCOOH + h\nu(490 \text{ nm})$$

This bioluminescent system has been widely used as reporter system for the detection of pollutants because of its high sensitivity. The co-substrate (FMNH2) and substrate (R-CHO) required for this reaction do not accumulate in the cell, but their levels depend on the level and balance of metabolism. Thus the bioluminescent reaction is directly linked to the cells metabolic activity, and so in models using *lux* as the reporter system, the light emission capacity is directly affected by factors such as changes in cell growth and dissolved oxygen concentration (Nelson and Lawrence, 1980).

## 'Green fluorescent protein' (from Aequorea victoria; gfp gene)

The green fluorescence protein (GFP) of the jellyfish *Aequorea victoria* is a photoprotein and has been used as reporter in eukaryotic and bacterial cells (Tsein, 1998). A photoprotein emits light when excited with light in a specific range of wavelength. Although the fluorescent reporter systems have not been widely used in the development of biosensors, they have a number of advantages. Owing its compact and closed structure, GFP is highly stable in eukaryote and prokaryote cells. These fluorescent proteins do not require cofactors or substrates. Moreover, GFP is not involved in bacterial metabolism and so is less sensitive to changes in the energetic state of the cell compared with the luminescent systems that depend on FMNH2 or ATP as reaction substrates. A further advantage of GFP is the development of mutant forms that alter the spectral properties of the protein to provide a range of systems for different applications. For example, the GFPmut1 has an excitation maximum at 488 nm and an emission maximum at 507 nm (Hakkila *et al.*, 2002). These structural alterations of the wild-type protein can produce less stable proteins. This can be advantageous since in contrast with the GPF wild-type, which is stable and accumulates in the cell, the mutant forms break down and reduce the background fluorescence, thus enhancing the sensitivity of the biosensor (Tauriainen *et al.*, 1999).

The selection of a genetic reporter can be difficult because of the wide range of genes available, and the scarcity of published comparative studies of reporter genes. Kurittu *et al.* (2000) compared the performances of two bioluminescent *Escherichia coli* strains that had been genetically modified so that they could be used for the detection of tetracyclines. The sensor bacteria containing the bacterial luciferase operon *luxCDABE* proved to be slightly more sensitive than the strain containing the *lucFF* operon. Hakkila *et al.* (2002) reported a comparison of the sensitivity achieved by using *lucFF*, *luxCDABE* and *gfp* reporter genes in the same constructs. *Escherichia coli* sensor bacteria were engineered to contain a reporter plasmid that carries the reporter gene under the control of mercury- (*mer* from Tn21) or arsenite- (*ars* from R773) responsive regulatory units. Characteristics of the strains were studied by using different arsenite or mercury concentrations and incubation times. The fluorescent system (*gfp*) requires longer incubation times than the luminescent proteins (*lucFF* and *lucCDABE*), and produces a detectable signal only at concentrations of target analyte that are about one hundred times higher than those necessary for the luminescent proteins. The former seem to be more appropriate for use in rapid tests and where environmental concentrations of the analyte are low. However, although the fluorescent reporter seems to be less sensitive, this is not the only factor that determines the overall sensitivity of the assay system, and the performance of the associated instrumentation must also be considered. For instance in some cases it is possible to provide enhanced sensitivity by integration of the signal over a long time period, and use of CCD cameras (Hakkila *et al.*, 2002).

### 3.4.3.3 Bioreporter Technology

In addition to the reporter systems described above some novel transductive techniques designed specifically to characterize molecular interactions between the pollutant molecule and its biomolecular target have been developed. In recent times there has been an increasing interest in the application of surface plasmon resonance (SPR) in this area. The attraction of this technique is that it offers the possibility of real time analysis using label free procedures. The SPR sensing mechanism is based on variations in the refractive index of the medium adjacent to the metal sensor surface during the interaction of an analyte with its corresponding recognition element that has been immobilized at the sensor surface (Homola, 2003). The interaction between the complementary biomolecules causes a change in the refractive index which is measured by a detector. Several SPR instruments are commercially available. Details of the characteristics of available instruments are available in the literature (Baird and Myszka, 2001; Mukhopadhyay, 2005; Gauglitz, 2005). SPR seems to have great potential as a reporter technology for use in environmental applications (Habauzit *et al.*, 2007) since it could allow a range of improvements that would be especially useful in environmental monitoring. These include: (1) miniaturization and use of optical fibre technology that would allow users to use the instrument on site, (2) possibility of multi-analysis systems that would support the study several interactions with the same system, (3) combination of the SPR system with other analytical techniques, such as mass spectrometry, electrochemistry, and amperometry.

The main practical environmental application of SPR has been as an immunosensor for the detection of phytosanitary products (Mullet, 2000). Recently, SPR immunodetection of 2,4D (Kim et al., 2007), DDT (Mauriz et al., 2006a, 2007), chlorpyrifos and carbaryl pesticides (Mauriz et al., 2006b) was found ten times more sensitive than ELISA methods, with detection limits in the ppt range, and an analysis time of less than 20 min.

SPR has been used for some time for the detection and characterization of biomolecular interactions, and this could have potential applications in environmental monitoring. Fukuda et al. (2007) characterized the interaction between the aryl hydrocarbon receptor (AhR) with several flavonoids (flavone, flavonol, and flavanone) that have an inhibitory effect of on ligand binding to the AhR. Results suggested that flavone, flavonol, and flavanone act as direct competitive antagonists of AhR, while catechin associates with the AhR and exerts its antagonistic effects by an indirect mechanism. SPR has been also been used to detect steroid hormones on the basis of their interaction with specific nuclear receptors: ERα and ERβ (McLachlan, 2001; Mangelsdorf et al., 1995). An understanding of the biological mechanism of interaction between ER and estrogens could lead to the development of biospecific and more sensitive quantification methods that could be applied for the quantification of estrogenic compounds in water. Applications of SPR in direct and indirect assays that use antibodies and receptor affinity interaction respectively have been described. For example, an assay system involving an anti-estrone-3-glucoronide (E3G) antibody as a recognition element for E3G bound onto ovalbumin has been developed for the quantification of estrogens. This assay allows the quantification of estrone-3-glucuronide in the range of 10 to 150 µg L$^{-1}$ in 70 minutes (Sesay and Cullen, 2001). The use of an indirect competition assay based on the interaction of ER with BSA-E2 bound on the sensor chip lowered the detection limit for estradiol to 0.2 µg L$^{-1}$ (Miyashita et al., 2005). Table 3.4.6 summarizes the different assays that use estrogen receptors (Habauzit et al., 2007). SPR based biosensors used for quantitative detection of estrogenic compounds shows low limit of detection that are comparable with those achieved by using classical methods based on chromatography (GC or HPLC). The SPR based methods have an advantage in that they do not require the preliminary sample treatment to concentrate the compound to be evaluated (usually by solid phase extraction) or the derivatization that is sometimes necessary for the quantification of some compounds by GC. However, the SPR method has not been validated for all known endocrine disruptor compounds, and further development will be necessary to enable the achievement of a global measure of estrogenic compounds that can be converted into the 17β estradiol equivalent concentration. Currently this prevents this method from providing a representative measure of the total quantity of estrogenic compounds in sample, and complementary analysis is still required.

**Table 3.4.6** Assay design based on the evaluation of estrogen receptor/estrogenic compound interactions (adapted from Habauzit et al., 2007). Estrogen receptor, E2-17PeNH, mAB, BSA-E2 complex, E2, estrogenic compound test chemical

| Assay characteristic | Compounds fixed on the chip | Target molecules | Figure | Concentration ranges ($\mu g\ L^{-1}$) | Reference |
|---|---|---|---|---|---|
| Direct binding assays | mAB anti-ER LBD | Estrogens : E1, E2, E3, EE2, DES, TAM. | | 20–200 | Rich et al., 2002 |
| | E2-PeNH | ER and E1, E2, E3, TAM, DES. | | 0.2–20 | Usami et al., 2002 |
| Indirect assays with competition | E2-BSA | ER and E2, DES | | 20 | Seifert et al., 1999; Pearson et al., 2001; |
| | E2-BSA | ER and E2, DES. mAB anti E2 | | 0.2–2 | Miyashita, 2005 |

E1: estrone, E2: estradiol, E3: estriol, EE2: ethynylestradiol, DES: diethylstilbestrol, TAM: tamoxiphen, BSA: bovine serum albumine, AB: antibody

*Transduction Phase*

The quartz crystal microbalance (QCM) technique has also been applied in this area where it has been used as an affinity-based biosensor. The principle is the use of mass-sensitive detector based on an oscillating piezoelectric quartz crystal that resonates at a fundamental frequency (Babacan *et al.*, 2000; Martin *et al.*, 2002). The technique has been applied to the detection of organophosphorus and carbamate pesticides (Karousos *et al.* 2002; Kim *et al.*, 2007). The detection is based on the chemisorption of a thiolated acetylcholine esterase over the QCM gold sensor surface and the measurement of a sum parameter of AChE inhibition. More recently, atrazine was shown to induce human aromatase gene expression via promoter II (ArPII) in a steroidogenic factor 1 (SF-1)-dependent manner. The binding affinity of atrazine to SF-1 was measured by using a 27-MHz Quartz Crystal Microbalance by immobilization of the target analyte onto a QCM electrode monitored continuously for QCM frequency change at 25°C. The SF-1 protein/atrazine interaction was detected by the evaluation of the subsequent time-course of frequency change (Fan *et al.*, 2007).

Detection of estrogen compounds has also been investigated with this powerful technique by using interaction with the estrogen receptor (Murata *et al.*, 2003; Carmon *et al.*, 2005). Generally, the receptor is immobilized on a piezoelectric quartz crystal via a single exposed cysteine, forming a uniform orientation on the crystal surface and the sensitive response of this biosensor to estrogenic substances results from changes in the structural rigidity of the immobilized receptor that occurs with ligand binding. This technique shows great selectivity and can detect the difference between agonistic and antagonistic compounds through difference in the behaviour of the ligand in response to their binding to the receptor. Substances that do not bind to the receptor are distinguished because they do not give a signal.

Kurosawa *et al.* (2006) investigated the potential of quartz crystal microbalance immunosensors for environmental monitoring by comparing the different available devices, and considering the biointerface of the QCM immunosensor, the reduction of nonspecific binding and the stabilization of immunologic activity that is achieved. The study was based on the quantification of dioxin and bisphenol A in environmental samples. The detection of dioxin showed a good relationship with the results obtained with both GC/MS and ELISA methods, and a detection level in the range of 0.1 to 0.01 µg $L^{-1}$ was reached for bisphenol.

Quartz Crystal Microbalance with Dissipation (QCM-D)-based techniques have also demonstrated to be relevant for understanding ligand binding effects although the mechanism of interaction is still under discussion. Peh *et al.* (2007) used QCM-D as a new alternative method for studying the conformational differences in protein-DNA complexes. Specifically, QCM-D was used in combination with surface plasmon resonance (SPR) spectroscopy to monitor the binding of ERα to a specific DNA (estrogen response element, ERE) and a nonspecific DNA in the presence of either the agonist ligand, 17β-estradiol (E2), the partial antagonist ligand, 4-hydroxytamoxifen (4OHT), or vehicle alone. Interaction between ERα with E2 and 4OHT affects not only the viscoelasticity and conformation of the protein-DNA complex but also the capacity of ERα to bind and immobilize ERE. The understanding of these fundamental effects promises to allow the application of these techniques for the quantitative detection of estrogens.

## 3.4.4 CONCLUSION

Due to the increasing number and variety of environmental pollutants and to the growing requirement for monitoring water quality (as described by the Water Framework Directive for surface water as example), there is an increasing need for methods that can provide an alternative to the classical standardized laboratory procedures. Biosensors that have been developed in recent years offer an answer to these needs. Biosensor development studies can be divided into two main fields: (1) the interaction between one pollutant (or a family of compounds) and a recognition element, (2) the characterization of this interaction. In the development of recognition element; artificial material (such as peptides, nucleic acids and molecularly imprinted polymers) and natural biological systems (such as whole organisms (including microorganisms), plant or animal cells, organelles, tissues, enzymes, antibodies, nucleic acids, receptors) have been studied and applied in biosensors.

Biosensors using biomolecular recognition systems (BRS) and in particular receptors, other proteins, and nucleic acids have become of increasing interest due to their potential in term of sensitivity, specificity, rapidity and the possibility of obtaining qualitative (the presence, the absence, the effect, the biological impact of a pollutant) and/or quantitative (concentration in the environment) information concerning the exposure to a contaminant. These biosensors are based on genetically engineered systems including genes coding for a reporter protein, placed under the control of a DNA sequence that recognizes the analyte of interest after its interaction with a BRS. The development of such biosensors combines the powerful specificity of a BRS for its substrate with a sensitive device to measure the response of ligand capture (for instance by way of reporter gene). While such biosensors are limited by the ability to control the biochemical systems involved, they have several advantages compared with conventional methods. In particular, they are able to detect the bioavailable fraction of the contaminant, to measure physiological effects (toxicity) caused by a contaminant as well as to give a concentration level of the targeted pollutant. They are faster and easier to use than other methods and they can also be more sensitive. Given the continuing scientific technical developments in this area, biosensors have the potential to become a common tool for routine on site water monitoring.

## REFERENCES

Asad N.R., Asad L.M., Silva A.B., Felzenszwalb I. and Leitao A.C., 1998. *Acta Biochim. Pol.*, **45**, 677.
Babacan S., Pivarnik P., Letcher S. and Rand A.G., 2000. *Biosens. Bioelectron.*, **15**, 615.
Baeumner A., 2003. *Anal. Bioanal. Chem.*, **377**, 434.
Baird C.L. and Myszka D.G., 2001. *J Mol. Recognit.*, **14**, 261.
Balaguer P., Francois F., Comunale F., *et al*., 1999. *Sci. Total Environ.*, **233**, 47.
Beck I.C., Bruhn R. and Gandrass J., 2006., *Chemosphere*, **63**, 1870
Behnisch P.A., Hosoe K. and Sakai S., 2001a. *Environ. Int.*, **27**, 413.
Behnisch P.A., Hosoe K. and Sakai S. 2001b. *Environ. Int.*, **27**, 495.
Belkin S., Smulski D.R., Vollmer A.C., Van Dyk T.K. and LaRossa R.A., 1996. *Appl. Environ. Microbiol.*, **62**, 2252.

# References

Belkin S., 2003. *Curr. Opin. Microbiol.*, **6**, 206.
Bilitewski U., Brenner-Weiß G., Hansen P.D., *et al.*, 2000. *Trends Anal. Chem.*, **19**, 428.
Blankenship A., Kannan K., Villalobos S., *et al.*, 1999. *Organohalogen Compound*, **42**, 217.
Brenner-Weiß G. and Obst U., 2003. *Anal. Bioanal. Chem.*, **377**, 408.
Brian I., Rissin D., Ron E. and Walt D., 2003. *Anal. Biochem.*, **315**, 106.
Byfield M.P. and Abuknesha R.A., 1994. *Biosens. Bioelectron.*, **9**, 373.
Carmon K.S., Baltus R.E. and Luck L.A., 2005. *Anal. Biochem.*, **345**, 277.
Coestier C., Lin L., Roig B. and Touraud E., 2007. *Anal. Bioanal. Chem.*, **387**, 1163.
Coldham N., Dave M., Sivapathasundaram S., McDonnell D., Connor C. and Sauer M., 1997. *Environ. Health Persp*, **105**, 734.
Combes R.D., 2000. *ATLA-Altern. Lab. Anim.*, **28**, 81.
Corbisier P., Van der Lelie D., Borremans B., *et al.*, 1999. *Anal. Chim. Acta*, **387**, 235.
Daunert S., Barrett G., Feliciano J.S., Shetty R.S., Shrestha, S. and Smith-Spencer W., 2000. *Chem. Rev.*, **100**, 2705.
DeLuca M. and McElroy W.D., 1974. *Biochem.*, **13**, 921.
Demple, B. and Harrison, L. (1994) *Annu. Rev. Biochem.*, **63**, 915.
Diaz E. and Prieto M.A., 2000. *Curr. Opin. Biotechnol.*, **11**, 467.
Dollard M.A. and Billard P., 2003. *J. Microbiol. Methods*, **55**, 221.
Drummond A.E., Britt K.L., Dyson M., *et al.*, 2002. *Mol. Cell. Endocrinol.*, **191**, 27.
Endo G. and Silver S., 1995. *J. Bacteriol.*, **177**, 4437.
Fahnrich K.A., Pravda M. and Guilbault G.G., 2003. *Biosens. Bioelectron.*, **18**, 73.
Fan W.Q., Yanase T., Morinaga H., *et al.*, 2007. *Biochem. Bioph. Res. Co.*, **355**, 1012.
Fang H., Tong W.D., Perkins R., Soto A.M., Prechtl N.V. and Sheehan D.M., 2000. *Environ. Health Persp.*, **108**, 723.
Fukuda I., Mukai R., Kawase M., Yoshida K. and Ashida H., 2007. *Biochem. Bioph. Res. Co.*, **359**, 822.
Gauglitz G., 2005. *Anal. Bioanal. Chem.*, **381**, 141.
Giesy J.P., Hilscherova K., Jones P.D., Kannan K. and Machala M., 2002. *Mar. Pollut. Bull.*, **45**, 3.
Gray L.E., Kelce W.R., Wiese T., *et al.*, 1997. *Reprod. Toxicol.*, **11**, 719.
Gray Jr. L. E., 1998. *Toxicol. Lett.*, **102-103**, 677.
Gutendorf B. and Westendorf J., 2001. *Toxicology*, **166**, 79.
Habauzit D., Chopineau J. and Roig B., 2007. *Anal. Bioanal. Chem.*, **387**, 1215.
Hakkila K., Maksimow M., Karp M. and Virta M., 2002. *Anal Biochem.*, **15**, 235.
Hamers T., Van Schaardenburg M.D., Felzel E.C., Murk A.J. and Koeman J.H., 2000. *Sci. Total Environ.*, **262**, 159.
Hidalgo E., Nunoshiba T. and Demple B., 1994. *Methods Mol. Genet.*, **3**, 325.
Hilscherova K., Machala M., Kannan K., Blankenship A.L. and Giesy J.P., 2000. *Environ. Sci. Poll. R.*, **7**, 159.
Hock B., Seifert M. and Kramer K., 2002. *Biosens. Bioelectron.*, **17**, 239.
Homola J., 2003. *Anal. Bioanal. Chem.*, **377**, 528.
Janosek J., Hilscherova K., Blaha L. and Holoubek I., 2006. *Toxicol. in Vitro*, **20**, 18.
Jungbauer A. and Beck V., 2002. *J. Chromatogr. B*, **777**, 167.
Justus T. and Thomas S.M., 1999. *Mutagenesis* **14**, 351.
Karousos N.G., Aouabdi S., Way A.S. and Reddy S.M., 2002. *Anal. Chim. Acta*, **469**, 189.
Kelce W.R., Lambright C.R., Gray J., Earl L. and Roberts K.P., 1997. *Toxicol. Appl. Pharm.*, **142**, 192.
Kenyon C.J., Brent R., Ptashne M. and Walker G.C., 1998. *J. Mol. Biol.* **160**, 445.
Kim S.J., Gobi V.K., Tanaka H., Shoyama Y. and Miura N., 2007. *Sens. Actuators B: Chem.*, In Press.

Kim N., Park I. and Kim D-K., 2007. *Biosens. Bioelectron.*, **22**, 1593.
Kizu R., Ishii K., Kobayashi J., *et al.*, 2000. *Mater. Sci. Eng., C*, **12**, 97–102.
Köhler S., Belkin S. and Schmid R.D., 2000. *Fresenius J. Anal. Chem.*, **366**, 769.
Kostrzynska M., Leung K.T., Lee H. and Trevors J.T., 2002. *J. Microb. Methods*, **48**, 43.
Kozubek S., Ogievertskaya M.M., Krasavin E.A., Drasil V. and Soska J., 1990. *Mutat. Res.*, **230**, 1.
Kramer K. and Hock B., 2003. *Anal. Bioanal. Chem.*, **377**, 417.
Kurittu J., Karp M. and Korpela M., 2000. *Luminescence*, **15**, 291.
Kurosawa S., Park J.W., Aizawa H., Wakida S., Tao H. and Ishihara K., 2006. *Biosens Bioelectron.*, **22**, 473.
Kuruto-Niwa R., Terao Y. and Nozawa R., 2002. *Environ. Toxicol. Phar.*, **12**, 27.
Lascombe I., Beffa D., Ruegg U., Tarradellas J. and Wahli W., 2000. *Environ. Health Persp.*, **108**, 621.
Lebail J.C., Marrefournier F., Nicolas J.C. and Habrioux G., 1998. *Steroids*, **63**, 678.
Lee H.Y., Choi S.H. and Gu M.B., 2003. *Bioprocess Eng.*, **8**, 101.
Lee H.J. and Gu M.B., 2003. *Appl. Microbiol. Biotech.*, **60**, 577.
Legler J., Brink C.E., Brouwer A., *et al.*, 1999. *Toxicol. Sci.*, **48**, 55.
Legler J., Dennekamp M., Vethaak A.D., *et al.*, 2002. *Sci. Total Environ.*, **293**, 69.
Machala M., Vondracek J., Blaha L., Ciganek M. and Neca J., 2001a. *Mut. Res.-Gen. Tox. En.*, **497**, 49–62.
Machala M., Ciganek M., Blaha L., Minksova K. and Vondraeek J., 2001b. *Environ. Toxicol. Chem.*, **20**, 2736.
Machala M., Blaha L., Lehmler H.J., *et al.*, 2004. *Chem. Res. Toxicol.*, **17**, 340.
McLachlan J.A., 2001, *Endocr Rev.*, **22**, 319.
Mangelsdorf D.J., Thummel C., Beato M., *et al.*, 1995. *Cell*, **83**, 835.
Mantovani A., Stazi A.V., Macri C., Maranghi F. and Ricciardi C.R., 1999. *Chemosphere*, **39** 1293.
Martin S.P., Lynch J.M. and Reddy S.M., 2002. *Biosens. Bioelectron.*, **17**, 735.
Matsumura A., Ghosh A., Pope G.S. and Darbre P.D., 2005. *J. Steroid Biochem. Mol. Biol.*, **94**, 431.
Mauriz E., Calle A., Montoya A. and Lechuga L.M., 2006a. *Talanta*, **69**, 359.
Mauriz E., Calle A., Manclús J.J., *et al.*, 2006b. *Sens. Actuators B: Chem.*, **118**, 399.
Mauriz E., Calle A., Manclús J.J., *et al.*, 2007. *Biosens. Bioelectron.*, **22**, 1410.
Meighen E.A. and Dunlap P.V., 1993. *Adv. Microb. Physiol.*, **34**, 1.
Min J.H., Kim E.J., LaRossa R.A. and Gu M.B., 1999. *Mutation Res.*, **442**, 61.
Minf W., Pascussi J-M. and Pillon A., *et al.*, 2007. *Toxicol. Lett.*, **170**, 19.
Miyashita M., Shimada T., Miyagawa H. and Akamatsu M., 2005. *Anal. Bioanal. Chem.*, **381**, 667.
Moore M., Mustain M., Daniel K., *et al.*, 1997. *Toxicol. Appl. Pharm.*, **142**, 160.
Mukhopadhyay R., 2005. *Anal. Chem.*, **77**, 291.
Murata M., Gouda C., Yano K., Kuroki S., Suzutani T. and Katayama Y., 2003. *Anal. Sci.*, **19**, 1355.
Murk A.J., Legler J., Denison M.S., Giesy J.P., Van de Guchte, C. and Brouwer A., 1996. *Fund. Appl. Toxicol.*, **33**, 149–160.
Murk A.J., Leonards P.E.G., Van Hattum B., Luit R., Van der Weiden M.E.J. and Smit M., 1998. *Environ. Toxicol. Phar.*, **6**, 91.
Mullett W.M., Lai E.P.C. and Yeung J.M., 2000. *Methods*, **22**, 77.
Nakai M., Tabira Y., Asai D., *et al.*, 1999. *Biochem. Bioph. Res. Co.*, **254**, 311.
Nakamura H. and Karube I., 2003. *Anal. Bioanal. Chem.*, **377**, 446.
Nelson P.O. and Lawrence A.W., 1980. *Wat. Res.*, **14**, 217.

Nunoshiba T. and Nishioka H., 1991. *Mut. Res.*, **245**, 71.
Paitan Y., Biran I., Shechter N., Biran D., Rishpon J. and Ron E.Z., 2004. *Anal. Biochem.*, **335**, 175.
Paris F., Balaguer P., Terouanne B., *et al*., 2002. *Mol. Cell. Endocrinol.*, **193**, 43.
Pearson J., Gill A., Margison G.P., Vadgama P. and Povey A.C., 2001. *Sens. Actuators B: Chem.*, **76**, 1.
Peh W.Y.X., Reimhult E., The H.F., Thomsen J.S. and Su X., 2007. *Biophys. J.*, **92**, 4415.
Quillardet P., Frelat G., Nguyen V.D. and Hofnung M., 1989. *Mutat. Res.*, **216**, 251.
Quillardet P. and Hofnung M. 1993. *Mutat Res.* **297**, 235–79.
Rich R.L., Hoth L.R., Geoghegan K.F., *et al*., 2002. *PNAS*, **99**, 8562.
Rogers J.M. and Denison M.S., 2000. *In Vitro Mol. Toxicol.*, **13**, 67.
Rosenstein R., Nikoleit K. and Götz F., 1994. *Mol. Gen. Genet.*, **242**, 566.
Saidi Y., Domini M., Choy F., Zryd J.P., Schwitzguebel J.P. and Goloubinoff P., 2007. *Plant Cell. Environ.*, **30**, 753.
Sassanfar M. and Roberts J.W., 1990. *J. Mol. Biol.*, **212**, 79.
Seifert M., Haindl S. and Hock B., 1999. *Anal. Chim. Acta*, **386**, 191.
Sesay A.M. and Cullen D.C., 2001. *Environ. Monit. Assess.*, **70**, 83.
Schrader T.J. and Cooke G.M., 2003. *Reprod. Toxicol.*, **17**, 15.
Stocker J., Balluch D., Gsell M. and Harms H., 2003. *Env. Science & Technology*, **37**, 4743.
Subrahmanyam S., Piletsky S.A. and Turner A.P., 2002. *Anal. Chem.*, **74**, 3942.
Tauriainen S., Karp M., Chang W. and Virta M., 1998. *Biosens. Bioelectron.*, **13**, 931.
Tauriainen S., Virta M., Chang W. and Karp M., 1999. *Anal. Biochem.*, **272**, 191.
Tauriainen S.M., Virta M.P.J. and Karp M.T., 2000. *Water Research*, **34**, 2661.
Tsein R.Y., 1998. *Annu. Rev. Biochem.*, **67**, 509.
Usami M., Mitsunaga K. and Ohno Y., 2002. *J. Steroid Biochem. Mol. Biol.*, **81**, 47.
Villeneuve D., Blankenship A.L and Giesy J.P., 1998. *In:* Denison, M.S.; Helferich, W.G. (eds.) *Toxicant-receptor interactions*. Taylor and Francis, Philadelphia, PA, USA, 69–99.
Vinggaard A.M., Hnida C. and Larsen J.C., 2000. *Toxicology*, **145**, 173.
Vollmer A.C., Belkin S., Smulski D.R., Van Dyk T.K. and LaRossa R.A., 1997. *Appl. Environ. Microbiol.*, **63**, 2566.
Vondracek J., Kozubik A. and Machala M., 2002. *Toxicol. Sci.*, **70**, 193.
Wood K.V., Amy Lam Y. and McElroy W.D., 1989. *J. Biolum.*, **4**, 289.
Zacharewski T., Harris M. and Safe S., 1991. *Biochem. Pharmacol.*, **41**, 1931.
Zacharewski T., 1997. *Environ. Sci. Technol.*, **31**, 613.

# 3.5
## Continuous Monitoring of Waters by Biological Early Warning Systems

Kees J.M. Kramer

3.5.1 Introduction
3.5.2 BEWS Systems
    3.5.2.1 Basic Requirements
    3.5.2.2 Commonly Used Systems
3.5.3 Applications
    3.5.3.1 Drinking Water Industry
    3.5.3.2 Surface Waters
    3.5.3.3 Groundwater
    3.5.3.4 Effluents
    3.5.3.5 Aquaculture
    3.5.3.6 Tuning Chlorination of Cooling Water
3.5.4 Quality Assurance Issues
3.5.5 Case Studies
3.5.6 Conclusion
Acknowledgement
References

## 3.5.1 INTRODUCTION

Monitoring surface water, ground water, seawater, effluents and drinking water for toxic compounds is traditionally carried out by discrete (spot) sampling that is followed by chemical analysis in the laboratory. This provides qualitative and quantitative information on specific analytes, in Europe often focused on the priority substances defined by the EC Water Framework Directive, WFD (European Commission, 2000). Although

*Rapid Chemical and Biological Techniques for Water Monitoring*    Edited by Catherine Gonzalez, Philippe Quevauviller and Richard Greenwood
© 2009 John Wiley & Sons, Ltd

this approach is common practice throughout the world, it may not always provide sufficient information for three main reasons:

- It will fail to identify contaminants which are not included in the routine analyses but which may nonetheless occur, albeit rarely, and which are toxic and/or in sufficiently high concentration that they impact aquatic environment.
- Since spot sampling is usually infrequent (e.g. 4 times per year) it may fail to identify sporadic discharges to the environment.
- The time lag between sampling and the availability of the analytical result, in case of a detected alarming concentration, usually does not allow timely action to prevent further harm.

There is an increasing need to obtain information on as many toxicants in the (drinking and surface) water as possible, preferably on a continuous basis and with immediate results to provide an early warning (Grayman et al., 2001). In general two approaches have been adopted that answer these needs:

- continuous physico-chemical monitoring;
- biological monitoring.

Continuous, or very frequent, automated measurement of individual or groups of toxic compounds or general physico chemical properties, can provide a rapid warning of changed environmental conditions. A few chemical methods (such as the LC SAMOS for polar compounds (Slobodnik et al., 1992)) may even detect the occurrence of high concentrations of specific contaminants of concern at the test site. In situ, online and on-site approaches are described in other chapters of this volume, but as yet, application is restricted by the limited availability of suitable, sufficiently sensitive and selective (bio)sensors and monitors for the analysis of (priority) pollutants that need to be monitored. Once available, they may provide (semi)-qualitative and quantitative information on the occurrence of compounds in the waters.

In contrast, biological monitoring techniques use organisms that 'sense' the toxicant(s). It will depend on the nature of the compound and its environmental concentration as well as on the organism whether a (toxic) reaction is induced and this will depend on the dose-effect relationship and organism specific sensitivity. Most organisms are, however, sensitive to more toxic compounds, than classical analytical methods that focus on standard sets of analytes that are included in the routine chemical analysis of aquatic monitoring programmes. Organisms are wide-range sensors. Unfortunately, organisms will never be able to identify the compound(s), and their response is usually only semi-quantitative. Their biological response will be a reaction to one or more of the pollutants from a wide range of chemical classes.

Examples of biological response measurements are ecotoxicological tests and the application of biological early warning systems (BEWS). Most eco-toxicological tests, or bioassays, are conducted batchwise, and take hours to days for acute toxicity tests, and many days for chronic toxicity tests. The 'OECD Guidelines for the Testing of Chemicals' are a collection of the most relevant internationally agreed testing methods.

They cover most groups of aquatic organisms like fish, daphnids and algae (OECD, 1998, 2004 and 2006, respectively). The methods may be applied to environmental water samples but the water may have to be treated to increase the concentration of the toxicants and thus increasing the chance of a measurable biological response.

In contrast, Biological Early Warning Systems (BEWS) fill the time gap between sampling and (toxicological) analysis: they provide a quantifiable response to a pollution incident within an hour, often within a half hour. BEWS have explicitly been developed to provide a rapid warning of the occurrence of contaminants at concentrations which could be of immediate threat to living organisms. BEWS are automated, continuous monitors which employ a biological organism or biological material as a primary sensing element (Baldwin and Kramer, 1994).

In contrast to the physico/chemical measurement systems, BEWS are sensitive to many toxic compounds, even to those that are not included in the routine monitoring programmes. They operate continuously (24/24h, 7/7d) and provide 'early' results. In the case of an accidental spill they should generate an alarm within minutes to one hour (van der Schalie *et al.*, 1999). BEWS have recently been included in the WFD Common Implementation Strategy Guideline 19 on surface water chemical monitoring as complementary method (European Communities, 2009).

This chapter provides an overview of the present situation regarding BEWS. The technical, financial and biological requirements of successful BEWS systems are discussed. Suitable organisms are identified and the needs for quality assurance and validation are highlighted. Some representative early warning systems that are successfully used in environmental monitoring are presented. Illustrations of typical field applications that focus on the requirements of environmental monitoring in the context of the WFD are included as case studies.

## 3.5.2 BEWS SYSTEMS

### 3.5.2.1 Basic Requirements

The basic principle of all BEWS is that suitable organisms are continuously exposed to the test water. This may be in situ or, more commonly, in an online lay out. The organism is the primary sensor. To allow for a fast response, a physiological or behavioural function of the organism has to be used as response parameter. This parameter must show a response to changing environmental conditions, notably to an increase in the concentration of one or more toxic compounds in the water. Behavioural responses that are used include for example: activity, locomotion, avoidance, positive rheotaxis and escape behaviour. Physiological parameters include: respiration rate, gill ventilation frequency, heart beat, pumping rate, bio-electric potential, photosynthetic activity, growth and bio-luminescence.

The response parameter is automatically and continuously recorded. This secondary sensing element may be electrical, electro-magnetic, optical, electro- or opto-chemical. Finally the data is evaluated. In most systems the current input information is compared with information obtained in the past, e.g. one hour before. When the changes in biological response significantly pass a predefined criterion, the biological early warning system will detect this and mark it as an alarming condition, and an alarm will be generated (Diamond *et al.*, 1988; Kramer and Botterweg, 1991; Brosnan, 1999). Many systems form part of a network and the signal will reach responsible staff immediately. If the alarm is used to induce sampling of the water, chemical analysis may provide the cause of the alarm by identifying the compound(s) quantitatively. To minimize the number of false positive or false negative alarms, today evaluation is performed by highly specialized software.

The choice of organism depends on practical and logistic aspects such as:

- availability (collected in the field or cultured);
- minor changes in appearance during the test period;
- insensitivity to handling and disease;
- maintenance requirements (feeding, replacement frequency); and
- a suitable response parameter that can be monitored and evaluated.

Organisms that are endemic for the test area have a preference as they will require minimal adaptation.

Lists of suitable organisms and their possible response parameters have been published (Baldwin and Kramer, 1994; Gruber *et al.*, 1994; Gerhardt, 2000). In recent years although the same groups of organisms are used, the response parameters have narrowed down to those that seem most practical and reliable. The following groups of organisms – with their response parameter(s) – are nowadays most commonly used in commercial BEWS:

- bacteria (e.g. *Vibrio fischeri, Photobacterium phosphoreum*): luminescence;
- algae (multi species): chlorophyll fluorescence;
- invertebrates (multi species, including e.g. *Gammarus pulex*): locomotor and ventilatory activity;
- crustacea (daphnids: *Daphnia* sp.): movement (and growth);
- bivalve molluscs (e.g. zebra mussel: *Dreissena polymorpha;* blue mussel: *Mytilus edulis*): valve movement;
- fish (e.g. gold ide: *Leuciscus idus*; blue gill: *Lepomis macrochirus*): positive rheotaxis, ventilatory activity.

Although organisms may 'sense' a wide range of pollutants, each group will have its specific sensitivities. No one biological sensor is likely to be equally sensitive to all potential toxicants of concern. It is obvious that algae will be more sensitive to herbicides than e.g. fish. Invertebrates such as daphnids and fish like bluegills appear to be relatively sensitive to volatile organic substances (Schorr and Gruber, 1991). Bivalves are sensitive to anti-fouling agents (copper, tributyl-tin) and often to organic solvents. Upstream industrial activities may give insight in the pollutants that may be expected in case of an accidental spill and thus an appropriate BEWS can be selected. A battery of several BEWS may be necessary to cover a reasonably large spectrum of toxicants, as is required for monitoring large river systems (van der Schalie *et al.*, 2006).

*BEWS Systems*

Clearly a biological testing system can be used only when the conditions of the water that is being monitored do not prevent normal life of the organism. Therefore, for example, temperature, oxygen content, pH, salt content and the background toxicity should be sustainable for the organisms used (i.e. should fall within the zone of tolerance of the test species). This may be critical when monitoring effluents, but is usually not a problem in surface waters.

### 3.5.2.2 Commonly Used Systems

As was discussed by Gruber *et al.* (1994) there is a considerable difference between test systems that are in the research and technical development phase in the laboratory and those that have reached maturity, and that have proven their functionality and reliability under realistic field monitoring conditions. The level of skill required by the operator is a further important consideration: if it takes a PhD student to operate the system then it is not likely to be of interest for the routine monitoring practice.

Successful early warning systems combine several factors (Gruber *et al.* 1994; von Danwitz *et al.*, 1998):

- organisms inexpensive, easy to acquire;
- sufficient sensitivity for various types of pollutants;
- rapid, reliable detection of adverse conditions;
- automated, stand alone, in situ or online, 24/24h 7/7d operation;
- appropriate data analysis, minimum false alarms;
- ease of use;
- high level of development for application under field conditions;
- validated;
- the price of the complete system;
- minimal time and cost of maintenance.

A successful BEWS is relatively inexpensive. In the calculation of costs the purchase price may not be the dominant factor, however. Running costs for maintenance and servicing, including consumables and staff costs, will in the end determine the cost effectiveness. Often the frequency of servicing and/or the replacement of organisms will be the dominant cost factor, especially if this is on a weekly basis.

Over the last three decades many different types of BEWS have been developed, using fish, bivalves, daphnids, other invertebrates, algae or bacteria as primary sensor (Kramer and Botterweg, 1991; Baldwin and Kramer, 1994; von Danwitz *et al.*, 1998; Gerhardt, 2000; Gerhardt *et al.*, 2006). As a consequence of the criteria for a successful BEWS it appears that only relatively few commercial BEWS systems have reached maturity and are used throughout the world for monitoring of the aquatic environment. Without trying to provide a complete account, several BEWS systems are introduced in the following section covering the range of sensor organisms, that are commercially available and have proven their functioning in practical monitoring of surface waters and effluents.

## Bacteria

Luminescent bacteria have been used in ecotoxicological studies since the 1980s, initially using *Photobacterium phosphoreum*. However, they have usually been applied in batch mode, e.g. in the Microtox® system (Bulich, 1979). One, the TOXcontrol® system (microLAN, Waalwijk, NL), meets the requirements of automated and stand alone operation. It makes use of freshly cultivated light emitting bacteria (*Vibrio fischeri*) as the biological sensor. The test uses the reduction in luminescent light as the indicator of toxic effect. The emitted luminescence is measured for the test water and a control sample, and the percentage of toxicity is calculated. The test organisms may be cultivated in a separate bioreactor. Maintenance is required once per week. This luminescent bacteria monitor is an automated version of the procedure as defined by the ISO 11348-1 standard (ISO, 2007). Applications claimed include drinking water protection (Appels *et al.*, 2007), and river and wastewater monitoring.

## Algae

In the Algae Toximeter® (bbe Moldaenke, Kiel-Kronshagen, DE) system the test water sample is almost continually pumped into the test chamber where the concentration and the activity of the naturally present algae – thus a mix of species – are determined using the photosynthetic activity of the chlorophyll in the green algae. A precisely defined amount of algae from a fermenter is then added to the measuring chamber. The activity of the added algae remains constant as long as no toxic substances are present. If a toxic compound is present, its interaction with the photosynthesis centre leads to an inhibition of algal activity (and oxygen production). This inhibition is quantified and an alarm is activated after a predefined threshold has been exceeded. The cultivation of the algae (e.g. *Chlorella* sp.) takes place in a fermenter which is regulated turbidostatically by an additional fluorescence measurement. Due to this regulation the algal concentration and their activity are kept in a steady state. An additional feature of the system is the possibility of identifying and quantifying algal groups (green, blue-green and brown algae, and cryptophyceae) by using coloured light emitting diodes (LEDs). Maintenance frequency is >7 days. Applications claimed include water supply systems, surface waters, water treatment plants and sewers.

## Invertebrates

Early warning systems involving the swimming behaviour of water fleas (*Daphnia* sp.) were first described by Knie (1978) as the 'Dynamic Daphnia test'. The principle consisted of a test vessel with infrared light beams where the blocking of the light indicated movement of the test organisms. It was used for many years to monitor surface waters (Gunatilaka *et al.*, 2001). Based on the Extended Dynamic Daphnia Test an improved system using video image analysis was developed: the Daphnia Toximeter® (bbe Moldaenke, Kiel-Kronshagen, DE). Sample water continuously flows through the measuring chamber containing the *Daphnia*. Live images recorded by a CCD-camera are evaluated online to analyse changes in the behaviour of the test organisms. If changes are statistically significant, an alarm is triggered. The method of image analysis enables a series of measurement methods to be applied. Based on measurements of each individual they include speed measurements (average speed, speed distribution), position in the water column, fractal dimension (measurements of turns and circling movements, curviness) and determination of *Daphnia* size (growth). An alarm is generated only if more than two of the measurands simultaneously show unusual behaviour. Maintenance frequency is seven or more days. Applications claimed include drinking water supply, surface waters (Lechelt *et al.*, 2000) and general environmental monitoring.

The measurement system of the MFB Multispecies Freshwater Biomonitor® (LimCo International, Ibbenbüren, DE) is based on the quadropole impedance conversion technology (Gerhardt *et al.*, 1994). The test chambers are flow-through cells with the two electrode pairs placed on the chamber walls. The MFB is a modular system and may contain 8–96 test chambers. Each cell contains usually one organism, but as the name of the instrument suggests, these may even be different species in one experiment (such as *Corophium volutator*, chironomid larvae, *Gammarus pulex, Daphnia* sp.); even small fish are an option. A cell with leaves but without organisms acts as control. An alternating current is applied between electrodes at opposite walls of the test chamber. Movements of the animals will change the conductivity and the electrical field between a second pair of electrodes and thus generate specific electrical signals, at different frequency ranges for different kinds of behaviour (swimming/locomotion, ventilation, inactivity) (Gerhardt *et al.*, 2007). Maintenance required is ~1 hour/week. Applications claimed include monitoring of surface waters, effluents and mixed sediment/water exposure, toxicity ($EC_{50}$) tests.

## Bivalve molluscs

The Musselmonitor® (Delta Consult, Kapelle, NL) operates with 8 bivalve molluscs, either freshwater mussels (e.g. zebra mussel (*Dreissena polymorpha*), clams (*Unio* sp., *Corbicula* sp.), or marine species (e.g. common or blue mussel (*Mytilus edulis*) or other mussel, clam or oyster species (*Perna* sp., *Crassostrea* sp.)). The principle, first applied in the 1980s (Kramer *et al.*, 1989), is that mussels have their shells open for respiration and feeding. Under stress they close their shells, and this is considered to be escape behaviour. In addition, under stress by (some) organic solvents like chloroform, they may increase the opening-closing frequency drastically, another behaviour change that is evaluated. The movement of the shells is the operational parameter which is measured by a high-frequency electromagnetic induction system using two small coils attached to each shell. For the evaluation the present valve movement behaviour of each mussel is compared with its own behaviour of about 1 h before. A significant change, e.g. when 5 mussels close during 5 min, results in an alarm (de Zwart *et al*., 1995; Kramer and Foekema, 2001).

The Musselmonitor was developed as an *in situ* system, but the flow-through version is more in demand by monitoring stations that already have a test-water supply installed. Maintenance involves weekly cleaning of the system (15 min); mussels need replacement only every three or more months. Feeding is not required: mussels are filter feeders and collect their food from the particles in the ambient water. Practical applications include surface water monitoring (freshwater and marine), ground water protection, drinking water industry, effluents, aquaculture and toxicity testing.

*Fish*

In the 1970s fish were the first organisms to be used in BEWS. Initially chemotaxis or avoidance were the behavioural changes used as indicators of exposure to toxicants. In these systems fish were allowed to choose between clean source water and the test water. In parallel, positive rheotaxis (the characteristic of swimming against the water flow), was used as a biological response (Juhnke and Besch, 1971), and was applied in for instance the Aqua-Tox-Control® (Kerren Umwelttechnik, Viersen, DE). In the case of bad water quality the fish try to escape and hit a touch screen at the rear of the test chamber. If the number of pulses exceeds a preset number, an alarm is generated. Fish typically used include in Europe the gold ide (*Leuciscus idus*). Weekly replacement of the fish is part of the maintenance. Applications include surface water monitoring and protection of drinking water.

The ventilatory behaviour of fish has also been used since the early 1980s (Gruber and Cairns, 1981). Fish that stand still in running water are used for the detection of the ventilation frequency; in Europe the rainbow trout (*Oncorhynchus mykiss*) and in the USA the bluegill (*Lepomis macrochyrus*) are commonly used. The BioSensor® (Biological Monitoring Inc., Blacksburg VA, US) is an online early warning system incorporating eight flow-through tanks. Each tank houses one fish, and signals are processed separately. Changes in ventilatory behaviour (and certain locomotor activities) are detected immediately through noninvasive electronic sensors in each of the test chambers. Computer-based algorithms continuously monitor and assess each fish's reaction to the water. When a significant number of fish respond simultaneously in an abnormal manner, an alarm is initiated (Mikol *et al*., 2007). Automated fish feeders reduce daily attendance, but weekly maintenance remains. Applications include monitoring the quality of drinking water, industrial and municipal effluents, and other water supplies.

In Germany, where many BEWS were installed in the river basins over the past decades, experience has taught that the fish-based systems applied in the monitoring of surface waters increasingly failed to detect adverse water conditions. This was attributed to a general improvement of the surface water quality and a lack of sensitivity of fish (von Danwitz *et al*., 1998). Obviously, in situations where a steep increase of (accidental) pollution is expected, such as in effluents, or terrorist attacks (drinking water protection), these systems remain of interest.

### 3.5.3 APPLICATIONS

#### 3.5.3.1 Drinking Water Industry

*Intake water*

Biological early warning systems have been used since the early 1970s initially by the drinking water industry to monitor the quality at their water abstraction points. In general water quality, e.g. in the river Rhine, was very poor in those days due to high loads of organic matter through untreated sewage (and consequently low oxygen content), and high concentrations of many pollutants were present (Gunatilaka and Diehl, 2001). Water quality has improved in many surface waters, but still (non)-accidental spills or accidents may occur. The Sandoz accident and subsequent spill of toxicants into the river Rhine near Basel (CH) in 1986 made all water managers very aware of the threats.

Since many water companies use surface water for the preparation of drinking water they long ago realized that in case of serious pollution of the water sources (rivers) the intake of water may have to be stopped for days to weeks, with a possible drinking water shortage as a result. In Figure 3.5.1 the number of days that the Dutch drinking water company WRK (now Waternet) had to stop or reduce intake at the water abstraction location in Nieuwegein (NL) because of the detection of too high pollution levels, confirmed by chemical analysis, is displayed for the period 1969–2006 (RIWA, 2007).

The contaminants observed changed over the years. In the 1980s the main cause was the high salt content of the river water, as a result of dumping by the potassium industry in the Alsace (1983, 1985, 1986 and 1989). The year 1986 also saw the Sandoz incident that led to 9 days with no intake. Most 'alarms' were caused by the presence of high concentrations of biocides. These were Endosulphan in 1969, 2,4-D in 1986, Isophorone and Mecoprop in 1988, and Metamitron in 1990. Since then Isoproturon has been the dominant factor that caused closing down of the water intake. This occurred in 1994, 1998, 1999, 2001 and 2002 (in the last two years Chlortoluron was also involved), and affected a total of 103 days in those five years.

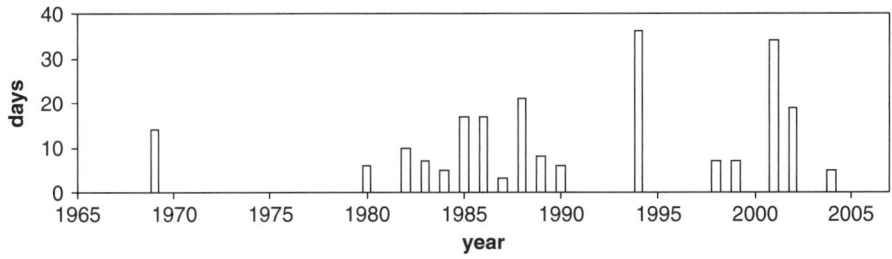

**Figure 3.5.1** Number of days/year that the drinking water company WRK/WCB stopped/reduced abstraction of drinking water from the River Rhine at Nieuwegein for the period 1969–2006 (RIWA, 2007)

Storage basins have been created in order to prevent water shortage during closure of abstraction sites and monitoring stations have been installed to prevent abstraction of polluted water. Measurement of the intake water quality is mostly by (partly on line) chemical analysis for a defined range of analytes, including pollutants. The implementation of BEWS to warn for other pollutants has become a rather common practice, as is illustrated e.g. for water from the Po river (Badino *et al.*, 2002) and from the River Rhine where the drinking water organization WRK started to use BEWS in 1988 at its abstraction point in Nieuwegein (NL) (see case study below).

*Drinking water*

Public safety in the field of drinking water has been an issue for a long time (Brosnan, 1999). However, drinking water security has become a more urgent issue after the terrorist attacks of 09/11 (Anon., 2003) since intentional spills now have to be considered as well as accidental spills. In the USA, the Homeland Protection Act of 2002 specifically calls for the investigation and use of Early Warning Systems (EWS) for water security applications (Green *et al.*, 2003; EPA, 2005; Appels *et al.*, 2007).

The challenge of monitoring tap water is that normally the water is purified and thus contains hardly any micronutrients or particles (food) that may be required to sustain life of the test organisms. In addition, sometimes for public health reasons drinking water is chlorinated; free chlorine is highly toxic to the test organisms and would thus prevent the successful monitoring of tap water by biological early warning systems. However, these problems have been overcome. On request of the Waterworks of Budapest (Budapest, HU) the Musselmonitor was adapted by the continuous addition to the inflowing water of a) an algal suspension (*Chlorella* sp.) as food source, and b) a solution of thiosulphate to neutralize the free chlorine. With these two additions it was possible to employ the same 8 zebra mussels (*D. polymorpha*) for more than 9 months (Baretto and Kramer, 2003). A dechlorination unit is also available for the BioSensor fish monitor (van der Schalie *et al.*, 2005).

### 3.5.3.2 Surface Waters

Notably in Europe, biological early warning systems have been used to monitor surface waters (Blübaum-Gronau *et al.*, 2001; Gunatilaka *et al.*, 2001; Gerhardt *et al.*, 2003). Monitoring stations that operate in trans-boundary rivers near the border are policy driven. Examples of such stations where BEWS are employed include in the River Danube 'Jochenstein' (DE) near the German-Austrian border, at the River Elbe 'Schmilka' (DE) near the Czech-German border, at the River Meuse 'Eijsden' (NL) near the Belgian-Dutch border and at the River Rhine 'Huningue' (FR) near the borders between Switzerland, France and Germany, and 'Bimmen' (DE) and 'Lobith' (NL) near the German-Dutch border. BEWS systems have also been installed in many other locations alongside rivers. In practice these systems have used either/or fish, mussels, invertebrates (daphnids) and algae as primary sensors (Bode and Nusch, 1999). Irmer (1994) and LAWA, the Ländergemeinschaft Wasser in Germany, were instrumental in evaluating BEWS, providing recommendations and serving as a think tank for improvements in applications (von Danwitz *et al.*, 1998).

### 3.5.3.3 Groundwater

Monitoring of groundwater may be for two reasons. The groundwater is a source of drinking water and thus has the same requirements as discussed above for drinking water. More important, however, is that when groundwater is extracted the water table may become lower, and to compensate for this surface water is introduced to the groundwater system. Once a pollutant in the surface water has entered the groundwater it can never be removed. For this reason the water companies monitor the water quality before it is infiltrated into the subsoil (Küster et al., 2004). For example, the 'Station d'Alerte de Huningue' (FR), a monitoring station alongside the River Rhine near the Swiss-French-German border is operated by APRONA. It is situated just upstream of a canal that carries water to the Alsace lowlands, where it infiltrates. The main objective of the station is to monitor River Rhine water and in case of an alarm to prevent infiltration of surface water that originates from the this river. In an emergency a lock may be closed downstream of the station. In the winter period the water company Vitens uses continuous monitoring, involving the Musselmonitor, to check on the quality of the surface water at their site near Epe (NL) before infiltration occurs.

### 3.5.3.4 Effluents

Effluents usually consist of untreated waters that may come from very diverse origins: household effluents and sewage, but also effluents from industrial complexes that may contain highly toxic compounds that can be released in accidents. In case of steep gradients in adverse water conditions, as is expected to occur in effluents, great confidence can be placed in the ability of BEWS to react, and applications have been reported in the literature since the 1970s (Morgan, 1977; Gruber and Cairns, 1981; de Zwart and Sloof, 1987; e.g. Shedd et al., 2001). There are, however, problems associated with the use of BEWS in effluent monitoring. The background quality (pH, temperature, salt content, dissolved oxygen concentration) of the effluent may not always be compatible with the healthy survival of the test organisms in the BEWS, and high loads of particles may interfere with the (optical) measurement system of the early warning sensor. Measuring at the source is of course the most direct approach, but deployments in the receiving water up- and downstream of the effluent line may yield measurements that are less affected by the harsh conditions of the raw effluent.

### 3.5.3.5 Aquaculture

Aquaculture in ponds and other enclosures, and mariculture in enclosures in seawater, produce shellfish, crustaceans and fish for human consumption. It is therefore of vital importance (for the organisms and for the consumers) that the water quality is optimal, and regular testing is essential to 'prove' that pollution is absent. BEWS are very suitable to carry out this task, as several test organisms belong to the same group as the cultured species: mussels, oysters, clams and fish. There seems no reason why other BEWS could not perform in freshwater aquaculture systems as well. BEWS not only monitor the quality of the ambient water in aquaculture farms, but may also be used to optimize production. Stressed animals are less productive in terms of e.g. growth rate and flesh quality, and also excrete increased levels of environmentally damaging products such as ammonia (Stewart *et al.*, 2004).

### 3.5.3.6 Tuning Chlorination of Cooling Water

Biofouling, the growth of organisms such as mussels, barnacles and bacteria on surfaces of e.g. cooling water pipelines and heat exchangers of power plants and industrial complexes is a major concern of the operators. The growth of organisms reduces the cooling water flow considerably. To allow sufficient cooling potential in the heat exchangers, the pumps must be geared up to allow the passage of sufficient water to export the excess heat. This involves a large power consumption that is very costly. In addition, cooling water systems may have to be dismantled and cleaned to remove fouling organisms. This involves a down time in the cooling/production process and serious maintenance costs.

A solution is often found in the addition of sodium hypochlorite to the cooling water, so-called chlorination (Anon., 1997). The free chlorine formed in the chlorination process is toxic to the fouling organisms and prevents them from attachment to the cooling surface walls. However, chlorine is also considered to cause serious environmental harm, directly and via chlorinated by-products that are formed. For these reasons attempts are made to reduce the use of chlorination, e.g. by applying only in the season when fouling occurs, by using intermittent chlorination, or by pulse-chlorination (Jenner *et al.*, 2004).

Dow Benelux, a petrochemical industry situated along the Scheldt estuary (Terneuzen, NL) used this intermittent chlorination technique. If bivalve molluscs at the exit of the cooling water lines could sense sufficient free chlorine to cause them close their shells, then this would imply that all organisms present in the line before this point would experience even higher concentrations, and therefore would be unable to attach themselves to the walls. Thus a Musselmonitor was installed to tune the minimum amount of chlorination to be effective. This proved to be very beneficial, both environmentally and financially (reduction of hypochlorite and pumping energy, less off time for maintenance) (Paping, 1991).

## 3.5.4 QUALITY ASSURANCE ISSUES

The EC WFD (European Commission, 2000) demands that monitoring data are of an appropriate quality, that they have been obtained by competent scientists, with appropriate (validated) methods and instruments. This requirement was originally formulated for the 'classical' chemical and ecological monitoring methods, but also applies to new (screening) techniques that explicitly have been included. This is enforced by a proposal for a Commission Directive adopting technical specifications for chemical monitoring and quality of analytical results in accordance with the WFD (legal act adopted by the end of 2008). Biological monitoring methods have to comply with these Directives.

For ecotoxicological studies, testing individual compounds for their biological (toxicological) effects upon specific organisms, the OECD has drafted a series of guidelines (e.g. OECD, 1998, 2004, 2006). Online biological early warning systems may be considered as 'sensors', albeit having a living organisms as primary sensor. One might expect that the ISO 15839 standard on online sensors/analysing equipment for water (ISO, 2003) is valid for BEWS. Unfortunately this is true only to a limited extent. Most sensors that are the target of ISO 15839 are able to quantify and to qualify specific compounds, and thus can be easily tested and validated. In contrast, BEWS can not identify individual compounds. Although they can show a response to one compound in a toxicity test, this is not the objective of a BEWS that is operated in the natural environment where a mix of toxicants and other interfering substances/conditions may occur. This makes validation less easy, and it is argued that there should be validation protocols for BEWS that differ from those for the more chemical analysis oriented sensors.

For BEWS few guidelines have been developed. For the fish monitors that use the ventilatory response there is the ASTM Standard E1768-95 (ASTM, 1995), while for the luminescent bacteria test there is the guideline ISO 13348, in three parts depending on the handling of the bacteria (ISO, 2007). These guidance documents detail the experiment and provide a good start to the validation of the methods in the monitoring practice.

It may be seen in Table 3.5.1 that there is a need for different approaches to validation.

Different groups of techniques are given, ranging from well defined, fully validated chemical methods that both qualify and quantify the compound of interest, via 'screening' methods that offer only semi-quantitative and semi-qualitative results, to BEWS, where there is no identification of the compound and the signal is only semi-quantitative. Each level requires a different approach for validation. A validation should be carried out under controlled conditions as well as in the monitoring practice in the field. Only too often, monitoring equipment, not only BEWS, works well in the laboratory, but fails in routine application in the environment.

In the last few years the Dutch KIWA Research Institute recognized a need for a better guidance for validation of BEWS and took the initiative to try to improve their standardization and to harmonize their use, to ensure better quality assurance, and raise awareness and acceptance of online biomonitoring. To this end they evaluated so far three BEWS that are operational in The Netherlands: Daphnia Toximeter, Algae

**Table 3.5.1** Overview of typical groups of chemical and biological methods applied in environmental monitoring, with an indication about their possibilities to (semi)qualify and (semi)quantify individual compounds, and the ease of their validation

| Type of analytical/screening method | Identification | Quantification | QA/QC |
|---|---|---|---|
| Spot sampling, lab analysis | ++ | ++ | ++ |
| Analyte specific (bio)sensors | ++ | + | ++ |
| Test kits | + | + | + |
| Analyte group screening | +/− | +/− | + |
| General parameters (e.g. TOC) | − | + | + |
| Eco-toxicological tests | − | +/− | +/− |
| Biological early warning system | − | +/− | +/− |

Toximeter and Musselmonitor, and have prepared three guidance documents on the standardization, quality assurance and data evaluation of online biological alarm systems (Wagenvoort *et al.*, 2005, 2008a; Penders *et al.*, 2006). Issues tackled included installation of the instrument, collection/procurement/breeding of the test organisms, quality assurance (inspection of test organism quality, their sensitivity, use of control samples and description of a validation test, maintenance plan), an evaluation of technical performance (up time, maintenance requirements) and the evaluation of measurement results (instrumental parameter settings, alarm generation).

### 3.5.5 CASE STUDIES

In order to give insight in the applications of BEWS in practical field operation, a selection of case studies is presented below.

**Case 1: Application of the Multispecies Freshwater Biomonitor MFB**

In view of support to the implementation of the EC WFD (European Commission, 2000) in situ biomonitoring was performed along the rivers Meuse (at the RIZA monitoring station in Eijsden, NL), Aller (DE) and Rhine (at the APRONA monitoring station in Huningue, FR) within the frame of the European Union-funded Project SWIFT-WFD. *Gammarus pulex*, a detritus feeder, was used as a test organism.

At the River Meuse, in a tank experiment using a through flow of spiked river water, *G. pulex* showed a biological response to pulse exposure of either a mixture of trace metals or of several organic xenobiotics, by showing up to 20% decreased locomotory activity in response to the 1st pulse and increased mortality during the 2nd or 3rd pulses only. At the River Aller, all *G. pulex* deployed in the MFB system survived (100%) and the biomonitoring did not result in any biological response towards chemical stress. In contrast, in a deployment at the monitoring station on the River Rhine the test organisms showed a clear biological response to the water composition (up to 20% decreased locomotory activity) (Gerhardt *et al.*, 2007).

**Case 2: Application of the Aqua-Tox-Control Fish Monitor**

The water company WRK (now Waternet) has used this fish monitor since 1988 at the water abstraction location Ir. Cornelis Biemond (WCB), in Nieuwegein (NL). Over the years several other BEWS have been added to supplement this monitor. In Figure 3.5.2 an illustration is given of the number and severity of the alarms generated by this monitor for the period July 2000–June 2001. In this period the occurrence of the herbicide Isoproturon caused major concern and closure of the water intake (Figure 3.5.1). The threshold level for an alarm was set at 60%, and it can be seen in the figure that this level was exceeded three, nearly four times (RIWA, 2003).

## Case Studies 211

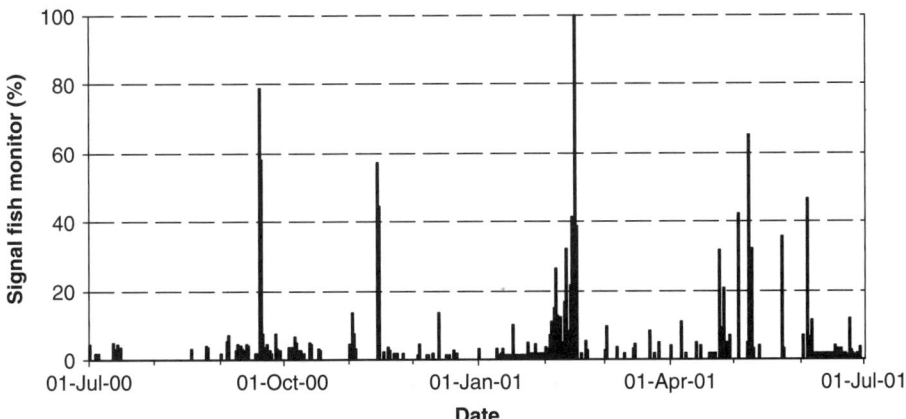

**Figure 3.5.2** Signal of the Aqua-Tox-Control fish monitor at the water company WRK water abstraction location Nieuwegein (NL) for the years 2000–2001; the threshold is at 60% (after RIWA, 2003)

### Case 3: Application of the Musselmonitor

The Musselmonitor took part in the same exposure trials for the EC Project SWIFT-WFD as discussed under Case 1. At the RIZA monitoring station Eijsden the in situ version of the early warning system was immersed directly in the River Meuse alongside the floating monitoring station during the period 5 April–20 May 2005; due to technical problems in data storage the evaluation was limited to a 4 weeks period: 22 April–20 May 2005. A set of 8 zebra mussels (*Dreissena polymorpha*) was used as primary sensor.

In the River Meuse trial one distinct alarm was recorded on 3 May 2005 (see Figure 3.5.3). The average valve opening of the eight mussels is presented in this figure. As usual, the baseline average is around 80–90% open, indicating that (nearly) all mussels are open for respiration and feeding. At 06:44 h the Musselmonitor generated a first alarm (D-alarm, indicating that the valve opening of the set minimum of 4 mussels had decreased by more than 20%), followed by the closure (C-) alarm at 06:52 h. The C-alarm was ended at 07:13 h, the D-alarm at 07:29 h. Neither the online LC SAMOS analyser nor the Daphnia Toximeter showed any response. Unfortunately the Musselmonitor alarm was not communicated in time to store water samples for further analysis.

In October 2006 the Musselmonitor, equipped with 8 zebra mussels (*Dreissena polymorpha*), installed at the water company Water Maatschappij Limburg (WML) along the River Meuse at Beegden (NL) showed an alarm. Chemical analysis of the water sample that was collected indicated the presence of the plasticizer N-butyl benzene sulfonamide (N-BBSA); its source remained unknown. In February 2007, as a result of this incident, the water company included this compound in the set of routine

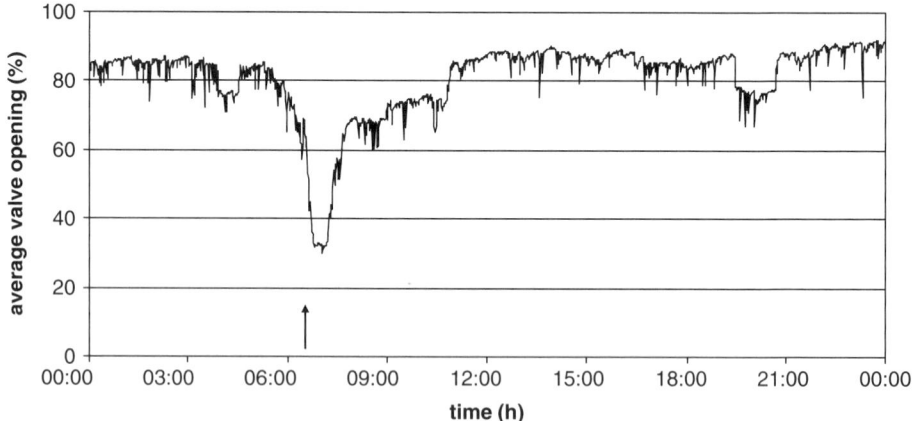

**Figure 3.5.3** Recording of 3 May 2005 of the average valve opening of 8 zebra mussels (*Dreissena polymorpha*) in the Musselmonitor exposed in situ at the transboundary monitoring station Eijsden (NL); 0% is fully closed, 100% fully open. The arrow indicates the time of the alarms that were generated by the Musselmonitor

measurements. Since then, N-BBSA has been chemically detected regularly, albeit in lower concentrations than when the alarm occurred (Wagenvoort et al., 2008c).

As part of the SWIFT-WFD trials at the 'Station d'Alerte de Huningue' alongside the River Rhine, near the borders of Switzerland, France and Germany and in the middle of industrial areas of these countries the flow-though version of the Musselmonitor was installed inside the monitoring station. Test organisms were zebra mussels (*D. polymorpha*). Unfortunately 2 mussels died early in the experiment, probably as a result of insufficient care in the sampling/transport and mounting of these animals. The system operated May–August 2006.

In Figure 3.5.4 the daily recordings of the average valve opening of the 6 remaining mussels are presented.

The top graph of 7 July 2006 shows typical baseline behaviour of mussels that feel well and feed continuously on the suspended particles in the water. The second graph is for 13 July, the next four consecutive days (14–17 July 2006). In this period the evaluation showed a strange phenomenon. From 13 July onwards, we see that from about midnight or slightly earlier, the average of the 6 mussels slowly drops to average values of 33%; the situation is restored at around 11:20 h. A priori this is not an abnormal behaviour; an (accidental) spill would show steeper closing and opening patterns. It becomes even more difficult to explain when we consider the days sequentially in the period 13–17 July. Each day at 23:00–24:00 h the average starts to drop, even down to around 15% open, while each day the normal situation returns to 90% and above around 11:00–12:00 h. All afternoon and evening the mussels stay fully open. This behaviour was seen every day over the period 11 July–2 August 2006. In the first instance it was suspected that external factors were affecting the valve movement response of the mussels in the first 12 hours of the day and we tried to find the origin of this phenomenon, but:

## Case Studies

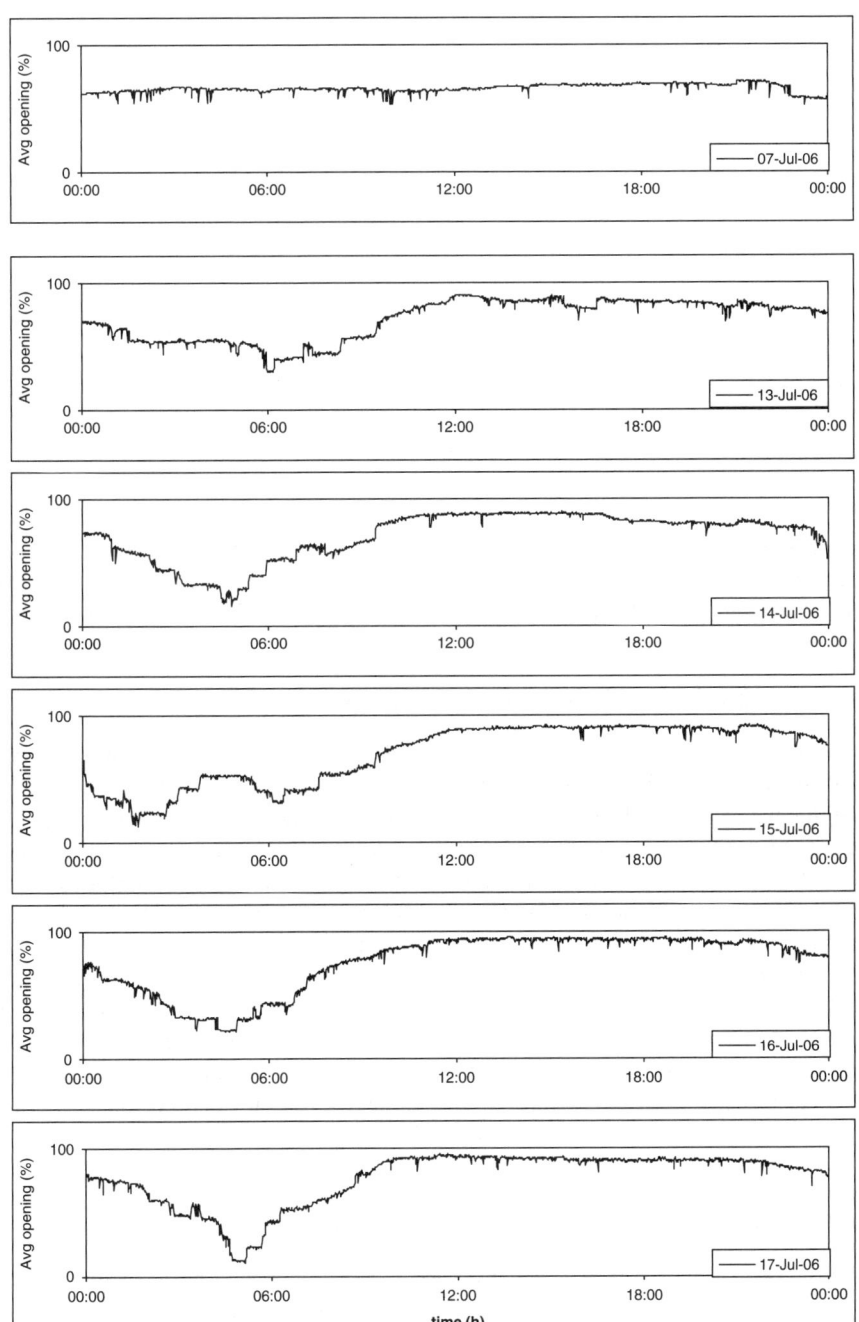

**Figure 3.5.4** 24-h recordings of 7 and 13–17 July 2006 of the average valve opening of 6 zebra mussels (*Dreissena polymorpha*) exposed in line in the Musselmonitor at the transboundary monitoring station Huningue (FR); 0% is fully closed, 100% fully open

- physical disturbance was ruled out: the station is remotely operated, and houses no staff, certainly not during the night;
- a daily pattern of activity was ruled out: in contrast to marine mussels, a daily biological rhythm has not been reported for zebra mussels;
- there was no evidence of a periodic change in the operation of the station between 0–12 h and 12–24 h: analyses are (semi) continuous during the entire day, and not focused on the 1st twelve hours;
- changes in characteristics of the water such as temperature, pH, conductivity, chlorophyll content, dissolved oxygen, suspended matter content (all obtained from the APRONA monitoring database) did not provide an explanation: none of these parameters showed a signal that was linked to the 0–12 h and 12–24 h cycle observed for the mussels, with a possible exception for oxygen. However, the minimum oxygen content during night was 7.5 mg/L which is well above the sustainability threshold (~4 mg/L).

As no cause for this effect could be identified (no physical disturbance, no macro parameters), there is suspicion that the cause may be that River Rhine water of lesser quality is passing by in these periods, possibly as a result of temporary emissions.

## Case 4: Application of the Algae Toximeter

In 2006 the Algae Toximeter operated at the RIZA transboundary monitoring station alongside the River Meuse in Eijsden (NL) recorded an alarm on three occasions. Twice the online liquid chromatographic analyser SAMOS (System for the Automated Monitoring of Organic substances in Surface waters, Slobodnik *et al.*, 1992) could identify the cause of the biological response: the herbicide Diuron. The third time no confirmation/identification could be obtained from the routine analyses. However, after samples collected during the alarm event were analysed it proved to be Terbutryn, a general use pesticide. After evaluation, it was decided that this compound should be added to the routine set of pesticide analyses, and since February 2007 analysis for this compound has been carried out twice daily in the River Meuse (Wagenvoort, 2008b).

This example shows that BEWS help to identify compounds that were not expected but appear to be present in the river water, sometimes at elevated concentrations, and how a routine monitoring program benefits from this finding.

## Case 5: Application of the Daphnia Toximeter

In spring 2004 the Daphnia Toximeter deployed at the RIZA monitoring station alongside the River Meuse in Keizersveer (NL) recorded several alarms. Routine chemical analysis did not reveal the identity of the offending compound(s). It took sophisticated chemical analyses, a combination of HPLC-DAD and Q-TOF MS techniques, to identify the so-far unknown contaminant as 3-cyclohexyl-1,1-dimethylurea. The maximum concentration of this compound in the River Meuse at the time of the alarm was estimated to be 5 µg/L. The response of the *Daphnia* to this compound was later confirmed with a short and a long-term toxicological test. The origin of the pollutant remained unknown (de Hoogh *et al.*, 2006). This result provides an excellent demonstration of the use of biological early warning systems. The biota detected elevated concentrations of a compound that was not detected by routine chemical methods. Analysis of samples collected at the alarm event revealed the identity of the toxic contaminant. In this example biological effect monitoring and specific compound analysis were complementary in providing insight in the identification of an unknown, potentially harmful substance in the river.

### 3.5.6 CONCLUSION

Biological early warning systems have developed into an invaluable tool in monitoring the aquatic environment. They operate continuously and detect adverse conditions in the aquatic environment. BEWS react to adverse environmental conditions in the water column by a physiological or behavioural response: they tell us that there is a problem with the quality of the water. Although biological sensors may detect many more compounds that those analysed by routine chemical measurements, they will never provide any information on the nature of the toxicant(s) and give only semi-quantitative information. Upon an alarm they can induce water sampling so that chemical analysis will identify the compound that caused the biological effect. As was demonstrated in the case studies, it is exactly in this that the strength of BEWS lies: they are complementary techniques (Gunatilaka and Diehl, 2001). BEWS are wide spectrum detectors, operating automatically around the clock (24/24 h, 7/7 d), but one needs chemical analysis for the identification and quantification of individual pollutants. All of the systems shown have matured into BEWS that have been tested and trialled in field operations. There are many proven applications that demonstrate the ability of this technology to support the implementation of the EC Water Framework Directive.

### ACKNOWLEDGEMENT

I would like to thank Almut Gerhardt (LimCo) and Arco Wagenvoort (AqWa) for their input and support, and the staff of the monitoring stations at Eijsden (RIZA) and Huningue (APRONA) for their hospitality and support.

### REFERENCES

Anon, 1997. *The Chlorination and Chloramination Handbook*. American Water Works Association, Denver CO, USA, 174 pp.

Anon, 2003. Protecting our water: Drinking water security in America after 9/11, *J American Water Works Association*, July 2003: 37–45.

Appels J., Küster E., van den Broeke J., Tangena B., de Zwart D. and Brandt A., 2007. Combination of an on-line biomonitor using light emitting bacteria and a UV spectrophotometer probe for Homeland Security and drinking water safety. In: *Proceedings of SPIE Conference*

on *Optics and Photonics in Global Homeland Security III*, 4 May 2007, TT Saito, D Lehrfeld, MJ DeWeert (eds), Vol. **6540**.
ASTM, 1995. Standard E1768-95. Guide for ventilatory behavioral toxicology testing of freshwater fish. American Assoc Testing Materials, West Conshohocken, PA, 14 pp.
Badino G., Meucci L. and Giacosa D., 2002. On line Po river water mussel monitoring, Int. Conf. 'Automation in Water Quality Monitoring', Vienna, Austria, 21–22 May.
Baldwin I.G. and Kramer K.J.M., 1994. Biological early warning systems (BEWS). In: *Biomonitoring of Coastal Waters and Estuaries*, KJM Kramer (ed) CRC Press, Boca Raton, pp. 1–28.
Barreto S. and Kramer K.J.M., 2003. Application of the Biological Early Warning System Musselmonitor® in monitoring of chlorinated drinking water. Poster presentation, SETAC-Europe, 28 April–1 May 2003, Hamburg, Germany.
Blübaum-Gronau E., Hoffmann M., Spieser O.H. and Scholz W., 2001. Continuous water monitoring. In: *Biomonitors and Biomarkers as Indicators of Environmental Change 2: A Handbook*; F.M. Butterworth, M.E. Gonsebatt-Bonaparte and A Gunatilaka (eds), Kluwer/Plenum, New York, pp. 123–41.
Bode H. and Nusch E.A., 1999. Advanced river quality monitoring in the Ruhr basin, *Water Science Technology* **40**: 145–52.
Brosnan T.M. (ed.), 1999. *Early Warning Monitoring to Detect Hazardous Events in Water Supplies*. International Life Sciences Institute, Risk Science Institute Workshop Report, Washington DC, USA, 45 pp.
Bulich A.A., 1979. Use of luminescent bacteria for determining toxicity in aquatic environments. In: *Aquatic Toxicology*. Marking, LL and RA Kimerle (eds) ASTM STP 667, pp. 98–106.
de Hoogh C.J., Wagenvoort A.J., Jonker F., van Leerdam J.A. and Hogenboom A.C., 2006. HPLC-DAD and Q-TOF MS techniques identify cause of *Daphnia* biomonitor alarms in the River Meuse, *Environ Sci Technol* **40**: 2678–85.
de Zwart D. and Slooff W., 1987. Continuous effluent biomonitoring with an early warning system. In: Bengston, Norberg-King and Mount (eds). *Effluent and Ambient Toxicity Testing in the Göta Älv and Viskan Rivers*, Sweden. Naturvardsverket report 3275.
de Zwart D., Kramer K.J.M. and Jenner H.A., 1995. Practical experiences with the biological early warning system 'Mosselmonitor', *Environ Toxicol Water Qual* **10**: 237–47.
Diamond J., Collins M. and Gruber D., 1988. An overview of automated biomonitoring – past developments and future needs. In: *Automated Biomonitoring: Living Sensors as Environmental Monitors*. Gruber, DS and JM Diamond (Eds), Ellis Horwood, Chichester, pp. 23–39.
EPA, 2005. Technologies and techniques for early warning systems to monitor and evaluate drinking water quality: A state-of-the-art review. US Environmental Protection Agency Office of Water Office of Science and Technology. Health and Ecological Criteria Division, Report EPA/600/R-05/156, 236 pp.
European Commission, 2000. Directive 2000/60/EC of the European Parliament and of the Council of 23 October 2000 establishing a framework for Community action in the field of water policy, *Off J Eur Comm* **L 327**: 1–72.
European Communities, 2009. Common implementation strategy for the Water Framework Directive (2000/60/EC). Guidance Document No. 19. Guidance on surface water chemical monitoring under the Water Framework Directive. Office for Official Publications of the European Communities, Technical Report 2009–25, Luxembourg, 222 pp.
Gerhardt A., 2000. Recent trends in online biomonitoring for water quality control. In: *Biomonitoring of Polluted Water*, A. Gerhardt (ed.). Trans Tech Publ, Uetecon-Zuerich, Switzerland, pp. 95–118.
Gerhardt A., Svensson E., Clostermann M., and Fridlund B., 1994. Monitoring of behavioral patterns of aquatic organisms with an impedance conversion technique, *Environ Int* **20**: 209–19.

Gerhardt A., de Bisthoven L.J. and Penders E., 2003. Quality control of drinking water from the River Rhine with the multispecies freshwater biomonitor, *Aquatic Ecosyst Health Managem* **6**: 159–66.

Gerhardt A., Ingram M.K., Kang I.J. and Ulitzur S., 2006. In situ on-line toxicity biomonitoring in water: recent developments, *Environ Toxicol Chem* **25**: 2263–71.

Gerhardt A., Kienle C., Allan I.J., et al., 2007. Biomonitoring with *Gammarus pulex* at the Meuse (NL), Aller (GER) and Rhine (F) rivers with the online Multispecies Freshwater Biomonitor®, *J Environ Monit* **9**: 979–85.

Grayman W.M., Deininger R.A. and Males R.M., 2001. *Design of early warning and predictive source-water monitoring systems*, American Water Works Association, Denver, USA, 328 pp.

Green U., Kremer J.H., Zillmer M. and Moldaenke C., 2003. Detection of chemical threat agents in drinking water by an early warning real-time biomonitor *Environ Toxicol* **18**: 368–74.

Gruber D. and Cairns J., 1981. Industrial effluent monitoring incorporating a recent automated fish biomonitoring system, *Water Air and Soil Pollution* **15**: 471–81.

Gruber D., Frago C.H. and Rasnake W.J., 1994. Automated biomonitors – first line of defense *J Aquat Ecosyst Health*, **3**: 87–92.

Gunatilaka A. and Diehl P., 2001. A brief review of chemical and biological continuous monitoring of rivers in Europe and Asia. In: *Biomonitors and Biomarkers as Indicators of Environmental Change 2: A Handbook*. F.M. Butterworth, M.E. Gonsebatt-Bonaparte, A Gunatilaka (eds) Kluwer/Plenum, New York, pp. 9–28.

Gunatilaka A., Diehl P. and Puzicha H., 2001. The evaluation of 'Dynamic Daphnia Test' after a decade of use. In: *Biomonitors and Biomarkers as Indicators of Environmental Change 2: A Handbook*. F.M. Butterworth, M.E. Gonsebatt-Bonaparte A Gunatilaka (Eds) Kluwer/Plenum, New York, pp. 29–58.

Irmer U. (Ed), 1994. *Continuous biotests for water monitoring of the river Rhine. Summary, recommendations, description of test methods*. Umweltbundesamt Texte 58/94, UBA, Berlin, Germany, 30 pp.

ISO, 2003. ISO 15839:2003, Water quality – On-line sensors/analysing equipment for water – Specifications and performance tests. ISO, International Organisation for Standardisation, Geneva, Switzerland, 30 pp.

ISO, 2007. ISO 13348 1-3:2007, Water quality – Determination of the inhibitory effect of water samples on the light emission of Vibrio fischeri (Luminescent bacteria test) – Part 1: Method using freshly prepared bacteria. Part 2: Method using liquid-dried bacteria Part 3: Method using freeze-dried bacteria. ISO, International Organisation for Standardisation, Geneva, Switzerland, 23 pp.

Jenner H.A., Polman H.J.G. and van Wijck R., 2004. Four years experience with a new chlorine dosing regime against macro fouling, *VGB Powertech*, **84**: 28–30.

Juhnke I., Besch W.K., 1971. Eine neue Testmethode zur Früherkennung akut toxischer Inhaltsstoffe im Wasser, *Gewässer und Abwässer*, **50/51**: 107–14.

Knie J., 1978. Der dynamischen Daphnietest – ein automatischer Biomonitor zur Uberwachung von Gewässern und Abwässern, *Wasser und Boden* **12**: 310–12.

Kramer K.J.M., Jenner H.A. and de Zwart D., 1989. The valve movement response of mussels: a tool in biological monitoring, *Hydrobiol* **188/189**: 433–43.

Kramer K.J.M. and Botterweg J., 1991. Aquatic biological early warning systems: an overview. In: *Bioindicators and Environmental Management*. D.W. Jeffrey and B. Madden (eds), Academic Press, London, pp. 95–126.

Kramer K.J.M. and Foekema E.M., 2001. The 'Musselmonitor®' as biological early warning system: The first decade. In: *Biomonitors and Biomarkers as Indicators of Environmental Change: A Handbook*; Vol **II**; F.M. Butterworth, M.E. Gonsebatt-Bonaparte and A Gunatilaka (eds). Kluwer/Plenum, New York, pp. 59–87.

Küster E., Dorusch F., Vogt C., Weiss H. and Altenburger R., 2004. On line biomonitors used as a tool for toxicity reduction evaluation of in situ groundwater remediation techniques, *Biosens Bioelectron* **19**: 1711–22.

Lechelt M., Blohm W., Kirschneit B., et al., 2000. Monitoring of surface water by an ultra-sensitive Daphnia Toximeter, *Environ Toxicol* **15**: 390–400.

Mikol Y.B., Richardson W.R., van der Schalie W.H. and Shedd T.R., 2007. An on-line real-time biomonitor for contaminant surveillance in water supplies, *J Amer Water Works Assoc* **99**: 107–15.

Morgan W.S.G., 1977. Biomonitoring with fish: an aid to industrial effluent and surface water quality control, *Prog Wat Tech* **9**: 703–11.

OECD, 1998. Guideline for Testing of Chemicals No 212. Fish, short-term toxicity test on embryo and sac-fry stages. Organization for Economic Co-operation and Development, Paris, 20 pp.

OECD, 2004. Guideline for Testing of Chemicals No 202. Daphnia sp, Acute Immobilization Test. Organization for Economic Co-operation and Development, Paris, 12 pp.

OECD, 2006. Guideline for Testing of Chemicals No 201. Freshwater Alga and Cyanobacteria, Growth Inhibition Test. Organization for Economic Co-operation and Development, Paris, 26 pp.

Paping L.L.M.J., 1991. Procesbeschrijving Westerschelde – doorpompkoelings and conditioneringsregime, Dow Benelux, Terneuzen, Netherlands, 23 pp.

Penders E., Wagenvoort A., de Hoogh C., Frijns N. and Kamps R., 2006. *Standardisation, quality assurance and data evaluation of on-line biological alarm systems. 2. Algae Toximeter* (BBE-Moldaenke, Kiel, D). KWR report BTO 2008.036, Kiwa Water Research, Nieuwegein, Netherlands, 57 pp.

RIWA, 2003. Jaarverslag 2001-2002, De Rijn. Vereniging van Rivierwaterbebrijven, RIWA-Rijn, Nieuwegein, Netherlands, 170 pp.

RIWA, 2007. Jaarrapport 2006, De Rijn. Vereniging van Rivierwaterbebrijven, RIWA-Rijn, Nieuwegein, Netherlands, 95 pp.

Schorr P. and Gruber D., 1991. Selecting a biological early warning system. In: Proc AWWA Water Quality Technol Conf, San Diego, November 11–15, 1991 AWWA, pp. 1491–1514.

Shedd T.R., van der Schalie W.H., Widder M.W., Burton D.T. and Burrows E.P., 2001. Long-term operation of an automated fish biomonitoring system for continuous effluent acute toxicity surveillance, *Bull Environ Contam Toxicol* **66**: 392–9.

Slobodnik J., Brouwer E.R., Geerdink R.B., Mulder W.H., Lingeman H. and Brinkman U.A.Th., 1992. Fully automated liquid chromatographic separation system for polar pollutants in various types of water, *Anal Chim Acta* **268**: 55–65.

Stewart S., Dick J. and Laming P., 2004. Development of Automated Biomonitors in the optimization of aquaculture production and environmental protection. Fourth SETAC World Congress, November 19, 2004, Portland OR, USA.

van der Schalie W.H., Gardner H.S., Bantle J.A., et al., 1999. Animals as sentinels of human health hazards of environmental chemicals, *Environ Health Perspect* **107**: 309–15.

van der Schalie W.H., Trader D.E., Shedd T.R., Widder M.W. and Brennan L.M., 2005. A residual chlorine removal method to allow drinking water monitoring by biological early warning systems. USACEHR Technical Report 0501, US Army Center for Environmental Health Research, Fort Detrick MD, USA.

van der Schalie W.H., James R.R. and Gargan T.P., 2006. Selection of a battery of rapid toxicity sensors for drinking water evaluation, *Biosens Bioelectron* **22**: 18–27.

von Danwitz B., Blübaum-Gronau E., Diehl P., et al., 1998. Recommendation on the deployment of continuous biomonitors for the monitoring of surface waters. Working Group of the Federal States on Water Problems (LAWA), Berlin, Germany, 46 pp.

Wagenvoort A., de Hoogh C., Penders E., Frijns N. and Kamps R., 2005. *Standardisation, quality assurance and data evaluation of on-line biological alarm systems. 1. Daphnia Toximeter* (BBE-Moldaenke, Kiel, D) Report BTO 2005.15, KIWA, Nieuwegein, The Netherlands, 55 pp.

Wagenvoort A., de Hoogh C., Kramer K., Engels P., Penders E. and Pikaar K., 2008a. *Standardisation, quality assurance and data evaluation of on-line biological alarm systems. 3. Mosselmonitor®* (Delta Consult, Kapelle, NL). KWR report BTO 2008.040, Kiwa Water Research, Nieuwegein, The Netherlands, 51 pp.

Wagenvoort A., de Hoogh C., Frijns N., Scaf W. and Pieper G., 2008b. *Alarm Evaluation of the bbe Daphnia Toximeter and the bbe Algae Toximeter. Examples of alarm events in 2006 and 2007.* In: KWR report BTO 2008.041, Kiwa Water Research, Nieuwegein, The Netherlands.

Wagenvoort A., de Hoogh C., Engels P. and Koppelaar S., 2008c. *Mosselmonitor® detects discharges and pollution in river Meuse. Alarm evaluation of the Mosselmonitor® at the abstraction point at Beegden.* In: KWR report BTO 2008.042, Kiwa Water Research, Nieuwegein, The Netherlands.

# 3.6
# Biological Markers of Exposure and Effect for Water Pollution Monitoring

Josephine A. Hagger and Tamara S. Galloway

3.6.1 Introduction
3.6.2 Case Study 1. Tributyltin (TBT) and Imposex/Interesex
3.6.3 Case Study 2. Fluorescence of Aromatic Hydrocarbon Metabolites
3.6.4 Integrating Biomarkers of Exposure and Sublethal Effects
3.6.5 Conclusion
References

## 3.6.1 INTRODUCTION

The use of biomarkers (functional biological measures of exposure to stressors expressed at the suborganismal, physiological, or behavioural level) (McCarty and Munkittrick, 1996) as surrogate measures of biological impact within laboratory and field studies has been prevalent for many years. However, the incorporation of biomarkers into regulatory legislation for environmental risk assessment (ERA) has generally been lacking. The focus for ERA has continued to be on chemical measurements as standard practice, made in the context of environmental quality standards (EQSs) for risk assessment of e.g. hazardous substances.

In general, biomarkers may be classified into three types: biomarkers of exposure, effect and susceptibility (Chambers et al., 2002). Biomarkers of exposure indicate that an organism has experienced exposure to a toxicant or other stressor; however, the change in the biomarker is not necessarily related directly to the toxicant's specific mechanism of action and may not be predictive of the degree of adverse effects on

---

*Rapid Chemical and Biological Techniques for Water Monitoring*   Edited by Catherine Gonzalez, Philippe Quevauviller and Richard Greenwood
© 2009 John Wiley & Sons, Ltd

either the organism or on higher levels of biological organization (e.g. population or community level) (Chambers *et al.*, 2002). Biomarkers of exposure provide quantitative and qualitative estimates of exposure to various compounds, and with particular reference for water pollution monitoring, exposure biomarkers may have the potential to offer an alternative to some chemical analyses or to measure effects of short-lived chemicals as well as giving a more biologically relevant indication of exposure (Walker *et al.*, 1996).

Biomarkers of effect are associated specifically with the toxicant's mechanism of action and are sufficiently well characterized to relate the degree of biomarker modification to the degree of adverse effects (Chambers *et al.*, 2002). Thus, in terms of ERA, biomarkers of effect can elucidate qualitative aspects of hazard identification by not only demonstrating that hazard is occurring but also probable mechanisms of action. These biomarkers can provide insights into both the causal factors of the hazard and its ecological consequences, depending on the degree of specificity of the biomarker for the pressure and, the degree to which its expression to higher order effects is understood (e.g. imposex in gastropods as both a biomarker of organotin exposure, reproductive impairment and population level consequences). In contrast to biomarkers of effect or exposure, biomarkers of susceptibility do not represent stages along the dose-effect continuum, but reflect an increase in the rate of transition between steps along that continuum (Figure 3.6.1) (Schlenk, 1999). They indicate the inherent or acquired ability of an organism to respond to the challenge of exposure to a specific xenobiotic substance, and include genetic factors and changes in receptors that alter the susceptibility of an organism to exposure. Therefore, biomarkers of susceptibility can be used in ERA to provide a characterization of variability which can be used in defining uncertainty factors (Schlenk, 1999). However, it is worth noting that no single biomarker can unequivocally measure environmental degradation and that the ability to differentiate between clean and polluted sites would be at best incomplete using a single biomarker approach (Galloway *et al.*, 2004a). A suite of endpoints, at

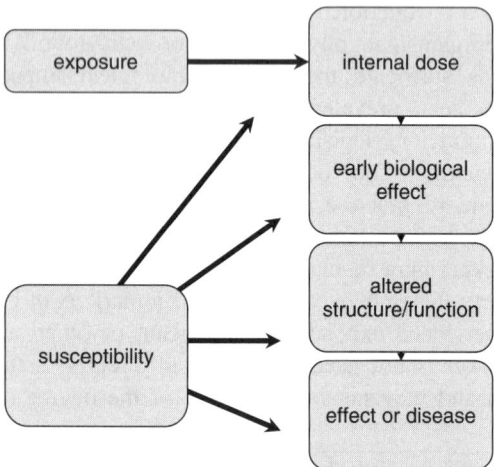

**Figure 3.6.1**  The biomarker continuum

# Introduction

different levels of biological organization, allows for a better evaluation of the hazard and allows for a weight of evidence approach to be adopted which minimizes the influence of natural variation and allow for the discrimination of clean/healthy and polluted/unhealthy sites to be carried out successfully.

Figure 3.6.2 illustrates how, ideally, biomarkers can be integrated into an ERA programme in conjunction with chemical analysis, toxicity testing and biological endpoints to identify water bodies at risk from pollution. The application of suites of biomarkers as a component part of the risk assessment process can provide hugely informative, descriptive information to characterize the relationship between contamination and the health status of exposed organisms. This characterization is achieved by consideration of individual biomarkers which may diagnose exposure to certain classes of contaminants or by assessing integrated biomarker responses. Table 3.6.1 illustrates some of the biomarkers that can be used for monitoring water pollution.

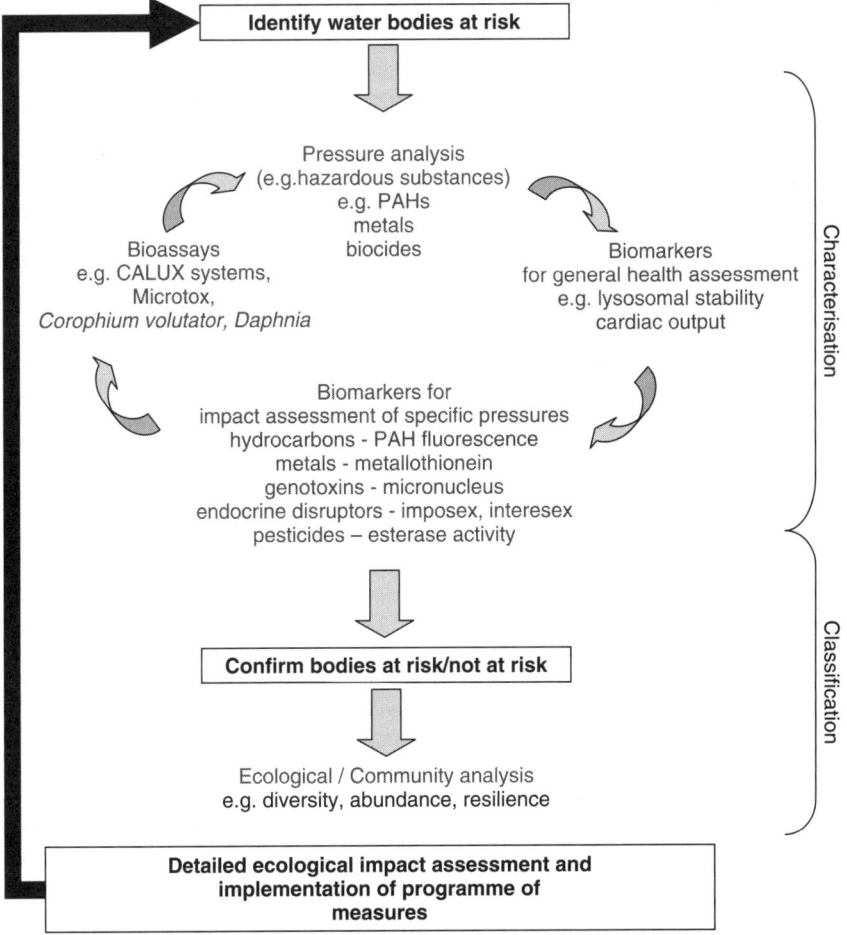

**Figure 3.6.2** An example of a hierarchal approach to risk assessment

**Table 3.6.1** Biomarkers of exposure and of sublethal effect. The examples chosen are relatively simple to perform and have been applied to a wide range of species in environmental monitoring programmes

| Biomarker | Chemical class | Details |
|---|---|---|
| Inhibition of acetylcholinesterase activity | Pesticides (organophosphates and carbamates) | Exposure to these pesticides causes inhibition of the enzyme which can be detected using a simple colour change assay in samples of blood or tissue from exposed animals (Mineau, 1991, (Galloway et al., 2002a)) |
| Fluorescence of aromatic hydrocarbon metabolites | Polyaromatic hydrocarbons | A fluorescence assay can be used to detect pyrenes and other PAHs and their metabolites, in fish bile and in the urine and blood of selected invertebrates (Aas et al., 1998), (Watson et al., 2004b) |
| Induction of metallothionein | Trace metals | Induction of metallothionein and other similar metal binding proteins, can be measured in the tissues (liver and kidney of vertebrates, digestive gland, hepatopancreas of invertebrates) of many different organisms from bivalves to humans (Viarengo et al., 1997) |
| Induction of vitellogenin | Estrogenic compounds | Vitellogenin is an egg yolk precursor protein which can be induced in both males and females following exposure to estrogens or estrogen mimics. It is measured in the plasma of fish using an immunoassay (Sumpter and Jobling, 1995) |
| Micronucleus formation | Genotoxins (e.g. ionising radiation, dioxins, PAHs) | Damage to chromosomes can cause small, cytoplasmic micronuclei to be formed at the end of cell division. They can be detected using a microscope (Wrisberg and Rhemrev, 1992) |
| Imposex/intersex | Organotin compounds | Imposex is a pseudo-hermaphroditic condition in female gastropods (snails) caused by TBT and manifested by the development of a false penis (Bryan et al., 1986; Gibbs et al., 1988). Intersex-affected females exhibit male features on female pallial organs (inhibition of the ontogenetic closure of the pallial oviduct), or female sex organs are supplanted by the corresponding male formations (Bauer et al., 1997) |
| Phagocytosis | Immunotoxins (e.g. pesticides, PCBs, PAHs, metals, organotins) | Phagocytosis of foreign particles by immune cells can be monitored using a microscope. This can provide a measure of the immune competence of the animal and also its general well-being (Galloway and Handy, 2003) |

Case Study 1. Tributyltin (TBT) and Imposex/Interesex    225

Table 3.6.1  (continued)

| Biomarker | Chemical class | Details |
| --- | --- | --- |
| CAPMON: computer aided heart rate monitoring | Many chemical and environmental impacts | Heart rate provides a general indication of an animal's metabolic status. The CAPMON uses infra-red sensors to monitor cardiac activity non-invasively in mussels and crabs (Depledge and Andersen, 1990) |
| Lysosomal stability | Many chemical and environmental impacts | Lysosomes are subcellular organelles involved in the metabolism of toxicants. Destabilisation of the lysosome can be measured using a microscope and has been used as indication of poor general health (Lowe et al., 1995) |
| Burrowing behaviour | Many chemical and environmental impacts | When amphipods are placed on contaminated sediment, characteristic changes to their burrowing behaviour can occur and have been used as an indication of toxicity (PARCOM, 1995) |

Source: Environment Agency (http://www.environment-agency.gov.uk/commondata/acrobat/tc_tbt_t_v2_1008150.pdf)

The aim of this chapter is to show how biomarkers can be used as simple, rapid cost-effective techniques to aid in an integrated approach to environmental management and risk assessment. Specifically, we use the Fal and Helford Estuaries located in the south west of England as a case study to highlight the application of biomarkers and to illustrate how they can aid ERA.

## 3.6.2  CASE STUDY 1. TRIBUTYLTIN (TBT) AND IMPOSEX/INTERESEX

The induction of a pseudohermaphroditism in the gonochoristic marine gastropod, the dogwhelk *Nucella lapillus*, remains the best characterized example of endocrine disruption in wildlife and the only example where an unequivocal causal association with a single class of chemical has been proven. Smith (1971) first named the superimposition of male genitalia in female gastropods as imposex and subsequent research has undisputedly linked the condition in *N. lapillus* to exposure to tributyltin (TBT) (Gibbs and Bryan, 1986; Bryan and Gibbs, 1991; Schulte-Oehlmann et al., 2000). In its most severe form, imposex measured as Vas Deferens Sequence Index (VDSI) prevents the reproductive (egg laying) capacity of affected females, leading to sterility and thus the ecologically significant decrease in population (Bryan et al., 1986; Gibbs and Bryan, 1986). Figure 3.6.3 illustrates a stage 5 imposexed female dogwhelk where the vaginal opening is blocked by the vas deferens which is connected to a pseudopenis.

Due to the unprecedented link between the presence of imposex and the decline in dogwhelk populations, the presence of imposex has been included as a biological

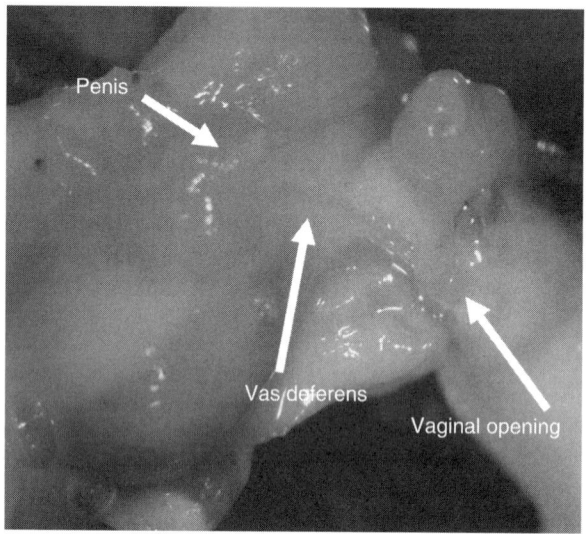

**Figure 3.6.3** Imposexed female dogwhelk (VDSI stage 5)

effects technique in many national and international regulatory monitoring programmes. Table 3.6.2 illustrates the OSPAR monitoring guidelines for the dogwhelk *Nucella lapillus* as well as the corresponding classes that are used to define impact of TBT within the Commission (EC) of the European Union (EU) Water Framework Directive (WFD) (2000/60/EC). It is also worth noting that imposex is the only biological effects measure that has been advocated by the UK Technical Advisory Group (UKTAG) on the WFD due to its unprecedented link with TBT exposure. Both exceedance of EQS for TBT and VDSI are used together in an integrated approach to characterize risk for this pressure and, in this sense, the use of the biomarker provides a measure of validation, reducing uncertainty in the risk assessment process associated with diffuse pesticide pressures in transitional and coastal waters.

Historically, the Fal and Helford Estuaries (Figure 3.6.4) have been monitored for pressures from organotin contamination (Bryan, 1980; Bryan and Gibbs, 1983). In particular the Falmouth area has experienced high levels of TBT with the principal source being Falmouth Dockyard, although sediment hotspots of TBT have been found at Mylor Creek and, more recently, at Porth Navas in the Helford Estuary (Langston et al., 2003). It has been reported that TBT concentrations are diluted rapidly with distance from the major sources at Falmouth, though levels above the EQS (2 ng $l^{-1}$) may be sustained nearby, as in Penryn Creek (mean 12.8 ng $l^{-1}$, 2004–2005) and are reflected in body burdens in species such as mussels which have been recorded as exceeding OSPARs upper ecotoxicological assessment criteria (Galloway et al., 2007).

*Nucella lapillus* populations continue to be absent from most of the Fal Estuary due to historical contamination by TBT. Observations on neogastropods just outside the mouth of the Fal confirm that imposex levels remain high and have not changed since studies began in the 1980s (Bryan et al., 1986). Hence, due to the extinction of

*Case Study 1. Tributyltin (TBT) and Imposex/Interesex*

**Table 3.6.2** Assessment classes for *Nucella lapillus* Vas Deferens Sequence Index (VDSI) used in the OSPAR monitoring guideline (OSPAR agreement 2004-15) and the equivalent impact category for assessment in the Water Framework Directive (WFD)

| OSPAR Assessment Class | WFD Impact Category | *Nucella* VDSI | Effects and impacts |
|---|---|---|---|
| A | None | <0.3 | The level of imposex in the most sensitive gastropod species is close to zero (0 – ~30% of females have imposex) indicating exposure to TBT concentrations close to zero, which is the objective in the OSPAR strategy of hazardous substances. |
| B | None | 0.3–<2.0 | The level of imposex in the more sensitive gastropod species (~ 30 – ~100% of the females have imposex) indicated exposure to TBT concentrations below the EAC derived for TBT. E.g. adverse effects in the mores sensitive taxa of the ecosystem caused by long-term exposure to TBT are predicted to be unlikely to occur. |
| C | Low | 2.0–<4.0 | The level of imposex in the more sensitive gastropod species indicates exposure to TBT concentrations higher than the EAC derived for TBT. E.g. there is a risk of adverse effects, such as reduced growth and recruitment, in the more sensitive taxa of the ecosystem caused by long-term exposure to TBT. |
| D | Moderate | 4.0–5.0 | The reproductive capacity in the populations of the more sensitive species, such as *Nucella lapillus*, is affected as a result of the presence of sterile females, but some reproductively capable females remain. E.g. there is evidence of adverse effects, which can be directly associated with the exposure to TBT. |
| E | High | >5.0 | Populations of the more sensitive gastropod species, such as *Nucella lapillus*, are unable to reproduce. The majority, if not all the females within the population have been sterilised. |
| F | High | – | The populations of the more sensitive gastropod species, such as *Nucella lapillus* and *Ocinebrina aciculate*, are absent/expired. |

*Source:* OSPAR agreement 2004-15 available from www.ospar.org

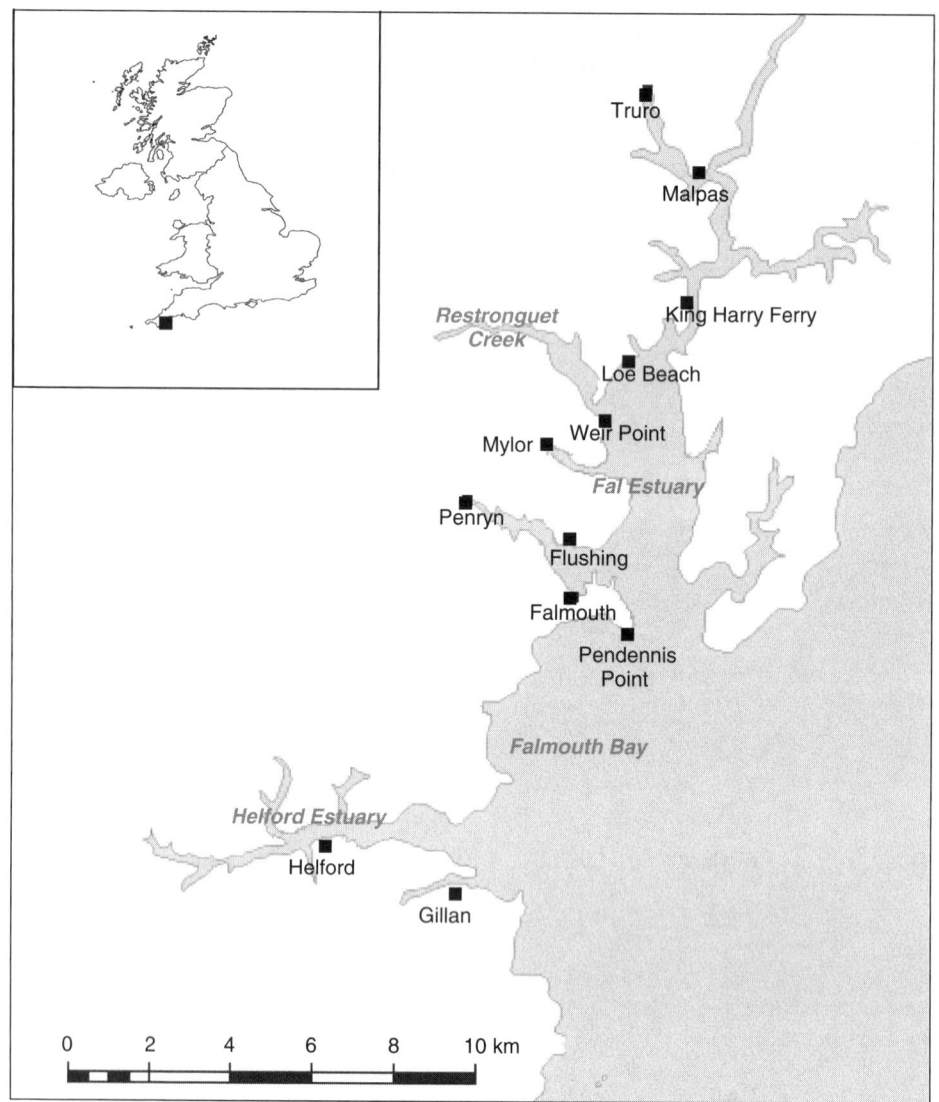

**Figure 3.6.4** Map of Falmouth and Helford Estuaries in South West England

populations of the highly sensitive *Nucella lapillus* from TBT exposure, OSPAR have also recommended the use of other relevant species as illustrated in Table 3.6.3. For example, Bauer *et al*. 1995 noted that although female periwinkles, *Littorina littorea*, did not display imposex following TBT exposure, malformations of the pallial genital tract were observed, and they termed this phenomenon intersex.

The intersex phenomenon is a gradual transformation of the female pallial tract and can be described by a progressive scheme with five stages (Bauer *et al*., 1997).

**Table 3.6.3** Biological-effect assessment criteria for TBT prepared and agreed during the technical TBT workshop, 6–7 November 2003, The Hague. Assessment criteria for imposex in *Nucella lapillus* are presented alongside equivalent VDSI/ISI values for sympatric populations of other relevant species (OSPAR agreement 2004-15 www.ospar.org)

| Assessment class | *Nucella* VDSI | *Nassarius* VDSI | *Buccinum* ~ PCI | *Neptunea* # VDSI | *Littorina* ISI |
|---|---|---|---|---|---|
| A | <0,3 | <0,3 | <0,3 | <0,3 | <0,3 |
| B | 0,3 – <2,0 | | | 0,3 – <2,0 | |
| C | 2,0 < 4,0 | 0,3 < 2,0 | 0,3 < 2,0 | 2,0 – 4,0 | |
| D | 4,0 – 5,0 | 2,0 – 3,5 | 2,0 – 3,5 | 4,0^ | 0,3 – <0,5 |
| E | >5,0 | >3,5 | >3,5 | | 0,5 – 1,2 |
| F | – | | | | >1,2 |
| | | Stroben *et al.*, 1995 | Stroben *et al.*, 1995, and field evidence that *Neptunea* has similar sensitivity as *Nassarius*, | # field evidence that *Neptunea* has similar sensitivity as *Nucella*, ^ highest value possible | Oehlmann, 2002 ASMO 02/4/8 |
| | | | ~ No correlation established | ~ No correlation established | |

Stage 0 represents a normal female without any pathomorphological changes. In stage 1, the female genital opening is enlarged by a proximal slit and the bursa copulatrix is split ventrally which may lead to a loss of sperm during copulation and, consequently, a reduction in reproductive success may occur. In stage 2, the entire pallial oviduct is split ventrally, forming an open structure with the internal lobes of the female glands exposed to the mantle cavity. In stage 3, the female pallial organs are totally or partially supplanted by a male prostate gland. Additionally, in stage 4, a seminal groove and a small penis occur. Female snails in stages 2–4 are classified as sterile because oocytes and the capsular material leak into the mantle cavity (in stage 2), or the glands responsible for extraembryonic nourishment and/or the formation of the egg capsules are missing (in stages 3 and 4) (Bauer et al., 1997).

For measurement of intersex intensities, the mean intersex stage of all females is calculated to provide the intersex index (ISI). This index provides the best results for assessing the reproductive capability of a population (Bauer et al., 1995; Bauer et al., 1997). A value of 0 indicates that only normal females occur, and no restrictions of the reproductive capability are to be expected. ISI values above 0 show that intersex-affected females can be found and that reproductive success may be reduced. An ISI above 1 indicates the presence of some sterile females, and values above 2 indicate that most females in a population are sterilized due to intersex development.

The periwinkle, *Littorina littorea*, unlike the dogwhelk *Nucella lapillus*, is present throughout the Fal and Helford Estuaries (Figure 3.6.4). The ISI values from sites within the Fal and Helford Estuaries are illustrated in Figure 3.6.5, and were calculated from approximately 20 female periwinkles as per the method of Bauer et al. (1997) described above. Periwinkles from Malpas, Pendennis point, Maenporth, Gillan and Helford exhibited either no or very low levels of intersex and would be classified as A in the OSPAR assessment class which corresponds to the 'no impact' category under the WFD (as illustrated in Table 3.6.2). Loe Beach and Flushing exhibited ISI values of

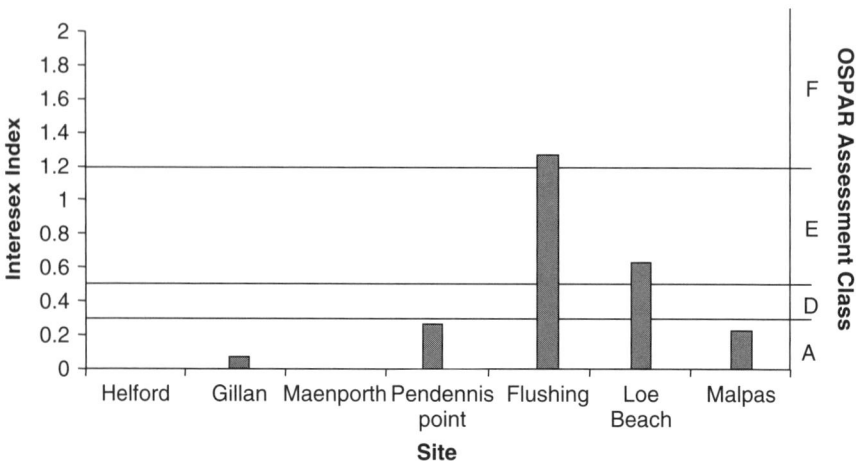

**Figure 3.6.5** ISI values and corresponding OSPAR class for *Littorina littorea* from the Fal and Helford Estuaries

## Case Study 1. Tributyltin (TBT) and Imposex/Interesex

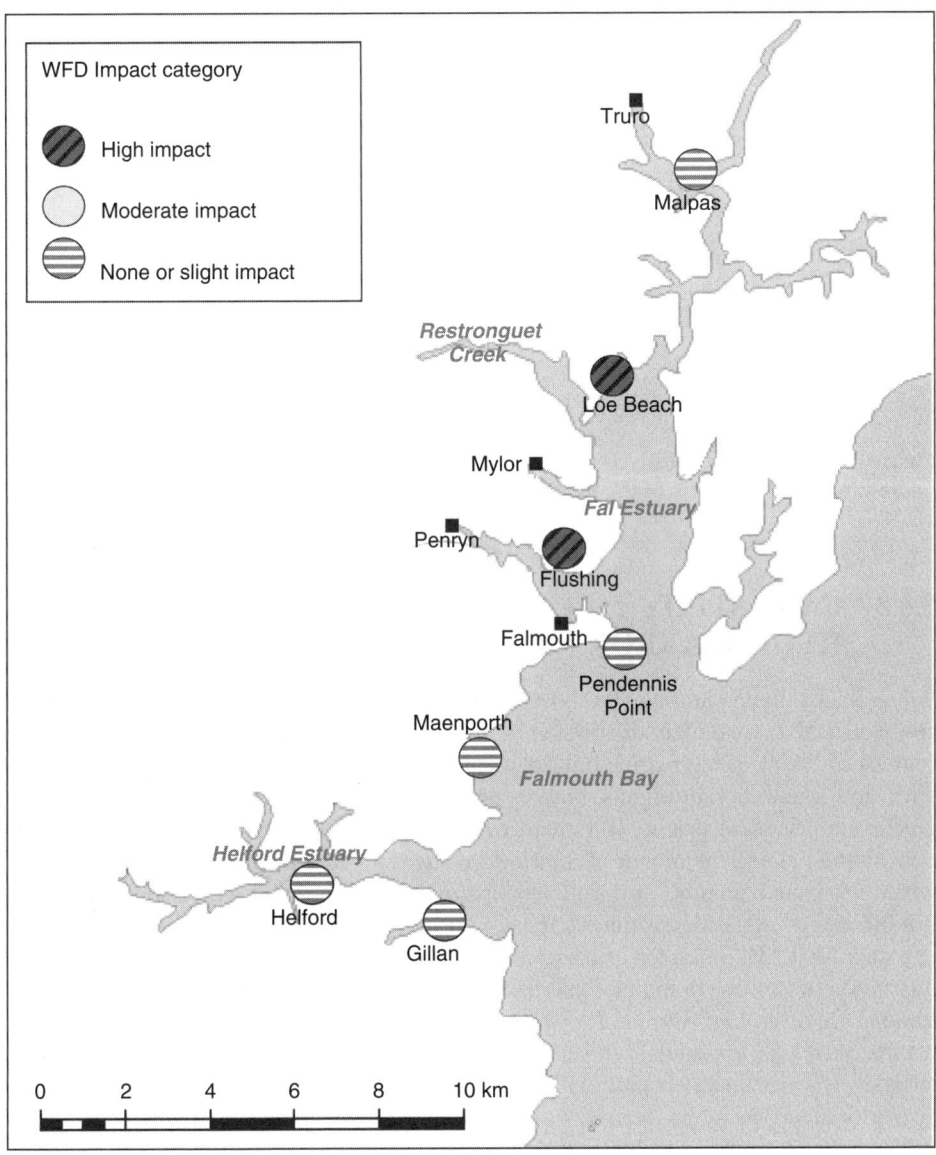

**Figure 3.6.6** Water Framework Directive impact classes for TBT based on the incidence of ISI in *Littorina littorea* from the Fal and Helford Estuaries

0.63 and 1.26 respectively, corresponding to 'highly impacted' for TBT (Figure 3.6.6) as defined by the WFD. Figure 3.6.7 illustrates the relationship between ISI and TBT body burdens for *Littorina littorea* obtained from the Fal and Helford Estuaries as well as the Tamar Estuary and Southampton Waters located along the south coast of England.

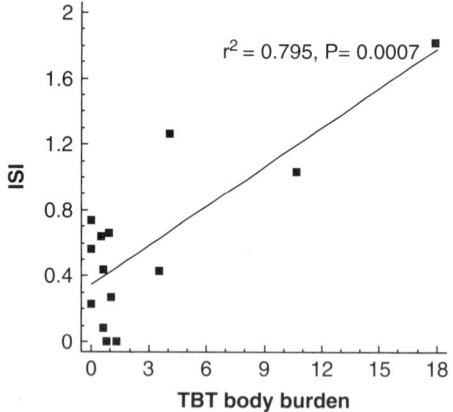

**Figure 3.6.7** Regression analysis of ISI and TBT body burdens (µg Sn g$^{-1}$ dry wt.) in *Littorina littorea* from four estuaries in the south of England

### 3.6.3 CASE STUDY 2. FLUORESCENCE OF AROMATIC HYDROCARBON METABOLITES

Polyaromatic hydrocarbons (PAHs) are persistent, toxic and sometimes carcinogenic compounds that are widely distributed in the aquatic environment. There are three main sources of PAHs in the marine environment: combustion of fossil fuels (pyrolytic origin), slow maturation of organic matter (petrogenic origin) and degradation of biogenic precursors (diagenic origin) (Baumard *et al.*, 1998). Historical records for hydrocarbons in the Fal and the mouth of the Helford suggest general contamination in the area which originates, possibly, from the dockyard and shipping, but may also be related to run-off and/or aerial deposition (Langston *et al.*, 2003). However, there are scarcely any data for PAHs in either estuaries, as the presence of PAHs in water is often very difficult to detect due to the fact that they are very sparingly soluble and hence concentrations are often low (Lin *et al.*, 1994). However, due to their hydrophobic/lipophilic nature PAHs due accumulate in sediments. Although, as with water analysis of PAHs, analysis of sediments is equally laborious and time consuming due to numerous clean up and extractions steps. Hence, when assessing PAHs it may be more pertinent to detect them in tissues and biological fluids of organisms. However, routine monitoring of parent aromatic hydrocarbons in tissue samples often reveal only small traces as many organisms are capable of rapidly metabolizing PAHs, which means that it is often more relevant to monitor concentrations of PAH metabolites. In addition, the determination of PAH metabolites is often more toxicological relevant, as it is often the biotransformed active metabolites that exert significant toxic, mutagentic and carcinogenic effects (James, 1989; Stroomberg *et al.*, 1996).

In the mid 1980s, Krahn and co-workers (1984) developed a moderately inexpensive, rapid screening method for estimating the relative amount of PAH metabolites present in fish bile. Their method used high-performance liquid chromatography with fluorescence detection (HPLC-F) for estimating bile metabolites. In brief, fluorescent peak areas for a fixed excitation/emission wavelength, excluding those from the early gradient (Lin *et al.*, 1996), are summed to estimate the exposure of fish to PAHs.

A simpler, less expensive method of estimating PAH biliary concentrations, that produces similar correlations with the HPLC-F method, is the synchronous fluorescent spectroscopic (SFS) method (Lin *et al.*, 1994). Furthermore, the fixed fluorescence (FF) technique was developed that measures the fluorescence response of diluted bile at a fixed excitation/emission wavelength pair without the need to perform any chromatographic separation, thus making it an even quicker method for detecting PAH metabolites. It is worth noting that although the results obtained with the FF technique cannot be compared directly with those obtained by HPLC-F, both methods discriminate equally well between sites impacted with PAHs and reference sites (Lin *et al.*, 1996). In an evaluation of both FF and SFS techniques, Dissanayake and Galloway (2004) recommended that the FF technique be used to conduct site comparisons using peak area quantification, whereas SFS analysis is more suited to illustrating differences between sites through spectral comparisons.

The use of PAH metabolites in routine monitoring has been supported by a number of studies that have shown correlations between the metabolites and PAH concentrations in sediments. Johnston and Baumann (1989) reported a correlation between relative levels of B[a]P equivalents in the bile of brown bullhead fish from the Black River, Ohio, and B[a]P levels in sediments. In addition to bilary metabolites being correlated to sediment concentrations of PAHs, it has been shown that fish with high levels of PAH metabolites also show signs of health problems. During a study of fish collected from Lake Erie tributaries, a significant association between the occurrence of barbell abnormalities and concentrations of naphthalene-type metabolites, and between the occurrence of raised lesions and the concentration of benzo(a)pyrene-type metabolites, was reported (Yang *et al.*, 2003; Yang and Baumann, 2006). Furthermore, techniques used to measure PAH metabolites have been adapted successfully to detect PAH exposure in polychaete worms (Giessing *et al.* 2003) and crustaceans (see

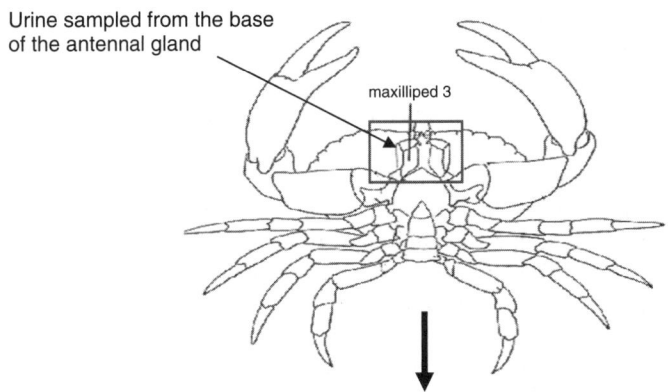

**Figure 3.6.8** Illustration of urine sampling from a crab for detection of PAH metabolites

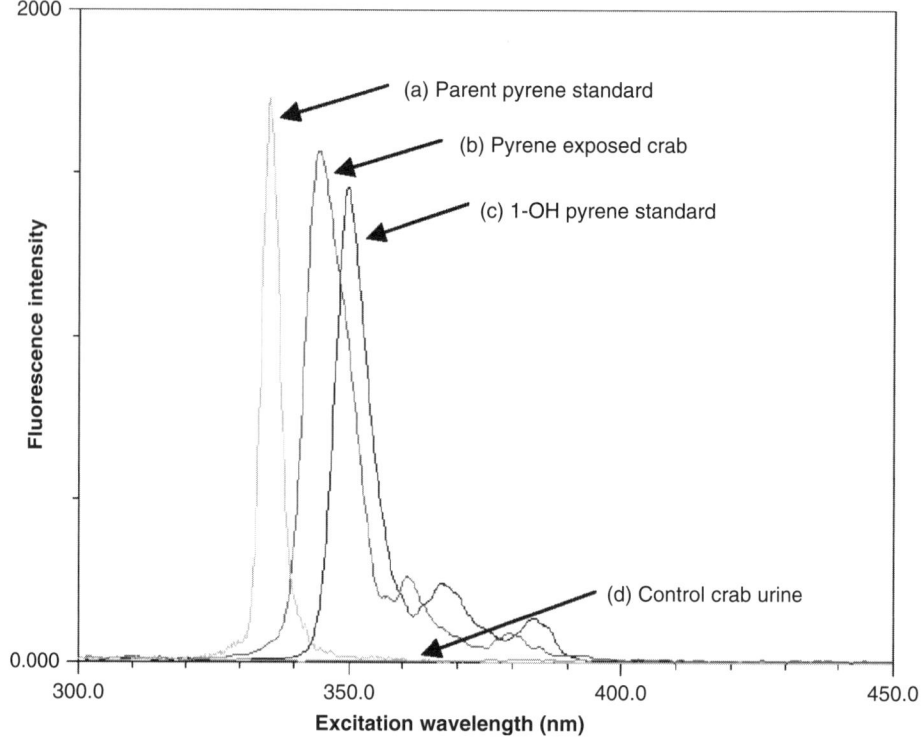

**Figure 3.6.9** Fluorescence (SFS, $\Delta\lambda$ 37 nm) spectra of crab urine (a) containing 100 μg $L^{-1}$ pyrene standard, (b) from an exposed crab (200 μg $L^{-1}$ pyrene), (c) containing 100 μg $Ll^{-1}$ 1–OH pyrene (metabolite of pyrene) and (d) a control crab

Figures 3.6.8 and 3.6.9) (Dissanayake and Galloway, 2004; Watson *et al.*, 2004a; Watson *et al.*, 2004b). Following laboratory validation of FF and SFS techniques using urine and haemolymph from crabs exposed to pyrene (Watson *et al.*, 2004a), these techniques have proven to be a reliable tool for assessing PAH contamination in environmental monitoring. Watson *et al.* (2004b) showed that PAH metabolites in crab urine could be used to discriminate between clean and contaminated sites as well as being sufficiently sensitive to detect gradients of PAH contamination along a Norwegian coastline. Galloway *et al.* (2004b) showed a moderately strong relationship ($R^2 = 0.751$) between PAH metabolites in crab urine measured using FF and PAH residues in sediments measured using GC-MS within Southampton Water (UK), which highlights the usefulness of this technique in conjunction with chemical analysis.

From Figure 3.6.10, it is apparent that pyrogenic PAH equivalents dominate the urine of crabs within the Fal and Helford Estuaries. This indicates that the source of the PAH exposure is less likely to be from ingestion of oil (such as might be expected after a fresh spillage) but more likely from larger 4–6 ringed pyrogenic products that accumulate in urban settings and harbours due to the combustion of petrols and residues from tyres from road traffic. This is particularly evident at Malpas, King Harry Ferry and at Flushing (opposite the Falmouth docks) in the Fal Estuary (Figure 3.6.10).

## Case Study 2. Fluorescence of Aromatic Hydrocarbon Metabolites

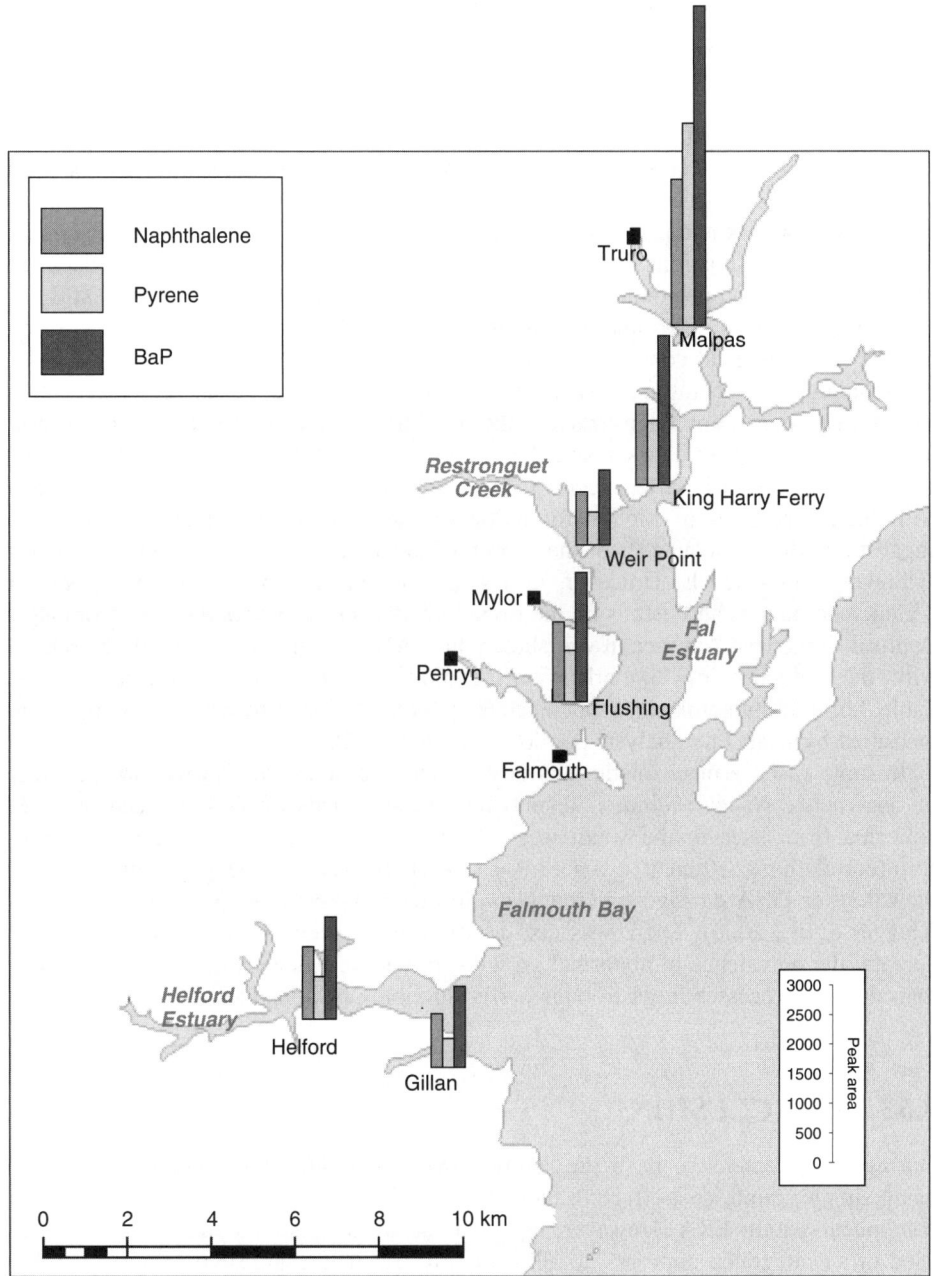

**Figure 3.6.10** Rapid qualitative screening of PAH metabolites in urine from *Carcinus maenas* collected in the Fal and Helford Estuaries

## 3.6.4 INTEGRATING BIOMARKERS OF EXPOSURE AND SUBLETHAL EFFECTS

As can be seen from the above discussion, biomarkers offer potential surrogate measures of chemical contamination and it is in this context that they have been most widely used. When incorporated into structured survey programmes, they can be used to help in the identification of pollutants responsible for environmental degradation, to identify sites at risk and prioritize their management. However, their greatest potential is in identifying situations in which exposure and adverse health consequences are linked. By combining biomarkers of exposure and sublethal effects (Table 3.6.1), it becomes possible to determine not just the presence of the pollutant in an inert form, but also the biological consequences of exposure.

In appropriate combinations (Figure 3.6.2), biomarkers can be used in a weight-of-evidence approach to show that organisms have been exposed to contaminants, or that exposure has resulted in a deterioration of health. For example, the relative tolerance of bivalve species to the toxic effects of PAHs has been associated with their low level of biotransformation of these compounds and has tended to suggest that the consequences of these contaminants are of more concern to organisms at higher trophic levels. However, in a study measuring exposure to and effects of organic compounds and metals on the ribbed mussel (*Geukensia demmissa*) from New Bedford Harbour in America, it was shown that PAH concentrations strongly correlated with the induction of biomarkers of genotoxicity (micronucleus formation – see Table 3.6.1), immunotoxicity (spontaneous cytotoxicity) and physiological impairment measured by heart rate analysis (Galloway *et al.*, 2002b).

In some cases, a mechanistic relationship can be established. For example, a study of dogwhelks, *Nucella lapillus*, displaying various stages of TBT-induced imposex collected from sites in the south west of England revealed a very strong relationship (correlation coefficient = 0.935, $P < 0.0001$) between the degree of imposex and the extent of DNA damage (micronucleus formation – see Table 3.6.1) in blood cells (Hagger *et al.*, 2006). There was also a strong association between TBT body burden and the prevalence of abnormal growths, providing evidence that TBT may affect reproductive processes in gastropods partly through DNA damage pathways.

## 3.6.5 CONCLUSION

During this chapter, we have provided a small snap shot of the potential value of using simple, rapid, cost-effective biomarkers within site assessment as an integrated component within ERA. However, as discussed, it is important that biomarkers are used in an integrated approach to ERA (as illustrated in Figure 3.6.2) where chemical data, toxicity testing and detailed ecology are measured, at the same time as the biomarkers, in order to provide a complete overview of the health of the ecosystem being investigated.

# REFERENCES

Aas E, Beyer J, Goksøyr A (1998) PAH in fish bile detected by fixed wavelength fluorescence. *Marine Environmental Research* **46**(1–5): 225–8.

Bauer B, Fioroni P, Ide I, Liebe S, Oehlmann J, Stroben E, Watermann B (1995) TBT effects on the female genital system of *Littorina littorea* – A possible indicator of tributyltin pollution. *Hydrobiologia* **309**: 15–27.

Bauer B, Fioroni P, SchulteOehlmann U, Oehlmann J, Kalbfus W (1997) The use of *Littorina littorea* for tributyltin (TBT) effect monitoring results from the German TBT survey 1994/1995 and laboratory experiments. *Environmental Pollution* **96**: 299–309.

Baumard P, Budzinski H, Michon Q, Garrigues P, Burgeot T, Bellocq J (1998) Origin and bioavailability of PAHs in the Mediterranean Sea from mussel and sediment records. *Estuarine, Coastal and Shelf Science* **47**(1): 77–90.

Bryan GW (1980) Recent trends in research on heavy metal contamination in the sea. *Helgoland Meeresuntersuchungen* **33**: 6–25.

Bryan GW, Gibbs PE (1983) Heavy metals in the Fal estuary, Cornwall: a study of long-term contamination by mining waste and its effects on estuarine organisms. *Journal of the Marine Biological Association of the United Kingdom*, Occasional Publication Number 2, Plymouth.

Bryan GW, Gibbs PE (1991) Impacts of low concentrations of tributyltin (TBT) on marine organisms: a review. In: Newman MC, McIntosh AW (eds) *Metal Ecotoxicology: Concepts and Applications*. Lewis Publishers, Boca Raton, Florida, pp. 323–62.

Bryan GW, Gibbs PE, Hummerstone LG, Burt GR (1986) The decline of the gastropod *Nucella lapillus* around southwest England – evidence for the effect of tributyltin from antifouling paints. *Journal of the Marine Biological Association of the United Kingdom* **66**: 611–40.

Chambers JE, Boone JS, Carr RL, Chambers HW, Straus DL (2002) Biomarkers as predictors in health and ecological risk assessment. *Human and Ecological Risk Assessment* **8**: 165–76.

Depledge MH, Andersen BB (1990) A computer-aided physiological monitoring system for continuous, long-term recording of cardiac activity in selected invertebrates. *Comparative Biochemistry and Physiology a-Physiology* **96**: 473–7.

Dissanayake A, Galloway TS (2004) Evaluation of fixed wavelength fluorescence and synchronous fluorescence spectrophotometry as a biomonitoring tool of environmental contamination. *Marine Environmental Research* **58**: 281–5.

Galloway TS, Millward N, Browne MA, Depledge MH (2002) Rapid assessment of organophosphorous/carbamate exposure in the bivalve mollusc Mytilus edulis using combined esterase activities as biomarkers. *Aquatic Toxicology* **61**(3–4): 169–80.

Galloway T, Handy R (2003) Immunotoxicity of organophosphorous pesticides. *Ecotoxicology* **12**: 345–63.

Galloway TS, Brown RJ, Browne MA, *et al.* (2004a) Ecosystem management bioindicators: the ECOMAN project – a multi-biomarker approach to ecosystem management. *Marine Environmental Research* **58**: 233–7.

Galloway TS, Brown RJ, Browne MA, *et al.* (2004b) A multibiomarker approach to environmental assessment. *Environmental Science & Technology* **38**: 1723–31.

Galloway TS, Hagger JA, Langston WJ, Jones MB (2007) The application of biological-effects tools to inform the condition of European Marine Sites. *Natural England*. Report 116 pp.

Galloway TS, Millward N, Browne MA, Depledge MH (2002a) Rapid assessment of organophosphorus/carbamate exposure in the bivalve mollusc *Mytilus edulis* using combined esterase activities as biomarkers. *Aquatic Toxicology* **61**: 169–80.

Galloway TS, Sanger RC, Smith KL, et al. (2002b) Rapid assessment of marine pollution using multiple biomarkers and chemical immunoassays. *Environmental Science & Technology* **36**: 2219–26.

Gibbs PE, Bryan GW (1986) Reproductive failure in populations of the dogwhelk, *Nucella lapillus*, caused by imposex induced by tributyltin from antifouling paints. *Journal of the Marine Biological Association of the United Kingdom* **66**: 767–77.

Gibbs PE, Pascoe PL, Burt GR (1988) Sex change in the female dog-whelk, *Nucella lapillus*, induced by tributyltin from antifouling paints. *Journal of the Marine Biological Association of the United Kingdom* **68**: 715–31.

Giessing AMB, Mayer LM, Forbes TL (2003) Synchronous fluorescence spectrometry of 1-hydroxypyrene: a rapid screening method for identification of PAH exposure in tissue from marine polychaetes. *Marine Environmental Research* **56**: 599–615.

Hagger JA, Galloway TS, Oehlmann J, Jobling S, Depledge MH (2006) Is there a causal association between genotoxicity and the imposex effect. *Environmental Health Perspectives* **114**: 20–6.

James MO (1989) Biotransformation and disposition of PAH in aquatic invertebrates. In: Varanasi U (ed.), *Metabolism of PAH in the Aquatic Environment*. CRC Press, Boca Raton, Florida, pp. 69–91.

Johnston EP, Baumann PC (1989) Analysis of fish bile with HPLC – fluorescence to determine environmental exposure to Benzo(a)Pyrene. *Hydrobiologia* **188**: 561–6.

Krahn MM, Myers MS, Burrows DG, Malins DC (1984) Determination of metabolites of xenobiotics in the bile of fish from polluted waterways. *Xenobiotica* **14**: 633–46.

Langston WJ, Chesman BS, Burt GR, Hawkins S, Readman JW, Worsfold PJ (2003) *Characterisation of the South West European Marine Sites: The Fal and Helford cSAC*. Marine Biological Association of the UK, Occasional Publication No. 8, Plymouth.

Lin ELC, Cormier SM, Racine RN (1994) Synchronous fluorometric measurement of metabolites of Polycyclic Aromatic Hydrocarbons in the bile of Brown Bullhead. *Environmental Toxicology and Chemistry* **13**: 707–15.

Lin ELC, Cormier SM, Torsella JA (1996) Fish biliary polycyclic aromatic hydrocarbon metabolites estimated by fixed-wavelength fluorescence: Comparison with HPLC-fluorescent detection. *Ecotoxicology and Environmental Safety* **35**: 16–23.

Lowe DM, Fossato V, Depledge MH (1995) Contaminant-induced lysosomal membrane damage in blood cells of mussels (*Mytilus galloprovincialis*) from the Venice lagoon: an *in vitro* study. *Marine Ecology-Progress Series* **129**: 189–96.

McCarty LS, Munkittrick KR (1996) Environmental biomarkers in aquatic toxicology: friction fantasy or functional. *Human and Ecology/Risk Assessment* **2**: 268–74.

Mineau P (1991) *Cholinesterase-inhabiting Insecticides – Impacts on Wildlife and the Environment*. Elsevier Science Publications, Amsterdam, Netherlands.

PARCOM (1995) A sediment bioassay using an amphipod Corophium sp. PARCOM protocols on methods for the testing of chemicals used in the offshore industry. Oslo and Paris Commissions, London, 35 pp.

Schlenk D (1999) Necessity of defining biomarkers for use in ecological risk assessments. *Marine Pollution Bulletin* **39**: 48–53.

Schulte-Oehlmann U, Watermann B, Tillmann M, Scherf S, Market B, Oehlmann J (2000) Effects of endocrine disruptors on prosobranch snails (Mollusca: Gastropoda) in the laboratory. Part II. Triphenyltin as a xeno-androgen. *Ecotoxicology* **9**: 399–412.

Smith BS (1971) Sexuality in the American mudsnail *Nassarius obsoletus* Say. *Proceedings of the Malacological Society of London* **39**: 377–8.

Stroomberg GJ, Reuther C, Kozin I, et al. (1996) Formation of pyrene metabolites by the terrestrial isopod *Porcellio scaber*. *Chemosphere* **33**: 1905–14.

Sumpter JP, Jobling S (1995) Vitellogenesis as a biomarker for estrogenic contamination of the aquatic environment. *Environmental Health Perspectives* **103**: 173–8.

Viarengo A, Ponzano E, Dondero F, Fabbri R (1997) A simple spectrophotometric method for metallothionein evaluation in marine organisms: an application to Mediterranean and Antarctic molluscs. *Marine Environmental Research* **44**: 69–84.

Walker CH, Hopkin SP, Sible RM, Peakall DB (1996) *Principles of Ecotoxicology*. Taylor & Francis Ltd, London.

Watson GM, Andersen OK, Depledge MH, Galloway TS (2004a) Detecting a field gradient of PAH exposure in decapod crustacea using a novel urinary biomarker. *Marine Environmental Research* **58**: 257–61.

Watson GM, Andersen OK, Galloway TS, Depledge MH (2004b) Rapid assessment of polycyclic aromatic hydrocarbon (PAH) exposure in decapod crustaceans by fluorimetric analysis of urine and haemolymph. *Aquatic Toxicology* **67**: 127–42.

Wrisberg MN, Rhemrev R (1992) Detection of genotoxins in the aquatic environment with the mussel *Mytilus edulis*. *Water Science and Technology* **25**: 317–24.

Yang X, Baumann PC (2006) Biliary PAH metabolites and the hepatosomatic index of brown bullheads from Lake Erie tributaries. *Ecological Indicators* **6**: 567–74.

Yang XA, Peterson DS, Baumann PC, Lin ELC (2003) Fish biliary PAH metabolites estimated by fixed-wavelength fluorescence as an indicator of environmental exposure and effects. *Journal of Great Lakes Research* **29**: 116–23.

# Section IV
Potential Use of Screening Methods and Performance Evaluation

# 4.1
## Monitoring Heavy Metals Using Passive Sampling Devices

Graham A. Mills, Ian J. Allan, Nathalie Guigues, Jesper Knutsson, A. Holmberg and Richard Greenwood

4.1.1 Introduction
4.1.2 Monitoring
4.1.3 Passive Sampling for Metals
4.1.4 Use of Passive Samples for Monitoring Metals in Surface Waters
    4.1.4.1 Tank Study
    4.1.4.2 Field Study
4.1.5 Monitoring of Organotins and Mercury
4.1.6 Conclusions
Acknlowledgements
References

## 4.1.1 INTRODUCTION

Metals are widely distributed in the aquatic environment and may be from either natural and/or anthropogenic sources. They are present in wide range of forms or species including freely dissolved metal, salts, organometalics, and bound to colloids and particulates. Factors such as pH and redox potential of the water and sediment, and presence of particulate material or organic matter, and microbial activity will determine in what form a particular metal is present in the environment. Since the different forms of a metal can have different bioavailabilities and toxicities, it is important to ensure that biologically relevant species are measured. Metals are generally bioavailable when present as free ions, bound to weak complexes or in a lipid soluble form (Hudson, 1998; Luider *et al*., 2004). Low concentrations of some metals, such as Cu and Zn,

---
*Rapid Chemical and Biological Techniques for Water Monitoring*    Edited by Catherine Gonzalez, Philippe Quevauviller and Richard Greenwood
© 2009 John Wiley & Sons, Ltd

are essential to living organisms, whilst others such as Cd and Pb, are toxic at low levels when present in a bioavailable form.

Metals can enter the aquatic environment through point and diffuse sources, and this applies to both natural and anthropogenic origins. In addition metals can be transported over long distances to remote areas through atmospheric deposition. Metal concentrations and speciation can vary widely across water bodies. In some cases, discharges with high pollutant concentrations may be mixed and become diluted, resulting in contaminants present at low (and in some cases trace) levels. The total concentration of a metal present in a sample is relatively straightforward to measure using techniques such as atomic absorption spectroscopy (AAS) or inductively coupled plasma mass spectrometry (ICP-MS). Currently available instruments are highly sensitive and can measure a suite of elements in a single analysis. However, where information on the individual species present is needed then more elaborate sample preparation, and/or analytical techniques are needed. These factors combine to present a challenge to those involved in monitoring the impact of metal contamination on the chemical and biological quality of environmental waters.

Under the European Union's Water Framework Directive (WFD, 2000/60/EC) the measured levels of priority pollutants are compared with environmental quality standards (EQS) defined within the legislation. Two types of EQS are used; namely an annual average concentration EQS (AA) that is the highest average concentration to which an aquatic ecosystem may be exposed without any likely adverse effects, and a maximum allowable concentration EQS (MAC) that is the lowest concentration that would be expected to cause adverse effects in a short-term exposure. The former is designed to protect against long-term (chronic) exposure to a pollutant, and the latter against short-term episodic exposure. These are set and refined on the basis of laboratory-based toxicological data. In some countries account is taken of variation in the properties (e.g. hardness, pH) of environmental waters, but detailed information on speciation is not used in compliance testing. Trace metals on the European WFD list of priority substances include Cd, Hg, Ni, and Pb, and the EQS values for these are based on dissolved concentrations (defined as the concentration in the filtrate obtained using a filter with a pore size of 0.45 µm, or equivalent pretreatment) (Lepom *et al*., 2008). For Cd it is mandatory to measure the water hardness since five levels of EQS are defined in terms of this variable.

## 4.1.2 MONITORING

Most regulatory monitoring of trace level pollutants depends on the use of bottle (spot or grab) sampling followed by instrumental analysis for the measurement of total or dissolved concentrations. Generally sample volumes are small (for most metals 10 mL is sufficient, and for Hg volumes >30 mL). The total metal fraction is the sum of all occurring species of a particular metal in a water sample. Where more detailed information is required the distribution between suspended solids and the water, then the most common practice is to filter the water sample through a membrane (usually with a nominal pore size of 0.45 µm), and to analyse the filtrate and the unfiltered sample. This filtration should be conducted in the field soon after sample collection in order to minimize artefacts due to water sample storage and transportation. The particulate fraction is defined by the pore size of the filter used. The dissolved fraction in the filtrate will be made up of free hydrated cations, and complexes with organic and inorganic ligands and colloids. The relative concentrations of these various forms will depend on the properties (including pH, hardness, and concentrations and properties of the dissolved organic material present) of the water sample and of the element being measured. Ultra-filtration can be used to obtain more information on the distribution of a metal between the species present in a filtrate (0.45 μm), but this is rarely used for regulatory purposes. Colloidal material (the size fraction between the truly dissolved species and the particulate material) is generally considered to be removed by use of a filter of nominal pore size 1–10 kDa MW. Colloids, organic and/or inorganic, are heterogeneous in composition and have a high surface area, and hence a large absorbance capacity for trace elements. In natural waters inorganic colloids generally consist of clay minerals and the organic colloids of humic substances. Colloidal matter is important since it influences the bioavailability and toxicity of elements (Luider et al., 2004; van Leeuwen et al., 2005).

Various sampling methods are available, and these can yield different information depending on the fraction or combination of fractions that they measure. In order to be able to interpret the information from bottle samples it is necessary to measure important variables (e.g. pH, Ca, Mg, and dissolved organic carbon) that determine the speciation pattern in natural waters. As outlined above the fraction measured by spot sampling will depend on the way in which the sample is treated prior to analysis. Other factors that can affect this include the method of sample storage during transport and in the laboratory. These can change the distribution of a metal between the various fractions found in the natural waters. Generally metals are stored in plastic containers, but for organotins storage in glass or aluminium vessels is preferred. Samples for analysis of Hg and methylmercury are stored in quartz or glass containers since this element diffuses through plastic. Usually samples are stabilized by the addition of preservative (e.g. potassium dichromate for Hg), and acidified, using e.g. nitric acid, to ensure that the pH is less than 2.

Analytical methods used in support of the regulations need to be fit for purpose, and have appropriate limits of quantification (LOQ) that should be no higher than one third of EQS value. If no analytical method is available that has the necessary sensitivity and that fulfils the required quality requirements, then the best available technique should be used with proviso that the cost is not excessive (Lepom et al., 2008). Where EQS values are set at very low levels on toxicological grounds (e.g. organotins, and Cd), then this poses a challenge for the analytical laboratories. For Cd, Ni and Pb ICP-MS methods provide adequate sensitivity. Atomic fluorescence spectrometry is used routinely for Hg. Specialized chromatographic and spectroscopic techniques are necessary to analyse tributyltin, and the methods that are routinely available in regulatory and commercial laboratories fail to meet the required levels of sensitivity. The EQS (AA) for tributyltin in surface waters is $0.0002\,\mu g\,L^{-1}$, whereas the LOQ is typically in the range 0.01 to $1\,\mu g\,L^{-1}$. In all cases a limiting factor can be the attainment of low levels of contamination in laboratory and field blank samples.

A great deal of emphasis has been placed on the analytical step where it is possible to enforce strict quality control procedures including the use of certified reference materials, and inter-laboratory proficiency testing schemes. Most laboratories involved in regulatory analysis are nationally accredited. This works well in compliance testing against EQS values, but where spot sampling is intermittent or undertaken with a frequency of unknown suitability the precise chemical measurements may not reflect average conditions that pertain in surface waters in the field. This is particularly problematic where concentrations fluctuate in time due to, for instance, sporadic discharges or weather events. It is possible to gain more representative data by continuous sampling, or repeated sampling procedures such as frequent spot sampling, weekly composite sampling or time or flow weighted 24 h collections. However, the equipment used for continuous and time-weighted monitoring is expensive and needs to be deployed in secure locations. Further the use of frequent spot sampling has implications for the costs of monitoring in terms of personnel time, transport and analysis. Other alternatives involve biomonitoring and passive sampling.

Biomonitoring offers another approach to the problem, and has been used routinely for many years in for the assessment of levels of organic pollutants and metals in surface waters. This may be particularly valuable for pollutants such as organotin where the EQS is below the achievable LOD and LOQ for routinely available methods. Biomonitoring involves deploying organisms such as caged fish, or more commonly bivalve molluscs at key sites of interest. Bivalve molluscs are sessile filter feeders and have advantages over the vertebrates in terms of using natural food sources, robustness, ease of handling, and ethical considerations, and have been used in mussel watch programmes in the Netherlands since 1990 (Smedes, 2007). Even depurated animals will have background levels of most common pollutants, and it is necessary to take a representative sample before deployment to measure this. Since the preparation of biological material for analysis is complex, then this adds significantly to the cost of this method. A further consideration is that biomonitors can be deployed only where conditions are suitable for maintaining the organisms in a healthy state. Different species have to be used in marine, brackish, and freshwater environments, and it is not possible to hold the biomonitors at sites close to discharge points where concentrations of pollutants could reach toxic levels. The information is usually expressed in terms of amounts in the organism normalized for either lipid content of the organisms or on a dry weight basis. This technique provides only qualitative or at best semiquantitative information on concentrations of pollutants in the environmental waters. It does, however, have the advantage of providing biologically relevant information, and can indicate trends in pollution over periods of years.

Passive samplers can provide an alternative to the use of living organisms. They have some advantages in terms of robustness, range of conditions in which they can be deployed, ethical considerations, ease of analysis, cost, and they can provide quantitative information on pollutant concentrations. This technology has been used routinely in support of regulations governing air quality since the late 1970s, but has only relatively recently been developed for use in aquatic environments. A range of devices is available for monitoring most classes of pollutants, including polar, and nonpolar organics, metals, and organometalics (Greenwood *et al.*, 2007).

## 4.1.3 PASSIVE SAMPLING FOR METALS

Passive samplers for all of classes of pollutants function in a similar manner and have the same essential components. These are a receiving phase with a high affinity for the analytes of interest and that maintains a concentration of the pollutant at it surface close to zero, and a diffusion limiting layer. The latter may be a physical barrier such as a membrane, and/or a boundary layer of water. The uptake kinetics can be described by an exponential approach to a maximum and hence two modes of use (kinetic and equilibrium) are available. Kinetic samplers, operate in the linear uptake region that is far from the thermodynamic equilibrium between sampler and water. In this regime the uptake rate is independent of the external concentration, and the samplers can provide time-weighted average (TWA) concentrations to which the samplers have been exposed over the deployment period. For long exposure times (in excess of the time to half maximum) where thermodynamic equilibrium between sampler and water is approached, sampling is no longer integrative, and the amount of analyte accumulated in the receiving phase reflects the capacity of the receiving phase. Under the latter circumstances the samplers no longer provide information on concentrations encountered earlier in the deployment and no longer yield TWA concentrations. A number of extensive reviews of the range of available designs have been published in recent years (Kot-Wasik *et al.*, 2007; Stuer-Lauridsen, 2005; Vrana *et al.*, 2005). Details of the structures of samplers for organic pollutants are described in Chapters 2.1 and 2.2 of this book. This chapter will focus on passive samplers that can quantitatively sequester metals and organometalics, and that have potential for use in a regulatory context.

The diffusion gradients in thin-films (DGT) was developed at the Lancaster University (UK) and has been in use since the mid 1990s for the *in situ* measurement of trace metals (Warnken *et al.*, 2007), and major cations (Ca and Mg), phosphate, and sulphide. A combination of configurations of this sampler can provide information on speciation of metals present in the water column. This sampler comprises a hydrogel layer (protected by a membrane prefilter) that acts as the diffusion layer, and an underlying binding hydrogel layer containing typically Chelex 100 as a receiving phase. The nature of the binding agent held in the receiving phase can be tailored to the sequestration of specific analytes (e.g. ferrihydrite for phosphates, and Spheron-Thiol for mercury). The pore size of the hydrogel can be varied to provide open pore (OP) and restricted pore (RP) samplers, and different thicknesses of hydrogel can be used. The OP DGTs sample a wider range of species than the RP variant. OP samplers are considered to sequester freely dissolved metal plus that bound in a wide range of complexes with inorganic and organic ligands where dissociation can occur in the time taken for them to traverse the diffusion layer. RP samplers are considered to sequester only free ions and small inorganic or organic complexes. A comparison of the data obtained with the RP and OP samplers provides information on the *in situ* speciation of the metals. The prefilter, diffusive hydrogel and receiving phase are assembled in a plastic body comprised of a base and cap (Figure 4.1.1).

After deployment the receiving phase is removed from the holder and extracted in a small volume of nitric acid. The resulting solution can be analysed directly using

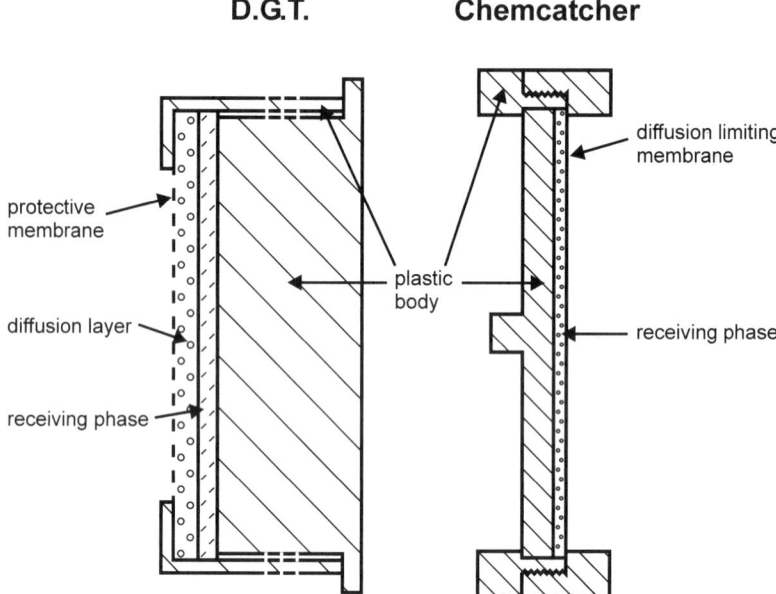

**Figure 4.1.1** Schematic diagrams of the DGT and Chemcatcher® passive samplers showing the key components

standard instrumental techniques. Calibration is based on diffusion coefficients of metal ions in the hydrogel, and these are available for a range of temperatures from the supplier. This design has a hydrogel diffusive layer that is thick relative to the thickness of the water boundary layer, and so uptake rates are minimally affected by changes in water turbulence during the deployment period (< 20%). However, the surface can become biofouled, and this can modify the uptake rate in a way that is difficult to predict. The study of the impact of biofouling is challenging since it is hard to create reproducibly even in the laboratory, and difficult to extrapolate laboratory results to the field situation. In addition, biofouling develops gradually during the deployment period and may impact metal uptake more significantly towards the end of the exposure. Some efforts have been made to use biocides to limit the growth of biofouling organisms on the sampler surface (Pichette *et al.*, 2007; Mills *et al.*, 2009). Samplers are normally deployed for relatively short periods so that they remain in the kinetic sampling region. Under these conditions calculations of TWA concentrations ($C_{DGT}$) in the water assume Fickian diffusion and are based on the mass of metal accumulated over a deployment period according to Equation 4.4.1.

$$C_{DGT} = \frac{M_{DGT} \Delta g}{D_e t A} \quad (4.1.1)$$

where ($M_{DGT}$) is the mass of metal accumulated during the exposure time ($t$), across a sampling surface area ($A$), through a diffusion gel thickness ($\Delta g$), where the diffusion coefficient for the metal of interest is ($D_e$).

The Chemcatcher® passive sampler (Figure 4.1.1) that was developed at the University Portsmouth (UK) and Chalmers University of Technology (Sweden) has been available since the late 1990s for monitoring organic pollutants, heavy metals and more recently for organotin compounds (Aguilar-Martínez, 2008). The variant used for monitoring heavy metals comprises a Teflon® watertight body that retains a 47 mm diameter 3M Empore™ chelating disk receiving phase overlaid with a cellulose acetate diffusion-limiting membrane (0.45 µm pore size; 0.135 mm thick). For organotins the same sampler body and diffusion-limiting membrane are used but the receiving phase is a $C_{18}$ 3M Empore™ disk. This sampler is likely to sequester free ions, and those complexes (both organic and inorganic) that are able to dissociate in the time taken to cross the diffusion layer. For metals the receiving phase is extracted with a small volume of acid, and the extract is analysed directly by routine methods. For organotins the receiving phase is extracted with methanolic acetic acid, and analysed by GC-ICP-MS or GC fitted with a flame photometric detector depending on the moiety being measured. Calibration data are in the form of uptake rates (expressed as mL h$^{-1}$) that can be considered as the equivalent volume of water cleared of analyte per unit time. Uptake rates are determined by a number of factors including the area of sampler available for diffusion, the properties of the diffusion-limiting layer (e.g. thickness and resistivity), and the properties (e.g. size and polarity) of the chemical. Uptake rates are measured in through flow plastic calibration tanks at known water temperatures and turbulence conditions (produced by rotating the samplers on a carousel at fixed speeds). Since uptake rates are affected by both temperature and turbulence it is important to obtain calibration data that represent the range of conditions found in the field. Where the deployment period does not exceed the kinetic sampling range of the samplers, calculation of the TWA concentration ($C_{\text{Chemcatcher}}$) in the water are, as for the DGT, based on the assumption of Fickian diffusion, and are obtained using Equation 4.1.2.

$$C_{\text{Chemcatcher}} = \frac{M_{\text{Chemcatcher}}}{R_s t} \qquad (4.1.2)$$

Where ($M_{\text{Chemcatcher}}$) is the mass of metal extracted from the receiving phase after a deployment time (t), and ($R_s$) is the uptake rate for the metal of interest. Like the DGT, these samplers may be adversely affected by biofouling of the membrane surface, and this may limit the length of deployment period that can be used. Unlike for the Chemcatcher® for nonpolar organic pollutants, it is not currently possible to use performance reference compounds to correct field uptake rates of the metals version of this sampler for the effects of environmental variables (temperature, turbulence and biofouling).

## 4.1.4 USE OF PASSIVE SAMPLES FOR MONITORING METALS IN SURFACE WATERS

Although passive samplers (and particularly the DGT) have been used by many different research groups to measure concentrations of metals in surface waters, and the work has been widely published, they have yet to be adopted within a regulatory context.

One of the main reasons for this is a lack of accepted validation procedures, particularly for the calibration stages, and the transfer of these to the field. Since this technology yields samples that are analysed by standard laboratory-based methods, this stage of the monitoring process is the same as for spot sampling. There is a need for validation studies using inter-laboratory trials, and the development of a reference site approach to validation and to quality control and quality assurance for the sampling phase. Tank and associated field trials within the EU-funded SWIFT-WFD project provided some insight into possible ways forward.

Most laboratory-based calibration experiments are conducted under controlled conditions, with constant concentrations of metals in clean water. This does not necessarily reflect field conditions, and there is uncertainty as to how laboratory estimates of calibration parameters will transfer to the field, where concentrations of metals can fluctuate over several orders of magnitude. Fluctuations are often a result of an accidental discharge or weather event, and they can be accompanied by changes in, for instance, levels of dissolved and particulate organic material, pH and redox conditions. It is difficult to reproduce these conditions in the laboratory, but the use of tank studies where conditions can be controlled, and defined patterns of changing pollutant concentration can be achieved could provide a solution, and provides a compromise between the use of routine laboratory-based calibration, and the use of a reference site.

### 4.1.4.1 Tank Study

A tank study carried out within the SWIFT-WFD project aimed to investigate the performance of two passive samplers (DGT and Chemcatcher®), speciation modelling, and frequent spot sampling in fluctuating concentrations of metals (Allan *et al.*, 2007). This study carried out at Eijsden (the Netherlands) used a 220 L Teflon tank system (Figure 4.1.2) filled with water from the River Meuse. The passive samplers were located on a carousel that was rotated at a fixed rate (30 rpm) to provide repeatable turbulence conditions. The water was spiked using aliquots of a stock solution of metals (Cd, Cu, Ni, Pb and Zn) at a concentration of $1 \text{ g L}^{-1}$. Fluctuating concentrations were produced by the addition of variable volumes (between 1 and 10 mL) of the spiking solution at fixed time intervals. Two peaks were produced over the 5 day experiment.

The first peak was introduced at 12 h after the start of the experiment (Figure 4.1.3). The nominal maximum concentrations were 70 to 140 $\mu\text{g L}^{-1}$, and the peak duration was 1.5 to 2 days. The concentrations of Cd used in this experiment simulated concentrations found in the Meuse during pollution events. In one such event in 2007, a peak level of 430 $\mu\text{g L}^{-1}$ (Cd) was observed. In this study Chemcatcher® and DGT samplers (with RP and OP hydrogels) were attached to the carousel during deployment, and frequent spot samples were taken to define the pattern of fluctuation in the actual concentrations of the various metals. Untreated water samples were analysed to determine total concentrations of metals; other samples were filtered (at 0.45 $\mu\text{m}$) and some of these then ultrafiltered (at 5 kDa) before analysis. The ultrafiltration and filtration procedures provide estimates of respectively the freely dissolved fraction, and the sum of the freely dissolved fractions and that containing metals bound to dissolved organic material. The various fractions were also predicted on the basis of modelling (Visual MINTEQ incorporating NICA-Donnan effects). The average metal

# Use of Passive Samples for Monitoring Metals in Surface Waters 251

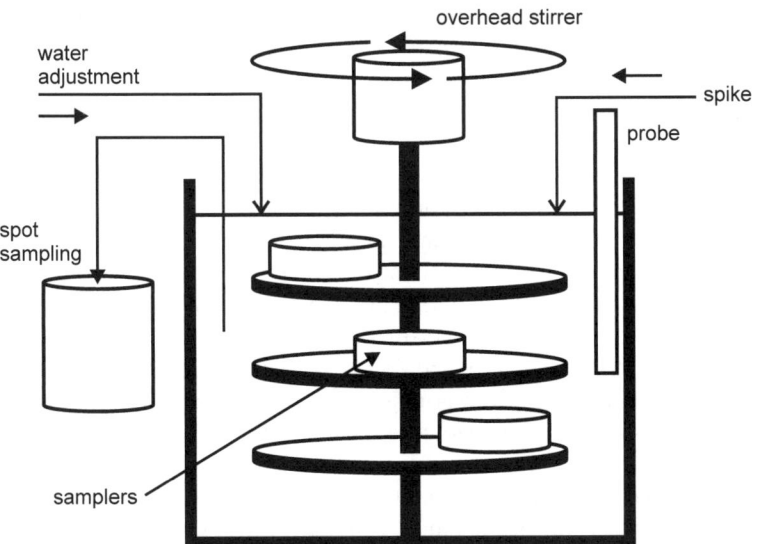

**Figure 4.1.2** Schematic diagram of the test tank showing the rotating carousel to which Chemcatcher® and DGT samplers were attached

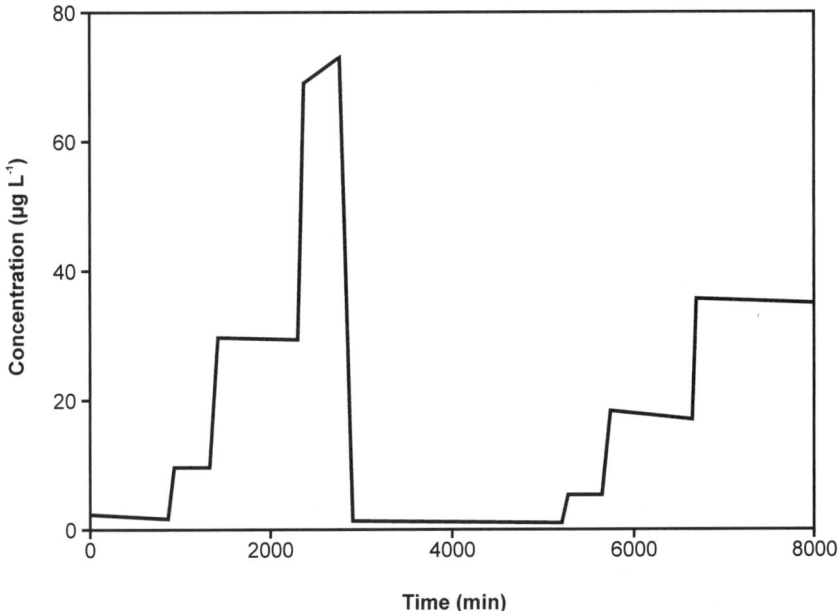

**Figure 4.1.3** Pattern of temporal variation in concentrations of metals produced by stepwise spiking in the test tank over the experimental period. The illustrative profile is for Cd

concentrations (total, filtered and ultrafiltered) in the tank water were calculated assuming linear changes in concentration between consecutive spot samples. The estimates of TWA concentrations in the water on the basis of the masses accumulated in the passive samplers were calculated using Equations (4.1.1) and (4.1.2) for the DGT and Chemcatcher® respectively.

The data obtained in this study (Allan et al., 2007) provide a useful comparison of the various methods of determining concentrations of metals in water, and help in the interpretation of the information that they provide. This study also provides a measure of the reliability of passive sampling relative to the use of infrequent spot sampling. As would be expected there was variation in the behaviour of the various metals in this study. On the basis of the spot samples treated in various ways 90% of the Cd was in free solution this contrasted with Cu and Pb where some 40% and 17% respectively was freely dissolved. This behaviour was reflected in the TWA concentrations estimated using the two passive samplers. However, the uptake of Pb by both samplers was relatively low, and especially so for the Chemcatcher® where there may be some binding of the metal to the diffusion limiting membrane. While there were differences in the fine detail of the estimates obtained with the various variants of passive sampler, there was generally a reasonable agreement between them, as exemplified by the data for Cd and Zn (Figure 4.1.4).

The TWA concentrations estimated with DGT had associated relative standard deviations in the range 1–5%, the equivalent values for Chemcatcher® were in the range 5–12%. There was a reasonable agreement between the TWA concentrations of Cd predicted by spot sampling (total, filtered, and ultrafiltered), Chemcatcher®, DGT, and modelling. This is consistent with the behaviour of this metal in this tank study where it was found to be mostly in the freely dissolved fraction. However, this may have been an artefact due to the high concentrations of Cd used, and in the field trial (see below). For Zn the filtered (0.45 μm) and ultrafiltered (5 kDa) fractions were 80 and 60% of the total concentration respectively. For this metal there was good agreement between the TWA concentrations measured by the DGT RP and the Chemcatcher® samplers, and a larger proportion was available for sequestration by the DGT OP (Figure 4.1.4). For Cu there was reasonable agreement between the estimates of TWA concentration based on filtered spot samples and Chemcatcher®, and total dissolved based on modelling; and between those based on ultrafiltered spot samples, DGT RP, and DGT OP. For the latter none of the concentrations of the fractions predicted by the modelling coincided closely with the DGT-based estimates. The differences between the samplers, and the uncertainties associated with the sampler-based estimates of TWA concentrations are small relative to the amplitude of the fluctuations, and if only one spot sample had been taken during the five-day period, the uncertainty associated with average concentration based on this would have been large.

When validating passive sampling it is difficult to select an appropriate benchmark with which to compare it since spot sampling concentrations are operationally defined, and fractions measured by filtration/ultrafiltration may not coincide exactly with fraction sequestered by passive samplers. The use of a tank trial where repeated spot samples can be taken to define the concentration profile over the deployment period shows promise, but in the study outlined above some problems were identified. There were indications that the dissolved organic material (DOM) in the finite volume

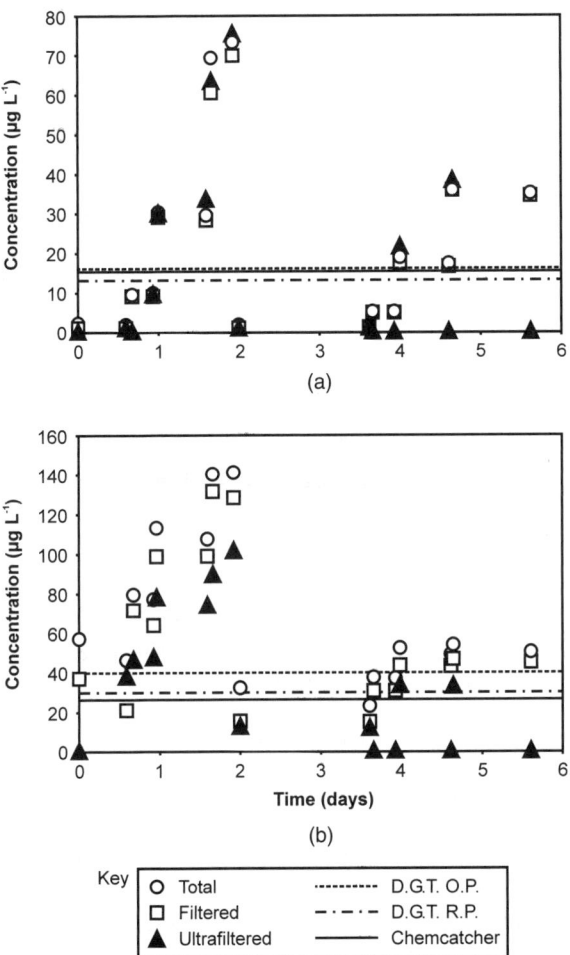

**Figure 4.1.4** Comparison of TWA concentrations ($\mu g\ L^{-1}$) of (a) Cd, and (b) Zn obtained using the three variants of passive sampling devices along side spot sampling with measurements of total, filtered (0.45 µm), and ultrafiltered (5 kDa) fractions. Samplers were deployed for six days in a through flow tank system using spiked water from the River Meuse

of river water may have approached saturation at the higher spiking levels. Further equilibrium between the various species of metals that show significant binding to the DOM may not have been established in the time scale of the additions of spiking solutions. If tank studies are to be used to validate the use of passive samplers in fluctuating concentrations of pollutants, then these issues will have to be addressed.

### 4.1.4.2 Field Study

The tank trial described above was followed by a field trial in the Meuse River (Allan et al., 2008) at the same location, and this was undertaken in order to determine whether

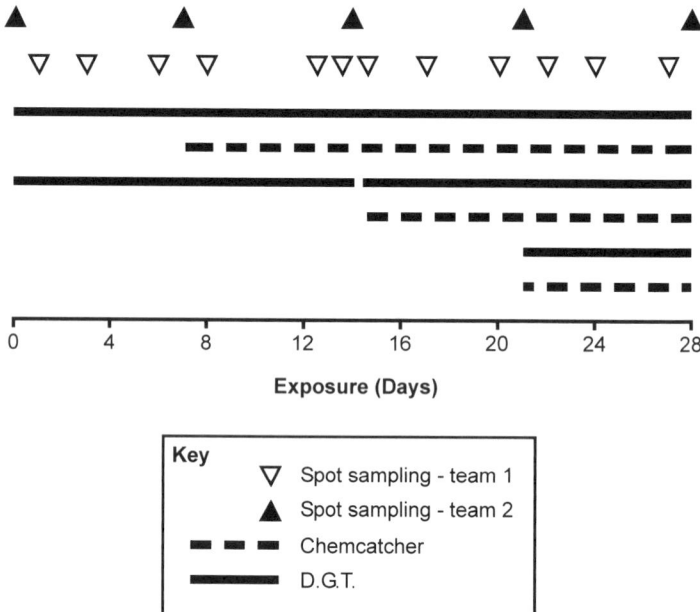

**Figure 4.1.5** Schematic diagram showing the pattern of deployment of the three variants of passive sampling devices, and the timing of the spot sampling events in the River Meuse field trial. Spot samples were collected independently by two teams to provide a check on the reproducibility of the sampling and analytical methods

the behaviour of the samplers in surface water was consistent with that found in the enclosed, controlled system. It aimed to assess the reliability and representativeness of passive sampling in monitoring TWA concentrations of metals (Cd, Cu, Ni, Pb and Zn), and their potential for inclusion in regulatory monitoring programmes. One way in which the samplers could be used is to reduce the uncertainty demonstrated to be associated with the estimates of average metal concentrations based on infrequent spot sampling when the concentrations of the pollutants are fluctuating in time. As in the tank trial both DGT (OP and RP) and Chemcatcher® samplers were deployed, and frequent spot samples were taken and analysed for both total metals and following filtration. Again predictions of equilibrium speciation patterns were made using the Visual MINTEQ speciation code (incorporating the NICA-Donnan model).

DGT (OP and RP) and Chemcatcher® samplers were exposed for overlapping/consecutive periods of 7, 14, and 21/28 days (Figure 4.1.5) at a depth of 1 m below the surface of river. The use of overlapping deployment periods provides an opportunity to assess the impact of biofouling.

Passive sampling protocols were based on BSI PAS-61 (British Standards Institute, 2006), and triplicate field blanks were used to assess possible contamination during manufacture, deployment, retrieval and transport. In field trials it is not possible to produce controlled fluctuations in concentrations of pollutants, but by chance in this work concentrations of most metals were higher during the first 14 days of the 28 day trial and two peaks of concentration of were observed in the frequent spot samples

*Use of Passive Samples for Monitoring Metals in Surface Waters* 255

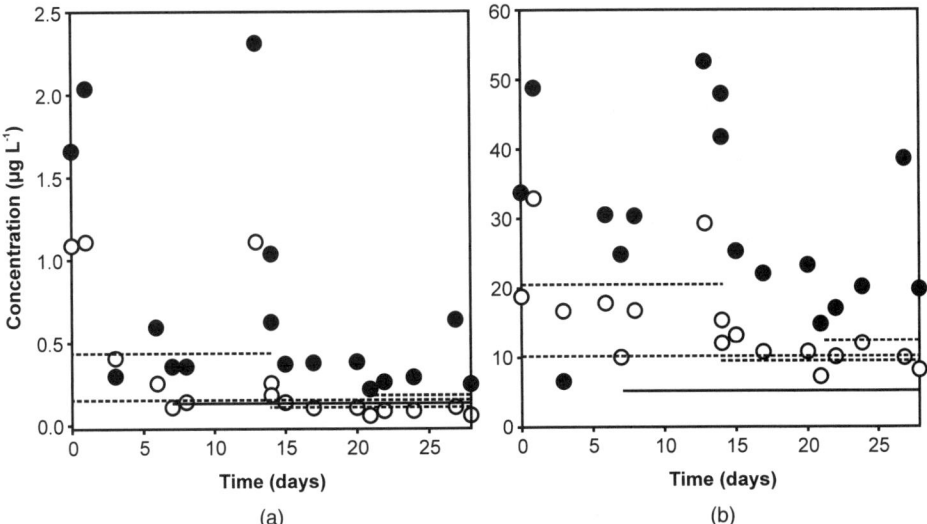

**Figure 4.1.6** Comparison of TWA concentrations (µg L$^{-1}$) of (a) Cd, and (b) Zn obtained using the DGT OP, and Chemcatcher® passive sampling devices deployed in the River Meuse (Eijsden, Netherlands) along side spot sampling with measurements of total (•) and filtered (0.45 µm) (o) fractions. DGT OP samplers (dashed line) were exposed for overlapping period of 7, 14, 14, and 28 days. Chemcatcher® samplers (solid line) were exposed for 21 days

(illustrated for Cd and Zn in Figure 4.1.6). The TWA concentrations of Cd measured by DGT OP and Chemcatcher® samplers were similar over the longer exposure times, and were close to the filtered concentrations. The DGT OP exposed during the first 14 days reflected the higher peak levels of Cd over this period. Over the period of this trial the TWA concentrations of Cd approached the AA-EQS value (0.25 µg L$^{-1}$) and some of the filtered concentrations in individual spot samples approached the MAC-EQS (1.5 µg L$^{-1}$). Even higher concentrations of Cd have been observed in spot samples from the Meuse during routine continuous monitoring (Allan *et al.*, 2008).

The sampling data were largely consistent with the concentrations of the various metals measured routinely at the Eijsden site, and were consistent with the expected behaviour of the metals. The filtered fraction in spot samples were on average 29 ± 11, 59 ± 6, 81 ± 4, 16 ± 14 and 47 ± 12% of the total concentrations of Cd, Cu, Ni, Pb and Zn, respectively. The samplers were in good agreement with the filtered spot samples for Cd and Zn where most of the filtered material is labile (Figure 4.1.6). However, for Cu, Ni, and Pb where a significant proportion of the filtered material is not labile, the concentrations estimated from the samplers are consistently lower than those from the filtered water samples (Figure 4.1.7).

From the results it appears that most of the Cd in the filtrate is labile and there was a good agreement between the three variants of sampler (DGT OP, DGT RP, and Chemcatcher®) and the spot samples. The estimates of concentrations of labile Cu and Ni were similar for all three samplers, and these were consistent with the output from the Visual MINTEQ model. Whilst these results for Cu and Ni are consistent with

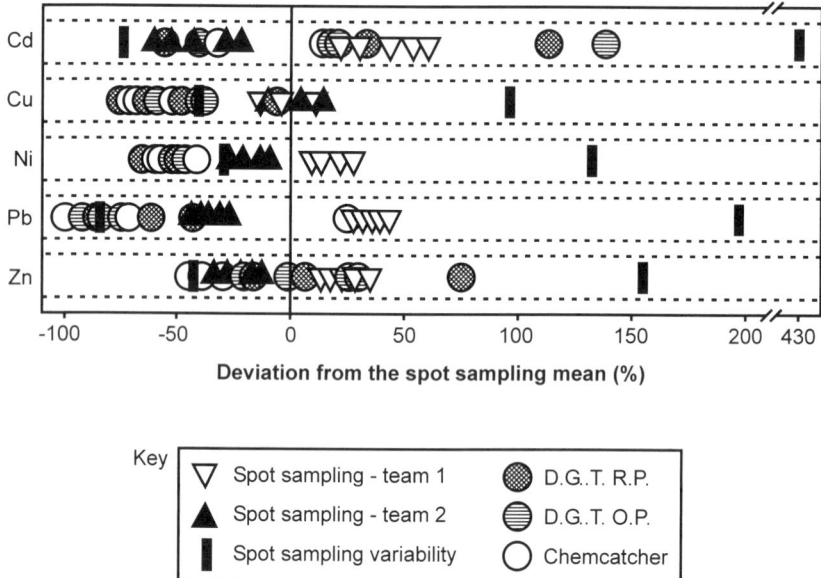

**Figure 4.1.7** The average filtered (0.45 μm) concentration based on all spot samples is plotted as a vertical line at zero deviation. The deviations of all other samples from this overall average are plotted for each of the five metals. The differences between the spot samples taken by the two monitoring teams over the 28 day trial indicate the reliability of these measurements. The deviations of the concentrations derived using the three variants of passive sampler represent values based on the overlapping 7, 14, 21 and 28 day deployment periods. The vertical bars indicate the range of concentrations found in the spot samples over the trial. This figure is adapted from Allan *et al.*, 2008

findings in previous studies in the River Meuse (Cleven *et al.*, 2005), those for Ni differ from those found in the tank study with Meuse water. This difference between tank and field may have been due to either saturation of available binding sites or failure to reach equilibrium between free ion and suspended solids and DOM during the spiking process as discussed above. There was a small difference between the DGT samplers and the Chemcatcher® for Zn, where the latter appeared to sequester the free ion fraction (predicted by the speciation model), and the former most of the metal present in the filtered fraction and this is consistent with observations in previous DGT studies in fresh water (Meylan *et al.*, 2004). The authors (Allan *et al.*, 2008) concluded that the differences between the two samplers for Zn may reflect the differences in diffusion limiting layer in relation to the dissociation kinetics of the fraction of this metal that is bound to DOM. For both samplers the rates of uptake of Pb were low (reflecting the low concentrations of labile metal).

The results of this field trial helped to establish possible limitations to extended deployment times. If the sampling rate were constant throughout the deployment, as it should be providing that the samplers remain in the linear uptake mode, then the average of the concentrations estimated from two consecutive 14 day deployments should be similar to those from the 28 day deployment. However, in this study (Allan *et al.*, 2008) the concentration of Cd estimated from the former was twice that derived from the latter. Since the receiving phase has a high capacity for the metals being sequestered in this study, it is likely that the samplers remained in the linear uptake phase throughout the trial, and the observed reduced estimate of concentration derived from the longer exposure may be due to biofouling of the sampler surface. This could increase the resistance to diffusion by increasing the length of the diffusion path, restricting the surface area of pores available for diffusion, or reducing the pore size. Whilst it is difficult to estimate the impact of biofouling in a repeatable and quantitative way, this warrants further investigation. Attempts to protect the DGT by the incorporation of antibiotics into the protective membrane were partially successful, but better results were obtained in long-term deployments using CuI and AgI. However, CuI was found to interfere with metal measurements and it not feasible to use this to provide protection when samplers are used to monitor metals (Pichette *et al.*, 2007). However, this approach is a potential solution for samplers for organic pollutants.

Some differences were found between the results of the tank and field trials carried out by Allan and co-workers (Allan *et al.*, 2007, 2008). Uncertainties associated with $C_{DGT}(OP)$ varied between 20 and 81% RSD (n = 3), and are consistent with those (28, 17 and 71% for Cu, Ni and Pb, respectively) observed for DGTs deployed in the River Meuse by Cleven and co-workers (Cleven *et al.*, 2005). The average uncertainties (RSD) for $C_{Chemcatcher}$ were 49, 24, 7.6 and 26% for Cd, Cu, Ni and Zn, respectively. For both samplers the variability is higher in this trial than in the tank trial, where the uncertainties were in the range of 1–5% and 5–12% for DGT and Chemcatcher® respectively. This may be due to for instance more homogeneous and consistent turbulence, lower biofouling because of the rotation of the samplers, and the use of higher concentrations in the tank water than were found in the river. This work indicates that even where river water is used in a tank trial, the sampling data do not necessarily provide a good measure of the variability found under field conditions, and provides an argument for the use of a reference site for validation of samplers.

The tank and field trials described above are unusual in that they combined the investigation of the performances of DGT and Chemcatcher® passive samplers alongside frequent spot sampling and speciation modelling to provide insight as to what fractions measured by various samplers and spot sampling and how they might be validated. Both the tank and field trials showed the same important feature: where the concentration fluctuates in time the uncertainty in the estimate of average concentrations associated with infrequent spot samples is large (Figures 4.1.4 and 4.1.7). This can be reduced by more frequent spot sampling or the use of passive sampling to complement infrequent spot sampling. There is a need for long-term investigations of the reliability of infrequent spot sampling and passive sampling, and potential strategies to combine these in obtaining representative data to support regulatory monitoring.

## 4.1.5 MONITORING OF ORGANOTINS AND MERCURY

The work described above covers many of the common trace metals, but two areas of concern not dealt with in those studies are metallic Hg, and organotins. Monitoring for these is difficult but a necessary part of regulatory monitoring. Due to its high toxicity the EQS (AA) value of tributyltin in surface waters is set very low (0.0002 μg L$^{-1}$), and this concentration is below the LOQ (typically in the range 0.01 to 1 μg L$^{-1}$) of most analytical methods used routinely in laboratories involved in regulatory monitoring. In order to attain the required detection limits large volumes of water would need to be collected and preconcentrated prior to instrumental analysis. An alternative approach is to use passive sampling devices that have a high uptake rate ($R_s$) and to deploy these for extended periods. Standard sized semi-permeable membrane devices (SPMDs) (see Chapter 2.2) have been used for this purpose in a number of studies (Følsvik et al., 2000, 2002). Følsvik et al. (2000) deployed SPMDs in Oslofjord (Norway) for 4 weeks. Spot samples of water were also collected over this period. The samplers accumulated both dibutyltin and tributyltin (concentrations for both analytes ranged from 3 to 220 ng tin SPMD$^{-1}$) but not monobutyltin. The concentration of tributytin in the water samples varied between 0.4–10 ng tin L$^{-1}$. This study demonstrated the potential utility of the samplers in monitoring organotins. However, as $R_s$ values were not available for these compounds, it was not possible to make a direct comparison with the concentrations measured in the spot samples of water. More recent work by Harman et al. (2008) measured the uptake kinetics of organotins for SPMDs over a four week period. Only tributyltin was accumulated significantly. Under the exposure conditions used, the calculated theoretical detection limits were in the low pg L$^{-1}$ range, which is below the EQS (AA) value for tributyltin. A variant of the Chemcatcher® sampler has been developed for organotins. This device uses a 47 mm Empore™ disk as the receiving phase that is overlaid by a thin cellulose acetate diffusion membrane. For tributlytin, the calculated detection limits for the device for a 14 day deployment under the different exposure conditions used were in the range 0.3–2.1 ng L$^{-1}$. These limits approach the EQS (AA) value and the device is therefore more appropriate for use in investigative monitoring.

Hg is toxic, and regarded as an important pollutant. However, its analysis is not straightforward, and in comparison with other trace metals its speciation in natural waters has been poorly studied. A recent study (Black et al., 2007) has shown that in oxygenated freshwater nearly all of the inorganic (Hg(II)) mercury present is very strongly bound by ligands present in the water. Some 85–90% of this Hg is complexed with hydrophobic ligands. The dissociation kinetics of these are very slow, and it was postulated that in this fraction it would be only the complexes with relatively small organic solutes that would be bioavailable. Measuring total Hg does not provide information on the bioavailable fraction. It may be possible to use passive sampling to provide some relevant information, but it is not straightforward to use existing passive sampling technologies. Dočekalová and Diviš (2005) applied the DGT samplers in monitoring Hg in aquatic systems. Hg was found to bind covalently to the amide nitrogen groups of the polyacrylamide, and so the polyacrylamide hydrogel was replaced by one made of agarose. These modified DGTs were found to sequester mainly inorganic ions and labile species of Hg, and not that complexed with inert organic species and colloids. A Chemcatcher® passive sample fitted with a chelating disk as receiving phase separated from the external medium by a porous polyethersulphone membrane was used in a preliminary investigation (Aigular-Martínez et al., 2008) of the use of passive sampling for monitoring concentrations of Hg in natural waters. The sampling rate for inorganic Hg was in the region of 0.1 L day$^{-1}$. This indicates that over a deployment period of 14 days the samplers will contain the equivalent amount of inorganic Hg that was dissolved in 1.4 L, and the limit of detection was of the order of 1.7 ng L$^{-1}$. In order to reach this sensitivity it is necessary to remove Hg from the receiving phases before deployment, and to use laboratory blanks. In a trial in Alicante Harbour (Spain) levels of Hg found in unfiltered spot samples were found to be in the range of 180–260 ng L$^{-1}$. The concentrations of labile Hg were estimated to be in the region of 34 ng L$^{-1}$ on the basis of the material accumulated in the passive samplers. In order to understand the significance of the fraction available to the variants of passive samplers, work is needed to identify the biologically important species of Hg in natural waters.

## 4.1.6 CONCLUSIONS

In natural waters metals can form complexes with organic and inorganic compounds. The extent and nature of the complexing varies between elements, but also with the physicochemical parameters (e.g. pH, redox conditions) and with the concentrations and nature of ligands that are present. The properties of large organic contaminants such as humic and fulvic acids change with age, and hence between run-off from peat deposits and discharges from sewage treatment plants (Tusseau-Vuillemin *et al.*, 2007). This makes it difficult to interpret the biological significance of concentrations of metals measured in the filtered (0.45 µm) spot samples used for regulatory purposes. Changes in spot samples during sampling, storage and sample preparation can modify results that may vary with water quality and are difficult to predict. Passive samplers avoid this problem since they provide in situ sampling in the raw unmodified water. Further, where concentrations and water quality can fluctuate unpredictably in time due to sporadic discharges and weather events, infrequent spot sampling may not provide a useful representative picture of the chemical quality of a body of surface water. Passive sampling may provide extra, useful information to provide the regulator with information that will allow the assessment of the quality of the data from routine spot sampling. In field measurements it is important to provide representative data, and less uncertainty would be associated with data obtained from passive samplers than from infrequent spot samples. However, in order for this to be accepted it is essential to have levels of uncertainty that are well defined and repeatable for passive sampling, and to have a benchmark that does not depend on low frequency spot sampling since the outcome of this is affected by the time of sampling, and by the way in which samples are treated prior to analysis.

There is an urgent need for longer-term studies similar to those carried out by Allan *et al.* (2007, 2008), and described above, in order to promote the acceptance of passive sampling in a regulatory role. These help to establish the validity and relevance of the technique alongside spot sampling. Most work in this area uses one type of passive sampler to investigate a few pollutants at limited sites, and there have been few systematic comparative studies such as those carried out within the SWIFT-WFD project. Currently no agreed international quality control protocols are available to allow those involved in passive sampling to validate their technologies and methodologies. In part this is hampered by a lack of suitable reference materials for use in laboratory calibration systems. There is an urgent need to develop quality assurance and quality control procedures for this technology. The development of reference materials, the use of large-scale tank studies at reference sites, or the use of in situ validation at reference sites are all possibilities. The use of tank trials with natural water has a number of advantages when compared with field deployments. It is possible to simulate a range of concentration profiles of pollutants under defined temperature and turbulence regimes, and this may provide a means of running quality assurance trials such as inter-laboratory calibration exercises. However, further work is necessary before this can be used with confidence. Studies are needed to establish the behaviour of the system during spiking experiments, especially where high concentrations are used, and it will be important to ensure that equilibrium positions for binding to DOM are reached in the spiking process.

Within a regulatory framework it is necessary to be able to assess the risk associated with concentrations of a range of chemicals in various divisions of the aquatic environment. In order to develop risk assessments on a river basin scale it is necessary to have reliable and representative data that are comparable and accepted as fit for purpose at national and international levels. If passive samplers were to become acceptable within the regulatory framework it would be necessary to define EQS values in terms of the fraction sequestered by passive samplers, and to work to international standards for their use. A further requirement of regulatory monitoring is the ability to detect trends in water quality, and in order to do this representative information on average levels is needed, even where there is short-term temporal variation in concentrations of individual pollutants. Since a lot of information is needed to meet these various regulatory requirements, it is important to use technologies and methodologies that are cost-effective in terms of the price of the technology and the costs of personnel and travel to and from monitoring points. Passive sampling has the potential provide reliable information within available budgets, but in the face of a natural reluctance to change current practice, further work will be necessary to convince regulatory authorities of this.

## ACKNLOWLEDGEMENTS

The authors are grateful to Joe Greenwood for the production of the figures in this chapter.

## REFERENCES

Aguilar-Martínez R., Palacios-Corvillo M.A., Greenwood R., Mills G.A., Vrana B. and Gómez-Gómez, M.M., 2008. *Anal. Chim. Acta*, **618**, 157.
Allan I.J., Knutsson J., Guigues N., Mills G.A., Fouillac A-M. and Greenwood R., 2007. *J. Environ. Monit.*, **9**, 672.
Allan I.J., Knutsson J., Guigues N., Mills G.A., Fouillac A-M. and Greenwood R., 2008. *J. Environ. Monit.*, **10**, 821.
Black F.J., Bruland K.W. and Flegal A.R., 2007. *Anal. Chim. Acta*, **598**, 318.

# References

British Standards Institute (BSI), 2006. *Publicly Available Specification: Determination of Priority Pollutants in Surface waters using Passive Sampling* (BSI PAS-61).
Cleven R., Nur Y., Krystek P. and Van den Berg G., 2005. *Water, Air, Soil Pollut.*, **165**, 249.
Dočekalová H. and Diviš, P., 2005. *Talanta*, **65**, 1174.
European Commission, 2000. Directive 2000/60/EC of the European Parliament and of the Council of 23 October 2000 establishing a framework for Community action in the field of water policy, *Official Journal of the European Communities* **L 327**, 22.12.2000, p. 1.
Følsvik N., Brevik E.M. and Berge J.A., 2000. *J. Environ. Monit.*, **2**, 281.
Følsvik N., Brevik E.M. and Berge J.A., 2002. *J. Environ. Monit.*, **4**, 280.
Greenwood R., Mills G. and Vrana B. (Eds), 2007. *Passive Sampling Techniques in Environmental Monitoring*, Elsevier, Amsterdam.
Harman C., Bøyum O., Tollefsen K.E., Thomas K. and Grung M., 2008. *J. Environ. Monit.* **10**, 239.
Hudson R.J.M., 1998. *Sci. Total Environ.*, **219**, 95.
Kot-Wasik A., Zabiegała B., Urbanowicz M., Dominiak E., Wasik A. and Namieśnik J., 2007. *Anal. Chim. Acta*, **602**, 141.
Van Leeuwen H.P., Town R.M., Buffle J., et al., 2005. *Environ. Sci. Technol.* **39**, 8545.
Lepom P., Brown B., Hanke G., Loos R., Quevauviller Ph. And Wollgast J., 2008. *J. Chromatogr. A*, **1216**, 302.
Luider C.D., Crusius, J., Playle R.C. and Curtis, P.J., 2004. *Environ. Sci. Technol.*, **38**, 2865.
Meylan S., Odzak N., Behra R. and Sigg, L., 2004. *Anal. Chim. Acta*, **510**, 91.
Mills G.A., Greenwood R., Allan I.A., et al., 2009, in: J. Namieśnik and P. Szefer (eds), *Analytical Measurements in Aquatic Environments*. Taylor & Francis, London.
Pichette C., Zhang H., Davison W. and Sauvé S., 2007, *Talanta*, **72**, 716.
Smedes F. 2007, in: R. Greenwood, G. Mills and B. Vrana (eds) *Passive Sampling Techniques in Environmental Monitoring*, Elsevier, Amsterdam.
Stuer-Lauridsen F., 2005. *Environ. Pollut.* 2005, **136**, 503.
Tusseau-Vuillemin M-H., Gourlay C., Lorgeoux C., et al,, 2007. *Sci. Total Environ.*, **375**, 244.
Vrana B., Allan I.J., Greenwood R., et al., 2005 *TrAC: Trends Anal. Chem.*, **24**, 845.
Warnken K.W., Zhang H. and Davison W., 2007, in: R. Greenwood, G. Mills and B. Vrana (eds), *Passive Sampling Techniques in Environmental Monitoring*, Elsevier, Amsterdam.

# 4.2
## On-site Heavy Metal Monitoring Using a Portable Screen-printed Electrode Sensor

Catherine Berho, Nathalie Guigues, Jean-Philippe Ghestem, Catherine Crouzet, Anne Strugeon, Stéphane Roy and Anne-Marie Fouillac

4.2.1 Introduction
4.2.2 Characteristics of the Screen Printed Electrode Sensor: Principle and Performance Criteria
4.2.3 Potential Functions of the Screen Printed Electrodes (SPEs)/PalmSens Device in the Frame of the WFD with Regard to the Performance Criteria
    4.2.3.1 Mapping of Water Bodies under 'Average' Quality Conditions to Assess the Spatial Variability of Metals
    4.2.3.2 Monitoring of Metals Under a Simulated Accidental Pollution Event: Assessment of the Temporal Variability
4.2.4 Conclusions
Acknowledgements
References

### 4.2.1 INTRODUCTION

The EU Water Framework Directive (WFD) is probably the most significant legislation in the water field that has been introduced on an international basis for many years. The directive has a broad view of water management and aims to ensure the prevention of any further deterioration of all water bodies, as well as the protection and enhancement of the status of aquatic ecosystems. Its implementation based on the

*Rapid Chemical and Biological Techniques for Water Monitoring*    Edited by Catherine Gonzalez, Philippe Quevauviller and Richard Greenwood
© 2009 John Wiley & Sons, Ltd

design of monitoring programmes for all EU Member States involves a wide range of measurements of several ecological, hydromorphological, physico-chemical and chemical quality elements (European Commission, 2000). The most commonly used method for measuring levels of chemical pollutants is spot (bottle) sampling coupled with laboratory analysis. In face of the increasing need of monitoring to underpin WFD, this approach can become too costly with growing numbers of samples being submitted for quantitative analysis especially when water bodies fail to achieve the necessary quality (Roig et al., 2007).

In this context, there are thus new opportunities to use field methods which can present well-known advantages such as the provision of information at lower cost and/or more quickly, the absence of artefacts caused by samples handling, transportation and storage, the possibility of rapid and real time analysis and the estimation of spatial and temporal variability (Buffle and Horvai, 2000).

For chemical monitoring, a list of priority substances has been established that includes metals such as cadmium, lead, and nickel. As far as metals are concerned, voltammetric techniques and more precisely electrochemical stripping analysis has long been recognized as a powerful technique in environmental samples. In particular, anodic stripping voltammetry (ASV) coupled with screen-printed electrodes (SPEs) is a great simplification in the design and operation of on site heavy metal determination in water, for reasons of cost, simplicity, speed, sensitivity, portability and simultaneous multi-analyte capabilities. The wide applications in the field for heavy metal detection were extensively reviewed (Honeychurch and Hart, 2003; Palchetti et al., 2005).

The aim of this work is to demonstrate how the screen printed electrodes (SPEs) can be used for on site heavy metals monitoring in surface waters in the frame of the WFD. The sensors used consist of mercury-coated screen-printed electrodes coupled with square wave anodic stripping voltammetry (SWASV) (Palchetti et al., 1999). Three metals Cu, Cd, and Pb which are classically analysed in water matrices have been considered. Moreover, Cd and Pb belong to the priority substances list of the WFD. Performance criteria of the device are first established to evaluate the level of confidence of the method. The potential use of the device and its main advantages are then highlighted through two illustrative field applications.

## 4.2.2 CHARACTERISTICS OF THE SCREEN PRINTED ELECTRODE SENSOR: PRINCIPLE AND PERFORMANCE CRITERIA

The sensor developed by the University of Florence (I) (Palchetti et al., 2005) consists of three screen-printed electrodes (SPEs) (Figure 4.2.1):

- a graphite working electrode modified by the deposition of a mercury salt entrapped in a cellulose derivative;
- a silver reference electrode;
- a graphite counter electrode.

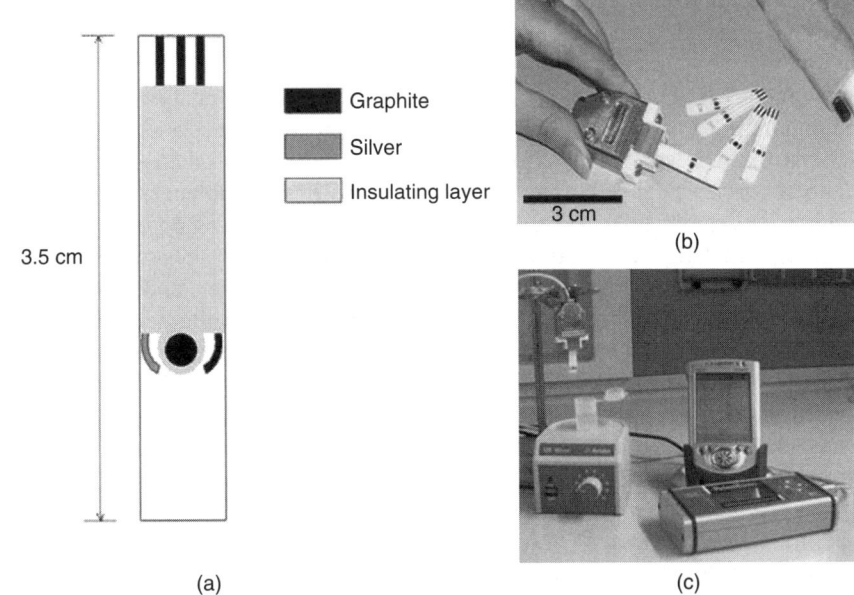

**Figure 4.2.1** (a) (b) Screen-Printed Electrodes (SPEs); (c) (SPEs) in combination with SWASV using a PalmSens portable instrument

It is used in combination with square wave anodic stripping voltammetry (SWASV) using a PalmSens portable instrument (Palm Instrument BV, Houten, The Netherlands) for the measurement of metals such as Cu (II), Cd (II) and Pb (II) (labile metallic complexes and free metals) in water. These disposable sensors require no calibration for use in the screening mode, so, many samples may be tested for the presence or the absence of metals in water. The quantification can also be performed using the standard addition method in less than 15 min.

A first evaluation of performance criteria in laboratory conditions based on the XPT 90–210 (French standard, 1999) was performed in order to characterize the method, to demonstrate its reliability and to define its capabilities to answer WFD requirements. The evaluation of the performance for the analysis of Cu, Cd and Pb in water samples concerns the following aspects:

- linearity range/calibration curve;
- limit of quantification;
- precision (repeatability and reproducibility);
- bias.

Specificity which is classically evaluated in method validation was neglected due to the high specificity of stripping voltammetry and the use of standard addition method. In this work, robustness is not taken into account since standard addition method is considered to reduce the influence of parameters (temperature, ionic strength ...) on the measurement. Analytical parameters characterizing the sensor performances are reported in Table 4.2.2.

The linearity test based on the XPT 90-210 (French Standard, 1999) was checked by performing a homogeneity test of variances based on the analysis of 5 standard solutions (ranging from 25 to 122 µg/L) using five different electrodes for each metal Cu, Pb and Cd. The application of the linearity test which has taken into account the five calibration curves gave correct results and revealed no problem of linearity.

In this work, the evaluation of the limit of quantification was done by the validation of a chosen target concentration (French Standard, 1999). There are also other approaches for this estimation, such as for example 10 times the standard deviation of the blank. The test used is based on trueness criteria and repeatability criteria estimated on the measurement of 10 sub samples of a spiked solution at the target concentration. The trueness criteria consist of checking that there is not significant difference between the target value and the measured value of limit of quantification. The precision criteria consist of checking that the repeatability is below 20%. The targeted value for Cu was chosen at 5 µg/L. For Cd and Pb, the targeted values of 2 µg/L close to 30% of the quality standard values of the European drinking water standard (3 µg/L for Pb and 1.6 µg/L for Cd) were chosen (European Commission, 1998). These tests result in the validation of the target values as limits of quantification (i.e 5 µg/L for Cu and 2 µg/L for Pb and Cd).

Repeatability and reproducibility were estimated by repeated analyses of spiked natural mineral water at three concentration levels for each metal covering the working range (from LOQ level to about 100 µg/L). Each spiked solution was analysed 3 times under repeatability conditions (same day, same operator, same sensor used at each level of concentration) and 3 times under reproducibility conditions (same operator, different days, different sensors at each day). The repeatability and the internal reproducibility were estimated at each level of concentration (Table 4.2.1). The relative standard deviation (RSD) of repeatability was close to the relative standard deviation of reproducibility at each level of concentration. The variability of the measurement is thus mainly due to the repeatability parameter. The higher value (24%) was obtained for Cd at a concentration level corresponding to the limit of quantification. In conclusion, we can consider that the precision of this method is of approximately 20% of the whole working range.

The variation of sensitivity between different sensors was also checked. Calibration curves with five different sensors were performed. A Relative Standard Deviation of 13, 13 and 42% of calibration slopes (sensitivity) were obtained for Cu, Pb and Cd respectively. These variations should have limited consequence on bias and precision when the standard addition method is used. However, for Cd, variations in the limit of quantification between two electrodes could be expected. Finally, the accuracy of the method was evaluated by the measurement of a SWIFT reference material used during the 2nd SWIFT-WFD Proficiency Testing exercise (Table 4.2.2). The reference value was chosen as the consensus value of the selected data population obtained after excluding the outliers. The performances of the device were estimated according to the Z-score (Z) calculation. Based on this score, results obtained with the SPEs/PalmSens method were consistent with those obtained by all methods for Pb and Cu ($|Z| \leq 2$) while the result was less satisfactory for Cd ($2 \leq |Z| \leq 3$).

**Table 4.2.1** Precision results

| Metal | Concentration level (μg/L) | Repeatability RSD (%) | Reproducibility RSD (%) |
|---|---|---|---|
| Cu | 100 | 15 | 15 |
|  | 25 | 16 | 16 |
|  | 5 | 5 | 5 |
| Pb | 100 | 14 | 14 |
|  | 10 | 17 | 17 |
|  | 2 | 12 | 14 |
| Cd | 100 | 12 | 17 |
|  | 25 | 19 | 20 |
|  | 2 | 16 | 24 |

**Table 4.2.2** Summary of performance criteria of the SPEs/PalmSens device

| Performance parameters | Screen printed electrode device – Cu | Screen printed electrode device – Pb | Screen printed electrode device – Cd |
|---|---|---|---|
| Linearity/calibration range | 25 – 122 μg/L | 25 – 122 μg/L | 25 – 122 μg/L |
| Coefficient of determination ($R^2$) | 0.999 | 0.999 | 0.999 |
| Limit of quantification checked (LOQ) | 5 μg/L | Close to 2 μg/L | Close to 2 μg/L |
| Precision of the method in the working range | 20%[a] | 20%[a] | 20%[a] |
| Accuracy (Bias) SWIFT RM 05 (spring fortified water) $Cu^b$ : 78.8 ± 5.9 μg/L $Pb^b$ : 10.6 ± 1.7 μg/L $Cd^b$ : 5.7 ± 0.6 μg/L | 80.3 ± 6.0 μg/L[c] (Z-score[d] = 0.2) | 9.3 ± 1.3 μg/L[c] (Z-score[d] = −1.2) | 4.2 ± 0.6 μg/L[c] (Z-score[d] = −2.6) |

[a] Approximate value whatever the concentration value in the working range.
[b] Mean of inter-laboratory exercise: consensus value ($X_{REF}$) of the selected data population obtained after excluding the outliers.
[c] Mean ($X_{LAB}$) and standard deviation obtained from 3 measurements in repeatability conditions.
[d] $Z = |X_{LAB} - X_{REF}|/S$ with ($X_{REF}$) the consensus value, ($X_{LAB}$) Mean of 3 measurements and (S) selected deviation unit chosen at 10%.

In conclusion, The SPEs/PalmSens method proved to be suitable for the determination of Cu, Pb and Cd in a water sample for concentrations ranging from few µg/L to one hundred µg/L. The next section is dedicated to examples of applications of this tool for field metal monitoring.

### 4.2.3 POTENTIAL FUNCTIONS OF THE SCREEN PRINTED ELECTRODES (SPEs)/PALMSENS DEVICE IN THE FRAME OF THE WFD WITH REGARD TO THE PERFORMANCE CRITERIA

This section aims to provide some illustrative examples of how information provided by the SPEs/PalmSens in field conditions can be useful in the frame of the WFD. These results were obtained during field trials organized within the SWIFT-WFD project. The first example highlights the ability of the sensor to assess spatial variability of metal concentration in a river system of 'average' quality conditions. The second one aims to demonstrate how the device can help to assess the temporal variability in case of a simulated accidental metal pollution scenario.

#### 4.2.3.1 Mapping of Water Bodies under 'Average' Quality Conditions to Assess the Spatial Variability of Metals

The Aller river system is located in the Basin of the River Weser that is situated in central northern Germany, and is one of the 10 German river-basin districts (Figure 4.2.2). There are several considerable pressures on surface waters in this river system such as municipal sewage plants, direct industrial discharges, rainwater run-off combined with wastewaters which have been identified as significant point sources of pollution in the Aller river system. Even if historical data indicate that contamination with metals can be expected in the Aller and Oker rivers, at low concentrations only (<1 µg/L), higher concentrations upstream in the upper Oker can be found due to close old mining activity areas. In order to assess the spatial variability of metals, 11 sampling sites alongside Aller and Oker rivers were considered (Figure 4.2.2). On site measurements were carried out on unfiltered samples to estimate the concentration of Cu, Cd and Pb by using the SPEs/PalmSens device and compared with results obtained by laboratory analysis (inductively coupled plasma-mass spectrometry, ICP/MS) on the same unfiltered samples

All metal concentrations found were relatively low and close to the limit of quantification of the SPEs/PalmSens method. As an example, cadmium concentrations estimated by SPEs/PalmSens and ICP/MS are presented in Figure 4.2.3. Although concentrations were over estimated they were in the same order of magnitude as the ICP/MS data. Results obtained by the SPEs/PalmSens are thus semi-quantitative but can lead to a comparison of a sample from one to another. Indeed, this screening resulted in the discrimination of the Oker River from the Aller River leading to the spatial variability assessment of the system. A representative picture of

*Potential Functions of the Screen Printed Electrodes* 269

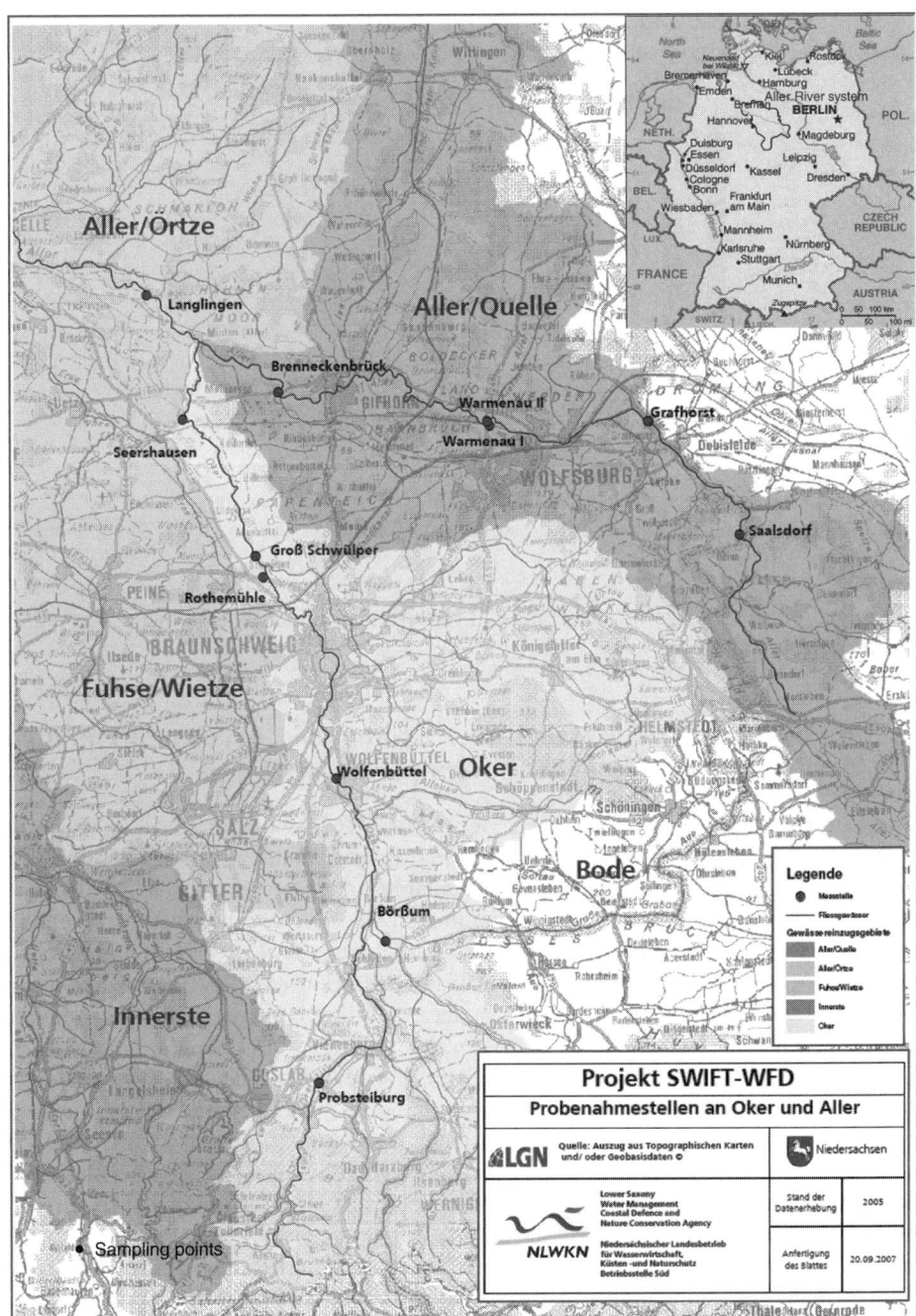

**Figure 4.2.2** Sampling point's location (See Plate 3 for colour representation)

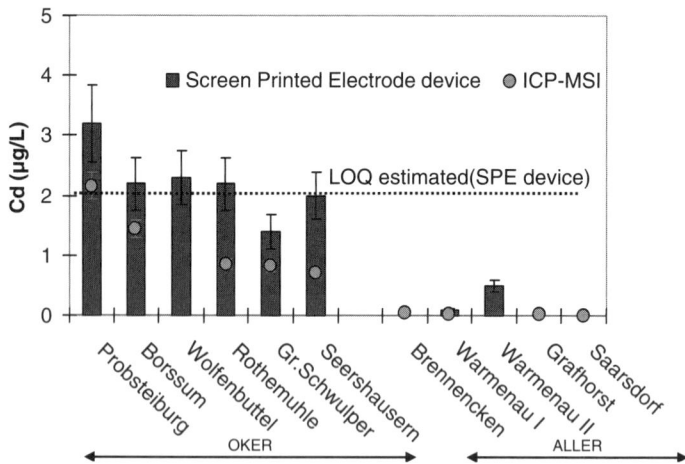

**Figure 4.2.3** Cadmium concentrations estimated by SPEs/PalmSens device and ICP/MS on unfiltered samples collected alongside the Aller and Oker rivers

contamination in this river system pointing out the sites under anthropogenic pressures was thus achieved.

### 4.2.3.2 Monitoring of Metals Under a Simulated Accidental Pollution Event: Assessment of the Temporal Variability

A field trial was conducted on the River Meuse at the RIZA monitoring station (Rijksinstituut voor Integraal Zoetwaterbeheer en Afvalwaterbehandeling) in Eijsden (NL) in order to test a number of commercially available or prototype tools selected in the toolbox under 'average' water quality conditions (Meuse river) such as those found in many European river basins (Allan et al., 2006a; Guigues et al., 2007). While contaminant levels prevailing in the river were generally those of an average large European river, additional tank experiments were conducted during five days in order to simulate higher contaminant concentrations for testing tools in case of accidental pollution. To demonstrate the potential use of the SPEs/PalmSens device in case of such event, water from the River Meuse was pumped into a plastic home-made tank through a recirculating system and spiked with a mixture of standard solutions of metals/heavy metals at various concentration levels in order to simulate two consecutive peaks of pollution in the working range of the method. A homogeneous stirring was maintained by a rotating carousel system. Sampling for metals was undertaken by manually pumping the water into a plastic bottle. In order to work in controlled conditions, physicochemical parameters (temperature, conductivity, dissolved oxygen, and pH) were monitored continuously during the experiment using a multi-parameter probe (YSI 6900). Sampling was performed in 16 occasions for SPEs/PalmSens measurements as well as ICP/MS measurements (Guigues et al., 2007).

Field measurements using SPEs/PalmSens device were carried out on unfiltered and filtered at 0.45-μm samples and compared with ICP/MS measurements on the

# Conclusions

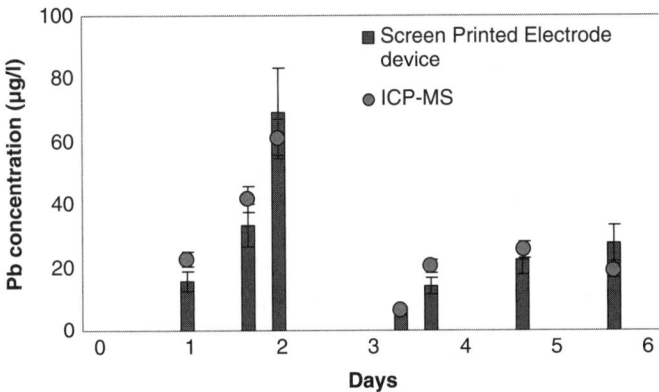

**Figure 4.2.4** Temporal variability of lead concentrations estimated by SPEs/PalmSens and ICP/MS during the 5-days tank experiments on raw samples

same samples. Metal concentrations in these samples varied over the working range of this instrument. There is a good agreement between the results obtained by the SPEs/PalmSens device instrument and the ICP-MS data for Cu, Cd and Pb. As an example, Figure 4.2.4 presents lead concentrations estimated by SPEs/PalmSens method and ICP/MS during the five-days tank experiments. The SPEs/PalmSens instrument was able to follow the changes in water concentration with reproducibility similar to that found using classical analysis (Allan et al., 2006b). This example demonstrates the usefulness of this device for temporal variability monitoring of metal concentration, in the case of an accidental pollution.

## 4.2.4 CONCLUSIONS

This study has successfully demonstrated the usefulness of the SPEs/PalmSens device for water monitoring in the frame of the WFD through two representative examples. Even if it may not be considered for monitoring at low level of metal concentration (<1 µg/L), the system is reliable, easy to use, safe, and could be used in a variety of situations to help environmental assessment and control:

- Rapid on site mapping of metal pollution associated with a minimization of delays and a reduction of costs in comparison with conventional spot sampling and laboratory analysis can be achieved.

- This tool can also screen samples, allowing the selection of few samples and thus reducing the number of samples being sent for more expensive and time consuming, but more precise and accurate conventional analysis.

- It can be extremely useful for the monitoring and the quantification of metal in waters under anthropogenic pressures. Hence, there are opportunities to deploy it to support operational (known pressures) or investigative monitoring (unknown pressures). As an example, in the case of an accidental pollution, the tool can be used to assess the spatial and the temporal variability of metal concentration in a water body in order to follow the spread of contamination and to implement remedial actions for instance.

According to the draft version of the Commission Directive 'implementing Directive 2000/60/EC concerning minimum performance criteria for analytical methods used for chemical monitoring and the quality of analytical results', the Chemical Monitoring Activity is likely to recommend the use of analytical methods for which the limit of quantification should be of 30% of the environmental quality standards (EQS) and the uncertainty at the EQS level of 50% (with k = 2) for surveillance monitoring (European Commission, 2008a). Concerning lead, the SPEs/PalmSens fulfils these requirements (EQS of 7.2 µg/L) whereas the tool is not able to reach the requested performances for cadmium (EQS from 0.08 and 0.25 µg/L depending on the hardness of water). It is also important to note that some measurements were performed on unfiltered samples. Indeed, when this work was performed, the draft version of the guidance on surface water chemical monitoring which recommends the measurement of metal concentration on filtered samples was not available (European Commission, 2008b).

The role of reliable and inexpensive field sensors in water monitoring may increase in the future, provided that they meet the criteria of analytical quality assurance in the same way as traditional methods do. It is obvious that conditions such as temperature, suspended matter in samples, proper to each site has an influence on the analytical results provided by sensors. Thus, these factors have to be taken into account through a robustness evaluation. Finally, validation and standardization of such methods would facilitate their use for routine monitoring by regulatory bodies in the frame of the WFD application. Indeed, to date very few organizations have addressed the problem of validation of such alternative methods. Initiatives such as the Environmental Technology Verification (ETV) Program of the US Environmental Protection Agency (USEPA) which verify the performance of innovative technologies and the method-validation program (designed specifically for test-kit methods) of AOAC International constitute a step forward to improve the acceptability of such tools for water monitoring.

## ACKNOWLEDGEMENTS

We acknowledge financial support from the Sixth Framework Programme of the European Union (Contract SSPI-CT-2003-502492; http://www.swift-wfd.com). We also thank Nel Frijns and the RIZA monitoring team in Eijsden (The Netherlands), the local Environmental Agency (in the Aller) for permission to work. Finally, we thank all the partners in the SWIFT-WFD consortium who carried out reference measurements, prepared the reference materials, participated in the proficiency testing and in field studies.

# REFERENCES

Allan I.J., Vrana B., Greenwood R., Mills G.A., Roig B. and C. Gonzalez, 2006a. *Talanta*, **69**, 302–22.
Allan I.J., Mills G.A., Vrana B., *et al*., 2006b. *Trends Anal. Chem.*, **25**, 704–15.
Buffle J. and Horvai G., 2000. *In situ monitoring of aquatic systems. IUPAC series on analytical and Physical Chemistry of Environmental Systems*. Vol. **6**. Chichester: John Wiley & Sons, Ltd.
European Commission, 1998. *Off. J. Eur. Commun.* **L 330** (1998) 32.
European Commission, 2000. *Directive 2000/60/EC of the European Parliament and of the Council of 23 October 2000 establishing a framework for Community action in the field of water policy*, *Off. J. Eur. Comm.* **L327** (2000) 1.
European Commission, 2008a. *Commission Directive 'implementing Directive 2000/60/EC concerning minimum performance criteria for analytical methods used for chemical monitoring and the quality of analytical results'*, in press.
European Commission, 2008b. *Draft version of the guidance on surface water chemical monitoring under the WFD*, Chemical Monitoring Surface Water Group.
French standard, 1999. XPT 90-210. *Water quality – protocol for the evaluation of an alternative quantitative physico-chemical analysis method against a reference method*.
Guigues N., Berho C., Roy S., Foucher J.-C. and Fouillac A.-M., 2007. *Trends Anal. Chem.*, **26**, 268–73.
Honeychurch K.C. and Hart J.P., 2003. *Trends Anal. Chem.*, **22**, 456–69.
Palchetti I., Cagnini A., Mascini M. and Turner A. P. F., 1999. *Microchim. Acta* **131**, 65–73.
Palchetti I., Laschi S. and Mascini M., 2005. *Anal. Chim. Acta*, **530**, 61–7.
Roig B., Valat C., Berho C., *et al*., 2007. *Trends Anal. Chem.*, **26**, 274–82.

# 4.3
## Field Monitoring of PAHs in River Water by Direct Fluorimetry on $C_{18}$ Solid Sorbent

Guillaume Bernier and Michel Lamotte

4.3.1 Introduction
4.3.2 Materials and Method
    4.3.2.1 Tank Experiment Design
    4.3.2.2 Chemicals
    4.3.2.3 Phase Preparation and Operational Conditions
    4.3.2.4 Front Face Fluorescence Spectra Measurements
4.3.3 Analytical Procedure
4.3.4 Results and Discussion
4.3.5 Conclusion
References

## 4.3.1 INTRODUCTION

Among the wide variety of anthropogenic chemicals, polycyclic aromatic hydrocarbons (PAHs) are of particular concern as widespread, persistent and toxic contaminants. (Neff, 1997; Manoli and Samara, 1999). The most widely used methods developed so far for the detection and quantification of PAHs in an aqueous matrix (Vo-Dinh, 1990) are based on separation techniques linked to a range of detection systems, and include high-performance liquid chromatography with fluorimetry (Kiss *et al.*, 1996; Manoli and Samara, 1999; Williamson *et al.*, 2002; Chen, 2004), gas chromatography (GC) with FID or MS detector (Manoli and Samara, 1999; Havenga and Rohwer, 2000; Santos and Galceran, 2002), supercritical fluid chromatography (SFC) (Bernal *et al.*,

*Rapid Chemical and Biological Techniques for Water Monitoring*    Edited by Catherine Gonzalez, Philippe Quevauviller and Richard Greenwood
© 2009 John Wiley & Sons, Ltd

1997; Manoli and Samara, 1999) or capillary electrophoresis (Martinez *et al.*, 1999). All these methods require sophisticated and expensive instruments which are difficult to transport and to adapt for on-site operation or for field monitoring. Moreover, these methods normally include an extraction step (liquid-liquid or solid phase extraction) for which a complex calibration process is needed to account for the appreciable loss of analyte (Thurman and Mills, 1998; Simpson, 2000) that occurs during the process. Further, they involve the use of organic solvents including halogenated compounds on whose use there are legal limitations (US EPA, 1998).

Some alternative, noninvasive and simpler methods mostly based on luminescence techniques have been proposed for the direct detection of PAHs in aqueous media. They include conventional, synchronous (Patra and Mishra, 2001; Miller, 1999) or laser excited fluorimetry (Karlitschek *et al.*, 1998; Giamarchy *et al.*, 2000; Whitcomb *et al.*, 2002; Burel *et al.*, 2003) or room temperature phosphorimetry (Vo-Dinh *et al.*, 1984; Hagestuen *et al.*, 2000) both eventually combined with the use of an enhancement agent such as micelles (Urbe and Ruana, 1997; Patra and Mishra, 2001) or cyclodextrins (Vo-Dinh and Gammage, 1978; Algarra and Hernandez, 1998; Yu *et al.*, 2003). Another approach based on front-surface fluorimetric detection on solid sorbent following solid phase extraction using $C_{18}$ Empore membranes has been proposed by Eastwood *et al.* (1994). Carr and Harris (1988) using alkylated silica adsorbent and Vilchez *et al.* (1994) using Sephadex G-25 gel have demonstrated the potential of this type of technique for the direct detection of PAHs in water. Other applications combining solid-phase extraction and spectrofluorimetry have also been evaluated (Arruda and Campiglia, 1999; Algarra *et al.*, 2000; Dmitrienko *et al.*, 2001; Whitcomb and Campiglia, 2001; Fernández-Sánchez *et al.*, 2003; Belfatmi *et al.*, 2005; Algarra *et al.* 2005). A similar approach was applied to the detection of BTEX (Wittkampt and Tilotta, 1995; Wittkampt *et al.*, 1997) or PCBs (Arruda and Campiglia, 1999; Belfatmi *et al.*, 2005), but to our knowledge none of these methods was applied for field monitoring.

In previous works various possibilities for the choice of the solid sorbent and the spectroscopic technique for detection were considered, and we paid particular attention to the use of tab-shaped fibre-glass $C_{18}$ extraction disks combined with synchronous fluorimetry using an available commercial instrument. This combination (Figure 4.3.1) was found to provide a simple, rapid and inexpensive analytical method for the detection of PAHs (Lamotte *et al.*, 2000; Algarra *et al.*, 2005) and PCBs (Lamotte *et al.*,

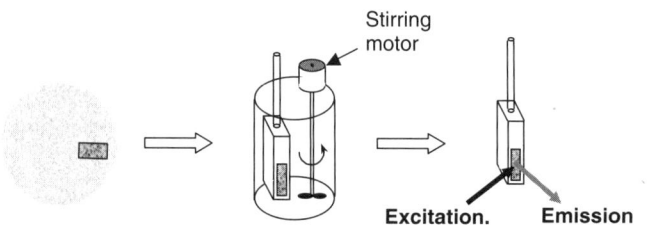

**Figure 4.3.1** Schematic diagram of the procedure used for direct front-face fluorimetric analysis of PAHs in fortified Maas river water after adsorption on a tab-shaped element cut from a $C_{18}$ extraction disk

2000; Belfatmi et al., 2005) in water and of PAH metabolites in the urine of PAH exposed persons (Algarra et al., 2000; Lamotte et al., 2003) at sub ppb level. In this chapter we describe the application of this method in a field monitoring trial conducted on the bank of the River Meuse, where river water was pumped into a tank and fortified by the addition of a controlled flow of a stock solution containing a mixture of PAHs. This test was conducted in Eijsden (NL) from 22 April to 27 April 2005 and was a part of a field trial that formed part of the European Union Project SWIFT-WFD (Screening method for Water data InFormation in support of the Implementation of the Water Framework Directive). This project involved the simultaneous testing of a number of commercially available and prototype monitoring tools under natural river conditions.

## 4.3.2 MATERIALS AND METHOD

### 4.3.2.1 Tank Experiment Design

A through-flow system with a continuous feed from two pump systems (illustrated in Figure 4.3.2) was used in order to minimize the rate of depletion of PAHs due to adsorption to the walls of the system, and to uptake by passive samplers deployed in the tank. The first pump provided a circulation of the Maas river water at a constant $10 \, L \, h^{-1}$ flow rate, and the second one added a solution of 16 PAHs (prepared by the dilution of a stock working solution containing equal concentrations of the test 16 PAHs) in acetonitrile at a rate of $0.1 \, mL \, min^{-1}$.

The theoretical variation of the PAH concentration in the tank was estimated from the programmed temporal concentration change in the spiking solution (Figure 4.3.3). This variation was derived by using the following expression:

$$C_i = C_{i-1} + (\Delta V_{PAH} \times C_{PAH})/V_{tank} - (\Delta V_{tank} \times C_{i-1})/V_{tank}$$

with:

- $C_i$ and $C_{i-1}$: expected PAH concentrations in the tank at $t = i$ min and $t = i - 1$ min respectively,
- $\Delta V_{PAH}$: flow rate of the PAH spiking solution added in the tank $= 0.1 \, mL \, min^{-1}$,
- $C_{PAH}$: PAH concentration in the spiking flowing solution,
- $\Delta V_{tank}$: flow rate of the Meuse river water through the tank $= 167 \, mL \, min^{-1}$,
- $V_{tank}$: water volume in the tank $= 215 \, L$

### 4.3.2.2 Chemicals

Essentially for spectroscopic reasons (see below) inherent in the method used, only 6 of the 16 polycyclic aromatic hydrocarbons contained in the spiking solutions could be monitored: benzo[k]fluoranthene (BkF), benzo[a]pyrene (BaP), chrysene (CHRY), benzo[a]anthracene (BaA), pyrene (PYR) and anthracene (ANTH). They were all purchased from Aldrich (Milwaukee, WI, USA). Solutions used for applying the

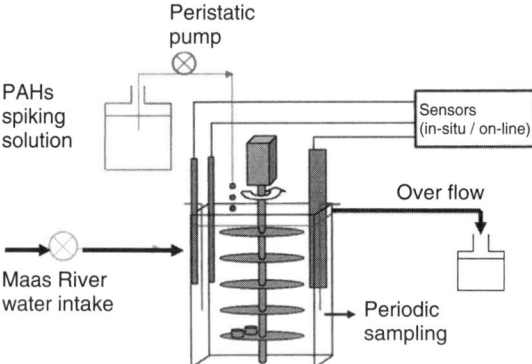

**Figure 4.3.2** Experimental configuration for tank experiment with Meuse river water fortified with PAH spiking solutions

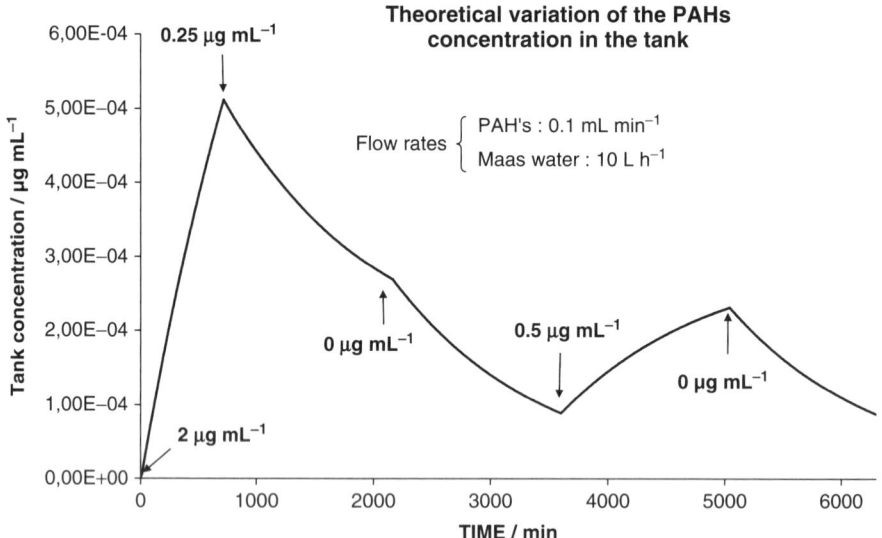

**Figure 4.3.3** Theoretical pattern of variation in PAH concentrations in the tank as predicted by the variational model (see text). The arrows indicate changes in the PAH spiking solution flow rate

standard addition method for their quantification were initially prepared in acetonitrile (SupraSolv®, Merck, Darmstadt, Germany). Pure water was prepared with a Milli-Q/Milli-Q2 system (Millipore, Bedford, MA).

### 4.3.2.3 Phase Preparation and Operational Conditions

The solid phase was prepared as 8 mm × 12 mm tab-shaped elements cut from a 90 mm SPE disk made of octadecyl silica phase enmeshed in a fibre glass support ($C_{18}$-ENVI-Disk™ from Supelco). The sorbent tab attached to the stainless steel

## Materials and Method

holder was then immersed into the water to be analysed for a fixed period of time (30 min) under stirring condition as illustrated in Figure 4.3.1.

Special attention was paid to avoiding adsorption onto the flask walls (Schaller et al., 1991) by using only stainless steel materials comprising a 200 mL beaker, the sorbent holder and a rod-like agitator driven by a small motor set above the beaker (Figure 4.3.1). The adsorption capacity of the extraction disks may vary between batches and so all of the measurements described in this work were made using sorbent elements cut from the same disk. A constant rotation rate of the stirring motor was maintained throughout the experiment by always using the same stirrer with 1.5 V batteries as a power source (Lamotte et al., 2003).

During the adsorption/concentration process, some fibres of the extraction disk were observed to be released from the sorbent surface upon stirring, and this was found to induce poor reproducibility of the signal. This problem was solved by prior treatment of the phase in which it was attached to its sampler holder and stirred vigorously in pure water for 30 min in order to eliminate the most fragile fibres from its surface.

An important source of error in the quantification of PAHs is the ubiquitous presence of humic substances (HS) in natural waters (Kile et al., 1994). These substances have been shown to form associations with PAHs and as such to play an important role in PAH transport and fate in the environment (McCarthy and Jimenez, 1985; Kile et al., 1994; Kopinke et al., 2001). These interactions are the basis of competitive complexation that reduces the uptake of PAHs by solid phase extraction (Sturn et al., 1998; Li and Kee Lee, 2001). This effect is illustrated in Figure 4.3.4 from Algarra et al. (2005), where the impact of humic substances is shown in a Stern–Volmer plot of the pyrene signal obtained by using the present method against the concentration of Suwanne River Humic acid. In addition to the humic substances many other organic substances and particulate matter may interact with PAHs and thus could significantly

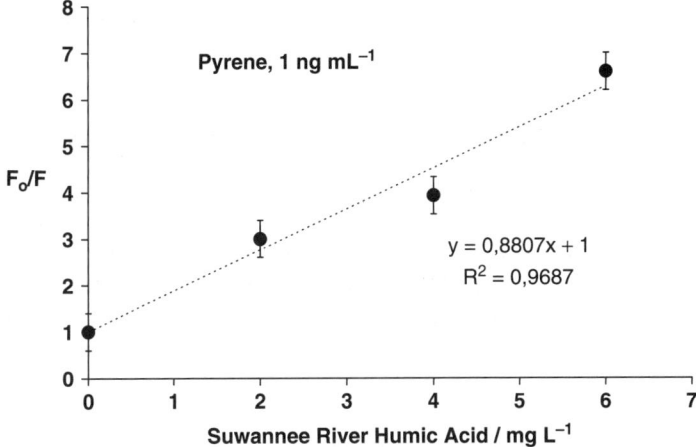

**Figure 4.3.4** Stern-Volmer type curve for pyrene fluorescence signal decrease upon increasing the concentration of Suwannee river humic acid in pure water (initial pH : 6.5). (From Algarra et al. (2005), with kind permission from Springer Science and Business Media)

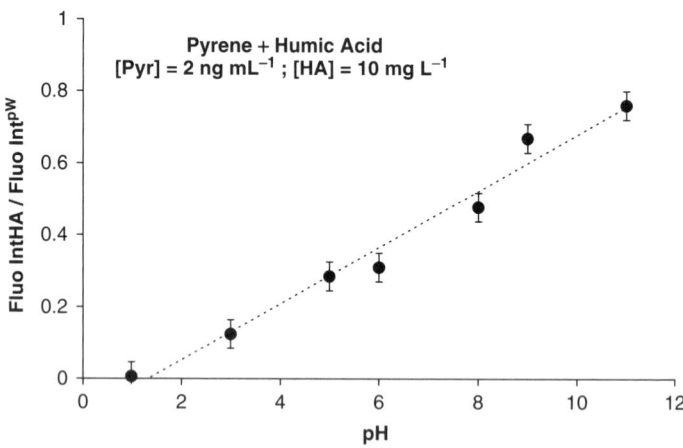

**Figure 4.3.5** pH effect on the fluorescence intensity signal of pyrene adsorbed from a water solution (C = $2\,\mu g.L^{-1}$) in the presence of Suwannee river humic acid (HA). The ratio of the fluorescence intensity in the presence of HA over the fluorescence intensity in pure water is plotted on the ordinate. (From Algarra *et al.* (2005), with kind permission from Springer Science and Business Media)

decrease the fluorescence signal of the PAHs present in the tank water. Some of these may have been released or created by other experiments that were carried out simultaneously in the test tank. One simple way to limit the negative effect of humic substances or natural organic matter in general was to increase the pH to 11 (Algarra *et al.*, 2005). Under these conditions the sequestering effect of these substances has been observed to be substantially inhibited. An illustrative example is provided by the case of the behaviour of pyrene in the presence of Suwannee River Humic acid (Figure 4.3.5). It shows that about 80% of the signal normally observed in pure water is recovered at pH 11 while it amounts less than 40% at pH 6. Accordingly, in the present work all measurements were performed at about pH 11 by addition of sodium hydroxide (400 μL of a 1N solution).

### 4.3.2.4 Front Face Fluorescence Spectra Measurements

Fluorimetric measurements were carried out with a commercial Hitachi F-4500 spectrofluorometer equipped with a 150 W xenon lamp and interfaced to a personal microcomputer for instrument operation and spectra processing. The slit-widths were adjusted at 2.5 nm both for excitation and emission and the photomultiplier (PMT R-3788) voltage set to operate at 950 V. The synchronous fluorescence spectra were corrected for variations with wavelength of the lamp intensity and photomultiplier sensitivity.

## 4.3.3 ANALYTICAL PROCEDURE

The first fortified sample was collected at an elapsed time of 3 h, and then subsequently every 6 h. For each sampling, about 1 L of the tank water was collected in a stainless steel bottle. A constant volume of 200 mL of this water was then transferred without further treatment (no filtering) into a 250 mL stainless steel Becher and sodium hydroxide (400 µL of 1N solution) added. The sorbent tab attached to the stainless steel holder was then immersed into this water sample for a 30 min under stirring conditions (Figure 4.3.1). After exposure the sorbent holder was placed in the sample compartment for front face synchronous fluorescence measurement with the sorbent surface oriented at 45° with respect to the excitation and emission beams. The PAH fluorescence synchronous spectrum is obtained by subtracting the blank signal recorded for the same experimental conditions from the sorbent tab previously immersed in a pure water sample for mechanical treatment. For each analysis, 3 corrected synchronous fluorescence spectra were recorded in an excitation scale between 240 and 400 nm with 3 different offsets $\Delta\lambda = \lambda_{em} - \lambda_{ex}$. It is important to note that such a method is not easily applicable for PAHs with low fluorescence quantum yield or whose absorption and fluorescence spectra are broad and do not exhibit clear characteristic and intense vibronic bands. Among the 16 PAHs used to artificially contaminate the Meuse river water in the tank, only 6 of them were found to be suitable for direct synchronous fluorescence determination in mixtures.

In order to simplify the procedure and to limit the time needed for each analysis, the following wavelength offsets were selected: $\Delta\lambda_1 = 18$ nm for ANTH and BaP, $\Delta\lambda_2 = 37$ nm for PYR and BaP and $\Delta\lambda_3 = 93$ nm for CHRY and BaA. For each determination, measurements are made on two aliquots (250 mL) from each of the periodic tank water samples (1 L).

- The first is made on the neat (unfortified) sample of tank water,
- The other is made after addition of a solution containing the PAHs to be analysed, all at the same concentration : $0.25\,\mu L\,L^{-1}$ for samples collected during the first 36 hours and $0.125\,\mu L\,L^{-1}$ during the rest of the exercise.

Some examples of the spectra obtained by using the 3 selected offsets are given in Figure 4.3.6. The concentrations were determined from the relative increase in fluorescence intensity following addition of the standard PAH mixture to the sample. The choice of using only 1 concentration for standard addition represents a compromise between the reduction in time necessary for one determination and the precision of the measurements.

## 4.3.4 RESULTS AND DISCUSSION

In total 18 water samples were analysed: the first sample contained plain Meuse river water without PAH addition. In this sample, no clear PAH fluorescence signal was detected indicating that during the period of this trial the PAH content of the Meuse river water was below our detection limit. The variations in concentration of the 6 PAHs in the 18 samples collected during the 6 days exercise are shown in Figure 4.3.7 and Table 4.3.1. It can be seen that while in samples T11 (3780 min) to T18 (6300 min) the measured concentrations follow approximately the expected pattern of change in

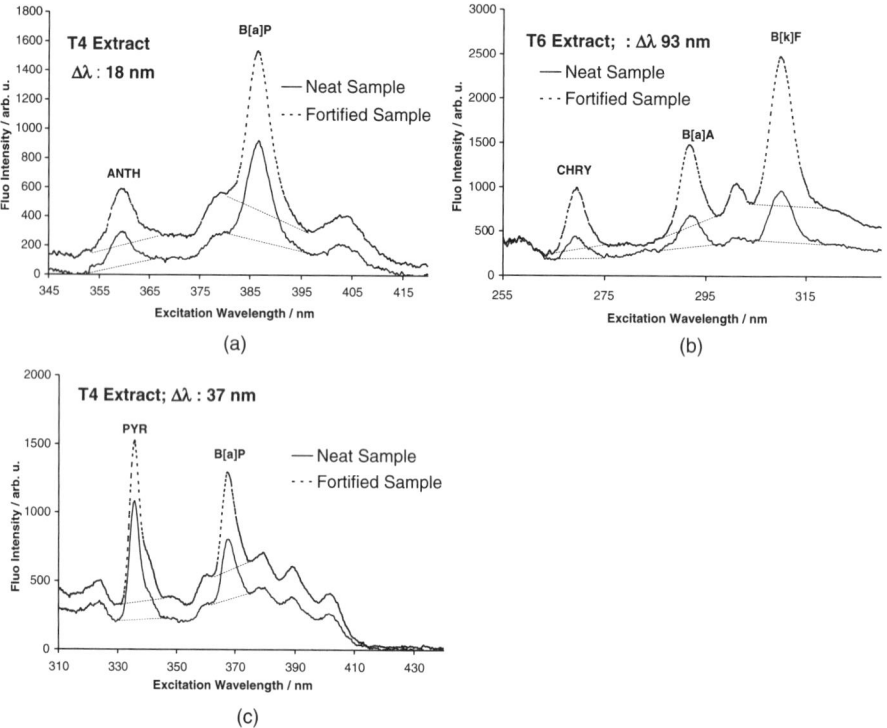

**Figure 4.3.6** Examples of front face synchronous fluorescence spectra obtained from tab-shaped $C_{18}$ sorbent elements after immersion in Meuse River water spiked with a mixture of PAHs. The marked fluorescence bands are those that were used for measuring the concentrations of the individual PAHs

concentration, this is not the case in samples T4 (1260 min) to T9 (3060 min) in which the observed concentrations were lower than expected. It is noteworthy, however, that in this time range ANTH, BaA, CHRY and PYR exhibit very similar changes in concentration, and this suggests that for these PAHs, the observed depletions in concentration are to a large extent dependent on the same cause. This may be explained by the observation made later around the time of sampling T6 that the peristaltic pump was not functioning properly and was suspected to delivered a lower PAH flow rate than expected. This probably affected the whole period following the first concentration change (from $2 \mu g L^{-1}$ to $0.25 \mu g L^{-1}$, Figure 4.3.3). This explanation is consistent with the fact that some time after having repaired the peristaltic pump (from about sampling T10, 3000 min) the changes in concentration followed more closely those expected.

Moreover, because of the presence in the tank of many types of materials (including natural organic particulates, polymeric devices, semi-permeable membranes and solid sorbents used for other types of measurements), a partial adsorption of PAHs leading to concentration depletion, at least at the beginning of the exercise, of the dissolved (free) PAHs is also highly plausible. This may also contribute to some extent to the low concentrations found for the four PAHs (ANTH, BaA, CHRY, and PYR) from samplings T4 to T9.

## Results and Discussion

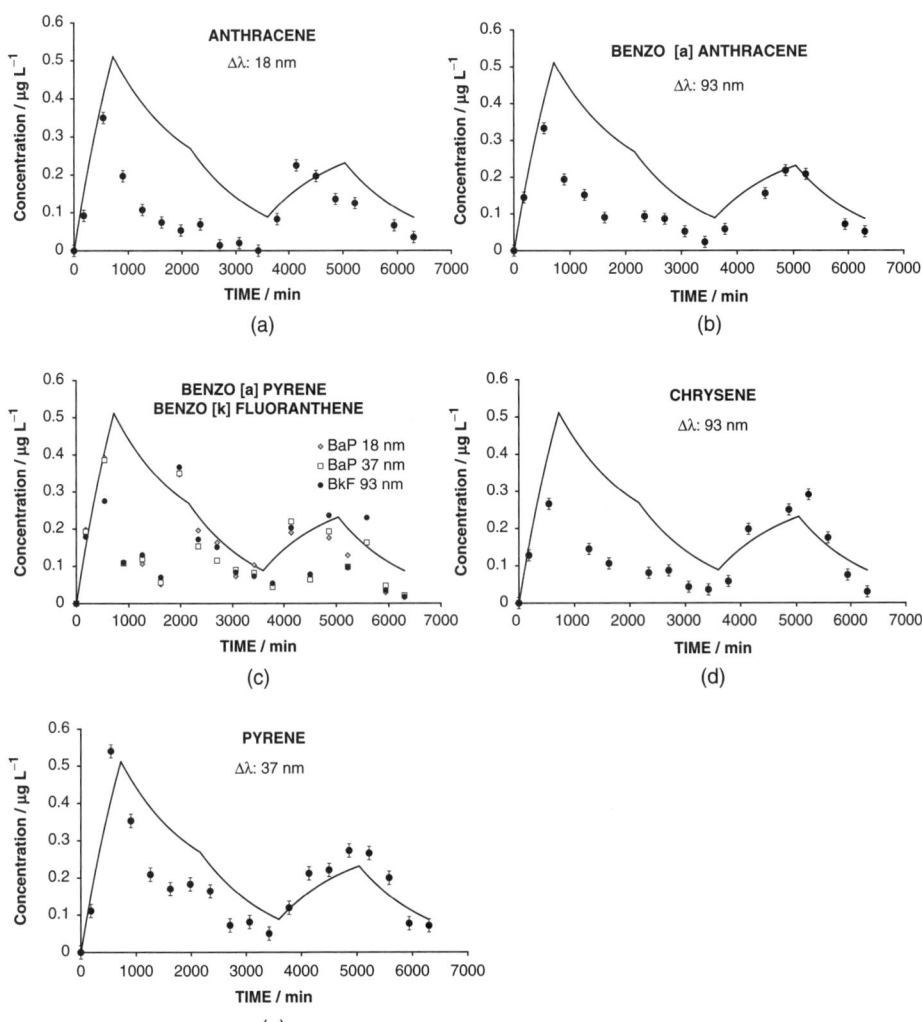

**Figure 4.3.7** Comparison of the measured changes in concentration of anthracene (a), benzo[a]anthracene (b), benzo[a]pyrene and benzo[k]fluoranthene (c), chrysene (d) and pyrene (e), during the present exercise, with the expected programmed changes in concentration (line)

A comparison of our results for BaA with those obtained by using a conventional chromatographic method is shown in Figure 4.3.8. A reasonable agreement between these results can be seen even at the beginning of the exercise. This lends further support to the idea that the low concentrations measured during this period did not result from some bias in the direct fluorimetric method but rather from an external source, probably the malfunctioning of the peristaltic pump and/or adsorption/desorption to/from the various materials present in the tank water. Similar agreement between the two sets of results was found for ANT, CHRY and PYR but not for BaP and BkF for which large discrepancies were observed between the expected

**Table 4.3.1** PAH concentrations in periodically collected water samples determined by front face synchronous fluorimetry on tab shaped $C_{18}$ sorbing elements after immersion in contaminated Meuse River water

| Extract | Time (h) | Time (min) | Concentrations (ng L$^{-1}$) ||||||
|---|---|---|---|---|---|---|---|---|
| | | | ANT | B[a]A | B[a]P | CHRY | PYR | B[k]F |
| T0  | 0   | 0    | 0     | 0   | 0   | 0   | 0   | 0   |
| T1  | 3   | 180  | 92    | 145 | 195 | 128 | 110 | 179 |
| T2  | 9   | 540  | 350   | 333 | 389 | 266 | 570 | 264 |
| T3  | 15  | 900  | 197   | 194 | 108 | 143 | 352 | 110 |
| T4  | 21  | 1260 | 107   | 152 | 112 | 145 | 222 | 130 |
| T5  | 27  | 1620 | 74    | 90  | 53  | 106 | 180 | 70  |
| T6  | 33  | 1980 | 53    |     | 348 |     | 185 | 367 |
| T7  | 39  | 2340 | 69    | 93  | 174 | 81  | 207 | 172 |
| T8  | 45  | 2700 | 14    | 86  | 139 | 87  | 77  | 151 |
| T9  | 51  | 3060 | 20    | 52  | 81  | 43  | 87  | 83  |
| T10 | 57  | 3420 | <LOD  | 23  | 92  | 36  | 84  | 73  |
| T11 | 63  | 3780 | 83    | 58  | 47  | 58  | 157 | 54  |
| T12 | 69  | 4140 | 225   |     | 205 | 198 | 212 | 203 |
| T13 | 75  | 4500 | 197   | 156 | 70  |     | 223 | 78  |
| T14 | 81  | 4860 | 136   | 218 | 184 | 250 | 248 | 236 |
| T15 | 87  | 5220 | 125   | 208 | 113 | 291 | 287 | 96  |
| T16 | 93  | 5580 |       |     | 81  | 175 | 230 | 230 |
| T17 | 99  | 5940 | 66    | 71  | 38  | 75  | 78  | 35  |
| T18 | 105 | 6300 | 35    | 51  | 18  | 30  | 87  | 17  |

*Notes*: Empty cells correspond to results which have been discarded in reason of obvious corrupted signals. <LOD : below the limit of detection.

and observed results throughout the experiment. These latter compounds have large molecular size and are strongly hydrophobic in character and the extent of their partitioning between the water phase and dissolved or suspended organic matter appears to fluctuate markedly. This partitioning is determined by environmental factors including temperature and pH. The fluctuations in the freely dissolved concentration of these PAHs during the experiment may reflect variations in magnitude of their interactions with humic substances and particulate matter due to changes in both environmental conditions and in the total amounts of organic materials present.

## 4.3.5 CONCLUSION

The results obtained in this work indicate that this fluorimetric method has a valuable potential for use in on site monitoring of PAHs in river water. Its application is, however, restricted to the determination of PAHs with suitable spectroscopic properties for synchronous fluorimetric detection in mixture. Nevertheless, with the exception of B[a]P and B[k]F, the PAHs that were monitored in the present work, are commonly present in most PAH emission sources and may be considered as suitable environmental markers for PAH contamination of river water.

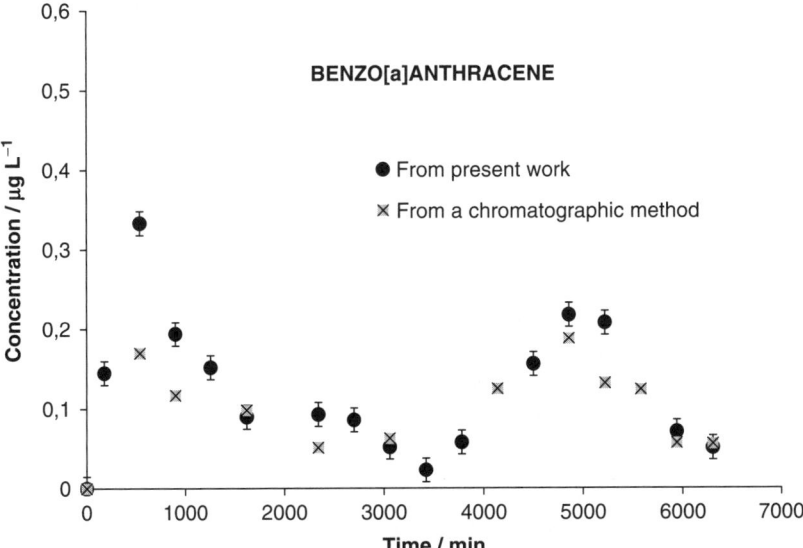

**Figure 4.3.8** Comparison of concentrations of B[a]A measured by direct fluorimetry with those obtained using a conventional chromatographic method

The practical limit of detection attained (LOD $\approx 20\,\text{ng}\,\text{L}^{-1}$) is very close to that determined during previous works (Lamotte *et al.*, 2003; Algarra *et al.*, 2005). This LOD value is in great part dependent on the fluorometer performance and of the efficiency of the extraction disk. A gain in sensitivity and reproducibility of the measurements could be expected by using a more sensitive fluorometer and by a possible improvement of the surface of the extraction disks since those currently available are not specifically devised for optically based measurements.

# REFERENCES

Algarra M. and Hernandez Lopez M., 1998. *Analyst*, **123**, 2217–21.
Algarra M., Jimenez V., Fornier de Violet Ph. and Lamotte M., 2005. *Anal. Bioanal. Chem.*, **382**, 1103–10.
Algarra M., Radin C., Fornier de Violet Ph., et al., 2000. *J. of Fluorescence*, **10**, 355–9.
Arruda A.F. and Campiglia A.D., 1999. *Anal. Chim. Acta*, **386**, 271–80.
Belfatmi R., Lamotte M., Ait-Lyazidi S. and Fornier de Violet Ph., 2005. *Chemosphere*, **61**, 761–9.
Bernal J.L., Nozal M.J., Toribio L., et al., 1997. *J. Chromatography A*, **778**, 321–8.
Burel L., Giamarchi P., Stephan L., Lijour Y. and Le Bihan A., 2003. *Talanta*, **60**, 295–302.
Carr J.W. and Harris J.M., 1988. *Anal. Chem.*, **60**, 698–702.
Chen H.-W., 2004. *Analytical Sci.*, **20**, 1383–8.
Dmitrienko S.G., Ya Gurariy E., Nosov R.E. and Zolotov Yu A., 2001. *Analytical Letters*, **34**, 425–38.
Eastwood D., Dominguez M.E., Lidberg R.L. and Poziomek E.J., 1994. *Analusis*, **22**, 305–10.

Fernández-Sánchez J.F., Carretero A.S., Cruces-Blanco C. and Fernández-Gutiérrez A., 2003. *Talanta*, **60**, 287–93.
Giamarchy P., Stephan L., Salomon S. and Le Bihan A., 2000. *J. of Fluorescence*, **10**, 393–402.
Hagestuen E.D., Arruda E.D. and Campiglia A.D., 2000. *Appl. Spectrosc.*, **52**, 727–37.
Havenga W.J. and Rohwer E., 2000. *Intern. J. Environ. Anal. Chem.*, **78**, 205–21.
Karlitschek P., Lewitzka F., Buenting U., Niederkrueger M. and Marowsky G., 1998. *Applied Physics B*, **67**, 497–504.
Kile D.E., Chiou C.T. and Brinton T.I., 1994. In: R.C. Averett, J.A. Leenheer, D.M. McKnight, K.A. Thorn (eds), *Humic substances in the Suwannee river, Georgia: interactions, properties and proposed structures*. U.S. Geological Survey Water Supply, Paper 2373, 21–32.
Kiss Z., Varga-Puchony J. and Hlavay J., 1996. *J. Chromatography A*, **725**, 261–72.
Kopinke F.D., Georgi A., and Mackenzie K., 2001. *Environ. Sci. Technol.*, **35**, 2536–42.
Lamotte M., Belfatmi R., Fornier de Violet Ph., Garrigues Ph., Lafontaine M. and Dumas C., 2003. *Anal. Bioanal. Chem.*, **376**, 816–21.
Lamotte M., Fornier de Violet Ph. and Garrigues Ph., 2000. *Spectra Analyse*, **222**, 27–31.
Li Nanqin and Kee Lee Hian, 2001. *J. of Chromatography A*, **921**, 255–63.
McCarthy J.F. and Jimenez B.D., 1985. *Environ. Sc. Technol.*, **19**, 1072–6.
Manoli E. and Samara C., 1999. *Trends Anal. Chem.*, **18**, 417–28.
Martinez D., Borrull F. and Calull M., 1999. *Trends Anal. Chem.*, **18**, 282–91.
Miller J.S., 1999. *Anal. Chim. Acta*, **388**, 27–34.
Neff J.M., 1997. *Polycyclic Aromatic Hydrocarbons in the Aquatic Environment: Sources, Fate and Biological Effects*, Applied Science, London.
Patra D. and Mishra A.K., 2001. *Talanta*, **55**, 143–153.
Santos F.J. and Galceran M.T., 2002. *Trends Anal. Chem.*, **21**, 672–85.
Schaller K.H., Angerer J. and Hausmann N., 1991. in: Ph. Garrigues and M. Lamotte (eds) *Polycyclic Aromatic Compounds, Proceedings of the 13$^{th}$ International Symposium on PAHs, Bordeaux*, Gordon and Breach, 1023–30.
Simpson N., 2000. *Solid-Phase Extraction: Principles, Strategies and Applications*, M. Dekker, New York.
Sturn B., Knauth H.-D., Theobald N. and Wünsch G., 1998. *Fresenius J. Anal. Chem.*, **361**, 803–10.
Thurman E.M. and Mills M.S., 1998. *Solid-Phase Extraction. Principles and Practice*, John Wiley & Sons, Inc., New York.
Urbe I. and Ruana J., 1997. *J. of Chromatography A*, **778**, 337–45.
US EPA Office of Solid Waste, draft PBT Chemical List, 1998.
Vilchez J.L, del Olmo M., Avidad R. and Capitan-Vallvey L.F., 1994. *Analyst*, **119**, 1211–24.
Vo-Dinh T. and Gammage R.B., 1978. *Anal. Chem.*, **50**, 2054–8.
Vo-Dinh T., 1984. In: P.J. Elving and J.D. Winefordner (eds), Kolthoff (Editor emeritus), *Chemical Analysis*. Vol. **68**, John Wiley & Sons, Inc., New York.
Vo-Dinh T., 1990. *Chemical Analysis of Polycyclic Aromatic Compounds*, John Wiley & Sons, Inc., New York.
Whitcomb J.L. and Campiglia A.D., 2001. *Talanta*, **55**, 509–18.
Whitcomb J.L., Bystol A.J. and Campiglia A.D., 2002. *Anal. Chim. Acta*, **464**, 261–72.
Williamson K.S., Petty J.D., Huckins J.N., Lebo J.A. and Kaiser E.M., 2002. *Chemosphere*, **49**, 717–29.
Wittkampt B.L. and Tilotta D.C., 1995. *Anal. Chem.*, **67**, 600–5.
Wittkampt B.L., Hawthorne S.B. and Tilotta D.C., 1997. *Anal. Chem.*, **69**, 1197–1203.
Yu J. C., Jiang Zi-Tao, Liu Ho-Yan, Yu Jiaguo and Zhang Lizhi, 2003. *Anal. Chim. Acta*, **477**, 93–101.

# 4.4
# Evaluation of the Field Performance of Emerging Water Quality Monitoring Tools

Catherine Berho, Nathalie Guigues, Anne Togola, Stéphane Roy, Anne-Marie Fouillac, Ian Allan, Graham A. Mills, Richard Greenwood, Benoît Roig, Charlotte Valat and Nirit Ulitzur

4.4.1 Introduction
4.4.2 Examples of Potential Applications for Emerging Tools
    4.4.2.1 Rapid Mapping of Nitrate Levels by On-site Based Monitoring Methods
    4.4.2.2 Use of Passive Samplers to Measure the Time-weighted Average Concentrations of Pollutants in Surface Waters
    4.4.2.3 Measurement of Time-weighted Average Concentrations of Heavy Metals using the Chemcatcher® and DGT Passive Sampling Devices
    4.4.2.4 Screening for the Presence of Emerging Pollutants Using Passive Sampling Devices
4.4.3 Use of Bioassays to Screen for Toxicity
4.4.4 Conclusions and Future Trends
Acknowledgements
References

## 4.4.1 INTRODUCTION

The implementation of the European Union's Water Framework Directive will require all Member States to design water monitoring programmes for the measurement of a range of ecological, hydromorphological, physico-chemical and chemical water quality elements (European Commission, 2000). This has increased the amount of monitoring

---

*Rapid Chemical and Biological Techniques for Water Monitoring*    Edited by Catherine Gonzalez, Philippe Quevauviller and Richard Greenwood
© 2009 John Wiley & Sons, Ltd

activity required to comply with the regulations, and this has cost implications since budgets are limited. There is therefore an urgent need to develop monitoring tools and methodologies that are able to provide the necessary information at a lower cost (Dworak *et al*., 2005). Since the WFD does not specify the methods that have to be used, there are opportunities to use new approaches, and/or to modify existing methods to obtain the necessary chemical and biological data. A significant research effort in the area of analytical chemistry is currently taking place to develop suitable field instrumentation and methods that can facilitate measurements at the field location of interest (Pawliszyn, 2006). However, before the methods can be used within the regulatory context it is necessary to demonstrate the quality of their performance and their utility. In order for them to be accepted by end-users full validation studies are needed. Since much of the developmental work on the newer methods has been laboratory based, the next step is to demonstrate their overall performance in realistic field situations.

In order to be able to compare and evaluate the burgeoning number of available methods it is helpful to have some form of classification. One promising approach is to group the methods on the basis of the relationship between the sampling and analytical processes. Most existing methods involve taking grab (bottle or spot) samples directly from a field site, and transporting them to a remote laboratory for chemical analysis or biological evaluation. Most regulatory monitoring is based on this approach. On-site methods involve similar manual or automatic collection of samples but the analysis is performed directly in the field location. This can be distinguished from *in-situ* methods where the analysis is carried out directly in the medium of interest (e.g. at the appropriate depth in a river or lake) (Buffle and Horvai, 2000). Methods can also be classified according to the type of information they provide. We can distinguish methods which are able to provide the same quantitative information as classical analysis from those that provide qualitative or semi-quantitative data: for instance, biological methods such as bioassays (Wadhia and Thompson, 2007). This type of method includes biological early warning systems (BEWS) based on the detection of changes in physiology or behaviour that can be used online or *in-situ* to indicate changes in water quality, and raise an alarm (de Zwart *et al*., 1995). Some methods, such as passive sampling, can be used to provide more representative information than can be obtained using classical spot sampling, but they are coupled with classical laboratory analysis. There are many of these alternative and or emerging methods in these various categories (Allan *et al*., 2006) but their potential for regulatory use within the context of the WFD is variable.

This chapter aims to give an overview of the performance and potential applications of a range of tools that were evaluated in the SWIFT-WFD field trials for water quality monitoring. The work was organized as a scoping trial on the Meuse River at the RIZA monitoring station in Eijsden (The Netherlands), followed by a series of trials at representative sites across Europe. The scoping study aimed to provide a preliminary evaluation of the performance of a range of chemical and biological tools. The following series of smaller scale trials provided more detailed information on the utility of methods selected on the basis of the outputs of the Eijsden trial. Some of these trials have been described in greater detail elsewhere (Allan *et al*., 2006; Roig *et al*., 2007; Allan *et al*., 2007; Togola and Budzinski, 2007; Togola, 2007).

## 4.4.2 EXAMPLES OF POTENTIAL APPLICATIONS FOR EMERGING TOOLS

This section describes a number of examples of the field deployment of a selection of potential monitoring tools for use within the WFD. Three types of tools (on-site based analytical methods, passive samplers and bioassays) are described. Each methodology was assessed by comparing its performance with that of the currently accepted monitoring regime (spot sampling combined with classical chemical analysis). In some cases the emerging methods provided information that was different from that provided by the classical methods.

### 4.4.2.1 Rapid Mapping of Nitrate Levels by On-site Based Monitoring Methods

This field trial was conducted in the Hardt catchment area in Alsace (France). A number of on-site analytical methods (multiparameter probe, immunoassays test kits, UV spectrophotometer) was evaluated for measuring the impact of anthropogenic pressures on water quality within the context of operational and investigative monitoring within the WFD. However, this section focuses on the use of a portable Pastel UV spectrophometer for the rapid mapping of concentrations of nitrate in a catchment.

The Pastel UV (Secomam) is an on-site analytical device (Figure 4.4.1) based on a spot sampling. This provide simultaneous measurements of up to six parameters

**Figure 4.4.1** Pastel UV portable measuring instrument (Secomam)

including chemical oxygen demand (COD), biological oxygen demand (BOD), total organic carbon (TOC), total suspended solids (TSS), nitrate ($NO_3^-$), and the surfactant dodecyl benzene sulfonate (DBS) by using a UV deconvolution method (Thomas et al., 1996). This tool was validated under both laboratory and field conditions for the measurement of nitrate concentrations in surface waters (Gonzalez et al., 2007). The instrument is easy to use, and the operator requires only a minimum of training. Other than dilution when concentrations are too high, no pretreatment of the sample is needed. The device has a short response time (ca. 5 minutes). It has a further advantage over standard laboratory based methods in that there is no need to preserve the sample for transport and storage prior to analysis for nitrate. Samples for laboratory analysis have to be analysed typically within 24 h to ensure optimum preservation.

The Pastel UV was selected to provide a rapid mapping of nitrate ($NO_3^-$) concentrations in the Hardt catchment area located in Alsace (France) (Roig et al., 2007). The City of Mulhouse stopped using wells to provide drinking water in the Hardt catchment area because nitrates and pesticides were found in high concentrations that in some cases exceeded the environmental quality standards (EQS). The hydrology of the catchment is such that surface waters infiltrate ground water through a series of gravel pits. Agricultural activities (mainly the growing of corn) in this area have a significant impact on ground water quality through the use of fertilizers and pesticides. Remedial actions are currently planned to improve the quality of the ground water in order to restore the use of wells in the region. However, the relative importance of a range of possible inputs into ground water is not well known. It was therefore necessary to assess the spatial variability of $NO_3^-$ in surface waters since this will give an indication of the main sources of nitrate contamination of the ground water. It will also aid the selection of representative sampling sites for use in routine water monitoring activities as required under the WFD.

On-site determinations of $NO_3^-$ concentrations were made using the Pastel UV at 37 sites in the Weiherbachgraben and Sauruntz sub-basins of the Hardt catchment. Both surface and ground waters were sampled over a two-day field campaign. In addition spot water samples were collected for analysis by ion chromatography in the laboratory. Immediately after collection an aliquot of the water sample was filtered through a 0.45 μm cellulose acetate filter (Millipore) and stored in a polyethylene bottle at 4–8°C until analysis.

The field measurements of concentrations of $NO_3^-$ obtained by the Pastel UV were consistent with the results obtained in standard laboratory-based analyses over the range 10–80 mg $L^{-1}$ $NO_3^-$ (Figure 4.4.2). One outlying point (indicated by a circle) was found, and this may be explained by high concentrations of suspended and organic matter in this sample.

The spatial variability of $NO_3^-$ concentrations determined using Pastel UV in surface and ground waters (wells) in the Hardt catchment area was mapped (Figure 4.4.3).

The $NO_3^-$ concentration in the surface waters of the Weiherbachgraben sub-basin are low at the sampling point in the vicinity of Dietwiller (D1 and D4 (26 and 44 mg $L^{-1}$ respectively)), compared with those at upstream and downstream sampling points (SLH, SLB, L) which varied between 50 and 60 mg $L^{-1}$ $NO_3^-$. However, the water quality parameters were markedly different between the up and downstream monitoring sites (Figure 4.4.4, indicated by the dashed border). Sites D1 and D2 were

## Examples of Potential Applications for Emerging Tools

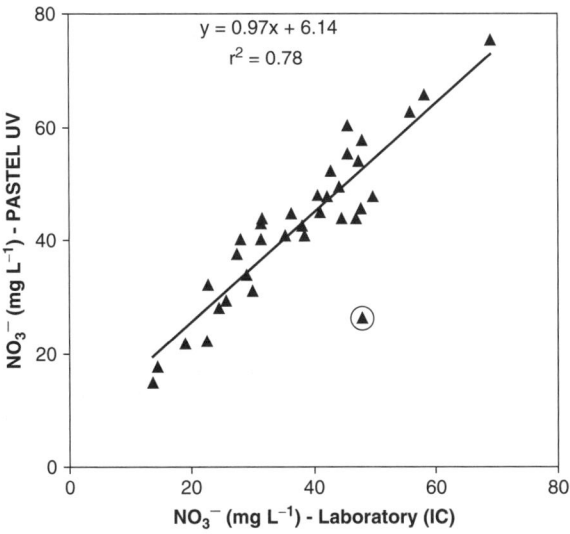

**Figure 4.4.2** Comparison of nitrate ($NO_3^-$) concentrations determined on-site using the PASTEL UV and in the laboratory using ion chromatography (IC)

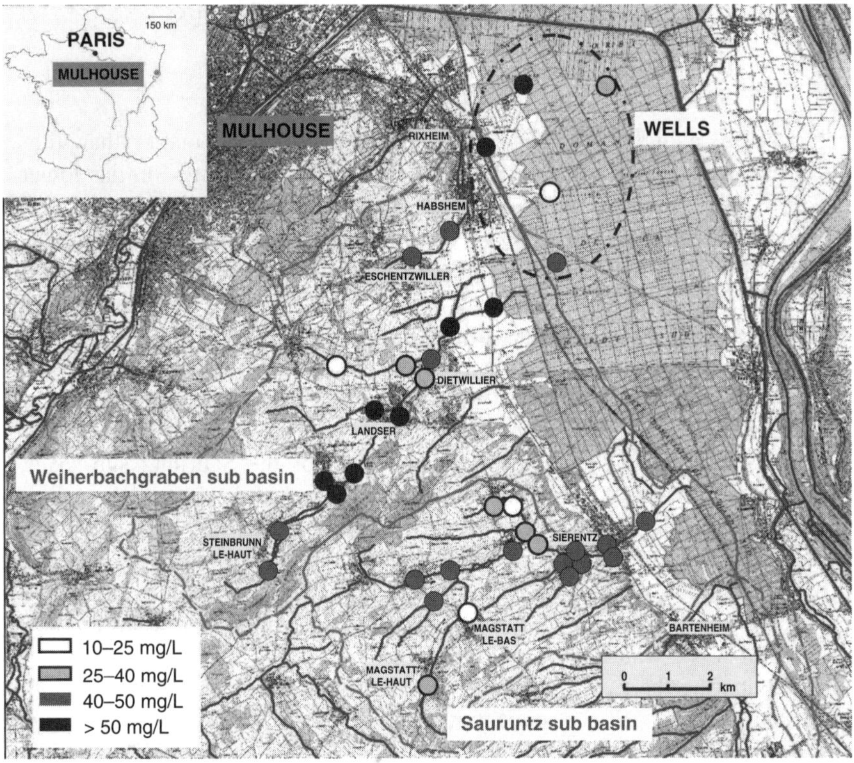

**Figure 4.4.3** Nitrate concentration estimated by Pastel UV in the Hardt catchment (Roig *et al.*, 2007) (See Plate 4 for colour representation)

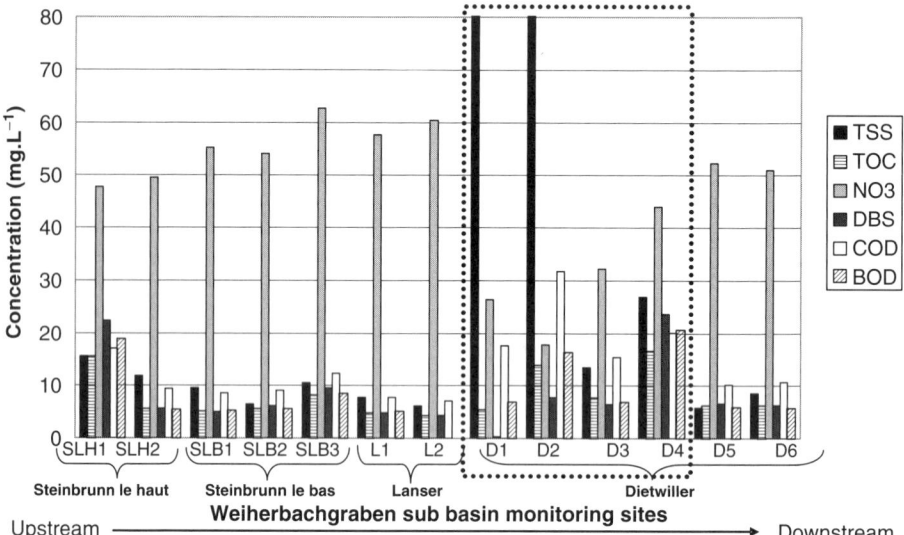

**Figure 4.4.4** Nitrate concentrations measured by Pastel UV for surface waters collected different locations within the Weiherbachgraben sub-basin. Key: chemical oxygen demand (COD), biological oxygen demand (BOD), total organic carbon (TOC), total suspended solids (TSS), nitrate ($NO_3^-$) and the concentration of dodecyl benzene sulfonate (DBS)

characterized by a higher content of organic and suspended matter than the other samples collected alongside the Weihergrabenbachgraben stream. Similar lower concentrations of $NO_3^-$ were also measured in the small stream running from the village of Bruebach (D2 and D3). The low $NO_3^-$ concentration and high chemical oxygen demand found in these samples are characteristic of urban sewage water inputs (Figure 4.4.4). Moreover, the deconvolution of the UV spectra for these four samples may have been influenced by the nature and the level of the concentration of organic matter present in the samples, reducing the accuracy of the $NO_3^-$ determinations compared with that of other samples. This especially marked for sample D1 that corresponds to the outlier in Figure 4.4.2.

The $NO_3^-$ concentrations in the surface waters of the Sauruntz sub-basin were generally lower than than those of the Weiherbachgraben sub-basin, with concentrations ranging from 15 to 50 mg $L^{-1}$. However, two point sources of $NO_3^-$ pollution were identified in the Sauruntz sub-basin: agricultural run-off near Koetzingue and Uffheim, where concentrations were roughly 45 mg $L^{-1}$. The organic content of most of the samples was rather high, especially the COD, and again this can be attributed to the input of urban waste water.

When comparing the two sub-basins, it appears that the Weiherbachgraben basin is characterized by higher $NO_3^-$ concentrations and lower COD concentrations whereas the Sauruntz basin shows higher COD and lower $NO_3^-$ concentrations. These results are consistent with the fact that waste waters of nearly all of the villages in the Weiherbachgraben sub-basin are collected into the sewage treatment plant (STP) whereas the waste waters of only two villages (Sierentz and Uffheim) in the Sauruntz sub-basin are collected to the same sewage treatment plant (STP).

The Pastel UV proved to be a suitable tool for the rapid (within two days) assessment of the spatial variability of $NO_3^-$ concentration in the Hardt catchment area. It demonstrated that most surface waters in the Weiherbachgraben and Sauruntz sub-basins were characterized by $NO_3^-$ concentration higher than 40 mg $L^{-1}$ and that $NO_3^-$ concentrations in ground water were above the EQS defined in the EU Drinking Water Directive 98/83/CE (European Commission, 1998). A further outcome of this study was the identification of critical pressure points in each sub-basin that may need further investigation and monitoring.

### 4.4.2.2 Use of Passive Samplers to Measure the Time-weighted Average Concentrations of Pollutants in Surface Waters

Passive sampling devices provide an alternative to repeated bottle or grab water sampling. The devices can be used to obtain time-weighted average (TWA) concentrations of pollutants over periods of exposure from days to weeks (European Commission, 1998; Pawliszyn, 2006). Samplers generally comprise a receiving phase with a high affinity for the pollutant to be monitored, separated from the bulk water phase by a diffusion limiting layer that can be a membrane. The TWA concentration to which the samplers have been exposed can be calculated from the mass of pollutant accumulated in the receiving phase over the deployment period. The chemicals accumulated in the receiving phase are extracted in the laboratory for subsequent instrumental analysis. Many types of samplers exist for the different classes of contaminants (polar organic, nonpolar organic, heavy metals and organo-metalic species) (Górecki and Namiesnik, 2002; Namiesnik *et al.*, 2005; Vrana *et al.*, 2005) and a more detailed account of their design and uses is given in Chapters 2.1 and 2.2 of this book.

A wide range of designs of passive samplers was tested tank and field trials conducted in April 2005 on the River Meuse at the RIZA monitoring station (Rijksinstituut voor Integraal Zoetwaterbeheer en Afvalwaterbehandeling) in Eijsden (Netherlands). However, this section focuses on a subset to illustrate potential applications for routine monitoring of selected heavy metals, and for the screening for presence or absence of some emerging organic pollutants. The results are discussed in the context of the WFD.

### 4.4.2.3 Measurement of Time-weighted Average Concentrations of Heavy Metals using the Chemcatcher® and DGT Passive Sampling Devices

The diffusive gradient in thin films (DGT) and Chemcatcher® passive samplers provide an alternative approach to monitoring concentrations of heavy metals in water. They measure the labile species, but in addition the DGT can provide information on the other species of an individual metal present in the environment (Górecki and Namiesnik, 2002; Vrana *et al.*, 2005; van Leeuwen *et al.*, 2005; Guigues *et al.*, 2007). The latter is determined by using two sets of DGT samplers, one with an open pore gel layer, and the other with a restricted pore gel layer. The difference between the uptake by these two sampler variants can yield information on the extent of speciation of the metals.

Chemcatcher® and DGT samplers were deployed in the Meuse River for varying periods up to 28 days and for the measurement of TWA concentrations of a range of metals (Cd, Cu, Ni, Pb and Zn) (Allan *et al.*, 2008). In order to evaluate the consistency of these TWA concentrations of labile metal, they were compared with total and filtered concentrations measured in spot samples of water taken at relatively high frequencies using standard spot and composite sampling procedures. Concentrations of metal species were predicted by equilibrium speciation modeling using Visual MINTEQ. Concentrations of most metals measured in spot samples were higher during the first 14 days of the trial and two peaks of concentration were observed. This natural variation in a field situation provided a good test of the utility of the technology in direct comparison with regulatory procedures.

The concentrations of Cd and Zn measured in filtered water samples and the passive sampling devices were generally similar, but concentrations of Cu and Ni were generally underestimated. This is consistent with the predicted speciation patterns of these metals. A large proportion of the Pb is associated with organic compounds in the water, and only a small proportion of the total load is labile. This resulted in a low uptake to and accumulation in the samplers, and so resulting estimates of TWA concentration were associated with a large uncertainty. The performance of the samplers is exemplified by the uptake data for Cd. For this element the TWA concentrations measured using both DGT and Chemcatcher® were reasonably consistent with the average of frequent spot sampling (samples filtered at 0.45 µm) over the 28 days of the trial. Despite the speciation modelling predicting a proportion of the filtered Cd associated with the humic/fulvic acid fraction, filtered Cd appeared mostly labile and available for uptake by Chemcatcher® or DGT. These observed levels of Cd are consistent with those from other studies in the River Meuse. During this trial the concentrations of Cd approached both the annual average EQS value ($0.25 \mu g\ L^{-1}$) and the maximum allowable filtered concentration EQS ($1.5 \mu g\ L^{-1}$).

The performance of the samplers is exemplified by the uptake of Cd by the DGT with an open pore gel layer (Figure 4.4.5). The samplers were able to detect the higher average levels of Cd over the first 14 days of the trial; however, the samplers exposed for 28 days underestimated the overall average filtered concentration. The latter effect may be due to the biofouling of the diffusive layer (Figure 4.4.6). This may reduce the uptake rate by either increasing the diffusion distance and/or increasing the resistance to diffusion. The most appropriate sampler deployment time needs to be defined in terms of the performance of the device and the objectives of the monitoring programme.

Successive shorter term (14 days) exposures may be suitable for investigative monitoring, but for routine monitoring purposes a balance needs to be struck between the costs associated with frequent sampling and the quality of the information obtained. Where concentrations fluctuate in time, then representative measurements of water quality can be obtained by spot sampling but only by using unrealistically high sampling frequencies with associated high costs. Deployment of passive samplers offers an alternative, and in this work the 14 day exposure yielded representative measurements of TWA concentrations of metals. If these samplers were used in conjunction with

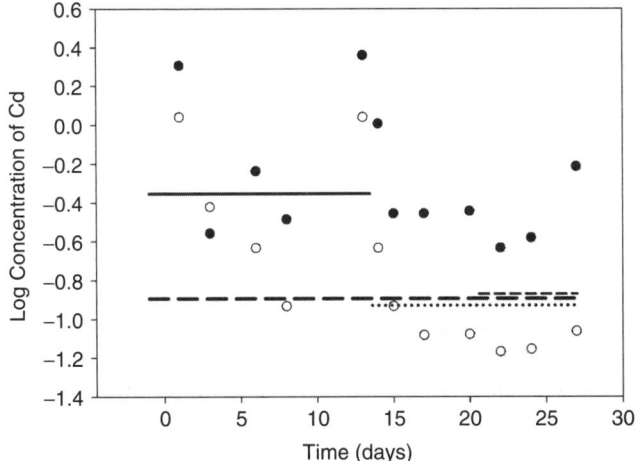

**Figure 4.4.5** Concentration of Cd (log scale) in frequent filtered (○) and unfiltered (total) (●) spot samples of water compared with time weighted average concentrations measured using the DGT open pore sampler deployed for overlapping periods of 7, 14 and 28 days in the River Meuse (NL)

**Figure 4.4.6** Typical biofouling: DGT (left) and Chemcatcher® (right) after a two-week deployment in the Meuse River at the RIZA monitoring station, Eijsden (NL)

infrequent spot sampling it would increase the reliability and confidence in data from routine WFD monitoring campaigns and associated risk assessments.

### 4.4.2.4 Screening for the Presence of Emerging Pollutants Using Passive Sampling Devices

Recent studies indicate the potential widespread occurrence of low-level concentrations (ng-µg $L^{-1}$) of emerging pollutants such as pharmaceuticals, hormones, personal protection products, household chemicals and their metabolites in the aquatic environment (Halling-Sørensen et al., 1998; Boyd et al., 2003; Sigg et al., 2006), particularly in wastewater. Some of these compounds are of particular concern because of their high levels of biological activity. Many of these substances are present at only very low levels, and conventional sampling methods (i.e., spot sampling) often require the collection and extraction of large volumes of water in order to detect these compounds at trace levels. This presents a challenge to the routine analytical laboratory.

Integrative passive sampling offers a solution to this since passive sampling devices accumulate pollutants over an extended deployment period, and effectively sequester quantities equivalent to those present in several litres of environmental medium. These devices can potentially provide information on TWA concentrations of pollutants even where concentrations fluctuate in time. Currently few calibration data are available for the passive samplers (POCIS, and Chemcatcher®) suitable for monitoring many of the polar organic compounds of concern (Tixier *et al.*, 2003; Petty *et al.*, 2004). However, these devices can provide screening data by indicating the presence or absence of individual trace pollutants. Since accumulation in the samplers is driven by the concentration of the freely dissolved (biologically available) fraction they provide estimates of potential exposure of aquatic organisms to these complex mixtures of contaminants.

In order to evaluate the technology in these roles, a five-day field trial was carried out in the River Meuse at Eijsden (NL) using the POCIS to detect the presence of pharmaceutical compounds. The results from the passive sampling campaign were compared with those obtained by using spot samples (1 L) that were extracted on-site using solid-phase extraction cartridges and analysed using gas chromatography-mass spectrometry.

After five days of exposure, the samplers accumulated a range of pharmaceutical compounds including stimulants, anti-psychotic, anti-inflammatory, and bronchodilatory drugs (Table 4.4.1). The amounts accumulated varied depending on the

Table 4.4.1 Comparison of the pharmaceutical compounds detected by the POCIS passive sampler (5-day deployment) and by simultaneous spot sampling (1 L) in the Meuse River. The detection limit for the GC/MS procedure are mentioned in brackets (Togola, 2007)

| Pharmaceutical compound | Clinical use | POCIS followed by GC-MS | Spot sampling followed by GC-MS |
|---|---|---|---|
| **Caffeine** | **Stimulant** | + | + |
| **Amitryptiline** | **Anti-psychotic drugs** | + | $- (<2\,\text{ng L}^{-1})$ |
| **Doxepine** | | + | $- (<2\,\text{ng L}^{-1})$ |
| **Imipramine** | | + | $- (<1\,\text{ng L}^{-1})$ |
| **Carbamazepine** | | + | + |
| **Diazepam** | | + | + |
| **Nordiazepam** | | + | + |
| **Aspirin** | **Anti-inflammatory drugs** | + | + |
| **Ibuprofen** | | + | + |
| **Paracetamol** | | + | $- (<2\,\text{ng L}^{-1})$ |
| **Gemfibrozil** | | + | + |
| **Naproxen** | | + | + |
| **Diclofenac** | | + | + |
| **Ketoprofen** | | + | + |
| **Terbutaline** | **Bronchodilatators** | + | $- (<2\,\text{ng L}^{-1})$ |
| **Salbutamol** | | + | + |

physicochemical properties of the analyte, and were equivalent to the contents of 1–12 L of river water. In comparison, some drugs (e.g. Amitryptiline, Imipramine, Paracetomol, and Terbutaline) from each class that were detected by the POCIS fell below the level of detection (1–2 ng L$^{-1}$) of the spot sampling method. However, with the currently available calibration database it was not possible to obtain quantitative data for the compounds detected with the passive sampler. Further work is urgently needed to obtain sampling rates for the two passive samplers available for monitoring polar organic pollutants, and to determine how these are affected by field variables such as temperature, turbulence and biofouling of the polyethersulphone membrane (Togola and Budzinski, 2007). This would enable their use in a quantitative manner such as is the case for many of the samplers used for monitoring nonpolar organic pollutants (see Chapters 2.1 and 2.2 on passive sampling).

### 4.4.3 USE OF BIOASSAYS TO SCREEN FOR TOXICITY

Bioassays are biological tools for the determination of the effects (positive or adverse) of a substance or a mixture of substances on living organisms or their component parts (Alvarez *et al.*, 2005). Whole organism bioassays have been used for many years to test the acute and chronic toxicity of aqueous samples or extracts from soils and sediments. Some of these have been designed to provide direct, rapid, sensitive and cost-effective assessments of chemical stress. Some can be used for high through-put determination of toxicity, and can be used to provide screening (Farré *et al.*, 2005).

A wide range of bioassays is available commercially. In this trial two whole cell bioassays, (ToxScreen II (CheckLight Ltd, Israel) and Metal Detector (CheckLight Ltd, Israel)) were tested under field conditions. The ToxScreen II test measures changes in light output from the naturally luminescent bacterium *Photobacterium leiognathi*. The test included the use of two assay buffers, one favoured the detection of heavy metals (Pro-Metal Buffer) while the other enhanced detection of organic contaminants (Pro-Organic Buffer). The Metal Detector is based on the inhibition of luminescence of *Escherichia coli* that have been modified to express the luminescence system of *Vibrio fischeri* in the presence of metallic compounds. In these assays, toxicity is measured as the minimal concentration (% dilution) of the environmental water sample that produces 50% inhibition of luminescence. A further well-established, general (nonspecific) assay system for toxicity is the Microtox® test (based on inhibition of luminescence in the bacterium *Vibrio fischeri*) was used alongside the more specific screens.

These bioassays were used to screen for toxicity in the River Ribble estuary (UK) (Roig *et al.*, 2007). The Ribble estuary is located in the Northwest of England and extends from Preston to Lytham. The estuary is known to fail the EQS values for a number of metals and certain organometalic compounds (e.g. tributyltin), particularly during tidal events, and this has been related to the region's industrial history. This is a complex system which is difficult to monitor because of the diverse pressures (agricultural run-off and sewage slurry, Preston docks, sewage-works effluents), large tidal flow, and fluctuations in the fresh water input depending on weather conditions. Water samples for use in the bioassays were taken at five points down the estuary between

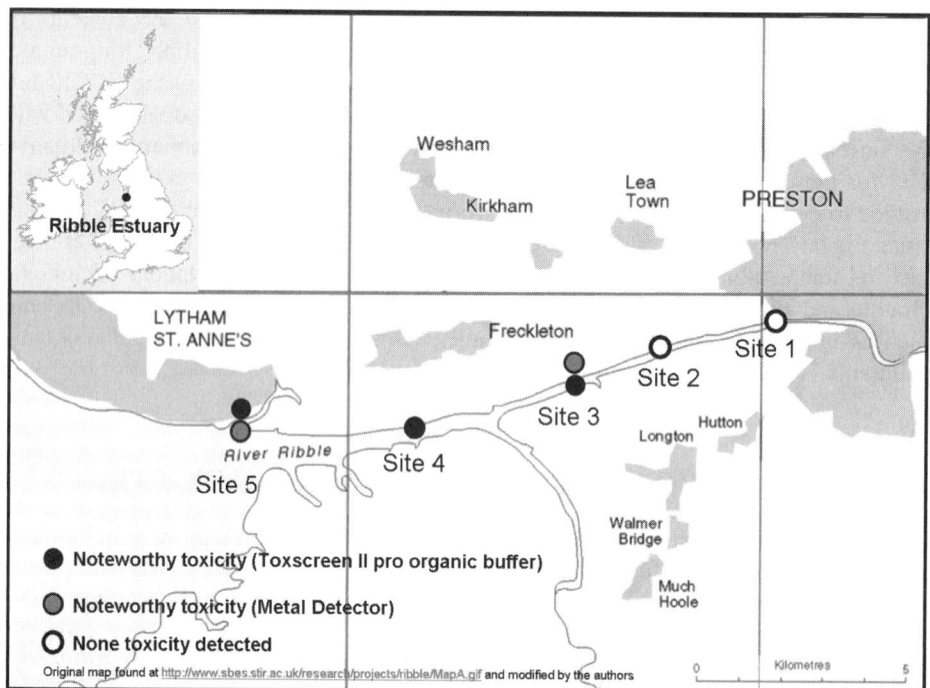

**Figure 4.4.7** Indication of the distribution of toxicity found by using ToxScreen II (pro organic buffer) and Metal Detector in the River Ribble estuary during the ebb tide. No toxicity was found with either ToxScreen II pro-metal buffer or with the standard Microtox® test (Roig *et al*., 2007)

Preston and the estuary mouth during the ebb and flood tides. Figure 4.4.7 shows the toxicity detected using the two screens during the ebb tide. The use of the ToxSceen II in combination with the Metal Detector permitted an association of observed toxicity effects with the type of pollutants present. The bioassay results indicated that the samples from sites 3 and 5 contained both heavy metals and organic toxicants whilst the sample from site 4 contained only organic toxicants. No toxicity was found using either the ToxScreen II pro-metal buffer or the standard Microtox® test.

These results demonstrate the utility of these methods for providing rapid on–site qualitative assessments of the presence of pollution in the water. They can be used for screening purposes in order to focus further monitoring activities and detailed analysis for the identification of the compounds responsible for the observed toxicity. They provide a mapping of the distribution of pollutants in a water course. This can reduce unnecessary monitoring efforts and associated costs. However, there is need for caution in the interpretation of the bioassay data using these bacterial screens. This is indicated by the failure to detect toxicity in any of the samples with either ToxScreen II pro-metal buffer or with the standard Microtox® test. This result highlights the variation in sensitivity between the various assay systems and the importance of using a battery of bioassays in any given monitoring campaign.

## 4.4.4 CONCLUSIONS AND FUTURE TRENDS

A range of emerging methods has been proposed for use in support of the monitoring required by the WFD. Three different modes of monitoring are defined in the WFD, and depending on the nature and quality of the information needed, some methods may be appropriate in only or in all of these modes. Many of the methods have been proved in laboratory studies, but there is an urgent need for demonstrating their utility in the field, and for field validation studies. In this chapter some of the possible roles for the emerging methods have been identified, and studies that demonstrate the potential advantages and limitations of these tools are described. This aims to identify areas where further work might be undertaken to increase the variety of tools available to regulatory organizations for obtaining reliable information on water quality at an affordable cost.

The on-site analytical and bioassay methods tested in some of these case studies provide quantitative data that can be comparable with results obtained from spot sampling linked to classical analysis, qualitative (screening) data respectively. The advantages of these methods is that the information is obtained in short timescale, and problems such as modification of samples associated with handling, storage, transport and preparation for analysis are avoided. They can provide a rapid mapping that would be helpful for instance after a pollution incident where they can provide timely information on the extent and rate of spread of a pollutant. They can also help in the selection of sampling sites, and identifying where water samples need to be taken and brought back to the laboratory for further analyses. This could lead to a more representative picture of water quality in a more cost-effective way than is achievable with current practice alone. Whilst many of the available on-site methods are not fully validated or as accurate as laboratory-based analyses, this may be addressed by present trends in instrumentation research. The development of fully integrated, miniaturized, noninvasive spectroscopic methods based on advances in the fields of nanotechnology, electronic data analysis and wireless communication will provide field technologies with performances similar to those of laboratory instrumentation (Pawliszyn, 2006).

Passive sampling technology provides a more representative measure of the concentration of pollutants than is provided by the current practice of infrequent spot sampling. However, the fractions measured by these devices differ from those measured in spot samples. In the former it is the freely dissolved fractions that are accumulated, while in the latter total (without filtration) or a sub fraction defined by the method of filtration used (e.g. ultra filtration or filtration at $0.45\,\mu m$). Passive samplers thus provide a useful indication of the bioavailable fraction, and a measure of the TWA concentration even where the environmental concentrations of a pollutant fluctuate in time. Under the WFD, they can be used for monitoring long-term trends and for screening a large range of contaminants at very low concentrations. Future challenges for this technology include the improvement of calibration models, availability of measurements of uptake rates for pollutants of current or emerging interest, the impacts of biofouling on performance, and the development of quality assurance procedures. Recently guidance (British Standards Institution Publicly Available Specification (PAS-61)) on the field deployment of passive samplers has helped to widen the availability of this methodology.

In order for emerging methods to be acceptable for use in a regulatory context work is needed to validate them, and to develop QA/QC procedures for their use. There is also a need for dissemination of information on their performance and potential uses within the WFD. This is especially important since the type of information obtained with some tools is different from that provided by current practice. In some cases it may be appropriate to use a battery of tools in water quality management. In this case there is a need for a system to integrate the available information to provide a comprehendible set of data that can provide support for those responsible for risk analysis, and decision making in this area.

## ACKNOWLEDGEMENTS

We acknowledge financial support from the Sixth Framework Programme of the European Union (Contract SSPI-CT-2003-502492; http://www.swift-wfd.com). We also thank the Environment Agency of England and Wales, the City of Mulhouse, Alsace, France and RIZA, The Netherlands for their generous assistance in helping to facilitate the different field trials In addition we thank all the partners in the SWIFT-WFD consortium who carried out the field and reference measurements.

## REFERENCES

Allan I.J., Vrana B., Greenwood R., Mills G.A., Roig B. and Gonzalez C., 2006. A 'toolbox' for biological and chemical monitoring requirements for the European Union's Water Framework Directive, *Talanta* **69**, 302–22.

Allan I.J., Mills G.A., Vrana B., et al., 2006. Strategic monitoring for the European Water Framework Directive *Trends Anal. Chem.* **25**, 704–15.

Allan I.J., Knutsson J., Guigues N., Mills G.A., Fouillac A-M. and Greenwood R., 2007. Evaluation of the Chemcatcher and DGT passive samplers for monitoring metals with highly fluctuating water concentrations, *J. Environ. Monit.* **9**, 1–10.

Allan I.J., Knutsson J., Guigues N., Mills G.A., Fouillac A-M. and Greenwood R., 2008. Chemcatcher® and DGT passive sampling devices for regulatory monitoring of trace metals in surface water, *J. Environ. Monit.*, **10**, 821–29.

Alvarez D.A., Stackelberg P.E., Petty J.D., Huckins J.N., Furlong E.T., Zaugg S.D. and Meyer M.T., 2005. Comparison of a novel passive sampler to standard water-column sampling for organic contaminants associated with wastewater effluents entering a New Jersey stream, *Chemosphere* **61**, 610–22.

Boyd G.R., Reemtsma H., Grimm D.A. and Mitra S., 2003. Pharmaceuticals and personal care products (PPCPs) in surface and treated waters of Louisiana, USA and Ontario, Canada, *Sci. Tot. Environ.* **311**, 135–49.

Buffle J. and Horvai G., 2000. *In situ monitoring of aquatic systems*. IUPAC series on Analytical and Physical Chemistry of Environmental Systems. Vol. **6**. John Wiley & Sons, Ltd, Chichester.

de Zwart D., Kramer K.J.M. and Jenner H.A., 1995. Practical experiences with the biological early warning system Mosselmonitor, *Environ. Toxicol. Water Quality* **10**, 237–47.

Dworak T., Gonzalez C., Laaser C. and Interwies E., 2005. The need for new monitoring tools to implement the WFD, *Environ. Sci. Policy* **8**, 301–6.

European Commission, 1998. *Off. J. Eur. Commun.* **L 330**, 32.
European Commission, 2000. Directive 2000/60/EC of the European Parliament and of the Council of 23 October 2000: *Establishing a framework for Community action in the field of water policy*, *Off. J. Eur. Comm.* **L327**, p. 1.
Farré M. and Barceló D., 2003. Toxicity testing of wastewater and sewage sludge by biosensors, bioassays and chemical analysis, *Trends Anal. Chem.* **22**, 299–310.
Farré M., Brix R. and Barceló D., 2005. Screening water for pollutants using biological techniques under European Union funding during the last 10 years, *Trends Anal. Chem.* **24**, 532–45.
Gonzalez C., Prichard E., Spinelli S., Gille J. and Touraud E., 2007. Validation procedure for existing, and emerging screening methods, *Trends Anal. Chem.* **26**, 315–22.
Górecki T. and Namiesnik J., 2002. Passive sampling, *Trends Anal. Chem.* **21**, 276–91.
Guigues N., Berho C., Roy S., Foucher J.-C. and Fouillac A.-M., 2007. The use of tank experiments in assessing the performance of emerging tools in optimizing water monitoring, *Trends Anal. Chem.* **26**, 268–73
Halling-Sørensen B., Nors Nielsen S., Lanzky, F., Ingerslev P.F., Holten Lützhøft H.C. and Jørgensen S.E, 1998. Occurrence, fate and effects of pharmaceutical substances in the environment – a review, *Chemosphere* **36**, 357–93.
Namiesnik J., Kot-Wasik A., Zabiegaa B., Partyka M. and Wasik A., 2005. Passive sampling and/or extraction techniques in environmental analysis: a review, *Anal. Bioanal. Chem.* **381**, 279–301.
Pawliszyn J., 2006. Why move analysis from laboratory to on-site? *Trends in Anal. Chem.* **25**, 633–4.
Petty J.D., Huckins J.N., Alvarez D.A., *et al.*, 2004. A holistic passive integrated sampling approach for assessing the presence and potential impacts of waterborne environmental contaminants, *Chemosphere* **54**, 695–705.
Roig B., Valat C., Berho C., *et al.*, 2007. The use of field studies to establish the performance of a range of tools for monitoring water quality, *Trends Anal. Chem.* **26**, 274–82.
Sigg L., Black F., Buffle J., *et al.*, 2006. Comparison of analytical techniques for dynamic trace metal speciation in natural freshwaters, *Environ. Sci. Technol.* **40**, 1934–41.
Thomas O., Theraulaz F., Agnel C. and Suryani S., 1996. Advanced UV examination of wastewater, *Environ. Technol.*, **17**, 251–61
Tixier C., Singer H.P., Oellers S. and Müller S.R., 2003. Occurrence and fate of carbamazepine, clofibric acid, diclofenac, ibuprofen, ketoprofen, and naproxen in surface waters, *Environ. Sci. Technol.* **37**, 1061–8.
Togola A., 2007. *Occurrence and fate of pharmaceuticals in aquatic systems*, PhD thesis, University of Bordeaux, France, p. 301.
Togola A. and Budzinski H., 2007. Development of Polar Integrative Samplers for analysis of pharmaceuticals in aquatic systems, *Anal. Chem.* **79**, 6734–41.
van Leeuwen H. P., Town R. M., Buffle J., *et al.*, 2005. Dynamic speciation analysis and bioavailability of metals in aquatic systems, *Environ. Sci. Technol.* **39**, 8545–6.
Vrana B., Allan I.J., Greenwood R., *et al.*, 2005. Passive sampling techniques for monitoring pollutants in water, *Trends Anal. Chem.* **24**, 845–68.
Wadhia K. and Thompson K.C., 2007. Low-cost ecotoxicity testing of environmental samples using microbiotests for potential implementation of the Water Framework Directive, *Trends Anal. Chem.*, **26**, 300–7.

# 4.5

# Sampling Uncertainty and Environmental Variability for Trace Elements on the Meuse River, France

**Anne Strugeon-Dercourt**

4.5.1 Introduction
4.5.2 Assessment of Sampling Step in the Uncertainty
    4.5.2.1 Procedure Followed
    4.5.2.2 Sampling Exercise Performed in the Meuse River
4.5.3 Environmental Variability
    4.5.3.1 Procedure Followed
    4.5.3.2 Application
    4.5.3.3 Temporal Variability
4.5.4 Conclusions
Acknowledgements
Appendix A
References

## 4.5.1 INTRODUCTION

When a spot sample arrives in the laboratory for standard analytical measurements, scientists can more or less easily make an assessment of the concentrations, with associated uncertainties, of analytes in the sample. This is because for the analytical chemist the uncertainty of the laboratory measurement appears as the most important parameter that describes the quality of results. However, the process of taking the

---

*Rapid Chemical and Biological Techniques for Water Monitoring*   Edited by Catherine Gonzalez, Philippe Quevauviller and Richard Greenwood
© 2009 John Wiley & Sons, Ltd

sample by the particular sampling method used is not taken into account, although it contributes to the global uncertainty of the reported measurement results. Despite its importance, knowledge of the contribution of sampling to the global uncertainty is very limited whatever the field (food, soils, water).

In the SWIFT project (SWIFT-WFD, EU RTD FP6 Project, SSPICT–2003-502492), of which this work forms a part, a number of investigations were undertaken to identify and evaluate some sampling methods and technologies that could be used in monitoring surface waters in the context of the WFD. The methodology for the evaluation of the environmental variability of water bodies, taking into account both sampling and analytical (classical in laboratory) uncertainties described in this chapter, was important for most of the work packages in the project. The outcomes of the work of the SWIFT project have important implications for the implementation of the 6th action plan on the environment under the Common Implementation Strategy (CIS) of the Water Framework Directive (WFD) (European Commission, 2000). For example, if the variability of the medium observed is very small (not significant), as is found for instance in some lake systems, the performance criteria of the tools used for monitoring are not the same as where the variability of the medium is very large (e.g. in some river systems with multiple inputs from point and diffuse sources, or in tidal waters). Where the medium is homogeneous in space and time, then there is a low chance of making a false decision concerning the chemical or biological quality of the water body, whereas where the medium is highly variable, there is a great chance of reaching a false conclusion (at the same confidence level). Measures of the overall quality of the medium have associated uncertainties that include not only the actual variability (linked to natural processes) of the medium, but also the uncertainties associated with analytical tools used to estimate it, and the sampling procedure followed.

The work described in this chapter was carried out during one of the major field trials undertaken as part of the SWIFT project. This trial was conducted during the period 12 April–10 May 2005 at a transboundary monitoring station on the River Meuse (Eijsden, The Netherlands (50°46'46.10"N; 5°41'58.91"E)) situated downstream of the border with Belgium. The different steps of the sampling process that could affect the uncertainty associated with measurements of the chemical quality of a river were identified and investigated. Those factors included a range of stages including sampling, sample preparation, transport and analysis, and in-situ measurements (e.g. physico-chemical measurements, filtration, flask/tube, $HNO_3$ acidification). The experiments were conducted on 'real natural samples'; this involves testing factors such as matrix effects and temporal fluctuations of the composition of the river. In order to minimize temporal fluctuations of the chemical composition of the river during the sampling exercise, complementary tests were carried out to estimate the temporal and spatial variability of the river. It would not be possible to estimate this variability without carrying out tests to determine the uncertainty due to the sampling process itself. Without this it is not possible to allocate the variation in the data to the real variability of the river. This chapter describes the investigations made to assess both the uncertainty from sampling and the spatial and temporal variability of the Meuse River during the field trial.

## 4.5.2 ASSESSMENT OF SAMPLING STEP IN THE UNCERTAINTY

### 4.5.2.1 Procedure Followed

*Analytical uncertainty*

A range of protocols exists for the evaluation of analytical uncertainty, these include the Guide to Uncertainty in Measurement (GUM exercise) according to the French Standard NF ENV 13005 (1999); the interlaboratory approach according to ISO 5725 part 1 (1998a); and the validation process or the intra-laboratory approach (Feinberg, 2000). The approach used here to estimate the uncertainty of trace analytical measurements by ICP/MS was the intra-laboratory process from the Shewhard Control Chart analysis (French Standard, NF X 06–031 part 0, 1995) by means of results obtained from Reference Materials. The latter are used to monitor the performance of the analytical method via a Control Chart Process: they are a recognized means to check the accuracy of the analytical procedure in a laboratory. The uncertainties concern the reproducibility, the drift of the appliance, the uncertainty of the reference material, the trueness obtained versus Certified Reference Materials analysed. Data used concern measurements on three types of Reference Material whose commercial names are 'CRM1' or NIST 1640, 'CRM2' or TMRAIN 95 and 'CRM3' or TMDA 62. The differences between these three Reference Materials are listed below in Table 4.5.1.

Data used for the assessment of analytical uncertainty by ICP/MS were obtained over a time period of two years. The statistical procedure used to assess the uncertainty of the different elements by ICP/MS, taking into account the matrix effect and level of concentrations, is described below.

**Table 4.5.1** Comparison of the three RMs selected for the evaluation of the analytical uncertainty (words in italic are not certified for these elements, and are mentioned only as Reference Mass Fractions)

|  | Type of RM | Type of water | Elements referenced |
| --- | --- | --- | --- |
| CRM1 (NIST 1640) | Standard Reference materials | A natural fresh water | Al, Sb, As, Ba, Be, B, Cd, Cr, Co, Fe, Pb, Mn, Mo, Se, Ag, Sr, V, *Cu, Li, Ni, K, Rb, Zn, Ca, Mg, Silicon, Na* |
| CRM2 (TMRAIN 95) | Certified Reference Materials | A simulated rain sample (derived from a rainwater) | Al, Sb, As, Ba, Be, B, Cd, Cr, Co, Cu, Fe, Pb, Li, Mn, Mo, Ni, Se, Ag, Sr, Tl, U, V |
| CRM3 (TMDA 62) | Certified Reference Materials | A fortified standard for trace elements (derived from lake water) | Al, Sb, As, Ba, Be, B, Cd, Cr, Co, Cu, Fe, Pb, Li, Mn, Mo, Ni, Se, Sr, Tl, U, V |

The explanation of the different statistical techniques used and the terminology of specific statistical terms used are gathered in Appendix A. The procedure followed for each element is:

- *Check the normal distribution of values*  The goal is to understand the random variability that exists in each measurement of the data set. The analysis provides a way of determining whether uncensored data follow a normal or another type of distribution. In any case, the normality or non-normality of data has to be determined prior to any other statistical tests in order to avoid any misinterpretation of results.

- *Check the variance homogeneity*  Before using the variance obtained on the Reference Materials over the last two years, the variance homogeneity has to be checked; the variance being used as a measure of the spread or dispersion of the data. This analysis is not applicable to all data obtained for each element because sometimes there is no replication of measurements under repeatability conditions or there are too few data. Under these circumstances the application of this statistical analysis would not be relevant.

- *Detection of aberrant (outlier) or suspected values*  The Grubbs test is the statistical test used to identify if there are some aberrant (outlier) or suspected values, the risk taken is also 5% (Feinberg, 2001). Aberrant or suspected values can also be checked graphically through 'Box and Whiskers' plots.

- *Check if the bias can be neglected through a test of the trueness*  The definition of trueness (Prichard, 2005) is detailed in Appendix A. The criterion $C_{obs}$ is compared with 1. It is considered as acceptable if it is less than 1 at a confidence level of 95% (Feinberg, 2001). Where the bias found is judged to be nonsignificant, the uncertainty associated with the bias is simply the combination of the standard uncertainty on the CRM value with the standard deviation associated with the bias (Eurachem, 2000).

- *Estimation of analytical uncertainty*  The analytical combined uncertainty for each element is noted: $U_{element}$, expressed as the percentage at a 95% confidence level (Eurachem, 2000). The standard followed (NF XPT 90–220) (French Standard, 2003) recommends that the uncertainty U is expressed as the percentage (at 95% interval confidence) to the upper value from 0 to 5. That is the reason why the results of uncertainty are expressed from 0% to 5%.

## Uncertainty from sampling

Sampling is defined in the international standard ISO 5667 part 14 (ISO, 1998b) as the process of removing a portion intended to be representative of a body of water (or sludge or sediment) for the purpose of examination for various defined characteristics. As sampling involves manual procedures and the use of several different devices (e.g. flasks, pump), it will be associated with uncertainties. The terminology of sampling uncertainty is defined in Appendix A (IUPAC, 2005).

When end-users of data need to know the concentration of a pollutant in a river, their overall objective is most of the time to make a decision on water monitoring; e.g.

whether an observed event is due to accidental pollution, or which type of monitoring (i.e. surveillance, investigative or operational) is more appropriate with reference to the Water Framework Directive (European Commission, 2000). It is clear that end-users need the values with an estimate of the uncertainty associated with them, as is required by the international ISO/CEI 17025 standard (French Standard, 2005). Now adopted as European standard, this describes (in addition to technical aspects of the use of RM, interlaboratory programmes, and evaluation of uncertainty) the quality aspects that should be followed by laboratories. Moreover, the global uncertainty has two distinguishable components, one resulting from the analytical method used and the other from the sampling process (spot sampling) in the field. The sampling uncertainty could reveal the degree to which sample results represent actual conditions for the population sampled, neglecting the contribution of analytical uncertainty. Even if one executes the best protocols available to the highest standards that can be achieved, it is still not possible to obtain a perfect 'representative' sample (Eurachem, 2007), since samples never have exactly the same average composition because of the heterogeneous composition of the medium and also because of the contribution of some effects that have to be included in the global uncertainty of sampling. The latter is not the easiest to determine, even if everybody is aware of its importance.

Different approaches for the estimation of uncertainty of sampling can be proposed. The first one, the international standard ISO 5667-14, (ISO, 1998b), comes from the analysis of quality control results which are aimed at assessing the random error associated with different levels of the sampling process (analytical variance, sub-sampling/transport variance and total sampling variance). These tests are based on replicate quality control samples (Eurachem, 2007). The further steps consist of determining the contamination due to sampling containers and to processes such as filtration, and of identifying the influence of transport, stabilization and storage samples. All of these tests are performed on both deionized water and spiked samples. An experimental protocol for the assessment of sampling uncertainty is proposed below. It is based on the GUM standard (French Standard, 1999), which consists of listing and assessing the different influential factors. The main difference between this and the protocol explained above is that the tests are not performed on spiked samples.

## Process to estimate the sampling uncertainty

The sampling programme began by setting out the location of the sampling points, the required number and frequency of samples, and the methods and equipment to be used. The sampling programme followed was a compromise between the purpose of sampling and its statistical evaluation and practical and economic constraints. In practical terms, the number of samples required by theory is often not realistic.

$$\frac{L}{2s}\sqrt{n} \geq t$$

Where $L$ is a confidence limit, $s$ the standard error, $n$ the number of samples and $t$ is the Student's $t$ for $(n-1)$ degrees of freedom at the selected confidence limit as described in the international standard ISO 5667-14 (ISO, 1998b; Krajca, 1989). Thus, all samples had to be taken under repeatability conditions.

The assessment of the contribution of sampling to the global uncertainty could not be effected without considering the associated analytical measurement since the sampling uncertainty involves that derived from the analytical procedure. Consequently, the first step of the process involves the estimation of the analytical uncertainty. Here the analytical technique used for the trace measurements was ICP/MS. The second step is the separation of the uncertainty due to the sampling process itself from the global uncertainty, and the third is the identification of the contribution of each factor listed as suspected of influencing this uncertainty. There are two categories of factors in the global uncertainty of sampling:

- the uncertainty due to the sample processing itself (e.g. influence of pumping, tube, flow rates);

- the uncertainty due to physico-chemical sample preparation (e.g. improper handling, incomplete decontamination of sampling equipment).

*Design of the tests*

Tests were performed on samples of crude Meuse river water.

- **Step A: uncertainty due to the whole process of sampling**
  10 different samples were collected under repeatability conditions (with the same operator, at the same place, with the same appliances and within a short period). These 10 samples were taken at the 'same' sampling point, defined as 'a precisely defined location, within a source or locality, selected for sample collection' as described in the international standard ISO 5667-14 (ISO, 1998b). This step enables the determination of the uncertainty due to the sampling process itself. This means that it also includes the uncertainties due to the pumping system, tubings, filtration process, acidification process, transport, storage and analytical techniques used.

- **Step B: uncertainty due to the filtration process**
  This step was performed to evaluate the effect of filtration on sampling. One sample of 500 cc was collected and divided between 10 flasks (where each subsample was filtered). The flasks were stored and sent to the laboratory for analysis.

- **Step C: uncertainty due to subsampling in the field**
  This step was carried out to assess the possible effect of dividing a large volume of water into small volumes in the field. One sample of 250 cc was collected, filtered, acidified in the field and dispatched in different flasks of 15 cc, which were stored and sent to the laboratory for analysis.

- **Step D: uncertainty due to subsampling in the laboratory**
  A step similar to Step C was performed in the laboratory in order to detect any possible effect of subsampling in the laboratory before analysis. For these steps (C and D), the containers were the only materials coming into contact with the samples from the moment of time of collection in the field to the time of analysis in the laboratory.

- **Step E: uncertainty due to the contamination of water by different flasks**
  The objective of this step was to check the inertness of the equipment in relation to the samples and their matrix. Contamination of the samples due to the release of compounds incorporated in the flask materials is possible. This may have happened with the different types of flasks (such as PP, PS and PE) used in this study for the different volumes. In order to assess this possibility, samples of pure water milliQ stored in different types of flasks were analysed.

*Global uncertainty of sampling*

In this programme, it was assumed that there would be no changes in the state of the sample nor any significant deterioration during transport and storage prior to analysis. Consequently, this effect was not evaluated separately; and forms part of the unexplained uncertainty.

Since the factors (mentioned) are assumed to be independent, then the combined uncertainty can be obtained by summing the variances of the various effects tested.

$$U_{sampling} = \sqrt{\sum \text{var}_{effects(AtoE)}}$$

All analyses (to evaluate the effects of the different factors) were performed using ICP/MS. The uncertainties due to the analytical measurements are described above (see *Analytical uncertainty*, pp. 310–311).

*ANOVA*

Further statistical analyses can be used to determine the relative influence that any factor or set of factors has on the total variation (global uncertainty). One of these methods is the analysis of variance (ANOVA). This is an important technique for analyzing the effects of categorical factors on a response. However, the assumption of normality of the data has to be checked prior to the use of ANOVA to decompose the variability in the response variable between the different factors. Depending upon the type of analysis, it may be important to determine: (a) which factors have a significant effect on the response, and/or (b) how much of the variability in the response variable is attributable to each factor (as described in the statistical software STATGRAPHICS, Vs 5.0).

*Multifactor ANOVA*

When more than one factor is present and the factors are crossed, a multifactor ANOVA is appropriate. Both main effects and interactions between the factors may be estimated. In a graphical representation of the results of an ANOVA the points are scaled so that any levels that differ by more than the difference exhibited in the distribution of the residuals are significantly different.

## 4.5.2.2 Sampling Exercise Performed in the Meuse River

*Choice of the site*

In order to evaluate the uncertainty associated with screening methods against environmental variability, both temporal and spatial variability of the water body have to be characterized at the selected reference site. The sampling site on the Meuse River at Eijsden (NL) is adjacent to a transboundary monitoring station located at the border with Belgium. The station is fully equipped with laboratories and analytical instruments, and RIZA carries out (semi) continuous monitoring of water quality. This site was selected to conduct a 28-days field trial since it provides all that is needed in terms of of pollutants (wide variety of priority pollutants; both organics and inorganics) as well as local facilities and a secure location. Further details can be found in the Deliverable D16 of the European Union project (SWIFT, 2003–502492).

*Analytical uncertainty*

*Trace analysis results.* 90% of analytical uncertainties found are below 25% (Figure 4.5.1). This demonstrates the fitness for purpose of ICP/MS for the measuring of most of the trace elements in such aqueous matrices over a range of concentration levels. A high relative uncertainty of 60% is obtained for two elements (Mo and Al). Both of these elements are present at very low concentrations (0.16 µg/L and 1.69 µg/L respectively), and this provide a common reason for the high uncertainty value. However, further statistical investigation of the data for Mo showed that there was a lack of fit to the regression line, and thus the estimated concentration (0.16 µg/L) of Mo should be treated with caution. In contrast the data for Al were found to be statistically sound, and although there was a large dispersion of the observations, no aberrant values were detected.

One of the advantages of using three different types of RM is the comparison of the results obtained for different types of matrix and and where concentration levels

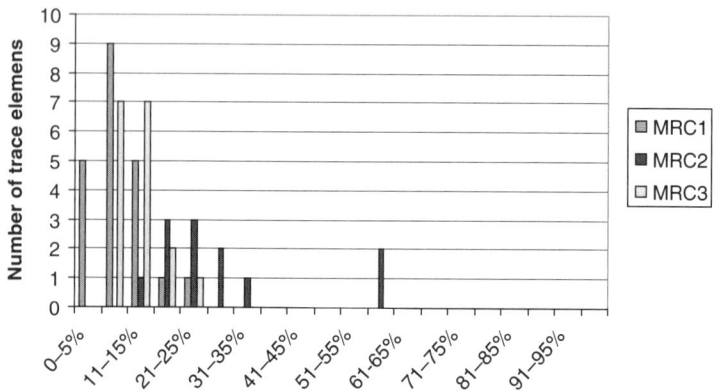

**Figure 4.5.1** Number of trace elements versus the analytical uncertainty associated for all three MRCs

are different for each trace element considered. In Figure 4.5.1, uncertainties obtained for all trace elements are classified into categories with a class increment of 5%. The lowest uncertainties are obtained for MRC1, where the concentrations are higher than for the other RMs.

The global analytical uncertainty selected for each element during the evaluation of sampling uncertainty takes into account, when possible, the concentration found in the medium. However, this analysis also indicated that sometimes the trueness found for some elements was large. This was found to be the case for V, Rb, Mo, Sb, Ba on the basis of the analysis of MRC1, for which the trueness of the method is very high. However, $C_{obs}$ was less than 1 for the other RMs analysed. The reason for this could not be attributed to the concentration levels of these elements since they were not far from the limit of quantification, and a more likely reason for the observations is the stability of the matrix. However, the small number of data (no replicates were available) does not allow further statistical tests such as Cochran or Grubbs or 'Gage linearity and accuracy'. The large uncertainty associated with these data precluded their use in the assessment of the contribution of sampling to global uncertainty. Another type of problem was encountered in considering the data for Cr, where the trueness was not acceptable, even though all of the statistical tests for normality, variance homogeneity and aberrant values gave results within acceptable bounds. All of this means that further investigations have to be performed such as experimental designs to test the ruggness of the method for the more problematic elements. The analytical uncertainties for the remaining elements described in this paragraph were used to assess the contribution of sampling to the global uncertainty.

*Uncertainty from sampling*

- **Step A: uncertainty due to the whole process of sampling.**
  All trace elements are represented in Figure 4.5.2, except for Sr because of its high concentration.

Box and whisker plots (Figure 4.5.3) are presented for all but Co and Cd elements, because no measurable variation was observed for these elements for any of the effects tested. Concentrations for these elements are very low. The variability due to the sampling process, including the analytical process, varies from 5% for Fe to 100% for Se at concentration levels of 13 µg/L and 0.2 µg/L respectively. In order to obtain the uncertainty associated with the sampling process itself, the uncertainty associated with the analytical processes (ICP/MS) was subtracted from the total. It was not possible to evaluate the uncertainty of sampling for most of the trace elements analysed (including V, Cr, Mn, Co, Ni, Cu, As, Rb, Mo, Sb, Ba, Pb and U) because no significant variation was observed between the 10 different samples. There are two main reasons for this: the low concentrations (close to the limit of detection) of these elements in the River Meuse, and the high level of uncertainty assessed from the Control Chart for these elements.

*Comparison of the values obtained with available data.* At this stage, in order to evaluate the water quality on the basis of these measurements, the concentration levels of traces were compared with projected Environmental Quality Standards (European

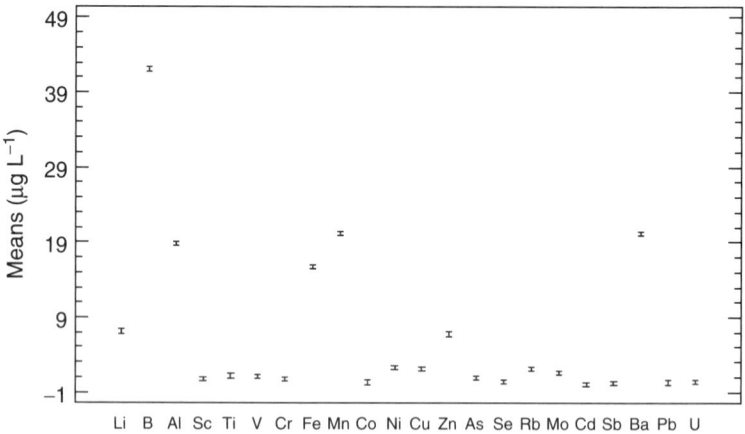

**Figure 4.5.2** Means and standard deviation (at 95%) of the concentrations (μg/L) of various trace elements (except Sr) measured under repeatability conditions on the Meuse River at Eijsden

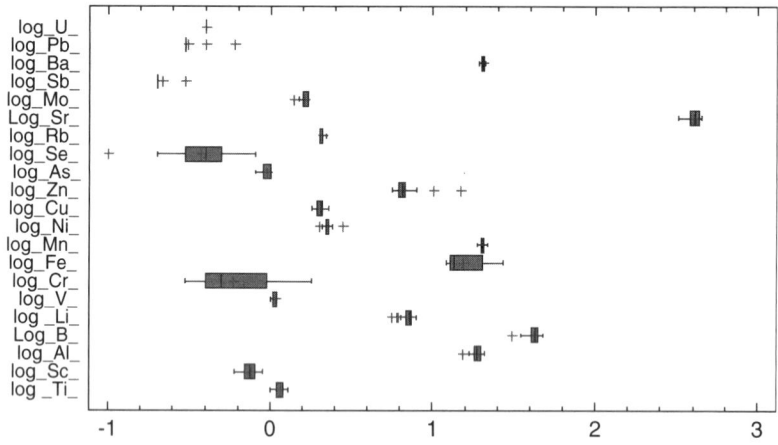

**Figure 4.5.3** Box and whisker plots of the log concentrations (μg L$^{-1}$) of various trace elements measured under repeatability conditions on the Meuse River at Eijsden

Commission, 2008). In this projected modification to the Directive, only 4 trace elements are listed: Cd, Ni, Pb and Hg. For these elements, the values obtained in this study were compared with the maximum allowable concentrations (MAC-EQS) for inland surface waters when available (Table 4.5.2). In France, a Circulaire (2007) defines NQEp (which are the French translation of the European MAC-EQS). However, since there is still a need to evaluate the geochemical background for many trace elements in surface waters, it has to be borne in mind when comparing water quality data with NQEp values, that these are not yet the definitive values. Finally, data were compared with limits fixed for drinking waters (Decree, 2001).

*Assessment of Sampling Step in the Uncertainty* 313

**Table 4.5.2** Comparison of the concentrations of some trace elements measured in the Meuse River at Eijsden with EQS values and drinking water limits

|  | Li | B | Al | Sc | Ti | Fe | Zn | Se | Sr | Cd |
|---|---|---|---|---|---|---|---|---|---|---|
| Concentration ($\mu g\ L^{-1}$) | 6.5 | 38.2 | 17.7 | 0.7 | 1.13 | 13.4 | 7.2 | 0.23 | 352 | 0.05 |
| Variance (%) | 14 | 19 | 7 | 15 | 10 | 5 | 82 | 100 | 9 | 12 |
| MAC-EQS ($\mu g\ L^{-1}$) | / | / | / | / | / | / | / | / | / | 0.45-1.5 |
| NQEp ($\mu g\ L^{-1}$) | / | 218 | / | / | 2 | / | 3.1–7.8 | 1 | / | 5 |
| Limit Values ($\mu g\ L^{-1}$) (drinking water) | / | / | / | / | / | / | 5 | 10 | / | 5 |

**Table 4.5.3** $U_{sampling}$ estimated for trace elements where there is a possible interpretation of data

| Range of $U_{sampling}$ | Elements |
|---|---|
| between 0 and 5% | Fe |
| between 6 and 10% | Al, Ti, Sr |
| between 11 and 20% | Li, B, Sc, Cd |
| between 21 and 50% | / |
| between 51 and 100% | Zn, Se |

The data in this study show the low level of pollution of the Meuse River by trace elements. A classification of these trace elements, with significant observed variations, into level of uncertainty due to the sampling process is given in Table 4.5.3.

The next step in the methodology consisted of explaining the sampling variance obtained for the different effects tested.

- **Steps B to E** The variance due to all selected effects (filtration, flask, blank) was calculated for each element in order to explain the contribution of each factor to the total variance and to estimate the part which is 'not explained' by these factors and that has to be correlated to the heterogeneity of the River Meuse.

For three elements, B, Zn and Sc, the explanation of the different effects tested was relevant. Figures 4.5.4 and 4.5.5 clearly show that the effects are not the same for each element studied at the levels of concentration found in the river.

Sc (Figure 4.5.5) shows a suprising distribution of effects. Whereas the sampling variance is around 100%, the variance obtained for the different factors tested is even larger. As the concentration of this element is very low, it is impossible to identify any trend in the distribution variance. All of the elements studied have in common a lack of influence of the blanks on the results; whereas the influence of the other effects tested varies from one element to another. Further statistical analyses were performed using STAGRAPHICS software, for the three elements (B, Sc and Zn) identified on the basis of the raw data without subtracting the analytical uncertainty. The results obtained are described below.

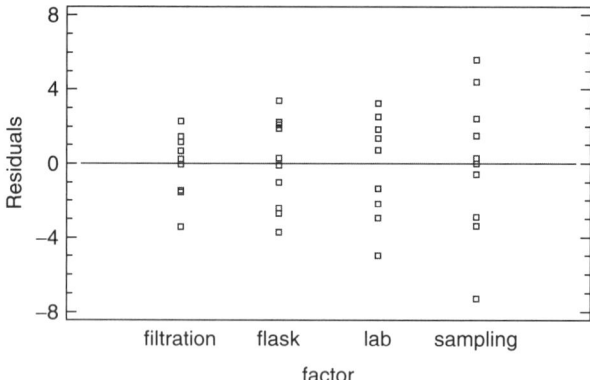

**Figure 4.5.4** Plot of residuals versus factor levels from an ANOVA of concentrations ($\mu g\ L^{-1}$) of boron

*Boron.* There was no evidence of significant deviation from normality (p = 98%) of the B data. This was not the case for Zn and Sc where a significant (p = 3.78 $10^{-10}$ and 1.25 $10^{-6}$ respectively) deviation from normality was detected. The multifactor **ANOVA** was performed on the boron data, and a plot of residuals versus factor levels plot indicates a lack of homogeneity of the variances across different factor levels as can be seen in Figure 4.5.4.

Results of the ANOVA analysis (the sum of squares type III was used to perform the analysis) are summarized in Table 4.5.4.

This analysis allows us to split the variability observed for B into contributions due to different factors. The probability (p-value) provides a measure of the statistical significance (at a confidence level of 95%) of each factor. Overall at least one of the factors has had a significant effect (p = 0.0001) on the measured level of B. This is in a good agreement with previous observations. Multiple range tests (Fisher's least significant difference (LSD)) was performed to determine which of the treatment means were significantly different from each other, and the results are summarized in Table 4.5.5.

The top portion of Table 4.5.5 identifies homogenous groups with columns of X that, within each column, indicates the group mean values for which there are no statistically significant differences. This shows that there is some discrimination between mean values observed between sampling and the other factors. If there were no discrimination

**Table 4.5.4** Analysis of variance for boron

| Source | Sum of squares | Degrees of freedom | Mean square | F | p-value |
|---|---|---|---|---|---|
| Effects | | | | | |
| A : factor | 234.585 | 3 | 78.1949 | 10.18 | 0.0001 |
| Residuals | 276.445 | 36 | 7.67903 | | |
| Total (corrected) | 511.03 | 39 | | | |

*Assessment of Sampling Step in the Uncertainty*

**Table 4.5.5** Multiple range tests for boron between factors, method 95% LSD (Fisher Least Significant DifferenceTest)

| Factor | Size | Means | Homogeneous group |
|---|---|---|---|
| Sampling | 10 | 38,2 | X |
| Filtration | 10 | 42,24 | X |
| Lab | 10 | 43,07 | XX |
| Flask | 10 | 44,8 | X |

| Contrast | Difference |
|---|---|
| Filtration – flask | *–2.56 |
| Filtration – lab | –0.83 |
| Filtration – sampling | *4.04 |
| Flask – lab | 1.73 |
| Flask – sampling | *6.6 |
| Lab – sampling | *4.87 |

*indicates a difference statistically significative.

between sampling and the other factors, it would not have been relevant to try to explain 'sampling' by these factors. The bottom portion of Table 4.5.5 shows the estimated difference between each pair of means. An asterisk indicates a statistically significant difference at the 95% confidence level. However, it is necessary to look carefully at statistically significant effects from the viewpoint of the analyst, and field monitoring scientist since these effects may not be chemically or environmentally significant, and some effects that do not reach statistical significance may be technically significant.

*Zinc and scandium.* The nonnormality of the distributions of Zn and Sc can be seen in the plots of residuals versus factors (Figure 4.5.5). Under these circumstances it is necessary to either apply a normalizing transformation, or to use nonparametric statistical procedures. Since the data distributions for Zn and Sc both demonstrate deviation from normality the statistical analyses of these two elements have been dealt with in the same paragraph. The box and whisker plots for these elements are presented in Figure 4.5.6 to provide further insight into their distributions.

Since the variances are not homogeneous, non parametric procedures, based on comparisons of medians rather than of mean values, are used. The dispersion observed for the Zn data indicates that the heterogenity of data can not be attributed to the factors tested here, and is more likely to be associated with the variability of the medium in this river. The results for Sc are different, and may be attributable to the extremely low concentrations (0.25 µg/L) compared with those of Zn and B (around 7 and 38 µg/L respectively). Further investigations should be carried out at higher concentrations of Sc, to check if the contribution of the factors associated with sampling and sample handling are the same.

These investigations do not allow us to provide accurate sampling uncertainties for each element, but they do illustrate the difficulties of explaining uncertainties in terms of the different factors involved in sampling. The advantage of working in a river is that the medium is a natural one, but the low concentrations and the heterogeneity of the

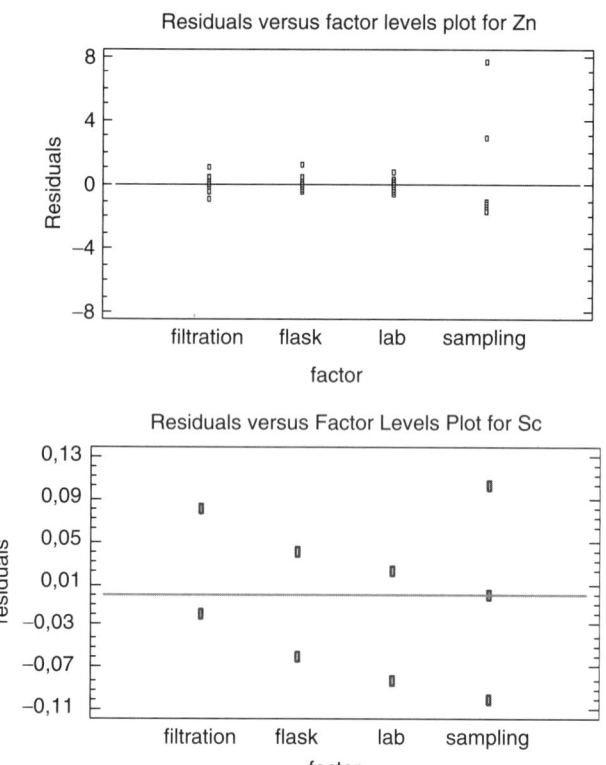

**Figure 4.5.5** Plot of residuals versus factor levels from an ANOVA of the concentrations (µg L$^{-1}$) of zinc and scandium

medium have a direct influence on the uncertainty due to sampling. The contribution of sampling uncertaintity to the overall data uncertainty is often significantly greater than that from analytical uncertainty because the former includes the heterogeneity of the river. Results could be totally different for organic compounds or major elements where concentration ranges are different.

## 4.5.3 ENVIRONMENTAL VARIABILITY

### 4.5.3.1 Procedure Followed

*Design of the four transects on the Meuse river*

Four transects (I, II, III, IV) on the Meuse River have been drawn perpendicular to the stream and are located upstream and downstream of Eijsden (see Figure 4.5.7).

The right side of the river is located in the Netherlands, whereas the left side is in Belgium. The stream runs from the south to the north on the map.

# Environmental Variability

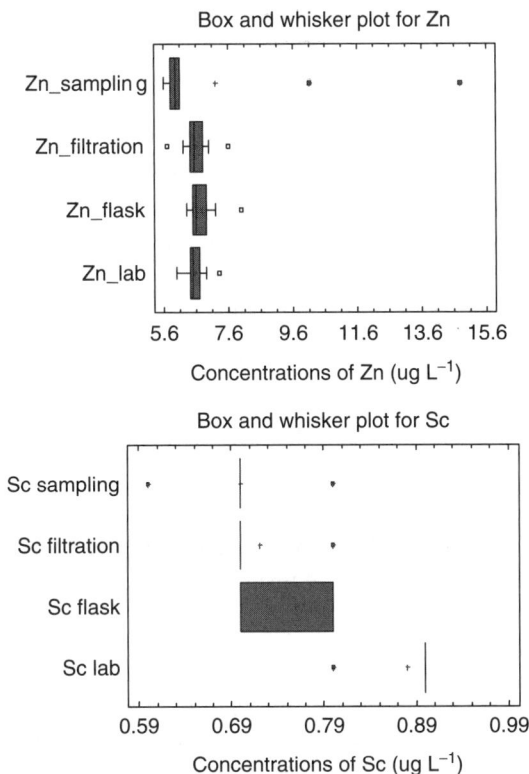

**Figure 4.5.6** Box and whisker plots for concentrations (µg L$^{-1}$) of zinc and scandium

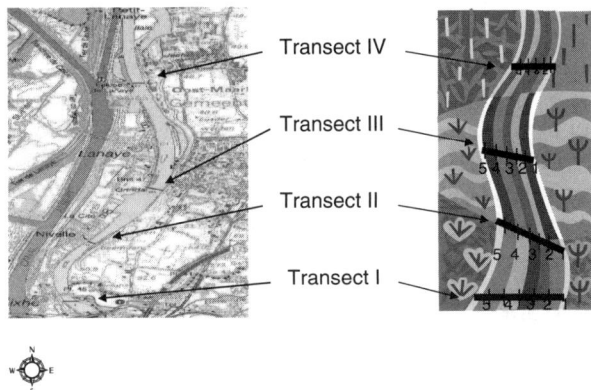

**Figure 4.5.7** Map of the Meuse River at Eisjden, including the 4 transects that were selected along the river

## Surveys

The procedure followed to estimate the variability of the medium was:

- to perform a *horizontal survey*, e.g. physico-chemical logs, of the river, from the Dutch to the Belgian bank, using a YSI multiparameter probe at a pre-defined depth (between 1 and 2 metres);
- to perform a *vertical survey* using the same YSI probe in order to check if there was any variation with depth; five vertical surveys were carried out alongside each transect, at approximately equidistant points;
- if there were no significant variations with depth, one sample was taken at mid depth;
- if a significant variation in physicochemical properties (e.g. pH, temperature, conductivity and turbidity) was observed with depth, then samples were taken at selected depths;
- in order to assess the temporal variability, these experiments were carried out twice with a two-day interval between campaigns.

## In situ measurements

All surveys were carried out using the same probe (YSI 6920 multiparameter probe) to avoid introducing a between probe effect. The parameters measured in situ with this probe were pH, conductivity, temperature, dissolved oxygen, redox potential, nitrate concentration, ammonium concentration, chloride concentration and turbidity.

## Water sampling and conditioning

Water samples were collected by manual pumping using a PTFE tubing attached to a 60 mL syringe. The procedure followed included rinsing the syringe, tubing, and containers (sterile PP 15 mL Falcon flasks) with river water twice prior to sample collection. Filtration was undertaken immediately using disposable 0.45 µm cellulose acetate filters. The first 20 mL was discarded and three Falcon flasks were filled with filtered river water. Ultrapure nitric acid was added (to bring the sample to pH<2) to stabilize trace elements and major cations in solution.

### 4.5.3.2 Application

*Total (spatial and temporal) variability*

*Vertical survey.* Analysis of data from the vertical surveys shows that the depth effect was not significant, and was eliminated from the sampling programme for the estimation of the spatial and temporal variability. The variations in the physico-chemical parameters, measured with the multiparameter probe, with depth are presented (for positions I-1, I-3 and I-5 along transect I) in Figure 4.5.8 for illustrative purposes. In

# Environmental Variability

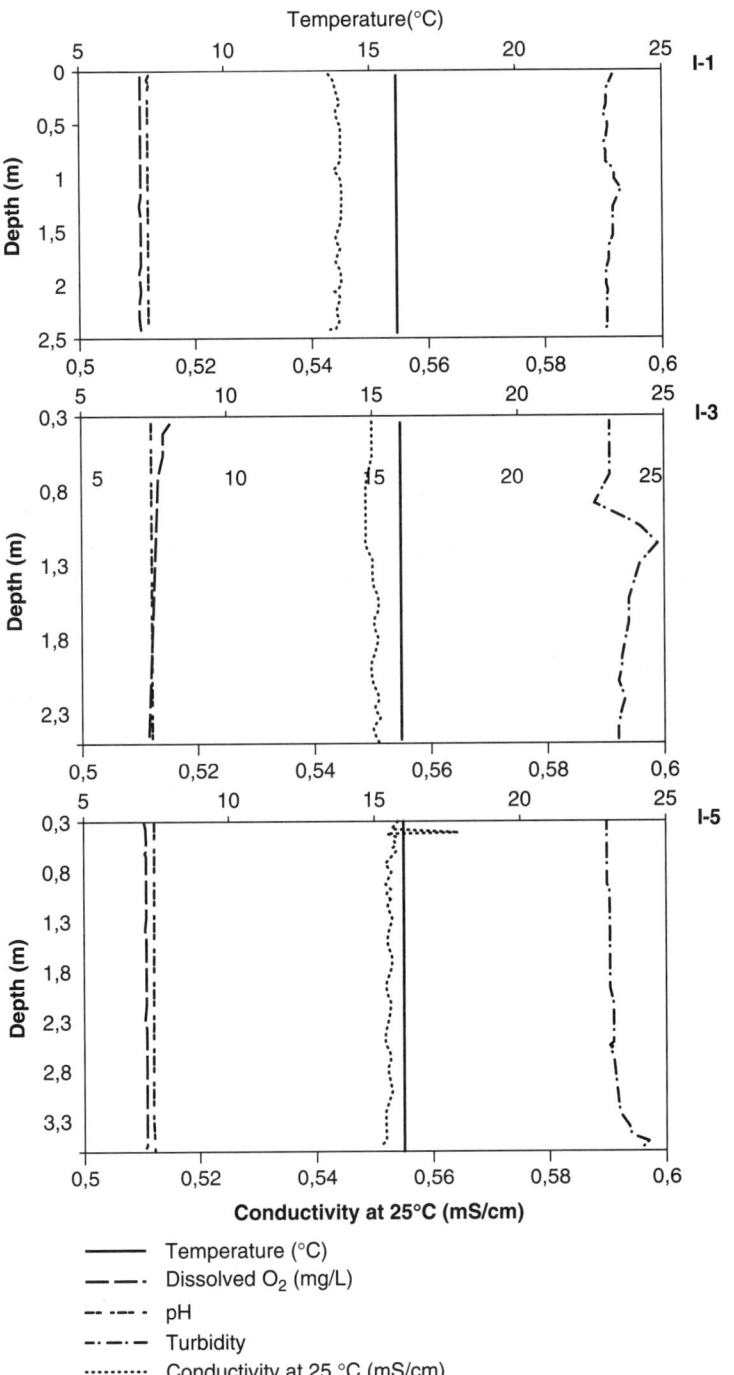

**Figure 4.5.8** Vertical surveys for three positions (transects I-1, I-3 and I-5) on transect I

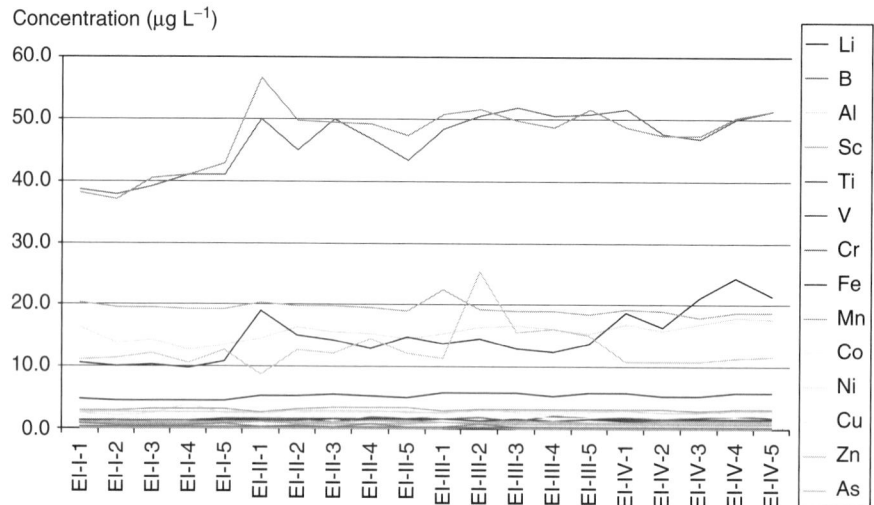

**Figure 4.5.9** Variations in concentrations (µg L$^{-1}$) of trace elements along the 4 transects in the Meuse River (See Plate 5 for colour representation)

view of the absence of an effect of depth in any of the vertical surveys, it was decided to collect only one sample for each depth profile, and this was taken at mid-depth.

*All trace elements.* All results obtained for trace elements are represented in Figure 4.5.9. Ideally the statistical analysis should take into account the effects of spatial location and time of sampling in order to analyze the total observed variation. However, spatial and temporal effects are not independent since it was not possible to collect all of the samples at the various locations simultaneously. It is clear that no significant variation was observed between sampling sites for some elements (e.g. Al, Cr, Cu) whereas trends can be observed for others (e.g. Fe, Mn, B). Concentrations of most trace elements were very low and this precluded any interpretation of results or any assessment of trends, and the statistical analysis focused on the same three trace elements (B, Sc and Zn) as those studied in the analysis of sampling uncertainty (see *Uncertainty from sampling*, above, pp. 311–316).

*Boron.* The scatter plots (Figure 4.5.10) obtained from the ANOVA of the B data show the concentrations for each position (1 to 5) across the four transects, and for all positions in each transect (I to IV).

The scatterplots facilitate the interpretation of the variability observed for each factor tested. Whereas there is no significant difference between the positions 1 to 5, the difference between transects I to IV can be clearly seen. The values in transect I are lower than those in the other three transects. An ANOVA (based on type III sums of squares) was used to investigate in greater detail some of the factors that could have an effect on the concentration of B (Table 4.5.6). This means that the contribution of each factor is measured after removing the effects of all of the other factors. In this table, there is a decomposition of the variability in concentration of B according to the

*Environmental Variability* 321

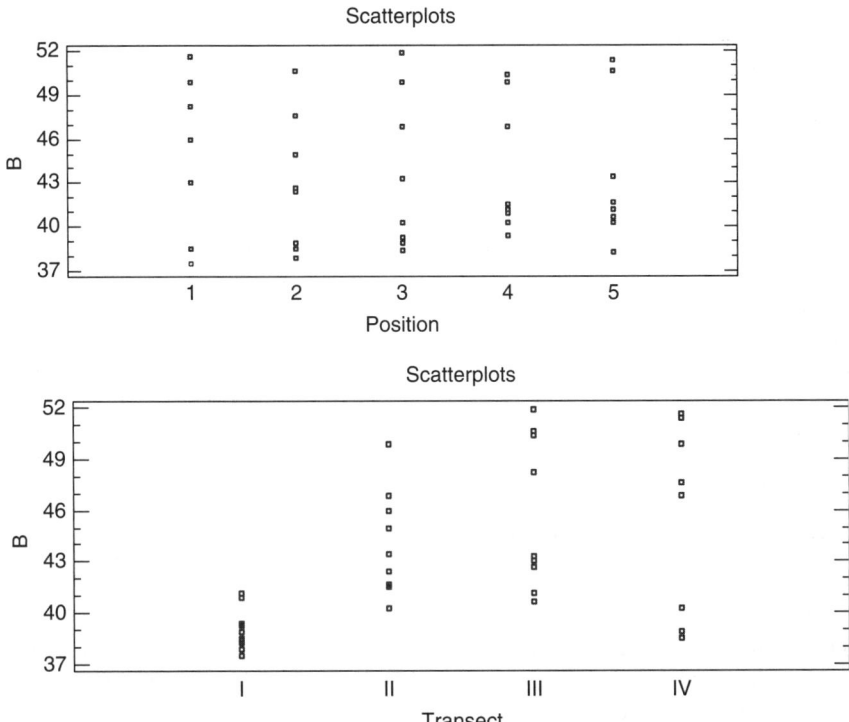

**Figure 4.5.10** Scatter plots of concentration (µg L$^{-1}$) of boron with respect to transect and position factors

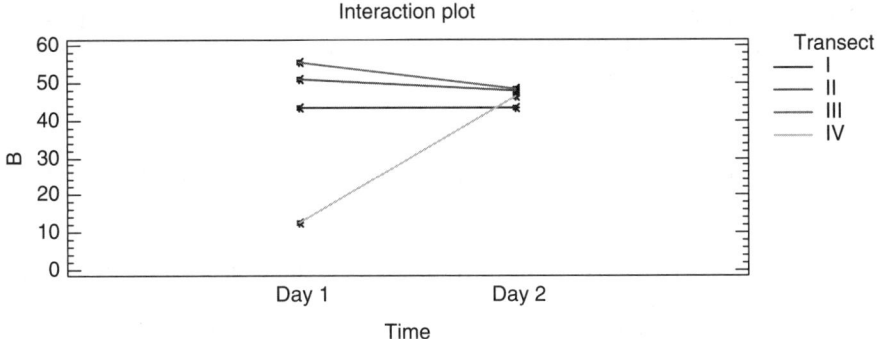

**Figure 4.5.11** Interaction (time and transect) for the concentration (µg L$^{-1}$) of boron

various factors tested. A p-value less than 0.05 indicates that a factor has a significant effect at a confidence level of 95%. There is an effect of transect, and an interaction between time (sampling replicate) and transect on the concentration of B during this trial in the Meuse River (See Figure 4.5.11). No other significant interactions were found between any of the other factors. This result is consistent with the scatter plots per transect.

**Table 4.5.6** Statistics for the main effects and one siginificant interaction in the ANOVA of the concentration of boron. These include the residual sums of squares, the degrees of freedom (df) for the residual, the residual mean square, an F-ratio calculated using the appropriate error term, and the p-value corresponding to the F-ratio

| Source | Sums of squares | Df | Mean square | F | p-value |
|---|---|---|---|---|---|
| **COVARIABLES** | | | | | |
| NO3 Ysi | 1.91241 | 1 | 1.91241 | 0.56 | 0.4619 |
| Conductivity | 1.12151 | 1 | 1.12151 | 0.33 | 0.5722 |
| pH | 3.7356 | 1 | 3.7356 | 1.09 | 0.3063 |
| **PRINCIPAL EFFECTS** | | | | | |
| A: Time | 0.809807 | 1 | 0.809807 | 0.24 | 0.6310 |
| B: Transect | 111.407 | 3 | 37.1356 | 10.84 | 0.0001 |
| C: Position | 14.3924 | 4 | 3.59809 | 1.05 | 0.4013 |
| **INTERACTIONS** | | | | | |
| AB | 46.3682 | 3 | 15.4561 | 4.51 | 0.0116 |
| RESIDUALS | 85.6129 | 25 | 3.42452 | | |
| TOTAL (corrected) | 861.348 | 39 | | | |

The variance due to the sampling process for B was first estimated as 19%. Although 10% was explained by the different contributing factors (e.g. flask), the remaining 9% was attributed to the variability of the medium. However, in this section, it is estimated as 21% (at the same level of concentration), including the uncertainty of chemical analyses (with RSD = 5.2%, k = 2). The relative standard deviation (RSD) is the standard deviation divided by the mean, expressed in percentage. According to this result the variability of the medium (excluding analytical process) should be about 20.3%, and this is greater than the estimate based on the sampling procedure. This discrepancy between results demonstrates that the interpretation of these results is not straightforward and that the uncertainty of the analytical process needs further investigation.

*Antimony.* The results obtained for Sb were not analysed statistically since the observed concentrations were very low (in the range 0.1 µg/L to 0.2 µg/L).

*Zinc.* The Zn data were analysed in a similar way to the B data (Table 4.5.7). Again there was a significant effect of transect. Only one (that between transect and time) of the possible interactions between factors had a significant effect on the concentration of Zn, and this is shown in the table. The concentration of nitrate measured with the YSI probe has a non-negligible association with the concentration of Zn. This association was not observed with B.

Even though the ranges of concentrations obtained for B and Zn are different, it can be seen that the concentrations found along transect IV are smaller for both elements than those found in the other transects, and that the variability is greater for transect IV (Figure 4.5.12). This was observed consistently for all elements (trace metals and major components).

# Environmental Variability

**Table 4.5.7** Statistics for the main effects and one siginificant interaction in the ANOVA of the concentration of zinc ($\mu g\ L^{-1}$). These include the residual sums of squares, the degrees of freedom (df) for the residual, the residual mean square, an F-ratio calculated using the appropriate error term, and the p-value corresponding to the F-ratio

| Source | Sums of squares | Df | Mean square | F | p-value |
|---|---|---|---|---|---|
| **COVARIABLES** | | | | | |
| NO3 Ysi | 33.715 | 1 | 33.715 | 9.84 | 0.0043 |
| Conductivity | 12.5066 | 1 | 12.5066 | 3.65 | 0.0676 |
| pH | 1.8581 | 1 | 1.8581 | 0.54 | 0.4683 |
| **PRINCIPAL EFFECTS** | | | | | |
| A: Time | 0.00503 | 1 | 0.00503 | 0.00 | 0.9697 |
| B: Transect | 59.0454 | 3 | 19.6818 | 5.74 | 0.0039 |
| C: Position | 14.4731 | 4 | 3.61828 | 1.06 | 0.3988 |
| **INTERACTIONS** | | | | | |
| AB | 67.2767 | 3 | 22.4256 | 6.54 | 0.002 |
| RESIDUALS | 85.6129 | 25 | 3.42452 | | |
| TOTAL (corrected) | 861.348 | 39 | | | |

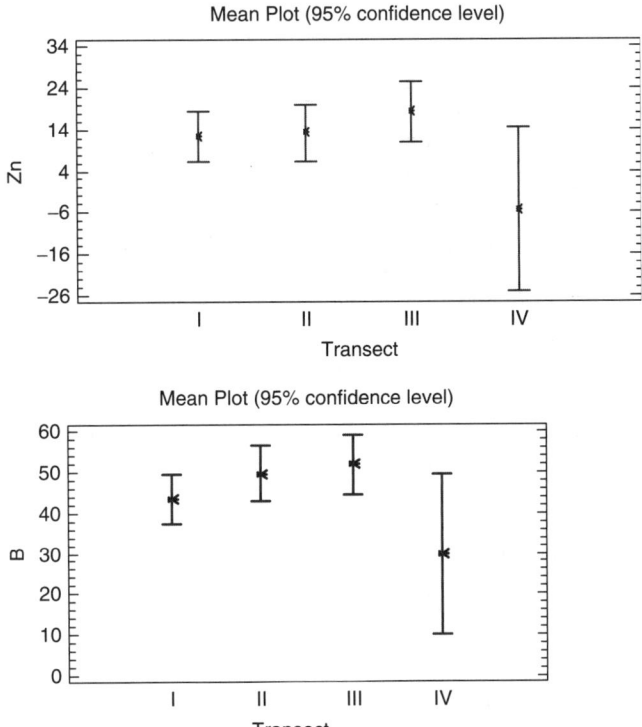

**Figure 4.5.12** Plots of the transect mean (with 95% confidence intervals) concentrations ($\mu g\ L^{-1}$) of Zn and B

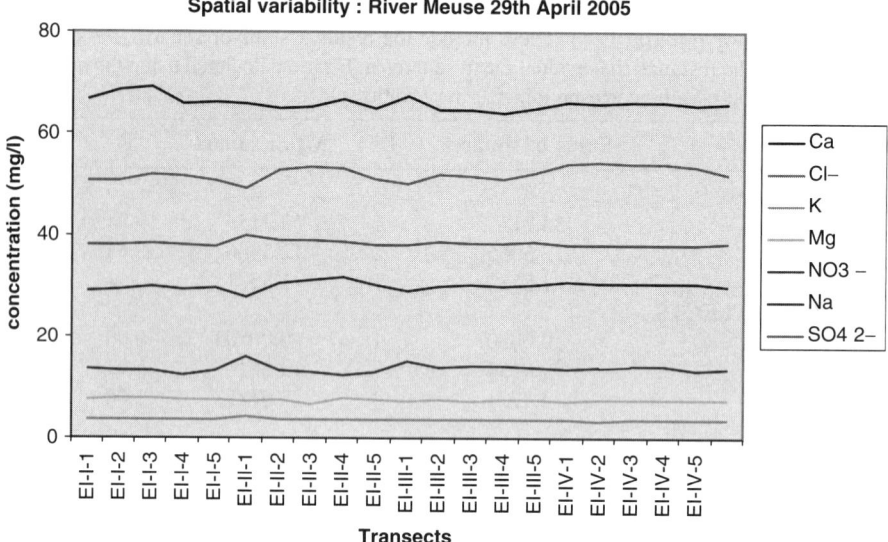

**Figure 4.5.13** Spatial variability in the concentrations (mg L$^{-1}$) of the major components (See Plate 6 for colour representation)

*Major components.* Major components were also measured for the four transects and the five positions identified (§ 4.5.3.1.1).

The concentrations of the major cations (Ca$^{2+}$, K$^+$, Mg$^{2+}$, Na$^+$) and anions (Cl$^-$, NO$_3^-$, SO$_4^{2-}$) are plotted for each position in each of the transects in Figure 4.5.13. There is no obvious variation between the concentrations of any of the ions either between transects or between positions within a transect, and no statistical analysis was undertaken. There was one exception, HCO$_3^-$, for which the variability between locations can be seen in Figure 4.5.14, and these data were analysed using an ANOVA (type III sums of squares) and the results are presented in Table 4.5.8. The results are consistent with the plot of means in Figure 4.5.14. There are significant differences between transects (p = 0.008), and a difference between replicate surveys in time that is approaching significance (p = 0.09). There were no significant differences between the positions in the transects, and no significant interaction effects.

### 4.5.3.3 Temporal Variability

In order to assess the temporal variability the sampling procedure described in (§ 4.5.3.1.1) was carried out twice with two day interval between surveys.

*Comparison between boron, zinc and antimony*

The data for B, Zn and Sb were analysed in a way similar to that described in *Design of the Four Transects on the Meuse River*, above, p. 316, and scatter plots of the concentrations (μg/L) of these elements in the two surveys are presented in Figure 4.5.15.

# Environmental Variability

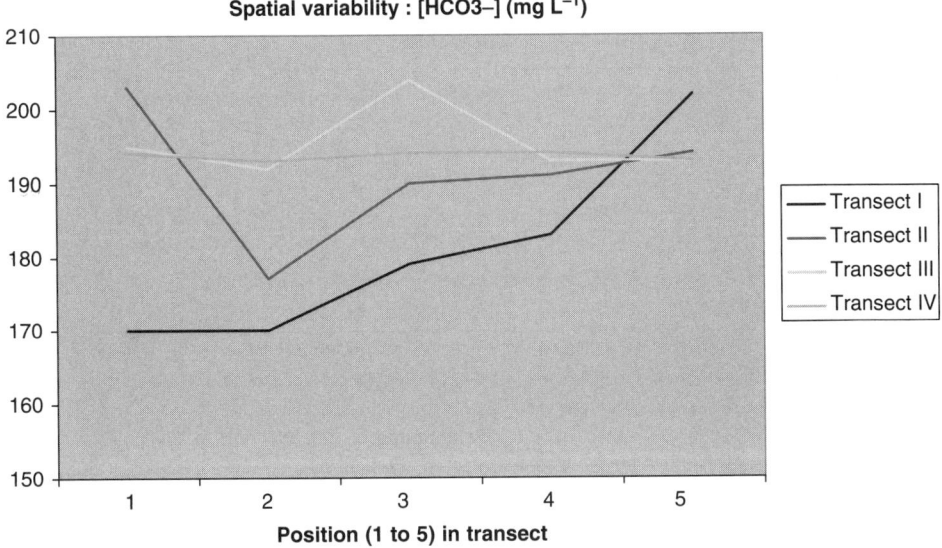

**Figure 4.5.14** Spatial variability in the concentration (mg L$^{-1}$) of HCO$_3^-$ across the transects

**Table 4.5.8** ANOVA statistics for HCO$_3^-$, main effects and interactions (sum of square type III)

| Source | Sum of square | Df | F | p-value |
|---|---|---|---|---|
| **PRINCIPAL EFFECTS** | | | | |
| A: Time | 140,625 | | 3,41 | 0,0896 |
| B: Transect | 782,475 | 3 | 6,32 | 0,0081 |
| C: Position | 167,65 | 4 | 1,02 | 0,4375 |
| **INTERACTIONS** | | | | |
| AB | 81,675 | 3 | 0,66 | 0.5922 |
| AC | 188,25 | 4 | 1,14 | 0.3836 |
| BC | 431,15 | 12 | 0,87 | 0.5925 |
| **RESIDUAL** | 494,95 | 12 | 41.2458 | |
| TOTAL (corrected) | 2286,77 | 39 | | |

Time is not a significant factor for Zn and B (see the results of the ANOVAs in Tables 4.5.6 and 4.5.7), though the variances are not homogeneous. Even though there appears to be a significant difference in the concentration of Sb between the two surveys in the scatter plot, it is not statistically significant. This is because the mean concentrations are very low (0.1 and 0.2 µg/L respectively). Whilst in this trial no significant differences were found between the concentrations of B, Zn, Sb on the two sampling occasions, it does not mean that in general temporal variability can be neglected, even where measurements are carried with only a short interval between

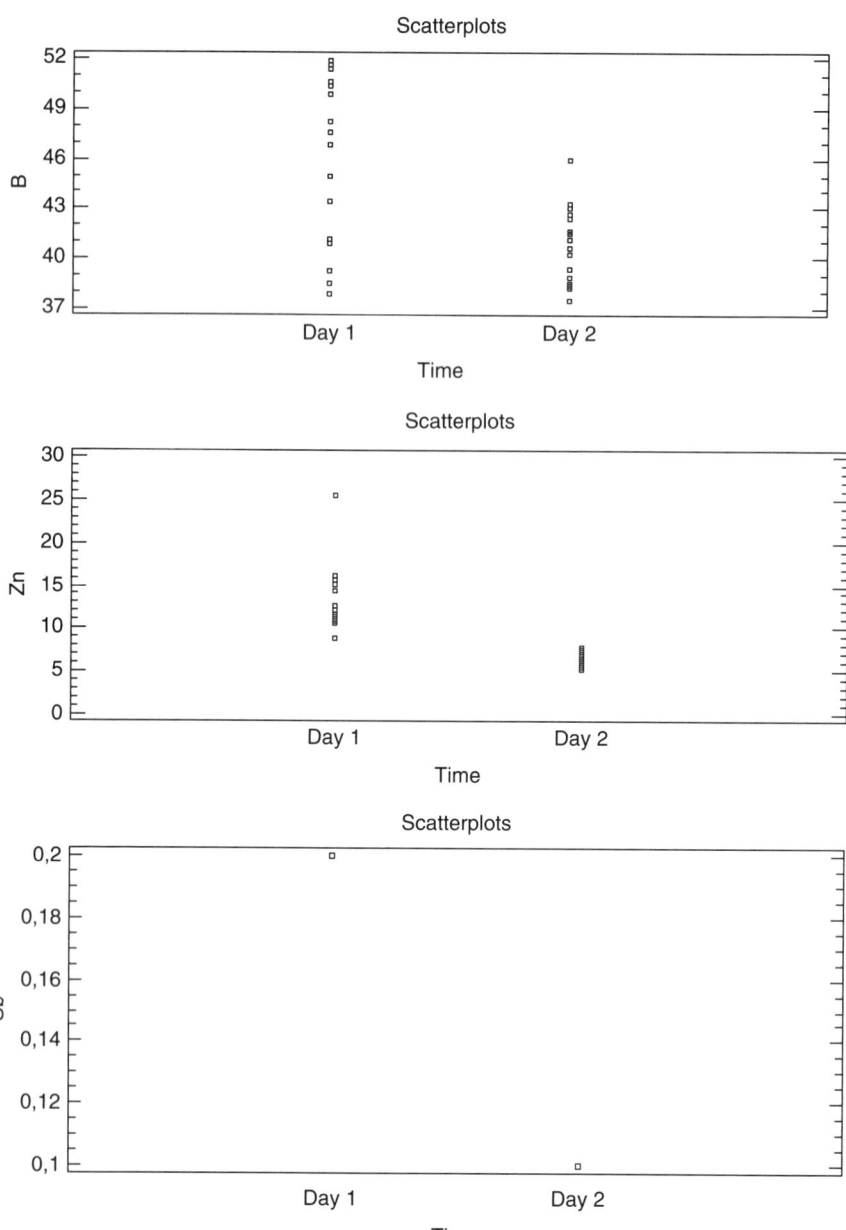

**Figure 4.5.15** Scatter plots of the concentrations (µg L$^{-1}$) of boron, zinc and antimony found at all positions on all transects on the two surveys carried out with a separation of two days

*Conclusions* 327

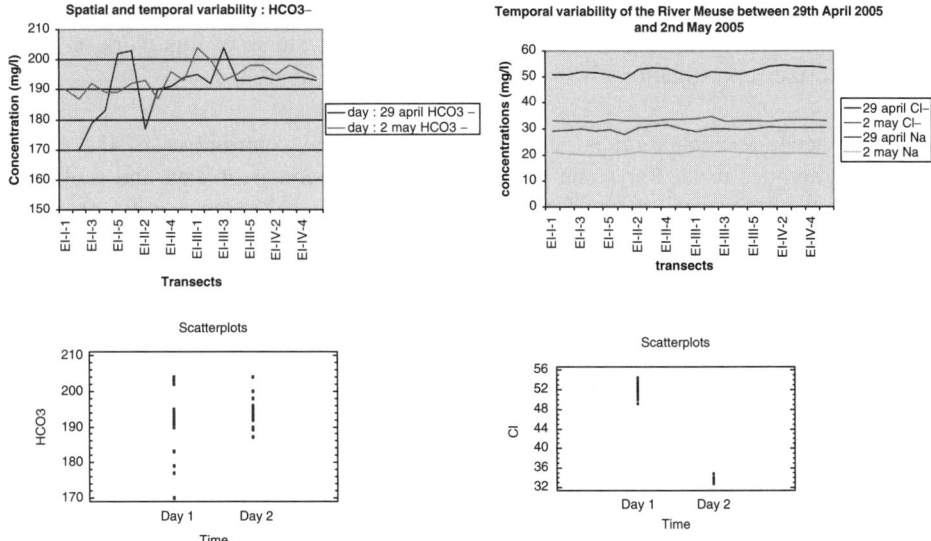

**Figure 4.5.16** Scatter plots of the concentrations (mg L$^{-1}$) of selected major elements (Na$^+$, Cl$^-$, and HCO$_3^-$) along the transects on two sampling expeditions separated by an interval of two days

them. In this work it was not possible to separate the two surveys by a longer period because of constraints on time and funding.

*Major elements*

For some of the major elements (Na$^+$, Cl$^-$, and HCO$_3^-$) there are indications in the scatter plots of variation between the sampling events (Figure 4.5.16). For HCO$_3^-$ the difference approached significance (p = 0.08) (Table 4.5.8), but it was not possible to make comparisons of the concentrations of Na$^+$, and Cl$^-$ between the two surveys because the sampling uncertainty was not evaluated for these analytes. This means that it was not possible to extract the effects of survey from the results in order to estimate the sampling uncertainty (the analytical uncertainty was available and subtracted from these data).

## 4.5.4 CONCLUSIONS

Classical or standard monitoring is generally based on spot (bottle) sampling and chemical analysis performed in the laboratory. This is the approach adopted for use in this study for the assessment of the contribution of sampling to the whole process of measuring the chemical quality of a river, and to the overall variability observed in the quality of the Meuse River. In light of evidence presented in this study, the successful implementation of the WFD will rely on a number of factors. Performing representative monitoring programmes must be implemented in order to have a good assessment (qualitative and/or quantitative) of the status of environmental waters (surface as well as groundwater). Some lessons were learned during the course of this work, and the need for further investigations in a number of areas was identified.

- More accurate evaluation is needed of the analytical contribution of the different techniques used. For some elements the small number of data and replicates didn't allow any accurate evaluation of uncertainty and this can lead to erroneous conclusions,

- Further field tests are needed to extend this work at strategic points on both the same type of water body (e.g. another European River) and on different types of water (e.g. groundwater). A methodology for estimating the uncertainties linked to an entire chain for groundwaters has been developed (Roy and Fouillac, 2004). That work showed that natural changes in the medium could generate variations in the data that were much more significant than any variation due to human actions. However, that study was performed on polluted sites, and totally different conclusions may be reached in sites that are unpolluted or characterized by different sources of pollution. In the present study, low concentrations of the selected pollutants were found, and this makes it difficult to detect any effects on the measurements and to interpret the results obtained,

- Further tests are needed to evaluate the influence of suspended matters on the spatial and temporal variability of trace measurements. In this study a low variability was observed in the quality of the river water. However, temporal variability was estimated over a very short window of time, using only two sets of surveys separated by a 2- or 3-day interval. There is therefore great uncertainty concerning the variation in water quality over time, and the measured quality of the water could be unrepresentative of the overall average quality. For instance samples could have been taken when concentrations were temporarily below or above normal levels or in localized 'clean' or 'dirty' areas. Where there is high variability in the environmental quality of a water body then it is necessary to have a large number of sampling sites and/or to increase the frequency of sampling to obtain a representative picture of the overall quality (Gonzalez *et al*., 2008). However, because of the high cost of both spot sampling and laboratory analyses it is necessary to reach a compromise that balances reliability of the measurements against affordability.

The size of the uncertainty due to the sampling procedure itself varied greatly depending on what was being measured: uncertainties for the trace elements measured in this work varied between about 5% for Ba (at 20 µg/L) to 82% for Zn (at 7 µg/L). Low uncertainties are obtained when tests are performed under repeatability conditions and by a highly qualified field team. Recently, several guides (IUPAC, 1990; Eurachem, 2007; Krajca, 1989) dealing with uncertainty in relation to the sampling process have been published in different fields (including environment, and food). The estimates of the total uncertainty associated with measurements ranged from a few percent up to 84% depending on the measurand, and the contribution of the sampling stage varied from a few percent to 90% of the total measurement variance. This led us to the conclusion that there is a need for an important effort to be put into the sampling step, and this will involve an increase in the proportion of expenditure spent on this step. This contrasts with the chemical analysis where the QA/QC are ensured in particular through accreditation according to the ISO 17025 standard (French Standard, 2005). The uncertainty associated with the sampling step could be reduced through the development and application of the QC of sampling. This could be achieved through the use of measures such as inter-organizational sampling trials, and/or the use of reference sampling targets (Eurachem, 2007), and this would need to be underpinned by the appropriate validation and training of sampling personnel. In addition, routine application of the duplicate method (duplicate sample is defined in Appendix A) could be followed as a way of monitoring the on-going sampling quality, whatever the type of water (e.g. surface, groundwater, rain) being sampled. Proficiency testing exercises carried out in the field would be very useful in terms of evaluation of the quality of the results. However, for such proficiency tests to be worth while it would be important to ensure an appropriate selection of sampling sites and of pollutants to measure. Sampling protocols have to be adapted according to both the analytes to be measured, and the environmental conditions (e.g. river, lake, groundwaters) where there could be spatial (depth and width) effects.

## ACKNOWLEDGEMENTS

I acknowledge financial support from the Sixth Framework Programme of the European Union (Contract SSPI-CT-2003-502492; http://www.swift-wfd.com). I gratefully thank RIZA monitoring team at Eijsden (The Netherlands) for their technical support. I also thank the ICP/MS Laboratory of BRGM for performing all trace analysis. Finally special thanks are expressed to the field team of BRGM for their active and supportive contribution in the sampling programmes, and to Nathalie Guigues for her helpful corrections to the English.

## APPENDIX A

### Normal Distribution of Values

Normal Distribution is a continuous probability distribution that is useful in characterizing a large variety of types of data. It is a symmetric, bell-shaped distribution, completely defined by its mean and standard deviation and is commonly used to calculate probabilities of events that tend to occur around a mean value and trail off with decreasing likelihood. Different statistical tests are used and compared: the $\chi^2$ test, the W Shapiro-Wilks test and the Z-score for asymmetry. If one of the p-values is smaller than 5%, the hypothesis ($H_0$) (normal distribution of the population of the sample) is rejected. If the p-value is greater than 5% then we prefer to accept the normality of the distribution. The normality of distribution allows us to analyse data through statistical procedures like ANOVA. In the absence of normality it is necessary to use nonparametric tests that compare medians rather than means.

## Variance Homogeneity

The statistical test used is Cochran's test (risk α = 5 %). The value obtained ($W_{exp}$) is compared with the theoretical value (table of Cochran tests at 5% risk).

## Trueness

The trueness is the closeness of agreement between the average value obtained from a large series of test results and an accepted reference value. Trueness is normally expressed in terms of bias (Eurachem, 2007).

As the bias is the difference between the mean of the test results and the accepted reference value, it has to be calculated for each trace element and for each Reference Material used. This calculation allows us to determine if the bias obtained can be neglected or not.

Even if different bias effects can be detected, here the objective is to evaluate the overall bias effect as a measurement bias for results obtained using Certified Reference Materials. The criterion is:

$$C_{obs} = \frac{|\bar{x} - X_{crm}|}{2 \times \sqrt{\left(\frac{s^2}{n} - S^2\right)}}$$

With $\bar{x}$ and $s^2$: mean and variance of the n values obtained on the CRM,
$S^2$: variance on the certified value of the CRM,
$X_{crm}$: certified value of the component on the certificate of the CRM.

## Uncertainty

Uncertainty is defined as follows (Eurachem, 2007): Parameter, associated with the result of a measurement that characterizes the dispersion of the values that could reasonably be attributed to the measurand.

## Uncertainty from Sampling

Defined as the part of the total measurement uncertainty attributable to sampling.
  Note: Also called sampling uncertainty (French Standard, 1995).

## Representative Sample (IUPAC, 2005)

Sample resulting from a sampling plan that can be expected to reflect adequately the properties of interest in the parent population.

## Duplicate (Replicate) Sample (IUPAC, 2005)

One of the two (or more for replicate sample) samples or subsamples obtained separately at the same time by the same sampling procedure or sub-sampling procedure.

Note: Each duplicate sample is obtained from a separate 'sampling point' within the sampling location.

## REFERENCES

Circulaire 2007/23 (ministère de l'écologie et du development durable).
Decree 2001-1220 relatif aux eaux destinées à la consummation humaine, à l'exclusion des eaux minérales naturelles.
Eurachem, 2000. *Quantifying Uncertainty in Analytical Measurement*, Second Edition, EURACHEM/CITAC Guide CG 4, QUAM: 2000.1
Eurachem, 2007. *Measurement Uncertainty Arising from Sampling*, EURACHEM/CITAC, First Edition.
European Commission, 2000. Directive 2000/60/EC of the European Parliament and of the Council of the 23 October 2000.
European Commission, 2008. Proposal for a Directive of the European Parliament and of the Council on Environmental Quality Standards in the Field of Water Policy and Amending Directive 2000/60/EC.
Feinberg M., 2000. *La validation des methods d'analyses*, Max. Feinberg (ed.). Masson.
Feinberg, M. 2001. L'assurance qualité dans les laboratoires agroalimentaires et pharmaceutiques, Feinberg (ed), Lavoisier Tec. & Doc., Paris.
French Standard, 1995. NFX 06-031-1, Application of statistics, Control Charts – Part 1: Shewhart control charts by variables.
French Standard, 1999. NF ENV 13005, Guide to the expression of uncertainty in measurement.
French Standard, 2003. XPT 90-220 water quality – protocol for the estimation of the uncertainty associated with a result for a physico-chemical analysis method.
French Standard, 2005. NF EN ISO/CEI 17025 (September 2005) General requirements for the competence of testing and calibration laboratories.
Gonzalez C., Fouillac A.M. and Greenwood R., 2008. Screening methods for ground water monitoring. *In Groundwater Science and Policy: An International Overview*. P. Quevauviller (ed.), Royal Society of Chemistry, London, pp. 363–77.
ISO, 1998a. NF ISO 5725 – 1, Application of statistics – Accuracy (trueness and precision) of measurements methods and results – Part 1: General principles and definitions.
ISO, 1998b. ISO 5667-14 Guidance on quality assurance of environmental water sampling and handling.
IUPAC, 1990. Nomenclature for sampling in analytical chemistry (Recommendations 1990), prepared for publication by Horwitz W, *Pure and Applied Chemistry*, **62**, 1193–1208.
IUPAC, 2005. Terminology in soil sampling (IUPAC Recommendations 2005), prepared for publication by De Zorzi P, Barbizzi S, Belli M, *et al*. *Pure and Applied Chemistry*, **77**(5), 827–41.
Krajca, J.M., 1989. *Water Sampling*, Jaromil M Krajca (ed.), Ellis Horwood, Chichester, UK.

Prichard, 2005. Guidelines for laboratories carrying out measurements where the results will be used to implement the WFD (2000/60/EC), Elisabeth Prichard, LGC: Deliverable D12 of EU 6$^{th}$ Framework Project SWIFT-WFD (Contract no. SSPI-CT-2003-502492) http://www.swiftwfd.com/document.php?project = swift&locale = en&level1 = menu1_swift_d4d62076d3b01800_7&level2 = 1&doc = delivrable_public_access: Accessed 30.11.2008.

Roy, S. and Fouillac A.M., 2004. *Trends Anal. Chem.*, **23**, 185–93.

Statgraphics software version 5.0 (software application combining graphics and statistics) Manguistics, Inc., Rockville, Maryland, USA.

SWIFT, WFD EU RTD FP6 Project, SSPICT-2003-502492, Selection of representative sites used in the SWIFT project; available on the web site of the project ( http://www.swift-wfd.com).

# Section V
Quality Assurance and Validation Method

# 5.1
# Preparation of Reference Materials for Proficiency Testing Schemes

Angels Sahuquillo, Marina Ricci, Ofelia Bercaru, Hakan Emteborg, Franz Ulberth, Roberto Morabito, Claudia Brunori, Yolanda Madrid, Erwin Rosenberg, Klara Polyak and Herbert Muntau

5.1.1 Introduction
5.1.2 Design of the Quality Control Tools
5.1.3 Description of Sampling Sites and Sampling Details
5.1.4 Preparation of Quality Control Tools for Inorganic Analysis
    5.1.4.1 Water Matrix RMs
    5.1.4.2 Calibrants and Spiking Solutions
5.1.5 Preparation of Quality Control Tools for Organic Analysis
    5.1.5.1 Water Matrix RMs
    5.1.5.2 Standard and Spiking Solutions
5.1.6 Homogeneity and Stability Studies
5.1.7 Discussion
5.1.8 Conclusions
Acknowledgements
References

## 5.1.1 INTRODUCTION

The present framework for Community action in the field of water policy (Water Framework Directive, WFD) (European Commission, 2000) aims at maintaining and

---

*Rapid Chemical and Biological Techniques for Water Monitoring*    Edited by Catherine Gonzalez, Philippe Quevauviller and Richard Greenwood
© 2009 John Wiley & Sons, Ltd

improving the aquatic environment in the European Union by preventing its deterioration and ensuring a progressive reduction of pollution, among other actions and measures. For the first time, the WFD considers water quality management at an international scale and establishes a model for management issues throughout the European Union. The establishment of the ecological status of the different water bodies requires biological, hydromorphological and physico-chemical quality elements for its classification. Moreover, different programmes of monitoring (surveillance, operational and investigative) have to be already implemented in the Member States since December 2006, with suitable frequencies.

The success in the implementation of the WFD depends on the achievement of reliable monitoring data, of undoubted quality and comparability, which assure good managing strategies and potential remediation actions in water bodies at EU-level. To obtain reliable data, water analysis laboratories must develop and implement quality assurance (QA) systems (Namiesnik and Zygmunt, 1999). Laboratories have to design strategies for assuring not only internal quality control (QC), to provide evidences that data produced is fit for their intended purpose (Thompson and Wood, 1995), but also some kind of external quality control such as regular participation in proficiency testing (PT) schemes. The latter aspects are especially relevant for laboratories envisaging accreditation processes according to ISO/IEC guide 17025 (ISO, 2005) for their competence.

When a suitable certified reference material (CRM) is available, it has to be used in the validation process of an analytical method (Quevauviller *et al.*, 1998). However, a non-certified RM produced following rigorous technical requirements and fit-for-purpose homogeneity and stability characteristics, can be used for intermediate reproducibility evaluation as well as in the organization of PT schemes (Emons *et al.*, 2004; De Guillebon *et al.*, 2001).

The European Commission funded the project 'Screening method for Water data InFormation in support of the Water Framework Directive (SWIFT-WFD Project)' from January 2004 till March 2007 (SWIFT-WFD project, 2004). One of the activities undertaken in the SWIFT-WFD project, included in Workpackage 2, was the production of quality control tools not only for classical, but also for a range of screening/emerging test methods for ecological/biological and chemical water monitoring purposes.

In this chapter, the preparation of selected water reference materials, and other quality control tools such as calibrants and multi-component standard solutions, is described. The produced materials were used in three PT schemes with the participation of a total of 94 laboratories from 21 European Countries, as described elsewhere (Brunori *et al.*, 2007).

## 5.1.2 DESIGN OF THE QUALITY CONTROL TOOLS

The preparation of RMs was undertaken, under the guidance of Institute for Reference Materials and Measurements (IRMM), after three different sampling campaigns. The produced RMs have been then used in three PT campaigns and in the validation of the newly developed screening/emerging test methods, within the frame of the SWIFT-WFD activities. The design of the RMs, calibrants, spiking and/or standard solutions to be used in the PT schemes was carefully planned in technical project meetings.

The selection of the sampling points followed several criteria: relevance in terms of European river basins definition; existence of an extended monitoring database built by the responsible organization; suitable concentration levels of the target pollutants; logistic assistance by host organization; easy access to the sampling site.

Concerning the analytes, both inorganic and organic compounds were chosen among the WFD list of priority pollutants, taking into account an increasing difficulty for the participating laboratories in subsequent PT campaigns. In all sets of materials the analysis of trace elements and major components was considered, and in the second and third set, PAHs and pesticides were included, with 9 to 13 analytes for each category. The preparation of RMs for organic analysis is of paramount interest due to the very limited number of suitable materials in the field of water monitoring, mainly for pesticides. The main difficulty when dealing with the preparation of such a material is connected to the long-term stability (Bercaru *et al.*, 2003).

The concentration level for the analytes to be determined in the provided materials was also matter of concern in the technical meetings. It was agreed, when possible, to provide natural matrix materials. However, when the previous characterization of the bulk water sample revealed low concentrations of the target analytes, in some cases even under detection limits, different fortification strategies were put in place: performed by the participating laboratory following spiking protocols, or performed by the material producer through enrichment of the bulk material or just before dispatch. In the cases where the spiking procedure of the water matrix had to be performed by each of the participating laboratories, both spiking solutions and detailed protocols for the execution of fortification were distributed. The final decision about the characteristics of the RMs to be produced came as consequence of the compromise between the closeness of the materials to natural samples and the minimum concentration levels possible to be detected by the most frequently used analytical techniques in water monitoring practices which could ensure the homogeneity and stability of the final obtained materials. Especially for organic compounds, their maximum solubility in water had to be considered when establishing the final concentrations, in order to avoid potential instability with time.

Along with the matrix RMs, a series of other quality control tools such as calibrants and standard solutions was also produced. On the occasion of the first set of reference materials, calibrants for trace elements and major components were also prepared and sent to the laboratories in order to minimize the differences in results produced by the use of different calibrants. For the other two sets of materials, multi component standard solutions for PAHs and pesticides were prepared and distributed in order to check the analytical performance of the chromatographic methods used.

Under the SWIFT-WFD project, overall 16 RMs (including calibrants and standard solutions) were prepared as summarized in Table 5.1.1. Matrix water RMs, spiking solutions for RM14, RM15 and standard solution RM16, were prepared at IRMM. Document flow and document history was managed by a dedicated document management system being part of IRMM's ISO Guide 34 (ISO, 2000) and ISO 17025 accreditation (Linsinger *et al.*, 2007). Calibrants and the rest of spiking and standard solutions were prepared at the Laboratory of Preparation of Quality Control Materials (Mat Control) of the Department of Analytical Chemistry of the University of Barcelona, following the implemented quality system of the unit (Sahuquillo *et al.*,

Table 5.1.1 Quality control materials produced within the SWIFT-WFD Project

| Matrix | Material code | Envisaged analysis | Additional materials |
|---|---|---|---|
| River water (River Meuse, The Netherlands) | RM01 | Trace elements (natural levels) Al, As, Cd, Cr, Cu, Mn, Ni, Pb, Se, Zn | Calibrants for As, Cd, Cu, Ni, Pb, Zn |
| | RM02 | Major components (natural levels) Na, K, Ca, Mg, total P, $Cl^-$, $NO_3^-$, $SO_4^{2-}$, pH, conductivity | Calibrants for total P, Na, K, Ca, Mg, $Cl^-$, $NO_3^-$, $SO_4^{2-}$ |
| | RM03 | Trace elements (fortified levels – spiking solution). Same analytes as RM01 | Calibrants for As, Cd, Cu, Ni, Pb, Zn |
| | RM04 | Major components (fortified levels – spiking solution). Same analytes as RM02 | Calibrants for total P, Na, K, Ca, Mg, $Cl^-$, $NO_3^-$, $SO_4^{2-}$ |
| Spring water (Brevilles, France) | RM05 | Trace elements (low fortified levels – spiking bulk material). Same analytes as RM01 | – |
| | RM06 | Major components (natural levels). Same analytes as RM02 | – |
| | RM07 | PAHs (fortified levels – spiking solution) anthracene, benzo(a)pyrene, benzo(b)fluoranthene, benzo(ghi)perylene, benzo(k)fluoranthene, phenanthrene, fluoranthene, Indeno(1,2,3-cd)pyrene, naphthalene | RM09: standard solution for PAHs in acetonitrile |
| | RM08 | Pesticides (fortified levels – pills) alachlor, chlorfenvinfos, chlorpyrifos, diuron, endosulfan (alpha+beta), isoproturon, simazine, aldrin (HHDN), terbuthylazine, lindane (gamma-HCH), chlortoluron, atrazine, desethyl-atrazine, glyphosate | RM10: standard solution for pesticides in ethylacetate RM11: glyphosate solution in double deionised water |
| River water (Ruhr, Germany) | RM12 | Trace elements (natural levels). Same analytes as RM01 | |
| | RM13 | Major components (natural levels). Same analytes as RM02 | |
| | RM14 | PAHs (low fortified levels - spiked before dispatch). Same analytes as RM07 except phenanthrene | RM16: standard solution for PAHs in acetonitrile |
| | RM15 | Pesticides (low fortified levels – spiked before dispatch). Same analytes as RM08 | RM17 (re-labelled RM10) |

2004). The pesticide pills intended for spiking RM08 were provided by IRMM through the producer Institute Pasteur de Lille (IPL, France) which developed and patented this new technology (WARP project, 2002–2004). Detailed description of the preparation of the materials is given in the following sections. Full information on the three sets of SWIFT-WFD RMs is reported in public project deliverables (Deliverable 10, Deliverable 13 and Deliverable 14) that can be freely downloaded from the project website (www.swift-wfd.com).

## 5.1.3 DESCRIPTION OF SAMPLING SITES AND SAMPLING DETAILS

For the first set of materials, sampling was performed in June 2004 at River Meuse, Keizersveer Meetstation 1, The Netherlands (operated by the Institute for Inland Water Management and Waste Water Treatment, RIZA), located north-east of Breda and south-west of Rotterdam. The water was sampled using a permanently installed flow-through system. The water was taken around 10 m from shore, at a depth of approximately 3 m and pumped using a peristaltic pump connected to PTFE tubing into an on-site laboratory. The PTFE tubings were preflushed for about 1 h before sampling to assure minimization of external contamination. Two drums of 200 L were filled at the average pumping rate of $4.4 \text{ L min}^{-1}$. The water was filtered through PALL Versaflow™ capsule filters 0.45 µm equipped with an 8 µm pre-filter into the preflushed drums (HDPE). The filters had to be changed due to clogging effects after about 100 L. The pH value of the river water was 7.4. The water foreseen for trace elements analysis was acidified by addition of concentrated $HNO_3$ (Merck Suprapur) to a pH of approximately 2 and homogenized with the aid of a PTFE paddle. The content of the other container (for major components analysis) remained untreated. Figure 5.1.1a shows an image of the sampling point.

The second set of materials was prepared with water sampled from a spring, situated in the countryside south of Montreuil-sur-Epte (Brevilles, France) in February 2005. The sampling station is under the continuous supervision of the *Bureau de recherches géologiques et minières* (BRGM). The spring itself was set in a fenced area and fed a small brooklet. The spring setting, about 50×50 cm, was circa 20 cm deep and it was covered by flint stone pebbles of medium size. The water production of the spring was estimated to $4 \text{ L s}^{-1}$, and it was odourless, colourless and extremely clear at visual inspection. The water was collected through a funnel covered by a 1 mm double-folded net using a peristaltic pump connected with PTFE tubes. The funnel was fixed by stones in a permanent position for the entire pumping time at the centre of the spring setting just below the water surface, as shown in Figure 5.1.1b. The pumping line was equipped with a PALL Versaflow™ capsule filter of 0.45 µm (prefilter 8 µm) and the observed pumping rate ranged from 4.0 to $3.3 \text{ L min}^{-1}$. Since no clogging of the filters was

**Figure 5.1.1** Sampling points. (A) River Meuse, (B) Spring at Brevilles, (C) River Ruhr

observed, there was no need to change filter while filling the 200 L container. In total 600 L of water were collected without any disturbance, changing the filter, however, for every container.

River basin Rhine hosts a large number of sub-basins of contamination characteristics widely varying in space and time, the highest priority however goes to the Ruhr district. Rivers Ruhr and Emscher drain a rectangle of roughly $40 \times 60$ km, which for a century hosted the German coal and steel industry. Though the district is since some decades undergoing fundamental structural changes, the residues of cokeries and steel works are still characterizing the water-courses to some extent (Muntau, 2005). The permanent monitoring stations of Ruhrverband at Hattingen and Oelbach were selected for collecting the water for the third set of materials, the former for the observed high metal concentrations, due to a number of lead/zinc mines in the upper reaches of the river, and the latter for its particular contamination characteristics. The sampling campaign took place in October 2005. These sampling stations are situated east of the city of Essen and south of Bochum. The pumping line at Hattingen is installed at the river side about 4 m from the shoreline and the pumping funnel being fixed at about 2 m of height above river bottom. The line ends in the laboratory, some 20 m away from the sampling spot and yields about $6 \text{ L min}^{-1}$. Water samples were taken from the tap in the laboratory after extensive flushing of the stainless steel pumping line. The sample appeared slightly turbid and exhibited in 30 cm layer a slight yellow colour. pH was 7.84. The second sampling spot, Oelbach, is situated on a fenced area belonging to the water treatment station under the supervision of Ruhrverband. The water appeared slightly turbid and yellowish. A very faint odour of phenol was noted and the measured pH was 8.35. In both sampling stations, the water was not acidified on the spot. A view of the river Ruhr at both sampling sites is shown in Figure 5.1.1c

## 5.1.4 PREPARATION OF QUALITY CONTROL TOOLS FOR INORGANIC ANALYSIS

### 5.1.4.1 Water Matrix RMs

The general flow chart for the preparation of RMs materials devoted to inorganic analysis is shown in Figure 5.1.2. The 200 L drums containing the bulk material were filtered through 0.45 µm on the spot, and stored at 4°C in the dark, immediately after arriving to the IRMM facilities. Initial characterization analysis of the sample for the target analytes was undertaken. At one point in time during the storage period, the drums were bubbled with compressed air (50 L bottle at 200 bar, technical grade from Air Liquide), in order to accelerate precipitation/flocculation of iron hydroxides and organic matter (humic and fulvic acids). Before filling into individual bottles, the content of each drum was pumped via an in-line filter (PALL AcroPackTM 1000, Supor® Membrane 0.8/0.2 µm offering also bacterial retention) to another precleaned drum using a peristaltic pump (pump speed 3.5 L/min) and PTFE tubing (the flexible part inserted in the pump was made of silicone). One to two filters per drum had to be used. Before immediate use, the PTFE tubes (and the silicone connections) were cleaned with acidified Milli-Q water (1% $HNO_3$ Suprapur®) and subsequently extensively rinsed. Before all the operations of water-transfer, preflushing of the tubes and filters was always performed.

The water material for trace element analysis was then acidified to pH 2 (if not acidified at the sampling) and homogenization was accomplished with a PTFE paddle.

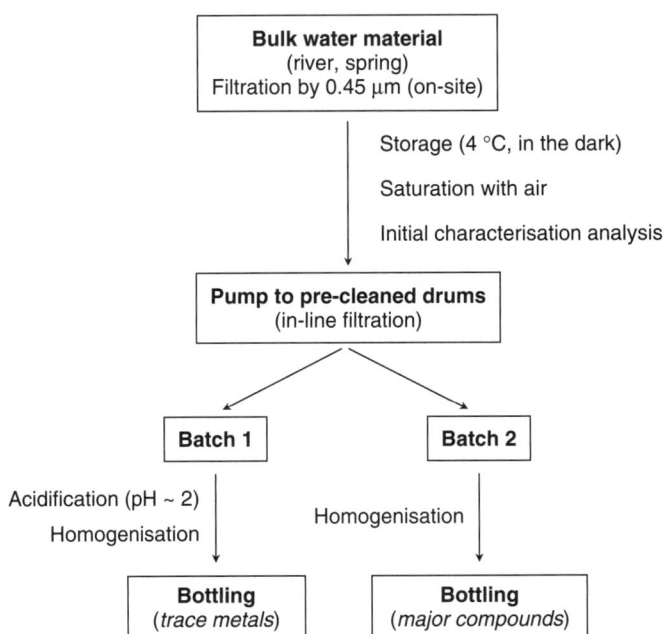

**Figure 5.1.2** Flow chart for the preparation of water RMs for inorganic analysis

At this point of the preparation, and due to the low concentration levels obtained at the Brevilles spring, the bulk material RM05 was fortified using a fit-for-purpose solution.

In all cases, the bulk material for the analysis of major components was not further pretreated.

The bottling was performed in a clean bench (clean air Voerden CLF490) at a pump speed of 1.3–1.5 L min$^{-1}$ with continuous flow, using 500 mL narrow mouth polypropylene bottles (Nalgene). An average number of 150–200 bottles was obtained per material, labelled and stored at 4°C in the dark until dispatch. Some of the produced units, picked up using a random stratified sampling scheme (Sample Number Assignment Program, SNAP), were set aside for homogeneity and stability testing.

The polypropylene bottles chosen for the preparation of these materials were previously assessed for blank levels. Five bottles as supplied by Nalgene, five bottles after having been washed in a lab-dishwasher and two PTFE bottles (Nalgene FEP) as reference, were filled with 0.01 M Suprapur® $HNO_3$. Checks were conducted with ICP-MS for Cd, Pb, Cu, Zn, Ni, As, Na, K, Ca, Mg. Comparing the blank data, the bottles as supplied by the company were found even cleaner than those washed, and therefore, it was decided to use them as received.

For the first set of materials, and with the aim of assessing the dispatch conditions, a short-term stability study was conducted at 40°C. The layout chosen for the stability study was the so-called isochronous scheme: samples were taken from the bulk, placed at 40°C and then moved back to the 'reference' temperature (4°C), after 1 and 2 weeks. Then, at the same time, the samples were analysed for major components and trace elements. The results, 3 time-points (0, 1, 2 weeks) and 2 units analysed per time-point, were evaluated by one-way analysis of variance ANOVA. As some parameters (especially As, Cd, Cu, and to a minor extent also Mn, pH) showed a statistically significant slope of the regression line, it was decided to assure the dispatch of the samples at 4°C (with cooling elements).

### 5.1.4.2 Calibrants and Spiking Solutions

Calibrants and spiking solutions for trace elements and major compounds were prepared following the procedure shown in Figure 5.1.3. In the cases where a reagent pro-analysis grade was used, the salt was previously dried for 3 hours at $100 \pm 5°C$, and directly weighed with an analytical balance. For trace elements and some major compounds, commercially available solutions CertiPUR® from Merck were used, all of them traceable to SRM from NIST. Individual solutions sealed in glass ampoules (10 or 15 mL) for each parameter were provided, with a final concentration of about 1000 mg L$^{-1}$. The solutions were prepared in $HNO_3$ for metals and some major components, and in double deionized water for anions and total P. The information provided together with the solutions is shown in Table 5.1.2.

All glass vials, except those used for nitrates, were previously cleaned with 10% $HNO_3$ (Hiperpur®, Panreac) by keeping them filled with the cleaning solution for 16 h. After rinsing with double deionized water, the vials were dried in an oven at 100°C. After cooling to room temperature they were stored in closed recipients until filling with the calibrant solution. The same cleaning procedure was used for the glass reactor.

A check was performed in order to discard any cross-contamination during preparation and to check the effectiveness of the cleaning procedure applied. In each calibrant

# Preparation of Quality Control Tools for Inorganic Analysis

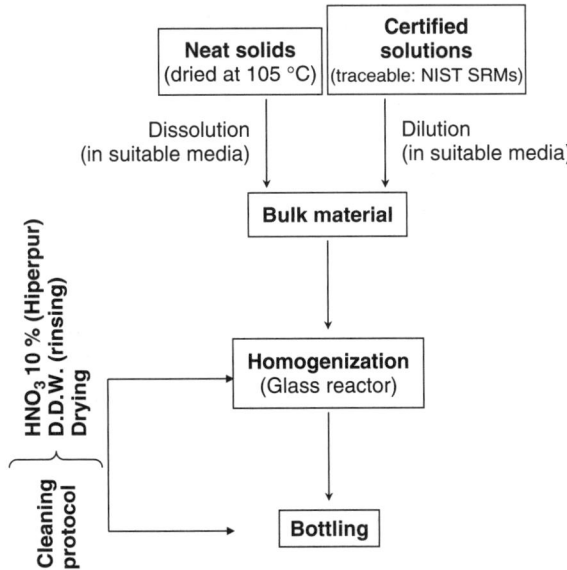

**Figure 5.1.3** Flow chart for the preparation of calibrants and spikes

**Table 5.1.2** Information provided with calibrants

| Compound | Concentration (mg L$^{-1}$) | u (combined uncertainty)[1] (mg L$^{-1}$) | U (Expanded uncertainty)[2] (mg L$^{-1}$) |
|---|---|---|---|
| As | 1000 | 8 | 16 |
| Ca | 1000 | 5 | 10 |
| Cd | 1000 | 11 | 22 |
| Cu | 1001 | 5 | 10 |
| K | 996 | 13 | 27 |
| Na | 1001 | 31 | 62 |
| Ni | 1001 | 7 | 13 |
| Mg | 1001 | 9 | 17 |
| Pb | 1001 | 4 | 9 |
| Zn | 1001 | 6 | 11 |
| P | 1000 | 8 | 15 |
| SO$_4^{2-}$ | 1001 | 40 | 79 |
| Cl$^-$ | 1001 | 8 | 16 |
| NO$_3^-$ | 1000 | 14 | 27 |

[1] As, Ca, Cd, Cu, K, Ni, Mg, Pb, Zn, P: determined by ICP-OES; Na: determined by flame emission spectroscopy (FES); SO$_4^{2-}$ and NO$_3^-$: determined by ion chromatography; Cl$^-$: determined by Mohr titration.
[2] For a level of confidence of 95%, k = 2.

solution, the presence of the remaining target elements was investigated at trace level. In all cases the concentration found by ICP-AES analyses was below the respective detection limits (for example, concentration values for Ca, Cd, Cu, K, Mg, Ni, Pb, Zn and P were below detection limits in the As calibrant solution).

For the preparation of the spiking solutions to be used either by the participants in their laboratories (to prepare RM03 and RM04) or by IRMM (to prepare RM05), the characterization values for the original bulk materials, as well as the spiking procedure to be performed, were taken into account. The amount of analytes in the spiking solutions was then designed to have concentration values around those indicated in the European Directive 98/83/EC (European Commission, 1998) for the analytes with low concentration in RM01 and RM02, or to have levels around five to ten times higher than their contents in the not fortified samples, for the rest of analytes.

In all cases the sealed ampoules were stored at room temperature until dispatch.

## 5.1.5 PREPARATION OF QUALITY CONTROL TOOLS FOR ORGANIC ANALYSIS

### 5.1.5.1 Water Matrix RMs

The first two materials for organic analyses from the Brevilles spring (RM07 and RM08) were prepared following the same protocol as described in Figure 5.1.2. Approximately 525 mL water were filled into polypropylene containers in order to ensure the necessary amount of 500 mL specified in the reconstitution protocols. The filled bottles were screwed tightly and the cap was further secured with a crimp-film. For these materials, due to the low concentration values found in the initial characterization analysis for the target analytes, a fortification process was envisaged as necessary, and it was performed directly by the participants using specific spiking solutions or pills and following a detailed spiking protocol, which yielded final concentrations between 0.5 and $1 \mu g L^{-1}$.

The preliminary characterization analysis on the bulk material from river Ruhr revealed also the necessity of a fortification step. However, for these materials, it was decided to opt for a low concentration level (down to around $0.1 \mu g L^{-1}$ per compound) achieved by fortifying the bottled materials just before dispatch of the samples (two days maximum in advance). Due to the chosen spiking procedure, a check of the integrity of the sample received by the laboratories was considered necessary and achieved by a control of the weight of the bottle. Additionally, and once again because of the spiking procedure used, for these materials participants were requested to analyse samples immediately after receipt.

Both RM14 and RM15, were bottled in 500 mL borosilicate glass bottles, previously flushed with 10 mL hexane and then dried at 40°C. Each bottle was weighed (without cap) right after filling on a calibrated Sartorius balance with registration of the weight on a computer. The bottles were filled up to the neck to ensure a minimum of head space. The samples were stored at 4°C in the dark, until spiking and subsequent shipment to the laboratories.

In the case of RM14 (PAHs determination), the filtration to a precleaned drum just before bottling was not performed. On the contrary, a PTFE paddle was placed in the

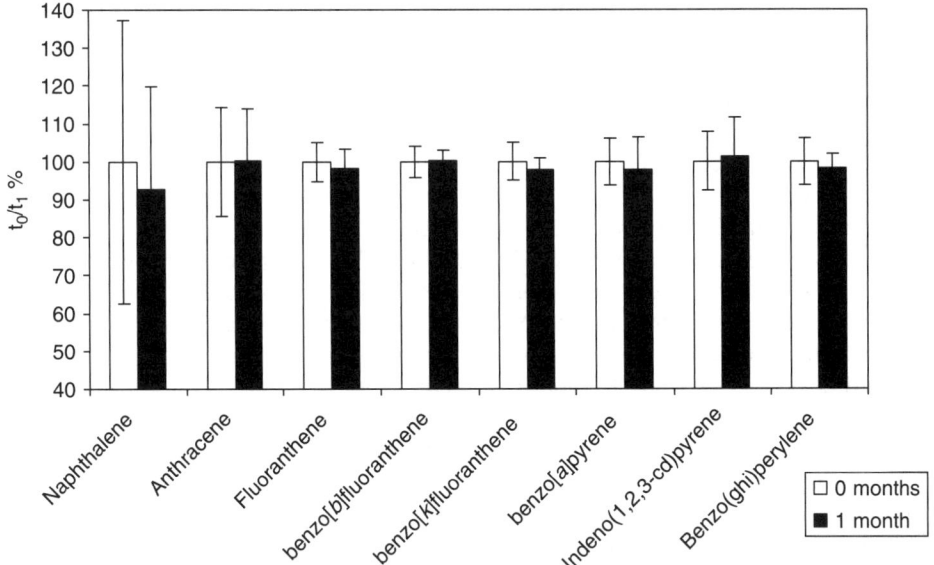

**Figure 5.1.4** Stability testing results for RM14 (PAHs at low concentration levels)

container in order to maintain the particulate matter present in the water in suspension during the bottling process. The decision of not removing the particulate matter in these samples (foreseen for PAHs analysis) was taken in the view of mimicking as close as possible a real river water sample.

Just before dispatch to the laboratories, each bottle (RM14 and RM15) was spiked with 1 mL of a spiking solution by means of a 1 mL dispenser. Before closing the bottle, an aluminium foil folded in four parts was placed in the cap in order to avoid adsorption phenomena as much as possible of the organic species on the inside of the plastic cap. The repeatability of the dispenser was checked beforehand and it was found to be better than 0.5% RSD.

The results from the one month-stability study revealed that this approach can be used for the preparation of water RMs at concentration levels down to $0.1\,\mu g\,L^{-1}$, for the eight PAHs (Figure 5.1.4) and 13 pesticides tested.

### 5.1.5.2 Standard and Spiking Solutions

The aim of providing PAHs as well as pesticides standard solutions of unknown concentrations to the participants was to check the performance of the chromatographic methods established in each laboratory. The solutions had to be analysed directly, or just after a dilution step in a suitable solvent compatible with the method of analysis. For this reason, after the selection of the compounds, the final concentrations of compounds in standard solutions were calculated considering on the one hand, a normal concentration factor of procedures for water analysis applied by laboratories (up to 500 fold), and, on the other hand, the recommended dilution factor to be performed on the solution.

The final solution medium for RM10, RM11 and RM09 was established according to the solubility requirements for assuring long-term stability at high concentration levels, that is ethylacetate for pesticides (except glyphosate which was prepared in double deionized water) and acetonitrile for PAHs.

The fit-for-purpose solutions used were accompanied by gravimetric certificates (Laboratory Dr. Ehrenstorfer) and they were prepared from individual solutions or from pure solid compounds depending on the final concentrations requested. The preparation procedure described in Figure 5.1.3 was followed.

The spiking solution for analysis of PAHs in RM07 was prepared also in acetonitrile. The robustness of the reconstitution protocol for this spring water was verified by the Organic Analysis Laboratory of the RM Unit at IRMM. Several samples were analysed following the protocol for liquid-liquid extraction given to the PT participants. The standard deviation for reconstitution was found to be lower than 10% for all 9 PAHs included in the study.

The PAHs spiking solution for the preparation of RM14 was prepared gravimetrically using the mass metrology service of IRMM. For that purpose, high purity crystalline substances provided by IRMM and Dr. Ehrenstorfer were subsequently diluted with acetonitrile to the required concentration. The same procedure was used for the preparation of the standard solution RM16.

The pesticides spiking solution for RM15 was prepared using pesticides reference materials (Pestanal®) from Riedel de Haen in form of neat crystals (apart from chlorfenvinphos which is liquid). Stock solutions were properly diluted with acetonitrile to the final concentrations. The stock solution for simazine was prepared in methanol.

## 5.1.6 HOMOGENEITY AND STABILITY STUDIES

The suitability of the quality control tools prepared under the SWIFT-WFD project was proved from the homogeneity and stability studies carried out on all materials. Table 5.1.3 summarizes the tested analytes for each material, the institution responsible for the analysis, and the analytical technique and sample intake used.

Analyses were carried out under repeatability conditions, that is in the same laboratory, by the same operator, applying the same instrumentation and analytical method on the same day, except for RM15 where, due to the extraction process, sample preparation was performed by two operators and samples were analysed on different days. In all cases, upon arrival in the laboratory, the samples were stored at 4°C before processing. Quantification was performed by external calibration graphs, using internal standards when ICP-MS and GC-MS were the analytical techniques.

For homogeneity testing in natural materials, the between-bottle variability ($S_{bb}$) was evaluated following the IRMM approach (Linsinger *et al.*, 2001), after the application of one-way analysis of variance (ANOVA) to the duplicates obtained in ten different units.

$$S_{bb} = \frac{\sqrt{\frac{MS_{between} - MS_{within}}{n}}}{\bar{y}} \tag{5.1.1}$$

where $n$ is the number of replicates per unit and $\bar{y}$ is the average value.

## Homogeneity and Stability Studies

**Table 5.1.3** Homogeneity and stability testing for quality control tools produced within the SWIFT-WFD Project

| Material | Center | Analysed elements (Analytical techniques) | Sample intake |
|---|---|---|---|
| RM01 | University of Veszprem | Ni, Al, Cr (GFAAS) | 1 ml |
|  | University of Veszprem | Mn (ICP-OES) | 10 ml |
|  | ENEA | As, Cd, Cu, Pb, Zn (ICP-MS) | 1 ml |
| RM02/RM06/RM13 | ENEA | pH, conductivity | 100 ml |
|  |  | Total P (UV-vis) | 50 ml |
|  |  | Na, K, Ca, Mg (ICP-OES) | 0.5 ml |
|  |  | $Cl^-$, $NO_3^-$, $SO_4^{2-}$ (Ion chromatography) | 2 ml |
| Calibrants | University of Barcelona | As, Cd, Cu, Ni, Pb, Zn, P, K, Ca, Mg (ICP-OES) | Dilution 1:1000 |
|  |  | Na (FES) | Dilution 1:1000 |
|  |  | $Cl^-$, $NO_3^-$, $SO_4^{2-}$ (Ion chromatography) | Dilution 1:100 |
| Spiking solution for RM03 | University of Barcelona | Cu, Zn (FAAS) | 1 ml |
|  |  | Se (ICP-MS) | 1 ml |
| Spiking solution for RM04 | University of Barcelona | Mg (FAAS); K (FES) | 1 ml |
|  |  | $SO_4^{2-}$ (Ion chromatography) | 2 ml |
| RM05/RM12 | University Complutense of Madrid | Al, Cr, Mn, Ni, Cu, Zn, As, Se, Cd, Pb (ICP-MS) | 1 ml |
| Spiking solution for RM07/RM09 | Vienna University of Technology | Anthracene, benzo[a]pyrene, benzo[b]fluoranthene, benzo(ghi)perylene, benzo[k]fluoranthene, phenanthrene, fluoranthene, indeno(1,2,3-cd)pyrene, naphthalene (HPLC-UV) | 10 µl |
| RM10/RM17 | Vienna University of Technology | Alachlor, chlorfenvinfos, chlorpyrifos, diuron, endosulfan (alpha+beta), isoproturon, simazine, aldrin (HHDN), terbuthylazine, lindane (gamma-HCH), chlortoluron, atrazine, desethyl-atrazine (GC-FID and HPLC-UV) | 1µl (GC-FID) 10 µl (HPLC-UV) |
| RM11 | Vienna University of Technology | Glyphosate (HPLC-UV after derivatisation) | 10 µl |
| RM14/RM16 | IRMM | Anthracene, benzo[a]pyrene, benzo[b]fluoranthene, benzo(ghi)perylene, benzo[k]fluoranthene, fluoranthene, Indeno(1,2,3-cd)pyrene, naphthalene (GC-MS) | 500 ml (1-1 extraction for RM14) |
| RM15 | Vienna University of Technology | Alachlor, chlorfenvinfos, chlorpyrifos, diuron, endosulfan (alpha+beta), isoproturon, simazine, aldrin (HHDN), terbuthylazine, lindane (gamma-HCH), chlortoluron, atrazine, desethyl-atrazine (GC-MS and HPLC-MS) | 500 ml (solid phase extraction) |

When $s_{bb}$ cannot be calculated (as $MS_{between} < MS_{within}$ in ANOVA results), the uncertainty contribution due to between-bottle heterogeneity is taken from the $u_{bb}^*$ value given in Equation 5.1.2 (where $df$ are the degrees of freedom). $u_{bb}^*$ represents the maximum heterogeneity that can be hidden by the method repeatability and it was considered as the minimum uncertainty contribution from heterogeneity in this approach.

$$u_{bb}^* = \frac{RSD_{method}}{\sqrt{n}} \sqrt[4]{\frac{2}{df_{MS within}}} \tag{5.1.2}$$

$$RSD_{method} = S_{within} = \frac{\sqrt{MS_{within}}}{\bar{y}} \tag{5.1.3}$$

As an example, the uncertainty contribution due to heterogeneity for RM01 (natural water) was quantified between 1.6 and 3.4% for most analytes, except Al and Cr, being 7 and 13% respectively. With regard to RM02 (major components) the uncertainty contribution due to the heterogeneity was between 1.5 and 2.5%, except for P (3.4%), probably due to the low content. The obtained results confirmed the homogeneity of the materials and their suitable use for PT schemes.

For materials RM14 and RM15, fortified at low concentration levels for organic compounds, a stability test was performed after one month of the spiking procedure in three bottles (one analysis each) stored at 4°C. The results obtained for RM14 are shown in Figure 5.1.4. The difference in percent obtained with respect to time 0, corresponding to the moment of the spiking procedure, resulted below 2% for all compounds, with the exception of naphthalene, for which the difference was 7%. For this analyte also the variability associated to the PT results was relatively higher compared to the other analytes, probably due to its high volatility.

## 5.1.7 DISCUSSION

The RMs for inorganic analytes were prepared following well established procedures, already in use since several years. On the contrary, novel approaches for the preparation of RMs for analysis of organic compounds in water (PAHs and pesticides) were explored and tested through the SWIFT-WFD PT campaigns, with a positive outcome of the results obtained. These data could serve for future developments in the field of RMs for organic compounds in water. Such quality control tools are, at present, missing because of still unsolved technical difficulties related to the stability of these substances in the water matrix.

In particular, in case of organic compounds the following was investigated: (1) the possibility to directly spike the materials immediately before their dispatch (3rd set of RMs); (2) the feasibility to leave the material unfiltered for PAHs analysis (3rd set of RMs); (3) the possibility to spike the material for pesticides using a newly developed technology by the Institute Pasteur de Lille (2nd set of RMs). The achieved results are promising: all the produced RMs, even in case of volatile analytes and low concentration values, proved to have satisfactory characteristics of homogeneity and stability (at least on a short-time scale) of the analytes and to be suitable for being used for the evaluation of analytical performances at European level. On the other hand, in many cases the produced RMs were not suitable for screening methods, for several reasons, but above all because much higher volume (tens of liters) of sample would be generally needed. The experience achieved along the project testifies that more efforts should be put in the development of new technologies for the production of RMs suitable for the quality control of screening methods and emerging tools.

The production of SWIFT-WFD RMs had a valuable impact in the evaluation of European laboratories performances using both classical and screening methods for four classes of analytes: trace elements, major components, PAHs, pesticides. The produced RMs allowed the evaluation of analytical performances at EU level at different levels of concentration, in different matrices (river water, spring water) with different composition.

## 5.1.8 CONCLUSIONS

The results obtained within the SWIFT-WFD project, dealing with the preparation of water reference materials, will contribute to the future developments of such materials, especially for the analysis of pesticides and PAHs at low concentration levels. The two different strategies employed, direct fortification in the laboratory by using spiking solutions following detailed protocols, and fortification of the material before dispatch, led to homogeneous and satisfactory stable reference materials for quality control purposes.

The production of calibrants and multi-component standard solutions, besides the matrix reference materials, allowed the organization of PT schemes under the step-by-step approach, by minimizing the effect of the use of standards coming from different sources and by allowing the evaluation of the performance of the instrumental analytical step.

The prepared reference materials were used in three different PT campaigns with the participation of European analytical laboratories involved in water monitoring, using not only classical methods but also the so-called screening methods and emerging tools, in spite of the different requirements of these latter methods (mainly compatibility of the matrix tested with the type of assay used and high sample volumes). Further developments would be nevertheless needed in this field.

## ACKNOWLEDGEMENTS

The European Commission is kindly acknowledged for the funding of the activity carried out under the SWIFT-WFD project, contract no: SSPI – CT – 2003 – 502492.

The staff from the organizations supervising the water sampling sites is acknowledged for their technical support. K. Kramer (Mermayde), A. Bau and Paul de Vos are also acknowledged for their contribution in the sampling campaigns.

The staff from RMs producers (IRMM and UB) involved in different steps along the production of materials provided, and E. Fernández-Díez from TU Vienna, and V. Pinto and E. Nardi from ENEA involved in some characterization analysis, are also kindly acknowledged.

## REFERENCES

Bercaru O, Gawlik BM, Ulberth F., Vandecasteele C., 2003. Reference materials for the monitoring of the aquatic environment. A review with special emphasis on organic priority pollutants. *J. Environ.Monitor.*, **5**, 697–705.

Brunori C., Ipolyi I., Pellegrino C., et al., 2007. The SWIFT-WFD Proficiency Testing campaigns in support of implementing the EU Water Framework Directive. *Trends Anal. Chem.*, **23**, 993–1004.

De Guillebon B, Pannier F, Seby F, Bennink D, Quevauviller Ph, 2001. Production and use of reference materials for environmental analyses: conclusions of a workshop. *Trends Anal. Chem.*, **20**, 160–6.

Emons H, Linsinger TPJ, Gawlik BM, 2004. Reference materials: terminology and use. Can't one see the forest for the trees? *Trends Anal. Chem.*, **23**: 442–449.

European Commission, 1998. Council Directive 98/83/EC of 3 November 1998 on the quality of water intended for human consumption. *Official Journal of the European Communities*, L **330**: 32–54.

European Commission, 2000. Directive 2000/60/EC of the European Parliament and of the Council of 23 October 2000 establishing a framework for Community action in the field of water policy. *Official Journal of the European Communities, L* **327** 1–72.

ISO Guide 34, 2000. *General requirements for the competence of reference material producers (E)*, International Organization for Standardization, Geneva, Switzerland.

ISO/IEC 17025, 2005. *General Requirements for the Competence of Calibration and Testing Laboratories*. ISO, Geneva, Switzerland.

Linsinger TPJ, Pauwels J, Van der Veen, Schimmel, Lamberty A, 2001. Homogeneity and stability of reference materials. *Accred. Qual. Assur.*, **6**, 20–5.

Linsinger TPJ, Bernreuther A, Corbisier Ph, et al., 2007. Accreditation of reference material producers: the example of IRMM's Reference Materials Unit. *Accred. Qual. Assur.*, **12**, 167–74.

Muntau H, 2005. SWIFT-WFD Project Third Sampling Campaign: River Basin Rhine–Sub-basins Ruhr and Emscher, 68 pp.

Namiesnik J, Zygmunt B, 1999. Role of reference materials in analysis of environmental pollutants. *Sci. Total Environ.* **228**, 243–57.

Quevauviller Ph, Cofino W, Cortez L, 1998. Use of reference materials in accreditation systems for environmental laboratories. *Trends Anal. Chem.*, **17**, 241–8.

Sahuquillo A, Carrasco E, Muntau H, Rubio R, Rauret G, 2004. Mat Control: a new laboratory for the preparation of reference materials at the University of Barcelona, Spain. *Accred. Qual. Assur.*, **9**, 272–7.

SWIFT-WFD project, 2004. Screening methods for Water data Information in support of the implementation of the Water Framework Directive. 6th Framework Programme, contract number SSPI-CT-2003-502492 (2004–2007).

Thompson M, Wood R, 1995. Harmonized guidelines for internal quality control in analytical chemistry laboratories (Technical Report). *Pure & Appl. Chem.*, **67**, 649–66.

WARP project, 2002. Preparation of a CRM: pesticides in water. Feasibility study. 5th Framework Programme, contract number G6RD-CT-2001-00607 (2002–2004).

# 5.2

Participation of Screening Methods and Emerging Tools (SMETs) to Proficiency Testing Schemes on the Determination of Priority Substances in Real Water Matrices Organized in Support of the Water Framework Directive Implementation

Claudia Brunori, Ildi Ipolyi and Roberto Morabito

5.2.1 Introduction
5.2.2 Description of the SWIFT-WFD PT Schemes
    5.2.2.1 Participation of SMETs in the SWIFT-WFD PT Campaign
    5.2.2.2 Description of the SMETs Participating in the PT Schemes
5.2.3 Data Evaluation
    5.2.3.1 Discussion of the Results Obtained by SMETs for the SWIFT-WFD PT Schemes
5.2.4 Conclusions
Glossary
References

---

*Rapid Chemical and Biological Techniques for Water Monitoring*  Edited by Catherine Gonzalez, Philippe Quevauviller and Richard Greenwood
© 2009 John Wiley & Sons, Ltd

## 5.2.1 INTRODUCTION

One of the short-term aims of the QA/QC activities within the SWIFT-WFD (*Screening methods for Water data InFormaTion in support of the implementation of the Water Framework Directive*) project funded by the European Commission (SWIFT-WFD project, ) was the evaluation of the state of the art of the analytical performances of screening methods in Europe, whereas one of the long-term aims was the establishment of a framework of ongoing proficiency testing (PT) activities supporting the European laboratories that are involved in the implementation of EU Water Framework Directive (WFD) either with classical methods and/or nonclassical' methods including screening methods, semi-quantitative methods, emerging tools, alternative or complementary methods, etc (SMETs).*

SMETs are tools and methods that are expected to provide better suited information within the WFD monitoring networks either in combination with classical techniques, or directly replacing them in their capacity e.g. to confirm the presence of certain compounds without tedious work required (semi-quantitative methods), to allow frequent monitoring (online systems) or to provide more representative data (passive samplers), and to provide data directly on-site (kits).

As such, SMETs assessed under SWIFT-WFD have the potential to fulfil diverse functions in the context of WFD monitoring. They can help supporting the design of monitoring programmes, e.g. by collecting information on current water status for grouping water bodies into groups that are homogeneous (in terms of pressures but also impacts) and for which one monitoring point per group only is required for complying with the requirements of the WFD as well as by applying screening methods and alternative methods to search for optimum locations and frequency for monitoring networks. SMETs can also be part of the monitoring programme per se for compliance checking (combined among each other/with existing traditional methods) or for enhancing the information base on water quality (e.g. with higher number of points/frequencies or 'new' information on water quality be it in terms of temporal variability of pollution or presence of new pollutants). Thus, they can help in supporting the selection of measures (operational monitoring), in assessing long-term trends in the overall status of the aquatic ecosystem (surveillance monitoring), or in establishing cause-effect relationships in water bodies that are considered being at risk but for which causes of risk are unknown (investigative monitoring). Even if these aspects can scarcely be measured in monetary terms, they often also result in both direct and indirect cost reductions (Greenwood and Roig, 2006; Dworak and Lügcke, 2006; Strosser and Graveline, 2006).

For this aim, three proficiency testing (PT) campaigns were organized within the SWIFT-WFD project, mainly dedicated to the determination of the list of priority pollutants annexed to the WFD. The SWIFT-WFD PT campaigns, one per year for three years, were organized on 17 reference materials (RMs) including 5 matrix RMs at natural concentration levels, 3 at low fortified concentration levels, 4 at fortified concentration levels, and 5 multi-analyte blind standard solutions, targeting the analysis of

---

* In order to facilitate writing and understanding, and exclusively only to achieve these reasons, the 'nonclassical' methods including screening methods, semi-quantitative methods, alternative or complementary methods, etc., are hereinafter referred to as 'screening methods – emerging tools', with the following acronym: SMETs.

trace elements, major components, pesticides and polyaromatic hydrocarbons. Laboratories active in measurement activities related to the implementation of the WFD, using either classical and/or nonclassical analytical methods (SMETs), from the EU member and associated states, were invited to participate in the SWIFT-WFD PT schemes. All together 94 laboratories from 21 target countries took part in the 17 exercises, with an average of 50 participants per exercises. The participation of SMETs in the PT schemes was significantly less than foreseen at the beginning of the project.

This chapter aims to discuss the results obtained in the SWIFT-WFD PT schemes by nonclassical methods. The representation of SMETs versus the overall participation in the PT schemes and the achieved analytical performances are introduced and evaluated by methods and analyte by analyte. The results are presented in the light of the state of the art performance of all the European laboratories on the field, on basis of the SWIFT-WFD PT experience.

## 5.2.2 DESCRIPTION OF THE SWIFT-WFD PT SCHEMES

The SWIFT-WFD PT exercises were organized in the frame of three campaigns, one per year for three years, from 2004 to 2006. More information on this activity is reported elsewhere (Ipolyi *et al.*, 2006a, 2006b, 2006c; Brunori *et al.*, 2008).

The matrix materials used as test materials in the SWIFT-WFD PT schemes were prepared within the project activities from four water batches of different origin in the three PT campaigns. The list of materials and the corresponding exercises is reported in Table 5.2.1; more details on the SWIFT-WFD RMs are reported in Chapter 5.1 of this book.

### 5.2.2.1 Participation of SMETs in the SWIFT-WFD PT Campaign

The 1st SWIFT-WFD PT schemes was primarily directed to laboratories applying classical analytical methods so that a reference of the state of the art of the analytical performance of European laboratories in trace element and major component determination can be established and to what the performance of SMETs could be compared. After the completion of the campaign, in December 2004, the laboratories using SMETs were also invited to analyse the first batch of SWIFT-WFD reference materials with their method(s). The 2nd and the 3rd SWIFT-WFD PT campaigns, conducted in 2005 and 2006, respectively, were immediately aimed at both groups of laboratories.

Table 5.2.2 shows the number of invitations sent out and the laboratories that applied to the PT schemes, according to the method used. The fifth column of the table presents the number of applications that arrived from SWIFT-WFD project partners.

All together 197 participants 5255 datasets were received and included in the statistical evaluation of the SWIFT-WFD PT schemes. Only 34 of these datasets were generated by SMETs, evenly distributed amongst the 3 PT campaigns. These numbers prove the participation of SMETs was very low, below 1%. In addition, most of the SMETs results were received from project partners and none of them arrived from laboratories from newly joined countries of the European Union and the Associated States.

**Table 5.2.1** List of exercises in the SWIFT-WFD PT schemes

| SWIFT-WFD PT campaign | Matrix | Material code | List of analytes | Range of concentration levels |
|---|---|---|---|---|
| First | river water | RM01* | Al, As, Cd, Cr, Cu, Mn, Ni, Pb, Se and Zn | natural levels (0.1 – 50 µg L$^{-1}$) |
| | | RM02* | Ca, Mg, K, Na, conductivity, pH, Cl$^-$, NO$_3^-$, SO$_4^{2-}$ and P$_{tot}$ | natural levels (0.1 – 60 mg L$^{-1}$) |
| | | RM03* | Al, As, Cd, Cr, Cu, Mn, Ni, Pb, Se and Zn | fortified levels (7 – 1000 µg L$^{-1}$) |
| | | RM04* | Ca, Mg, K, Na, Cl$^-$, SO$_4^{2-}$ and P$_{tot}$ | fortified levels (2 – 300 mg L$^{-1}$) |
| | | RM05 | Al, As, Cd, Cr, Cu, Mn, Ni, Pb, Se and Zn | low fortified levels (4 – 80 µg L$^{-1}$) |
| | | RM06 | Ca, Mg, K, Na, conductivity, pH, Cl$^-$, NO$_3^-$, SO$_4^{2-}$ and P$_{tot}$ | natural levels (0.02 – 110 mg L$^{-1}$) |
| | spring water | RM07 | benzo[a]pyrene, benzo[b]fluoranthene, benzo[k]fluoranthene, benzo[ghi]perylene, indeno(1,2,3-cd)pyrene, fluoranthene, anthracene, fenanthrene, naphtalene | fortified levels (0.15 – 1.4 µg L$^{-1}$) |
| Second | | RM08 | alachlor, chlorfenvinfos, chlorpyrifos, chlortoluron, diuron, endosulfan, isoproturon, simazine, aldrin, terbuthylazine, lindane, atrazine, desethylatrazine, glyphosate | fortified levels (0.2 – 1.4 µg L$^{-1}$) |
| | acetonitrile | RM09 | as RM07 | blind standard solution (10 – 80 mg L$^{-1}$) |
| | ethylacetate | RM10 | as RM08 (except glyphosate) | blind standard solution (30 – 80 mg L$^{-1}$) |
| | deionized water | RM11 | glyphosate | blind standard solution (~80 mg L$^{-1}$) |
| | | RM12 | Al, As, Cd, Cr, Cu, Mn, Ni, Pb, Se and Zn | natural levels (0.05 – 150 µg L$^{-1}$) |
| | river water | RM13 | Ca, Mg, K, Na, conductivity, pH, Cl$^-$, NO$_3^-$, SO$_4^{2-}$ and P$_{tot}$ | natural levels (0.05 – 1000 mg L$^{-1}$) |
| | | RM14 | as RM07 (without anthracene) | low fortified levels (30 – 85 ng L$^{-1}$) |
| Third | | RM15 | as RM08 (without desethylatrazine, glyphosate, terbuthylazine) | low fortified levels (90 – 170 ng L$^{-1}$) |
| | acetonitrile | RM16 | as RM14 | blind standard solution (70 – 130 µg L$^{-1}$) |
| | ethylacetate | RM17** | as RM10 | blind standard solution (30 – 80 mg L$^{-1}$) |

*Including calibrants for target analytes
**Re-labelled RM10

*Description of the SWIFT-WFD PT Schemes*

Table 5.2.2 Participation of laboratories in the SWIFT-WFD PT schemes

| SWIFT-WFD PT campaign | | Invited laboratories | Applications | SWIFT-WFD partners |
|---|---|---|---|---|
| First | SMETs | 15 | 1 | 1 |
|  | Classical methods | 75 | 58 | 19 |
| Second | SMETs | 35 | 5 | 3 |
|  | Classical methods | 95 | 76 | 16 |
| Third | SMETs | 35 | 4 | 3 |
|  | Classical methods | 95 | 63 | 18 |

## 5.2.2.2 Description of the SMETs Participating in the PT Schemes

As specified in Table 5.2.3, in the 1st PT campaign only 1 laboratory (project partner) using SMETs did submit results for some major components in both RM02 and RM04 exercises. Within the 2nd PT campaign four laboratories (two project partners and two invited external participants) submitted quantitative results for some target analytes in RM05, RM08 and RM10. In addition, one qualitative result was received for both RM08 and RM10 for the total content of pesticides, which could not be included in the statistical evaluation of the PT results. For the 3rd PT campaign three laboratories (all project partners) submitted results; the quantitative results comprised datasets for RM12, RM13, RM15 and RM17. In addition, two qualitative results were also provided, one for RM14 and one for RM15; as previously mentioned, these results could not be included in the statistical evaluation of the PT results.

A brief description of the applied SMETs is reported below.

*Analytical method for the determination of Ca, K, Na, chloride, pH and conductivity*

The results provided in the 1st PT campaign for pH, conductivity, Ca, K, Na and chloride were obtained by in-house fabricated microelectrodes (Chapter 4.1.4 of this book). The sensor used for pH determination is a pH-ISFET with silicone nitride membrane, the sensors for $Na^+$, $K^+$, $Ca^{2+}$ and $Cl^-$ are ISFETs with ion-selective membranes, while the sensors for conductivity is a 4-bar platinum electrode. All the circuits used for measuring with the sensors (ISFET meter and conductivity meter) were also developed in-house. The results for the ions ($Na^+$, $K^+$, $Ca^{2+}$ and $Cl^-$) were received being expressed in activity, therefore could not be compared with the other PT results.

*Analytical method for the determination of Cd, Cu and Pb*

The results for Cd, Cu and Pb in RM05 and RM12 were obtained with Palmsens – Screen Printed Electrode Voltametric Sensor (Simultaneous determination with square wave anodic stripping voltammetry – SWASV) with standard addition

Table 5.2.3  Participation and performance evaluation of SMETs in the SWIFT-WFD PT schemes

| SWIFT-WFD PT campaign | N° of laboratories | Type of results | N° of submitted datasets | Analytes | Z-score |
|---|---|---|---|---|---|
| First | 1 SWIFT-WFD partner | quantitative | 6 for RM02<br>4 for RM04 | pH, conductivity<br>Ca, K, Na, Cl$^-$ | n.d.<br>(results expressed in activity) |
|  | 1 SWIFT-WFD partner | quantitative | 3 for RM05 | Cd, Cu, Pb | 2 satisfactory datasets ($Z \leq |2|$)<br>1 doubtful dataset ($2 < |Z| \leq 3$) |
| Second | 1 SWIFT-WFD partner and<br>1 external laboratory | quantitative | 4 for RM08<br>3 for RM10 | atrazine, diuron,<br>isoproturon,<br>chlorpyrifos | 5 satisfactory datasets ($Z \leq |2|$)<br>2 unsatisfactory datasets ($|Z|>3$) |
|  | 1 external laboratory | qualitative | 1 for RM08<br>1 for RM10 | pesticides | n.d. |
|  | 1 SWIFT-WFD partner | quantitative | 3 for RM12 | Cd, Cu, Pb | 3 datasets < LoD |
| Third | 2 SWIFT-WFD partners | quantitative | 4 for RM13 | $NO_3^-$ | 2 satisfactory datasets ($Z \leq |2|$)<br>2 unsatisfactory datasets ($|Z|>3$) |
|  | 1 SWIFT-WFD partner | quantitative | 2 for RM15<br>2 for RM17 | diuron, isoproturon | 3 satisfactory datasets ($Z \leq |2|$)<br>1 unsatisfactory dataset ($|Z|>3$) |
|  | 1 SWIFT-WFD partner | qualitative | 1 for RM14 and<br>1 for RM15 | PAHs and<br>pesticides | n.d. |

*n.d. : not determined

*Description of the SWIFT-WFD PT Schemes*                                                                357

(Chapter 4.1.2 of this book). The principle of the technique is as follows:

deposition step (metal reduction) : $Cd^{2+} \rightarrow Cd^0(Hg)$

stripping step (metal redissolution) : $Cd^0(Hg) \rightarrow Cd^{2+}$

This method is suitable for the quantification of the free and labile complexed forms of Cd, Cu and Pb (and Zn is also possible); sample analysis takes ~ 15 min. (Roig et al., 2006a).

## Analytical method for the determination of atrazine, diuron and isoproturon

The results for atrazine, diuron and isoproturon in RM08, RM10 and RM15 and RM17 were gained by ELISA (enzyme-linked immunosorbent assays) in enzyme tracer format and coating antigen format, respectively (Chapter 3.1.3 of this book). The immunoassays were formatted into microtiter plates. In the in-house plate assay developed for isoproturon, two different monoclonal antibodies (mAbs) IOC 7E1 and IOC 10G7 were used, separately or together, depending on the analytical requirements. These mAbs have very similar selectivity, but different sensitivities to the main target isoproturon (Krämer et al., 2004). The ELISAs for diuron and atrazine were based on mAbs previously developed by Dr A. Karu (formerly UC Berkeley, CA, USA) and formatted into optimized immunoassays in-house. For diuron, mAb 481.3 was used together with an optimized in-house enzyme-tracer for this mAb. For the analysis of atrazine, an immunoassay was applied using mAb AM5D1 together with a formerly commercially available enzyme-tracer (Atrazine-HRP; Fitzgerald Industries International, Inc., Concord, MA, USA) (Krämer et al., 2007).

## Analytical method for the determination of chlorpyrifos

An enzyme sensor with amperometric detection was used to attempt to quantify chlorpyrifos in RM08. However, the test was reported not being sensitive enough to differentiate between chlorpyrifos and chlorfenvinfos, therefore these results were not included in the statistical evaluation of results. A new system was under development for the determination (selective detection and quantification) of the various organophosphorous compounds using recombinant enzymes.

## Analytical methods for the determination of nitrate

One of datasets obtained for $NO_3^-$ in RM13 was obtained with HACH Kit Colo (Nitraver 5) which is based on the reduction of cadmium (Roig et al., 2006a). All the necessary reagents are packaged in the Nitraver 5 reagent powder pillows (high and medium range). In this test cadmium is used to reduce nitrates ($NO_3^-$) to nitrites ($NO_2^-$) as follows:

$$NO_3^- + Cd + 2H^+ \rightarrow NO_2^- + Cd^{+2} + H_2O$$

The nitrite ions then react with sulfanilic acid (in an acidic medium) to form an intermediate diazonium salt:

$$NO_2^- + H_2N-\underset{\text{sulfanilic acid}}{\underline{\bigcirc}}-SO_3H + 2H^+ \longrightarrow HO_3S-\underset{\text{diazonium salt}}{\underline{\bigcirc}}-\overset{+}{N}\equiv N + 2H_2O$$

When coupled with gentisic acid (2,5–dihydroxibenzoic acid), an amber coloured solution is formed. Colour intensity of this compound is directly proportional to the nitrate concentration in the water sample.

For instrument calibration Merck (Anion multielements, Std II) CERTIPUR and the calibrants produced and distributed within the 1st SWIFT-WFD PT campaign were used.

Another dataset for $NO_3^-$ in RM13 was obtained with SECOMAM Pastel UV (a laboratory based portable UV spectrophotometer for the fast check of wastewater and natural water quality, for six analytes simultaneously) plus deconvolution (Roig *et al.*, 2006a).

The third dataset was obtained also by a laboratory based portable spectrophotometer, MERCK spectroquant kit, which functions according to the principles of colorimetry. For the determination of $NO_3^-$ the reaction is based on the formation of a red complex of nitrate with a derived benzoic acid in a sulphuric acid media and the absorbance of the samples is measured at 517 nm (Roig *et al.*, 2006b).

At last, the forth dataset for nitrate was obtained by YSI Multiparameter robe 6900 supplied with a specific electrode for nitrate measurements based on the principles of potentiometry (Roig *et al.*, 2006b).

### *Qualitative analysis for organophosphate and/or carbamate pesticides*

A set of qualitative results gained by a prediagnostic (qualitative) biosensor kit (OP-Prot sensor) developed for organophosphate and/or carbamate pesticide detection also arrived for RM08 and RM10. In case of the presence of pesticide mixtures, e.g. as it is the case in most natural samples, the kit detects unselectively the traces of all the present organophosphate and carbamate pesticides. The LOD of the applied kit is $0.1–10 \,\mu g \, L^{-1}$ depending on the pesticide.

### *Qualitative analysis for pesticides and PAHs – ecotoxicity tests*

The qualitative results obtained for RM14 and RM15, reported in Table 5.2.4, were generated by ecotoxicity tests (Rodriguez-Mozaz *et al.*, 2006; Martins *et al.*, 2007). With Microtox® test the percentage of inhibition of luminescense is the effect observed in the presence of toxic pollutants. According to the Microtox test protocol sodium chloride solution is added to the tested aliquot which effectively dilutes the samples in a 1:1 ratio. The two invertebrate tests, Daphtoxkit (*Daphnia magna* immobilization bioassay) and Thamnotoxkit (*Thamnocephalus platyurus* bioassay performed using), are based on the percentage mortality of the corresponding species. Both are performed

*Data Evaluation*                                                                                         359

**Table 5.2.4** Results obtained by ecotoxicity tests

| Sample | Microtox® | | *Daphnia magna* | | *Thamnocephalus platyurus* |
|---|---|---|---|---|---|
| | 5 min % Inhibition | 15 min % Inhibition | 24 h % Inhibition | 48 h % Inhibition | 24 h % Inhibition |
| RM14 | <5%* | 9.8% | <10%* | 10% | <10%* |
| RM15 | <5%* | <5%* | <10%* | 85% | 100% |

*Less than (<) value indicates limit of detection (LOD) for the particular test

on undiluted samples with pH adjusted to 6–8 if needed. For Daphnia test 24 h and 48 h durations were employed; while for *Thamnocephalus* only a single duration, 24 hours, was employed.

## 5.2.3 DATA EVALUATION

Data evaluation was carried out by the Tool4PT © Cortez & MERMAYDE, 2002–2004 software. The normality of average data was checked by the Kolmogorov-Smirnov test. The statistical outliers were identified by the application of the Hampel test (test of averages) (Davies, 1988; Linsinger et al., 1998) and Cochran test (test of variances) at 95% of significance level.

The evaluation of the analytical performances was performed by calculation of the Z-scores values characterizing the individual laboratories (ISO 13528:2005); these were calculated by using the consensus values as reference value ($X_{REF}$) and deviation unit (S), according to the following formula:

$$Z = (X_{LAB} - X_{REF})/S$$

where  $X_{LAB}$ = average result of the laboratory
$X_{REF}$ = reference value
S = deviation unit

The deviation units (S) were set as standard deviation of the selected data in the 1$^{st}$ SWIFT-WFD PT campaign; while in the 2$^{nd}$ and 3$^{rd}$ SWIFT-WFD PT campaigns the deviation units were based on the fit-for-purpose criteria. The Z-score classifies laboratory performance as follows: results obtaining $|Z| \leq 2$ are to be considered satisfactory, results with $2 < |Z| \leq 3$ are defined as doubtful and results obtaining $|Z| > 3$ are considered unsatisfactory.

### 5.2.3.1 Discussion of the Results Obtained by SMETs for the SWIFT-WFD PT Schemes

A summary of the SMETs performance evaluation is reported in the last column of Table 5.2.3. As evidenced in the table, the laboratories using SMETs did provide results only for some among the target analytes of the SWIFT-WFD exercises. In general the

**Table 5.2.5**  Results obtained for trace elements by SWASV

| | RM05 (low fortified levels) | | | | RM12 (natural levels) | | | |
|---|---|---|---|---|---|---|---|---|
| Analyte | SWASV ($\mu g\ L^{-1}$) | PT reference value ($\mu g\ L^{-1}$) | PT unit of deviation | Z-score | SWASV ($\mu g\ L^{-1}$) | PT reference value ($\mu g\ L^{-1}$) | PT unit of deviation | Z-score |
| Cd | $4.2 \pm 0.6$ | 5.7 | 10% Xref | −2.6 | <1 | $(0.07 \pm 0.05)$* | – | n.d. |
| Cu | $80.3 \pm 6.0$ | 78.8 | 10% Xref | 0.2 | <2 | 3.37 | 10% Xref | n.d. |
| Pb | $9.3 \pm 1.3$ | 10.6 | 10% Xref | −1.2 | <0.5 | $(0.68 \pm 0.25)$* | – | n.d. |

*Indicative consensus values of data population

number of satisfactory results was prevailing on that of the unsatisfactory results, but the low number of datasets does not allow making general comments.

## Results obtained for trace elements

In case of trace elements results were provided only for Cd, Cu and Pb (Table 5.2.5). In RM05 exercise (spring water) Z-score was calculated considering a fit-for-purpose unit of deviation corresponding to 10% of Xref. Satisfactory results were obtained for Cu and Pb and doubtful results for Cd; for this last analyte the result was identified as a 'Cochran-outlier' (variance) amongst the results submitted for RM05, but not as 'Hampel-outlier' (mean).

While having slightly higher standard deviation than the state-of-the-art, the SWASV technique proved to be suitable for the determination of the three trace elements in the test sample which contained the target analytes at low fortified concentration level.

However, the same elements, Cd, Cu and Pb, could not be quantified in river water RM12, which contained these analytes at natural concentration levels. The target concentration levels were lower than the LOQ of the method. Due to the large spread of results and the corresponding high CV% values, Z-score was not calculated for Cd and Pb in the corresponding exercise. The average of data population and corresponding standard deviation were calculated and presented in table 5.2.5. According to presented values, the '<LOD' results obtained by SWASW are in line with the average result of PT schemes for Cd and Pb, whereas for Cu the reported <LOD value is apparently a false negative.

## Results obtained for nitrate

The first two SMETs datasets evaluated for nitrate analysis in RM13 were obtained by the same laboratory with two different methods: 1, HACH Kit Colo (Nitraver 5) (laboratory code 18a) and 2, Pastel UV SECOMAM (Spectrophotometry UV + deconvolution) (laboratory code 18b). The two results are encircled on Figure 5.2.1.

Both results were part of the 'selected data' population, thus included in the calculation of consensus value. The performance of neither methods, on the other hand, was graded satisfactory with the statistical evaluation; and received Z-score + 4.2 and + 4.7, respectively (5% – preset deviation unit, fit-for-purpose criteria). As seen furthermore, the result gained by the HACH kit is characterized by a visibly very

*Data Evaluation*   361

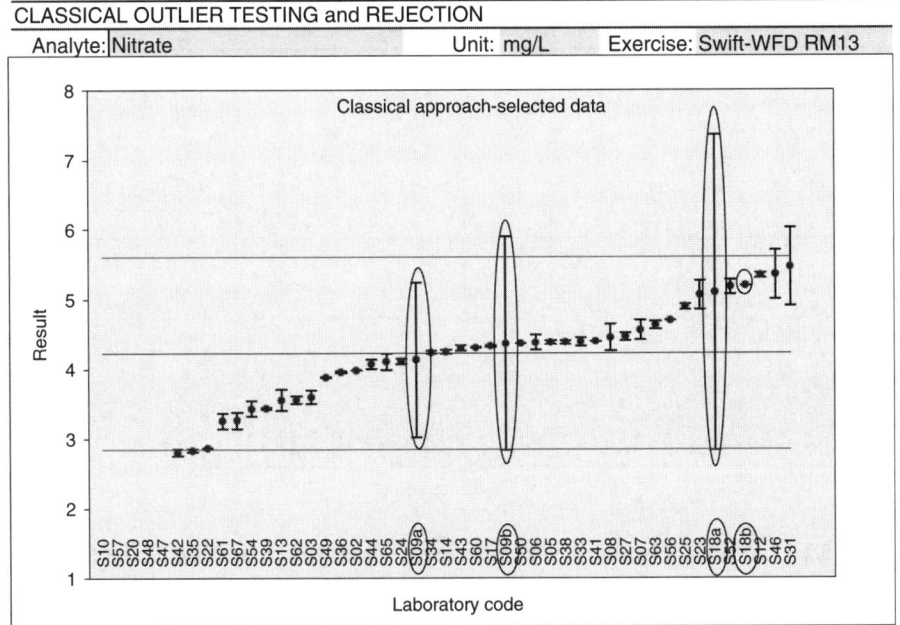

**Figure 5.2.1** Results obtained for nitrate in the 3$^{rd}$ SWIFT-WFD PT Schemes (SMET results encircled)

high standard deviation, significantly higher than the rest of the population. On the other hand, the standard deviation associated to the result obtained by Pastel UV SEC-OMAM is comparable with the standard deviation of the rest of the values in the population. On this basis, Pastel UV seems to be a promising alternative method for nitrate determination.

In addition to its participation in the SWIFT-WFD PT schemes, the performance of 'Pastel UV SECOMAM' was further evaluated under laboratory conditions, as well as within the frame of the river Tevere field trials organized within the project. All results are detailed in Chapter 5.3 of this book. Before the field trial the performance of Pastel UV was tested in field conditions (under various environmental parameters) with test sample RM13 used as quality control material. In addition, in order to keep the on site performance Pastel UV Secomam under control, test sample RM13 was used as a control sample within the field trials: one RM13 replicate was measured in parallel with each field measurement. The results obtained within the field trial are presented in Figure 5.2.2 with the name of the place where the field trial took place. The results obtained in the frame of the field trial are in the upper range of the population if treated together with the PT results, and according to the statistical evaluation should have received an unsatisfactory Z-score.

The other two SMETs datasets evaluated for nitrate analysis in RM13 were obtained by one other laboratory, again with two different methods: 1, MERCK Specktroquant Kit (Spectrophotometry – colorimetry) (laboratory code 09a) and 2, YSI Multiparameter Probe 6900 nitrate specific electrode (laboratory code 09b). The two results are presented encircled on Figure 5.2.1.

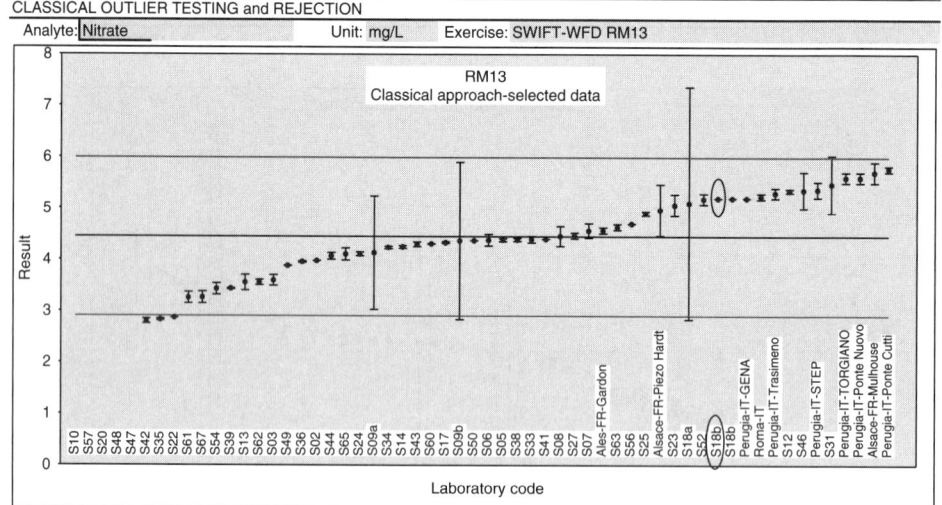

**Figure 5.2.2** Results obtained on field for nitrate in RM13 in comparison to the PT results

These results were also included in the calculation of consensus value. The performance of both methods was graded satisfactory with the statistical evaluation; received Z-score − 0.4 and + 0.7, respectively. However, both results are characterized by a significant standard deviation, close to that associated with the HACH kit result.

The overall evaluation of the methods suggests that Pastel UV is indeed characterized by a good repeatability and reproducibility, but its results might lead to an overestimation of the actual nitrate concentration even if comparable with other methods. At the same time, even though the Spectroquant kit and the nitrate specific electrode provided results with optimal Z-scores in this population, the standard deviation of the methods was much higher than that of the rest of the population, therefore staying behind in the state-of-the-art expectations.

## Results obtained for pesticides

Within the 2nd and 3rd SWIFT PT campaign, results obtained with SMETs for pesticides arrived from two laboratories.

The result for chlorpyrifos in RM08 was gained by an enzyme sensor with amperometric detection. However, since the sensitivity of the sensor is not sufficient to differentiate between chlorpyrifos and chlorfenvinfos, and both were present in the sample, the outlying position of the result (indicated on Figure 5.2.3 – laboratory code 69) in the chlorpyrifos population is understandable. This method is currently under improvement with recombinant enzymes to gain specificity for various organophosphorus pesticides.

The rest of the quantitative results received for pesticides within the 2nd and 3rd PT campaign by SMETs were gained by ELISAs by one participant. In the 2nd SWIFT-WFD PT campaign the participant used the enzyme-tracer format of the

# Data Evaluation

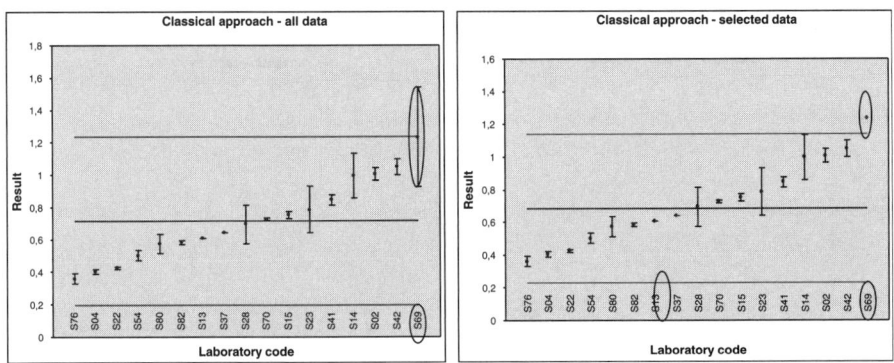

**Figure 5.2.3** Results obtained for Chlorpyrifos in RM08 – 'All data' and 'Selected data', respectively (SMET result encircled)

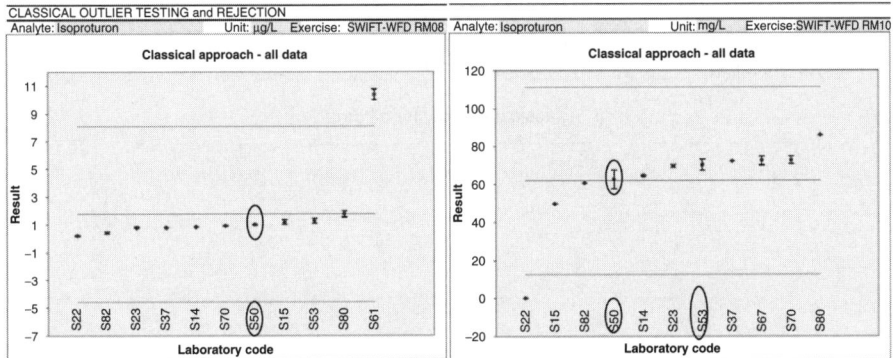

**Figure 5.2.4** Results obtained for Isoproturon in RM8 and RM10, respectively – 'All data' (*ELISA results encircled*) (from Krämer *et al.* (2007) with kind permission from Springer Science and Business Media)

ELISA, whereas in the 3rd PT campaign the coating antigen format was used in order to test both formats for their proficiency.

The results gained for isoproturon and diuron in test sample RM08 (spring water) and RM10 (blind solution) with ELISA in enzyme tracer format lay within the results obtained by all the methods (Figures 5.2.4 and 5.2.5, respectively; laboratory code 50), with calculated Z-scores for isoproturon and diuron in RM08: + 0.5 and -1.2, respectively; in RM10 -0.4 and +0.3, respectively; thus both lying in the category of satisfactory results ($|Z| \leq 2$).

At the same time the performance of the immunoassay for atrazine in RM08 demands more critical attention. The obtained Z-score was 23, therefore as it falls in the $|Z|>3$ category, means unsatisfactory result (Figure 5.2.6). It was about 6 times higher ($0.98 \pm 0.06\,\mu g\,L^{-1}$) than the consensus value ($0.17 \pm 0.02\,\mu g\,L^{-1}$) calculated from the results of 11 participants. At first glance the difference seems to be the typical and thus often described matrix effect in immunoassays. However, a more careful data analysis suggests that it might be due to the presence of other triazines in the sample.

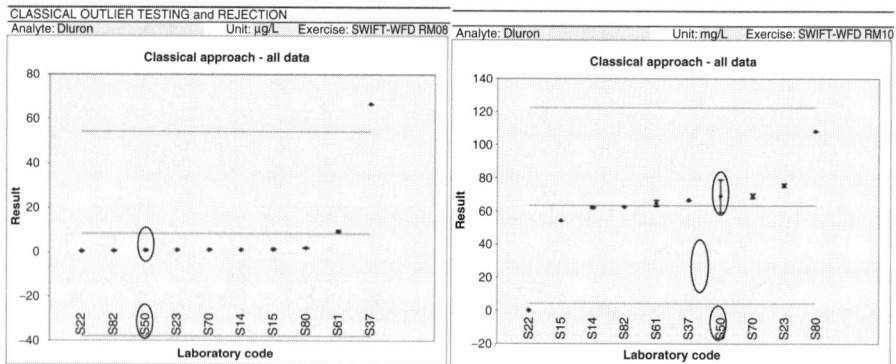

**Figure 5.2.5** Results obtained for Diuron in RM08 and RM10, respectively – 'All data' (ELISA results encircled) (from Krämer *et al.* (2007) with kind permission from Springer Science and Business Media)

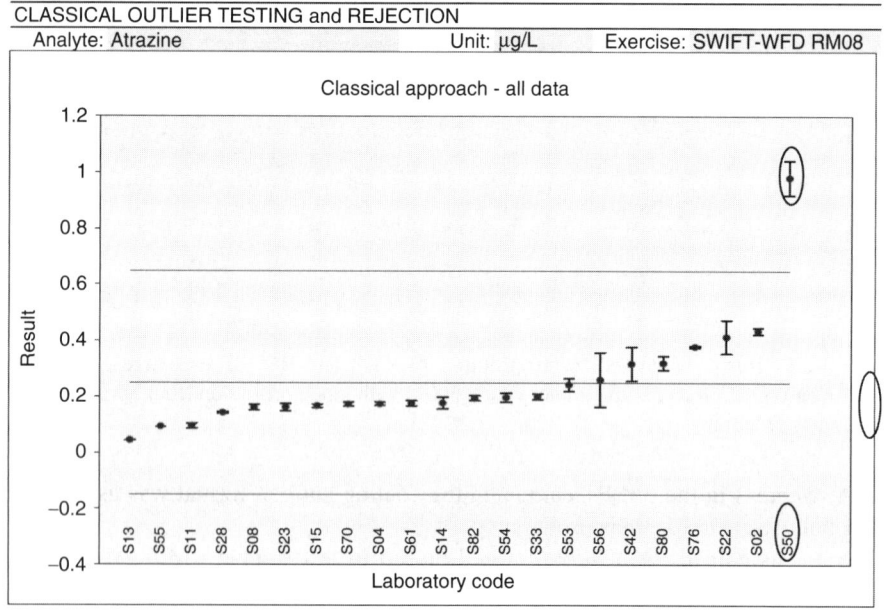

**Figure 5.2.6** Results obtained for atrazine in RM08 – 'All data' (ELISA result encircled) (from Krämer *et al.* (2007) with kind permission from Springer Science and Business Media)

The sum of some other triazines that show cross reactivity (CR) in this form of ELISA (mAb – monoclonal antibodies, AM5D1), such as simazine, desethylatrazine and terbutylazine determined by the reference methods in RM08 adds up to a total of 2.3 µg L$^{-1}$. Taking the CR of the monoclonal antibodies in the ELISA into account, however, the immunoassay should have given the value 0.48 µg L$^{-1}$ and not 0.98 µg L$^{-1}$ referring to about 20% of the sum of triazines present. It has to be noted that even though this mAb also shows a considerable CR for propazine (12%), in this

Data Evaluation 365

value propazine is not considered as it was not a target analyte of the exercise and therefore was not analysed by the reference methods. However, the value gained by ELISA (0.98 µg L$^{-1}$) is only twice the hypothetic reference value (0.48 µg L$^{-1}$) and not six times. This supposes a signal suppression which might be due to a typical matrix effect eventually, either the presence of humic acids in the test sample or in the unidentified matrix components of the spiking pills used for the fortification of the reference material (unusual effect, connected stricktly to the use of the spiking pills), or to the presence of the nonconsidered triazine, propazine, or additional, yet unknown cross reactant(s) of this assay (occurs only if the test sample is a natural sample).

The evaluation of the results obtained for the corresponding blind solution (RM10) confirms the same degree of the above mentioned recognition by the immunoassay. The sum of triazines (atrazine, simazine, desethylatrazine and terbutylazine on the basis of the accepted consensus value) in the sample is 198.7 mg L$^{-1}$. With the immunoassay 53.1 ± 6.7 mg L$^{-1}$ was obtained which refers to 26.7% of the summarized reference values. The percentage of the recognized triazines is not significantly different from the percentage recognized in RM08, especially if the dilution factor applied on the blind solution (1:25000) for the analysis is taken into consideration (dilution factor of this scale highly increases standard deviations). This shows that the Z-score of +0.3 gained for RM10 is only at a first look a satisfactory result: the amount of atrazine present in RM10 was by chance about 25% of the total triazines present in RM10, 50.2 ± 10.05 mg L$^{-1}$.

In the 3rd PT campaign, for isoproturon and diuron determination in both RM15 and RM17 the ELISAs were used in the coating antigen format which is usually less sensitive compared to the enzyme-tracer format.

For the determination of isoproturon in RM15 the value 105 ± 4 mg L$^{-1}$ was gained versus the consensus value of 125 ± 19 µg L$^{-1}$ and received a Z-score of −0.6, ($|Z|<2$, satisfactory). In RM17, the corresponding blind solution as well, the received Z-score was −1.5. However, for diuron the determined concentration was about 6 times higher, 946 ± 25 µg L$^{-1}$ than the consensus value 152 ± 11 µg L$^{-1}$. This indicates a clear matrix effect of the sample due to the presence of e.g.: humic acids or other yet unidentified compounds, especially in the view of the result obtained for RM17, the corresponding blind solution: the value gained was 59.1 ± 7.4 µg L$^{-1}$ versus the consensus value of 65.4 ± 4.3 µg L$^{-1}$, which means a Z-score −0.5, ($|Z|<2$, satisfactory). Figure 5.2.7 presents the diuron results in RM15 and RM17, respectively, with the ELISA results indicated.

Based on the results gained by the ELISA tests within the SWIFT-WFD PT schemes, these immunoassays are potentially very useful tools for the successful implementation of the Water Framework Directive. They are rapid in delivering results and have higher sample throughput than classical methods within the same time. In addition, in comparison to the classical methods used in the SWIFT-WFD PT schemes, the immunoassay-generated results were obtained in the water samples directly. The sensitivity of the method has also to be mentioned, the blind solutions had to be diluted generally 1:10000–50000, because the concentration range given was too high for immunochemical detection. They do not need sample preconcentration, nor sample pretreatment, are easy to use, therefore very cost-effective and environment-friendly.

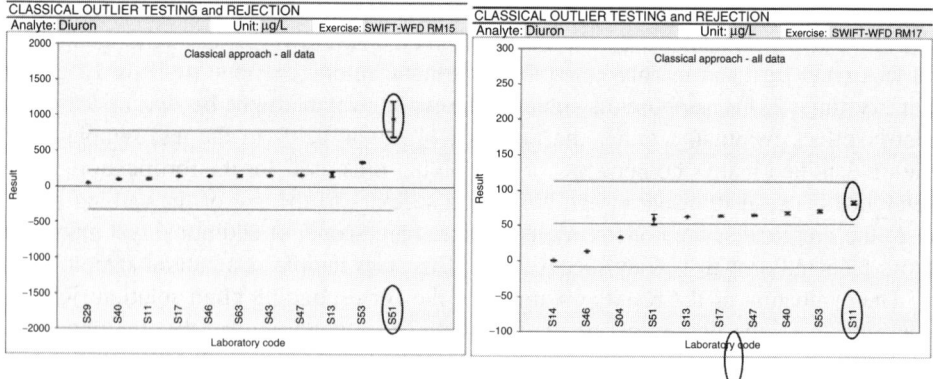

**Figure 5.2.7** Result obtained for Diuron in RM15 and RM17, respectively – 'All data' (ELISA results encircled)

Although the SWIFT-WFD PT schemes were not specifically designed for immunoassays, they clearly showed that immunoassays reach easily the required sensitivity for the WFD and even for the EU Drinking Directive. However, care is required for the recognition of matrix effects of certain water matrices. Although immunoassays are usually not group-selective, they can to a certain extent detect different targets of similar structure (e.g. triazines). These effect call for thorough evaluation before the methods are recommended for routine/monitoring purposes. The systematic comparison to conventional analytical methods is recommended, revealing the response factor of analytes by the immunoassay.

## Results of qualitative tests

In case of the qualitative test applied for the qualitative analysis of organophosphate and/or carbamate pesticides, the inhibition of the biosensor was observed both in RM08 and RM10, thus detected the presence of the organophosphate and/or carbamate pesticides. The applied test is simple, easy-to-use and rapid, thus suitable for field measurement. However, provides information only about the presence of the representatives of the organophosphate and carbamate groups and does not allow the identification of the compounds that are present.

The results gained with the ecotoxicity tests are presented in Table 5.2.4. With its standard application times, Microtox test showed positive results (inhibition 9.8%) only for PAHs, (RM14) after 15 min, which means that this tests is not sensitive to the presence of pesticides at the tests concentration ranges. The Daphnia test detected the presence of both target compound groups, however, only with 48 h of tests time. It proved more sensitive to present concentration of pesticides than PAHs. The Thamnocephalus test was the most sensitive out of the applied ones for pesticides, but did not react at all to the PAH concentration of RM14.

Ecotoxicity tests react to the presence of a wide range of toxic substances, therefore suitable for the overall evaluation of the ecological status of waters. However, not sensitive enough in comparison with laboratory techniques, and just as the above mentioned biosensor, not applicable for the identification of the toxic compounds and compound groups present. These tests should be applied for their dedicated purpose, for the evaluation of the overall ecological status of waters, but cannot at the moment be considered as alternatives to chemical laboratory methods.

## 5.2.4 CONCLUSIONS

One of the short-term aims of the quality assurance/quality control (QA/QC) activities of the SWIFT-WFD project was the evaluation of the state-of-the-art of the analytical performances of SMETs in use in Europe, whereas one of the long-term aims was the establishment of a framework of ongoing PT schemes activities supporting the European laboratories involved in the implementation of WFD applying either classical methods or SMETs.

In general, the establishment of sustainable PT schemes requires the careful evaluation of certain mandatory conditions. First of all, an adequate number of laboratories have to be sufficiently motivated to participate in the PT schemes to be organized (a). Then the applied methods must be under analytical control (b), otherwise performance in the PT schemes is not informative. Adequate reference materials must be available or easily producible upon request for the PT schemes (c), and finally, the PT schemes providers must be capable of organizing and managing the PT schemes properly (d). When considering commercial PT schemes providers this last condition is satisfied only if reasonable income or at least coverage of costs is achieved.

According to the experience gained from SWIFT-WFD PT schemes for SMETs, the following conclusions can be drawn on the four above mentioned issues.

The participation of SMETs in the SWIFT-WFD PT schemes was very low in comparison with the participation of classical methods, even though the participation of classical methods was limited by the invitation of a restricted number of selected laboratories per country. On the other hand, all the laboratories known to the SWIFT-WFD PT schemes organizers as SMETs users were invited to participate.

Evidently and as expected, the diffusion of SMETs is indeed still very low compared to the diffusion of classical methods. Therefore the primary need is to put strong efforts in the diffusion of SMETs, especially in the analytical laboratories involved in WFD related activities routinely. Furthermore, the willingness of SMETs applying laboratories to participate in PT schemes also has to be highlighted. The percentage of laboratories applying classical methods accepting to participate in the SWIFT-WFD PT schemes was around 75% with respect to the total of invited laboratories, while this percentage is down to about 15% in case of the invited SMETs laboratories. Together with analytical laboratories, SMETs manufacturers were also given the possibility to participate thereby offering a powerful tool for advertising the analytical performances of their produced SMETs; however, there was no positive response from their side. This testifies the scarce attitude in the 'SMETs world' to the application of quality control tools, expressly external quality control tools. In this view, the state-of-the-art of the 'SMETs world' now is comparable to that of 'classical methods world' about 20–25 years ago, when laboratories scarcely participated in PT schemes. However, in the past decades relevant results were obtained on large scale following the considerable efforts at a European level for the improvement of the QC/QA culture in the laboratories and the encouragement of their participation in PT schemes. Following this positive experience, a similar attitude is deemed necessary towards the 'SMETs world'. As seen, these efforts should be pursued without any further delay, possibly exploiting the SWIFT-WFD experience.

Finally, the scarce motivation of laboratories in the validation and external performance evaluation of SMETs also has to be mentioned. This may be due to the fact that these methods are not yet extensively applied in monitoring and that the Directives usually do not clearly suggest their application in monitoring. SMETs are applied in research laboratories, but not in routine laboratories. If the Directives included the possibility of the application of these methods in monitoring, it would certainly contribute to a rapid and diffuse spread of SMETs outside the academic world.

As stated above, the analytical methods must be under control before their participation in external quality control is rendered useful.

The SWIFT-WFD experience presents cases where SMETs applied in the PT schemes can be considered under control. ELISA and to a certain even Pastel UV are examples. These methods could from now be considered ready to participate in existing and not specifically designed PT schemes. While the limit of detection of SWASV appears to be the main constraint for application to natural matrices, but for the rest this method may also be considered under control.

However, except for these SMETs, validation and being under control seem to be far from being achieved. The main constraints are nonsatisfactory (too high) limit of detection, lack of specificity/selectivity, being prone to matrix effects and the scarce performance in terms of repeatability and reproducibility. Efforts must be put in reducing these constraints both in the manufacturing and in the method development phase.

Other methods cannot be evaluated by the existing PT schemes. In some cases the results cannot be compared for being expressed in different units than required by the PT schemes, in other cases they provide only qualitative or semiquantitative results or would require much larger sample volumes for analysis than what is those usually considered in classical PT schemes. For these methods, the design of specific PT schemes is inevitable if feasible.

Some SMETs provide only qualitative or semiquantitative results (yes/no, toxicity index, group of components or 'a family of chemicals', i.e. organotins, triazines, etc.). Some other SMETs need large volumes of test samples, even up to hundred liters (e.g. passive samplers). For these methods no RM is available. The up-to-date available RMs production technologies would allow the preparation of specific RMs in certain cases. This is the case for SMETs providing results for 'families of chemicals'. The production of RMs for qualitative responses, toxicity index, etc. still presents more problems. However, considerable motivation to formulate specific designs and adjust the available technologies would render it possible. For the third group of methods, e.g. passive samplers, the available RMs production technologies are not sufficient. The size of the required samples, influences not only the required logistics (batch reactors, homogenizers, containers, etc) but also the homogeneity and stability of the materials themselves. In this case, efforts to modify the technologies would be at present the only solution to overcome the difficulties and individuate the possible alternatives.

As already mentioned some SMETs are 'ready' to participate in classical PT schemes from the organizers point of view. To convince laboratories and pursue their participation is another matter.

For some SMETs there are no suitable PT schemes available at the moment. The low diffusion of SMETs, their scarce motivation and thus attitude to apply proper quality control measures results in a very low participation in PT schemes. Therefore commercial PT schemes providers are not motivated to offer PT schemes designed specifically for SMETs, being the risk of failure with consequent economic losses far to high to be taken.

In order to come out of this vicious circle, two possibilities can be proposed.

On the one hand, as already experienced in case of the analytical classical methods in the past years, also in case of SMETs at European level there should be substantial efforts in providing resources to support their diffusion in routine laboratories, to disseminate the correct QA/QC culture for these methods and contemporarily to support RM producers to shape-cut technologies to satisfy the test sample needs of SMET.

On the other hand, the above mentioned problems related to the organization of specifically designed PT schemes for SMETs might be overcome through the creation of a network of PT schemes providers, which would be able to enlarge the basin of potential SMETs participants in PT schemes, thus decreasing the risk of failure.

## GLOSSARY

| | |
|---|---|
| RMs | Reference Materials |
| PT | Proficiency Testing |
| SWIFT-WFD | Screening methods for Water data InFormaTion in support of the implementation of the Water Framework Directive |
| WFD | Water Framework Directive |
| SMETS | Screening Methods/Emerging Tools |
| ELISA | Enzyme-Linked ImmunoSorbent Assays |
| QC/QA | Quality Control/Quality Assurance |
| SWASV | Square Wave Anodic Stripping Voltametry |

## REFERENCES

Brunori C., Ipolyi I., Pellegrino C., Ricci M., Bercaru O., Ulberth F., Sahuquillo A., Rosenberg E., Madrid Y. and Morabito R., 2008. *Trends Anal. Chem.*, accepted for publication.

Brunori C., Ipolyi I., Pellegrino C., Ricci M., Bercaru O., Ulberth F., Sahuquillo A., Rosenberg E., Madrid Y. and Morabito R., The SWIFT-WFD Proficiency Testing Campaigns in Support of the EU Water Framework Directive Implementation Organized within the Frame of the SWIFT-WFD Project, 2007, *Trends in Analytical Chemistry*, **26**(10) 993–1004.

Davies P. L., 1988. *Fres. J. Anal. Chem.*, **331** 513–19.

Dworak T. and Lügcke H., 2006. Project SWIFT-WFD Policy Brief 10 – 2006, available at http://www.swift-wfd.com

Greenwood R. and Roig B., 2006. Project SWIFT-WFD Deliverable 5, Operational manual – A toolbox of existing and emerging methods for water monitoring under the WFD. Public report available at http://www.swift-wfd.com

Ipolyi I., Brunori C. and Morabito R., 2006a. Project SWIFT-WFD Deliverable 17, First report on intercomparisons on reference materials. Public report available at http://www.swift-wfd.com

Ipolyi I., Brunori C. and Morabito R., 2006b. Project SWIFT-WFD Deliverable 18, Second report on intercomparisons on reference materials. Public report available at http://www.swift-wfd.com

Ipolyi I., Pellegrino C., Brunori C. and Morabito R., 2006c. Project SWIFT-WFD Deliverable 19, Third report on intercomparisons on reference materials. Public report available at http://www.swift-wfd.com

ISO 5725-2, Accuracy (trueness and precision) of measurement methods and results – Part 2: Basic method for the determination of repeatability and reproducibility of a standard measurement method, http://www.iso.org.

ISO 13528:2005, Statistical methods for use in proficiency testing by interlaboratory comparisons, http://www.iso.org.

Krämer P. M., Kremmer E., Forster S. and Goodrow M. H., 2004. *J. Agric. Food Chem.*, **52**, 6394–6401.

Krämer P. M., Martens D., Forster S., Ipolyi I., Brunori C. and Morabito R., 2007. *Anal. Bioanal. Chem.*, **387**, 1435–48.

Linsinger T. P. J., Kandler W., Krska R. and Grasserbauer R., 1998. *Accr. and Qual Ass.*, **3**, 322–7.

Martins J., Oliva Teles L. and Vasconcelos V., 2007. *Environ. Intern.*, **33**, 414–25.

Rodriguez-Mozaz S., Lopez de Alda M. J. and Barceló D., 2006. *Anal. Bioanal. Chem.*, **386**, 1025–41.

Roig B., Gonzalez C., Berho C., *et al.*, 2006a. Project SWIFT-WFD Deliverable 44, Report of laboratory and field validations of screening tools in terms of performance criteria; Recommendations for their further applications in water monitoring. Public report available at http://www.swift-wfd.com

Roig B., Mills G., Greenwood R., A *et al.*, 2006b. Project SWIFT-WFD Deliverable 43, Report of performances and valuation of screening methods (field trials results). Available at http://www.swift-wfd.com

Strosser P. and Graveline N., 2006. Project SWIFT-WFD Brief – Assessing the impact of enhanced information on decision making, available at http://www.swift-wfd.com

SWIFT-WFD project. Screening methods for water data information in support of the implementation of the Water Framework Directive. 6th Framework Programme, contract SSPI-CT-2003-502492 (2003–2006).

# 5.3
# Traceability and Interlaboratory Studies on Yeast-based Assays for the Determination of Estrogenicity

Rikke Brix and Damià Barceló

5.3.1 Introduction
5.3.2 Types of Estrogen Assays
5.3.3 Relative Estrogenic Potency Determined by Yeast-based Assays
5.3.4 Relative Estrogenic Potentials Determined by Non-yeast-based Assays
5.3.5 Inter-laboratory Exercises
5.3.6 Yeast-based Assays May Be Applied with Different Objectives
5.3.7 Correlation with Chemical Results
5.3.8 Limits of Detection
5.3.9 Initiatives for the Future
5.3.10 Conclusions
Acknowledgements
References

## 5.3.1 INTRODUCTION

Since the 1962 publication of the book *Silent Spring* (Carson, 1962), describing health problems in the environment and linking them to the environmental use of chemicals such as pesticides, there have been a growing focus on the long-term effects of the chemicals by which we are surrounded. One outcome of this concern has been the discovery of the estrogen mimicking behaviour of a long list of synthetic chemicals produced for a variety of applications.

Currently there are several definitions on what estrogenicity is and what makes a compound estrogenic, here a few are presented; 'Estrogenicity is a physiologic response to a compound that induce estrus in vivo' (bringing an animal into heat) (Andersen *et al.*, 1999). Hertz proposed this definition: 'estrogens are substances, which elicit proliferative activity of the organs of the female genital tract' (Hertz, 1985). A third definition states that 'estrogens are substances that, by binding to their receptors, elicit the expression of genes that are controlled by estrogen-responsive elements' (Soto *et al.*, 2006a). Finally, the biological and biochemical aspects of estrogen action are combined in a fourth definition stating that 'estrogens are substances that elicit the proliferative activity and the control of expression of specific genes in tissues of the female genital tract' (Soto *et al.*, 2006a).

As further investigation into the field showed that negative reproductive effects weren't only produced by estrogens, but also by anti-estrogens, androgens, anti-androgens etc., the need arose for a more comprehensive description and the term 'endocrine disruptor' was taken into use at the Wingspread Conference (held in Racine, WI, in 1991) to cover multiple chemicals which, through different modes of action, disrupted the endocrine system (Schomberg *et al.*, 1999; Wang *et al.*, 2003; Soto *et al.*, 2006b).

It has been suggested that endocrine disruptors may play an important role in the observed decrease in human semen quality (Carlsen *et al.*, 1992; Giwercman *et al.*, 1993). Endocrine disrupters further influence the growth, differentiation and function of many target organs, such as the mammary glad, uterus, vagina, ovary, testis, epididymis and prostate. They also play an important role in bone maintenance, the central nervous system and in the cardiovascular system (Schomberg *et al.*, 1992; Couse *et al.*, 1993; Wang *et al.*, 2003). Sex-changes in fish (Purdon *et al.*, 1994) and higher animals such as alligators (Arthur, 1995) have been observed in several places as a consequence of endocrine disruptors.

The aim of this chapter is to give a brief overview of yeast-based assays; the comparability of the results and the different uses and future perspective.

## 5.3.2 TYPES OF ESTROGEN ASSAYS

Most of the *In vitro* assays for the determination of endocrine disrupting effect fall into one of four categories:

1. Cell proliferation assays. The most commonly known is the E-Screen (first described by Soto *et al.*, 1992), which is based on the proliferation of the cells of the MCF-7 human breast cancer cell line (Korner *et al.*, 1999).

2. Estrogen Receptor (ER) competitive binding assays, which measures the binding affinity of a chemical for the estrogen receptor (Fang *et al.*, 2000).

3. ER binding assays, an example is the ER-CALUX (first described by Legler *et al.*, 1999) the assay uses T47D human breast cancer cells stably transfected with an ER-mediated luciferase gene construct. The luciferase reporter gene activity can be easily quantified following short-term exposure to chemicals activating endogenous estrogen receptors.

4. Reporter gene assays that measure ER binding-dependant transcriptional and translational activity. The yeast-based assays fall within this group. The vast majority of yeast-based assays are based on *S. cerevisiae* containing a stably transfected human estrogen, receptor (hER), although there are several variations in strain and detection method (NICEATM, 2002)

The best known yeast-based assay is the YES assay (first described by Routledge *et al.*, 1996) where a DNA sequence of the human estrogen receptor (hER) is integrated into the yeast genome, which also contains expression plasmid carrying estrogen-responsive elements (ERE) controlling the expression of the reporter gene *lac-Z*, encoding the enzyme β-galactosidase. Thus, in the presence of estrogens β-galactosidase is synthesized and secreted into the medium, where it causes a colour change from yellow to red.

Another much used yeast-based assay is the yEGFP assay (first described by Bovee *et al.*, 2004). In this assay recombinant yeast cells are constructed that express the human estrogen reporter α (ER α) and yeast enhanced green fluorescence protein), the estrogenic potential is then measured by a Fluorometer.

## 5.3.3 RELATIVE ESTROGENIC POTENCY DETERMINED BY YEAST-BASED ASSAYS

The relative estrogenic potency of a compound has been estimated by many different biological assays. Some values found in the literature are shown in Tables 5.3.1 and Table 5.3.2. The values are all given as Estradiol equivalency factors (EEF), defined as $EC_{50}$ of the compound relative to the $EC_{50}$ of 17β-estradiol.

As shown in Tables 5.3.1 and 5.3.2, there is a large variance between the different estimations, because the assays use different estrogen receptors and different reporting elements. This means that the EEFs are not universe determinations related only to the compound, but so far there is a strong correlation to the assay used and the laboratory carrying out the assay.

This clearly underlines the necessity of the establishment of a traceability chain so that all measurements can be related to an internationally recognised standard.

Until this traceability chain has been established, it is strongly recommended that each laboratory estimates relative estrogenic potentials for all compounds in order to achieve a good correlation to chemical measurements. The variation coefficients are discussed further in the section about correlation with chemical analysis.

**Table 5.3.1** Relative Estrogenic Potency Of Selected estrogens

|        | EE*        | Estrone     | DES**      | Estriol      | Genistein     |
|--------|------------|-------------|------------|--------------|---------------|
| EEF(M) | 1.60 [2]   | 0.096 [21]  | 0.64 [22]  | 3.7E-3 [22]  | 4.9E-4 [21]   |
|        | 3.33 [2]   | 0.14 [23]   | 0.267 [2]  | 2.4E-03 [24] | 2.45E-7 [25]  |
|        | 1.19 [24]  | 0.38 [24]   | 0.40 [2]   | 0.37 [23]    | 1.27E-7 [25]  |
|        | 1.2 [26]   | 0.1 [26]    | 1.1 [27]   |              | 2.50E-7 [25]  |
|        | 0.96 [28]  | 0.02 [23]   | 0.05 [23]  |              | 3.13E-4 [29]  |
|        | 0.7 [27]   |             |            |              |               |
|        | 0.18 [23]  |             |            |              |               |
| Mean   | 1.309      | 0.147       | 0.491      | 0.125        | 1.6E-04       |
| Std    | 0.997      | 0.137       | 0.402      | 0.212        | 2.3E-04       |
| %CV    | 76.2       | 93.2        | 81.8       | 169.0        | 142.2         |

*Ethynyl Estradiol (EE), **Diethylstilbestrol (DES)
[2] Andersen *et al*., 1999
[22] Gaido *et al*., 1997
[23] Cespedes *et al*., 2004
[24] Rutishauser *et al*., 2004
[25] Dhooge *et al*., 2006
[26] Murk *et al*., 2002
[27] Folmar *et al*., 2002
[28] Segner *et al*., 2003
[29] De Boever *et al*., 2001

## 5.3.4 RELATIVE ESTROGENIC POTENTIALS DETERMINED BY NON-YEAST-BASED ASSAYS

This section presents relative estrogenic potential determined by E-SCREEN and ER-CALUX, two well recognized *in-vitro* assays for the determination of estrogenicity (see Table 5.3.3).

It is not very scientific to calculate averages of results from two different methods, which also show a large variation between then, but from the point of illustration, it was found acceptable and these averages have been plotted in Figure 5.3.1, along with averages and standard deviations from Tables 5.3.1 and 5.3.2 on yeast-based assays.

The figure shows that because the standard deviations are so large, a large part of the mean values of the non-yeast-based are included in one standard deviation of the results of the yeast-based assays. However, more than half the values are outside the range, showing that there is a low consistency between the results of the non-yeast and yeast-based assays.

## 5.3.5 INTER-LABORATORY EXERCISES

So far only two inter-laboratory exercises have been published by Andersen *et al*. (1999) and Dhooge *et al*. (2006). The first is comparing ten *in-vitro* assays and one *in-vivo* assay performed by ten different laboratories. The inter-laboratory consist of three laboratories performing the E-SCREEN assay, four laboratories performing

*Inter-laboratory Exercises*

**Table 5.3.2** Relative Estrogenic Potency Of Selected Compounds

|  | **Nonylphenol** | **NP1EO** | **Octylphenol** | **Bisphenol A** | **Tamoxifen** |
|---|---|---|---|---|---|
| EEF(M) | 5.0E-5 [21] | 7.4E-6 [30] | 3.0E-5 [21] | 5E-5 [21] | 4.7E-5 [21] |
|  | 8.9E-5 [31] | 7.8E-06 [24] | 1.0E-5 [26] | 1.40E-6 [25] | 6.67E-04 [2] |
|  | 2.05E-4 [22] | 4.0E-6 [26] | 4.9E-4 [28] | 9.3E-7 [25] |  |
|  | 2.33E-4 [30] | 1.3E-5 [23] | 1.6E-4 [23] | 6.36E-7 [25] |  |
|  | 2.5E-4 [24] |  |  | 3.7E-5 [23] |  |
|  | 5.7E-4 [26] |  |  | 6.6E-5 [22] |  |
|  | 7.2E-7 [27] |  |  | 8.00E-5 [2] |  |
|  | 4E-4 [23] |  |  | 4.0E-5 [2] |  |
|  | 2.5E-4 [32] |  |  | 1.1E-4 [24] |  |
|  |  |  |  | 1.0E-5 [26] |  |
|  |  |  |  | 8.1E-5 [28] |  |
|  |  |  |  | 1.0E-4 [21] |  |
|  |  |  |  | 7.9E-4 [29] |  |
| **Mean** | 2.28E-04 | 8.05E-06 | 1.73E-04 | 1.05E-04 | 3.57E-04 |
| Std | 1.89$^E$-04 | 3.71E-06 | 2.22E-04 | 3.34E-05 | 4.38E-04 |
| %CV | 83.2 | 46.1 | 128.6 | 31.7 | 122.8 |

[2] Andersen *et al.*, 1999
[21] Coldham *et al.*, 1997
[22] Gaido *et al.*, 1997
[23] Cespedes *et al.*, 2004
[24] Rutishauser *et al.*, 2004
[25] Dhooge *et al.*, 2006
[26] Murk *et al.*, 2002
[27] Folmar *et al.*, 2002
[28] Segner *et al.*, 2003
[29] De Boever *et al.*, 2001
[30] Garcia-Reyero *et al.*, 2004
[31] Metcalfe *et al.*, 2001
[32] Beresford *et al.*, 2000

**Table 5.3.3** Relative Estrogenic Potency Of Selected Compounds, Determined By E-SCREEN and ER-CALUX

| EEF(M) | E-SCREEN | | | ER-CALUX | | | Mean |
|---|---|---|---|---|---|---|---|
| Ref. | Folmar *et al* (2002) | Andersen *et al* (1999) | | Legler *et al* (2002) | Legler *et al* (1999) | Murk *et al* (2002) | |
| NP |  | 7.50E-8 | 7.00E-7 |  | 2,3E-5 |  | 3.88E-7 |
| OP |  | 8.57E-8 | 1.40E-6 |  | 1.4E-6 |  | 7.43E-7 |
| NP1EO |  |  |  |  |  | 3.8E-6 | 3,8E-6 |
| BisA |  | 3.00E-6 | 1.00E-5 | 7,8E-6 |  |  | 6.50E-6 |
| Gen |  |  |  | 6.00E-5 |  |  | 6.00E-5 |
| Estriol |  |  |  | 1 |  |  | 1.00 |
| EE | 1.9 | 0.88 | 3.33E-2 | 1 |  | 1.2 | 0.95 |
| DES | 2.5 | 8.75E-2 |  |  |  |  | 1.29 |
| Estrone |  |  |  | 0.22 |  |  | 0.22 |

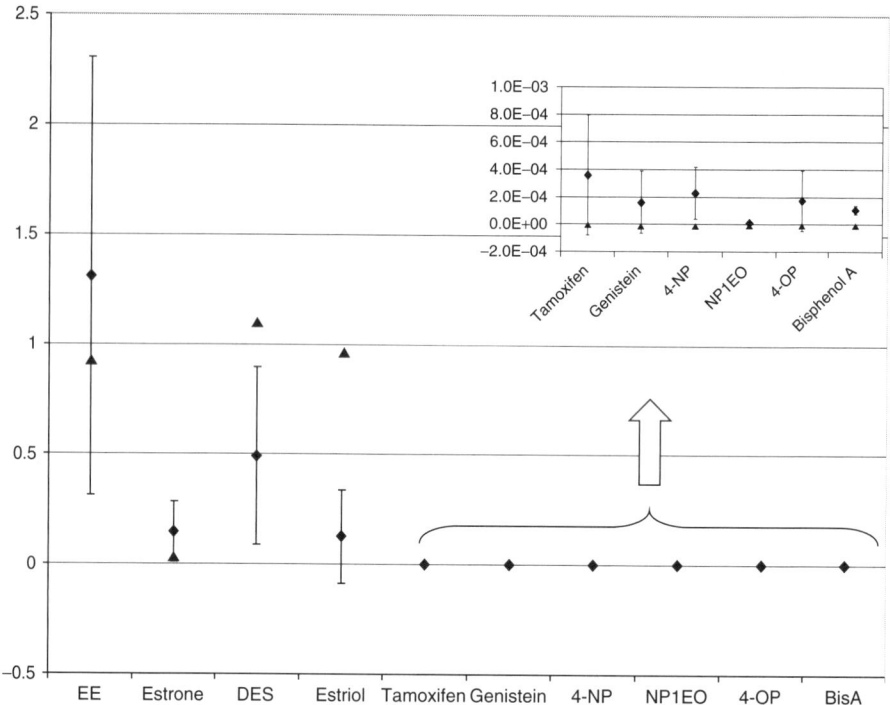

**Figure 5.3.1** Relative estrogenic potential determined by yeast-based assays (diamonds) and non-yeast-based assays (triangles)

a yeast-based assay, one laboratory performing one of the following assays; Direct competitive estrogen *in-vitro* binding assay, recombinant hER, Transcient gene expression assay in MCF-7 cells, *in-vitro* ER binding assay, rabbit uterine tissue and finally, Vitellogenin production in juvenile rainbow trout *in-vivo*.

Unfortunately, the study only reports the results from two of the four yeast-based assays, which are presented in Table 5.3.4.

The paper by Dhooge *et al*. (2006) is a validation/optimization process along with being an inter-laboratory exercise and contains several considerations regarding the optimal conditions for the assay. Regarding the inter-laboratory comparison, it presents the results of three laboratories performing the YES assay (see Table 5.3.5).

**Table 5.3.4** Relative Estrogenic Potentials From Inter-Laboratory By Andersen *et al*. (1999)

| EEF(M) | EE | DES | BisA | Methoxychlor |
|---|---|---|---|---|
| Lab 1 | 1.60 | 0.27 | 8.00E-05 | 2.00E-06 |
| Lab 2 | 3.33 | 0.40 | 4.00E-05 | 6.6667E-06 |
| **mean** | **2.47** | **0.33** | **6.00E-05** | **4.3333E-06** |
| **%CV** | **49.7** | **28.3** | **47.1** | **76.15** |

*Correlation with Chemical Results*

**Table 5.3.5** Relative Estrogenic Potentials From Inter-Laboratory By Dhooge *et al.* (2006)

| EEF(M) | Bis A | Genistein | Methoxychlor |
|---|---|---|---|
| Lab 1 | 1.40E-06 | 2.45E-07 | 8.80E-06 |
| Lab 2 | 9.29E-07 | 1.27E-07 | 5.91E-06 |
| Lab 3 | 6.36E-07 | 2.50E-07 | 3.57E-06 |
| mean | 9.87E-07 | 2.08E-07 | 6.09E-06 |
| %CV | 38.8 | 33.5 | 43.0 |

For a biological assay, both studies are showing a good reproducibility, however, when comparing the results between the two studies a variation coefficient (%CV) of Bisphenol A of 143% is observed, confirming the large variations between different studies that were mentioned previously in this chapter.

## 5.3.6 YEAST-BASED ASSAYS MAY BE APPLIED WITH DIFFERENT OBJECTIVES

These types of assays may be used in several conceptually different approaches to environmental chemistry. One approach is using the *in-vitro* assays as a screening to select compounds with a potential for endocrine disruption for further studies to determine modes of action etc. with a battery of *in-vivo* and *in-vitro* assays.

The other, which is commonly utilized in environmental monitoring, is to screen unknown environmental samples for their total biological (toxicological) activity for a given endpoint and report these findings without knowledge of the presence of individual compounds. In this approach, the idea is to use the assay as a screening method, thus it should basically give yes/no response as to whether there are particular types of contaminants present or not. Previous studies (Pawlowski *et al.*, (2004); Pinto *et al.*, (2005); Schmitt *et al.*, 2005) have indicated that a yeast assay may be a highly reliable methodology for a first level screening to assess surface water quality in terms of estrogenic activity. The samples with a positive response can then be further analysed by chromatographic techniques to identify and quantify the active compounds. The purpose of this type of screening is to perform a fast and cheap elimination of samples having low estrogenic activities and thus allow for a more in-depth analysis of the samples that give a clear response in the bioassay.

## 5.3.7 CORRELATION WITH CHEMICAL RESULTS

Recent studies show that alkylphenols alone can largely account for the estrogenic activity seen in fresh water samples (Rutishauser *et al.*, 2004; Cespedes *et al.*, 2005; Quiros *et al.*, 2005). However, unpublished data from the authors' laboratory show that natural estrogens (in this case Estriol and Estrone) can be found in natural samples at concentrations high enough to account for 90% of the observed estrogenicity. So if yeast-based assays are used as prescreening for chemical analysis, it is important to keep in mind that the chemical analyses should include both alkylphenols and natural estrogens.

Acceptable variation coefficients are determined by the application of the assay, for prescreening purposes with a sufficiently low limit of detection, it is acceptable to have high variation coefficients. But when trying to establish a correlation between chemical data and results from yeast-based assays a variation coefficient on the relative estrogenic potential of more than 100% is quite problematic.

Thus, there is a clear need for reproducible determinations of relative estrogenic potentials from natural samples with no or very little estrogenicity spiked with known concentrations of the various compounds and mixtures of compounds.

## 5.3.8 LIMITS OF DETECTION

Many studies calculates limits of detection, two examples are: LOD Estradiol 0.88 μg/l (Bovee *et al.*, 2005) and LOD Estradiol 0.88 μg/l (Li *et al.*, 2004). In this situation there are two main considerations:

1. These LOD have been determined with standard solutions and may not give a realistic image of LODs of natural samples. Yet unpublished data from the authors' laboratory show that in natural samples the LOD is several orders of magnitude higher.

2. The second issue, when using the yeast-based assays as a prescreening for chemical analysis, is whether these limits of detection are low enough. Studies have shown that 1 ng/l ethynyl estradiol cause induction of vitellogenesis in carp (Purdom *et al.*, 1994), which means that the sample would have to be concentrated at least 1000 times and this often makes the sample toxic because of elevated levels of other compounds, e.g. LAS, which kills the yeast at high concentrations.

## 5.3.9 INITIATIVES FOR THE FUTURE

Both within OECD and US National Health Service work is in progress for a standardization of *in-vivo* and *in-vitro* assays for the determination of endocrine disruption. The OECD endocrine disrupter testing and assessment task force (EDTA) is performing validations of an array of assays and developing standard guideline protocols. At this point only a draft protocol is available for a yeast-based assay (OECD, 2006). More information on EDTA's work can be found on http://www.oecd.org/document/62/0,2340,en_2649_34377_2348606_1_1_1_1,00.htm

The US National Toxicology Program (NTP) Interagency Center for the Evaluation of Alternative Toxicological Methods (NICEATM) has in collaboration with Interagency Coordinating Committee on the Validation of Alternative Methods (ICCVAM) prepared comprehensive background review documents

(BRDs) to assess the validation status of *in-vitro* estrogen receptor (ER) and androgen receptor (AR) binding and transcriptional activation (TA) assays, which were proposed as screens to identify substances with potential hormonal activity. An independent expert panel review concluded that there were no adequately validated *in-vitro* ER- or AR-based test methods. Based on the expert panel's conclusions and recommendations, along with comments from the public, ICCVAM developed test method recommendations that included minimum procedural standards and a list of 78 reference substances that should be used to standardize and validate *in-vitro* ER and AR binding and TA test methods. These recommendations were made publicly available in the report: *ICCVAM Evaluation of the In Vitro Methods for Detecting Potential Endocrine Disruptors: Estrogen Receptor and Androgen Receptor Binding and Transcriptional Activation Assays* (NIH No: 03–4503). More information can be found on http://iccvam.niehs.nih.gov/methods/endocrine/end_TMER.htm

## 5.3.10 CONCLUSIONS

Yeast-based assays show a lot of potential for various applications, but especially as a cost-reducing prescreening tool for chemical analysis in environmental monitoring.

If *in-vitro* assays are to be used for total estrogenic load (as effects have been shown to be additive) and as such as a 'pre-filter' for LC-MS analyses, a better traceability in the results is needed. This can be obtained by the use of certified reference materials and participation in inter-laboratory exercises. Unfortunately, there are no commercially available reference materials (which should be natural samples) to date, nor is there any organization of open regular inter-laboratory exercises, which should be given priority in the future. Thus, there are issues which remain to be solved, but initiatives to solve these problems are on the way, so this is hopefully only a question of time.

## ACKNOWLEDGEMENTS

This work has been supported by the EU Project: Screening methods for water data information, in support of implementation of the Water Framework Directive (SWIFT WFD, Contract SSPI-CT-2003-502,492) and by the Ministerio de Ciencia y Tecnología (Contracts CTM2005-24255-E, BIO2005-00840, CTM2006-26227-E/TECNO, and CGL2007-64551/HID).

## REFERENCES

Andersen H.R., Andersson A.M., Arnold S.F., *et al.*, 1999. Comparison of short-term estrogenicity tests for identification of hormone-disrupting chemicals, *Environ. Health Persp.*, **107**, 89.

Arthur R.A.J., 1995. *Water & Environ. Int.*, **4**, 10.

Beresford N., Routledge E.J., Harris C.A. and Sumpter J.P., 2000. *Toxicol. Appl. Pharmacol.*, **162**, 22.

Bovee T.F.H., Helsdingen R.J.R., Koks P.D., Kuiper H.A., Hoogenboom R. and Keijer J., 2004. *Gene,* **325**, 187.
Bovee T.F.H., Heskamp H.H., Hamers A.R.M., Hoogenboom R. and Nielen M.W.F., 2005. *Anal. Chim. Acta,* **529**, 57.
Carlsen E., Giwercman A., Keiding N. and Skakkebaek N.E., 1992. *British Med. J.,* **305**, 609.
Carson R., 1962. *Silent Spring.* Houghton Mifflin, New York.
Cespedes R., Petrovic M., Raldua D., *et al.*, 2004. *Anal. Bioanal. Chem.,* **378**, 697.
Cespedes R., Lacorte S., Raldua D., Ginebreda A., Barcelo D. and Pina B., 2005. *Chemosphere,* **61**, 1710.
Coldham N.G., Dave M., Sivapathasundaram S., McDonnell D.P., Connor C. and Sauer M.J., 1997. *Environ. Health Persp.,* **105**, 734.
Couse J.F., Korach K.S., Keiding N. and Skakkebaek N.E., 1993. *Environ. Health Persp.,* **101**, 358.
De Boever P., Demare W., Vanderperren E., Cooreman K., Bossier P. and Verstraete W., 2001. *Environ. Health Persp.,* **109**, 691.
Dhooge W., Arijs K., D'Haese I., *et al.*, 2006. *Anal. Bioanal. Chem.,* **386**, 1419.
Fang H., Tong W., Perkins R., Soto A.M., Prechtl N.V. and Sheehan D.M., 2000. *Environ. Health Persp.,* **108**, 723.
Folmar L.C., Hemmer M.J., Denslow N.D., *et al.*, 2002. *Aquatic Toxicol.,* **60**, 101.
Hertz R., 1985. The estrogen problem. Retrospect and prospect., In: *Estrogens in the Environment II. Influences in Development*, McLachlan J.A. (ed), Elsevier/North Holland, New York.
Gaido K.W., Leonard L.S., Lovell S., *et al.*, 1997. *Toxicol. Appl. Pharmacol.,* **143**, 205.
Garcia-Reyero N., Requena V., Petrovic M., *et al.*, 2004. *Environ. Toxicol. Chem.,* **23**, 705.
Giwercman A., Carlsen E., Keiding N. and Skakkebaek N.E., 1993. *Environ. Health Persp.,* **101**, 65.
Korner W., Hanf V., Schuller W., Kempter C., Metzger J. and Hagenmaier H., 1999. *Sci. Total Environ.,* **225**, 33.
Legler J., Van Den Brink C.E., Brouwer A., *et al.*, 1999. *Toxicol. Sci.,* **48**, 55.
Legler J., Jonas A., Lahr J., Vethaak A.D., Brouwer A. and Murk A.J., 2002. *Environ. Toxicol. Chem.,* **21**, 473.
Li W., Seifert M., Xu Y. and Hock B., 2004. *Environ. Intern.,* **30**, 329.
Metcalfe C.D., Metcalfe T.L., Kiparissis Y., *et al.*, 2001. *Environ. Toxicol. Chem.,* **20**, 297.
Murk A.J., Legler J., Van Lipzig M.M.H., *et al.*, 2002. *Environ. Toxicol. Chem.,* **21**, 16.
NICEATM, 2002. Background Review Document: *Current Status of Test Methods for Detecting Endocrine Disruptors: In vitro Estrogen Receptor Transcriptional Activation Assays.*
OECD, 2006. Stably Transfected Transcriptional Activation (TA) Assay for Detecting Estrogenic Activity of Chemicals, *Draft OECD Guideline for the Testing of Chemicals.*
Pawlowski S., Ternes T.A., Bonerz M., Rastall A.C., Erdinger L. and Braunbeck T., 2004. *Toxicology in Vitro,* **18**, 129.
Pinto B., Garritano S. and Reali D., 2005. *Mar. Pollut. Bull.,* **50**, 1681.
Purdom C.E., Hardiman P.A., Bye V.J., Eno N.C., Tyler C.R. and Sumpter J.P., 1994. *Chem. Ecol.,* **8**, 275
Quiros L., Cespedes R., Lacorte S., *et al..*, 2005. *Environ. Toxicol. Chem.,* **24**, 389.
Routledge E.J. and Sumpter J.P., 1996. *Environ. Toxicol. Chem.,* **15**, 241.
Rutishauser B.V., Pesonen M., Escher B.I., *et al.*, 2004. *Environ. Toxicol. Chem.,* **23**, 857.
Schmitt M., Gellert G. and Lichtenberg-Frate H., 2005. *Water Res.,* **39**, 3211.
Schomberg D.W., Couse J.F., Mukherjee A., *et al.*, 1992. *British Med. J.,* **305**.
Schomberg D.W., Couse J.F., Mukherjee A., *et al.*, 1999. *Endocrinology,* **140**, 2733.
Segner H., Navas J.M., Schafers C. and Wenzel A., 2003. *Ecotoxicol. Environ. Safety,* **54**, 315.

# References

Soto A.M., Lin T.M., Justicia H. and Silvia R.M., 1992. *Adv. Modern Environ. Toxicol.*, **21**, 295.

Soto A.M., Maffini M.V., Schaeberle C.M. and Sonnenschein C., 2006a. *Best Pract. Res. Clin. Endocrin. Metab.*, **20**, 15.

Soto A.M., Maffini M.V, Schaeberle C.M. and Sonnenschein C., 2006b. *Best Pract. Res. Clin. Endocrin. Metab.*, **20**, 15

Wang L., Andersson S., Warner M. and Gustafsson J.A., 2003. *Proc. Nat. Academy Sci. of the United States of America,* **100**, 703.

# Section VI
Integration of Screening Methods in Water Monitoring Strategies

# 6.1

# Assessing the Impacts of Alternative Monitoring Methods and Tools on Costs and Decision Making: Methodology and Experience from Case Studies

Helen Lückge, Pierre Strosser, Nina Graveline, Thomas Dworak and Jean-Daniel Rinaudo

6.1.1 Introduction
6.1.2 The Use of SMETs in the Context of Costs and Better Information
6.1.3 The Costs of Using SMETs
    6.1.3.1 Costs of Sensors and Portable Instruments – Replacement of Traditional Methods
    6.1.3.2 Passive Samplers – Obtaining Additional Information
6.1.4 Summary and Conclusion
References

## 6.1.1 INTRODUCTION

Monitoring programmes play a key role in the implementation process of the Water Framework Directive (WFD) as they provide the basis for establishing programs of measures tackling water pollution. To develop effective programs of measures, monitoring networks need to deliver representative information on water quality and to quickly identify new or potential pollution sources which might have negative effects on human health and the environment. The new requirements that arise from the WFD however put considerable financial pressure on water management authorities. Several Member States are supposed to adjust their monitoring networks by increasing monitoring frequency, density or the measurement of new parameters under constant or even reduced budgets. Other Member States need to install completely new monitoring programmes in order to fulfil the WFD requirements.

The classical approach of spot sampling and laboratory analysis has up to now been the dominating monitoring method. On one hand, this approach is often linked to high costs which will rise proportionally with adjustments in frequency, density or the measurement of new parameters. On the other hand, the WFD asks for additional information in some areas that cannot be obtained by using classical approaches (e.g. information on average water quality).

Screening Methods and Emerging tools (SMETs) might help to answer some of the above mentioned issues. Some of these methods and tools might have a considerable potential to perform equally compared to classical methods but at lower costs. Other tools obtain different, additional and more representative information on water quality that can support an improved decision-making process and reduces the risk of wrong decision making and costs that are associated with such wrong decisions.

The European Framework project Screening methods for Water data InFormation in support of the implementation of the Water Framework Directive* carried out five case studies, assessing the potential impact SMETs can have on costs of monitoring and impact on decision making. The experiences gained and lessons learned from these case studies are described in the subsequent sections.

## 6.1.2 THE USE OF SMETs IN THE CONTEXT OF COSTS AND BETTER INFORMATION

The Water Framework Directive requires that an integrated monitoring programme is established within each river basin district. These monitoring programmes will in many cases be extensions or modifications of existing programmes and will enable the collection of physical, chemical and biological data necessary to assess the status of surface and groundwater bodies in each river basin district. In other cases there is a need to collect additional information in order to base decisions on tackling water pressures on a solid basis. In order to assess the need and advantageousness of additional information on water quality, the benefit of having this additional information needs to be compared to the costs that come along with obtaining it.

As stated in the WFD CIS monitoring guidance (European Commission, 2003), 'it is likely that there will have to be a balance between the costs of monitoring against the risk of a water body being misclassified'. In other words the cost for information has to be weighted against the cost of wrong decisions. When entering this discussion there is a need to distinguish between three aspects:

---

* See http://www.swift-wfd.com

1. Potential costs savings from the replacement of classical tools with SMETs: The comparison between actual costs of monitoring and the costs of monitoring if traditional methods would be replaced by new methods is highly relevant for supporting discussions and decisions on the development and use of new methods. Indeed, the cost difference would represent (part of) the expected benefits one might get from the accreditation of new methods and tools that would remove a significant constraint to their use.

2. Additional cost for introducing SMETs to fulfil new requirements under the WFD which cannot be performed with classical tools: The monitoring requirements for successfully implementing the WFD will directly depend upon available measurement techniques of demonstrated quality, which will be able to deliver reliable data at an affordable cost. Besides the necessary 'traditional' laboratory analyses, SMETs might play a key role in the WFD implementation. For example in order to assess the magnitude of the chemical pressure to which bodies of surface water are subject Member States shall monitor for all priority substances and other pollutants discharged in significant amounts. Several of these substances cannot be measured with classical methods. SMETs can provide a sufficient opportunity to monitor these substances, but this will lead to additional total costs of monitoring.

3. Avoided costs due to better information for better decision making: There has been an increasing recognition that emerging tools have a key role to play in WFD monitoring for ensuring better information on the state of the environment. In this respect, SMETs would be used in complement to traditional methods that are accredited and that comply with existing legal requirements. SMETs can deliver additional information compared to classical methods, leading to considerable additional costs in monitoring. This means that the costs for complying with the basic monitoring requirements of the WFD by using alternative methods will be higher than the costs of traditional techniques alone. These additional monitoring costs can however lead to a better decision making and thus in the medium to long-run prevent wrong decisions. Beside potential damages to human health and the environment that come along with such wrong decision, costs for cleaning up these damages might arise. In other words the follow-up benefits (or avoided costs) of using SMETs providing additional information can thus outweigh the initial cost of using them.

While points (1) and (2) follow the methodology of a straightforward cost-assessment, point (3) is more complex as it needs to deal with some issues that are difficult to grasp in monetary terms. While information on costs of programs of measures is partly available, the benefit of preventing wrong decisions and their potential impact on human health and the environment are more difficult to assess. Also, within assessing the impact on decision making uncertainties of methods have to be taken into account.

For assessing the costs of using SMETs within future monitoring networks and their impact on decision making, the SWIFT case studies were build on a common methodology. Based on a gap assessment of existing monitoring programmes, potential functions of SMETs in future monitoring networks have been specified.

The tools and devices subject to this book can play diverse functions in the context of the WFD monitoring. These tools can help supporting the design of monitoring programmes, e.g. through collecting information on current water status for grouping water bodies into homogenous groups or for the selection of representative monitoring sites. Some of the tools can also be a part of the monitoring programme per se, be it for compliance checking (combined among each other/with existing traditional methods) or for enhancing the information base on water quality (e.g. with higher number of points/frequencies or 'new' information on water quality be it in terms of temporal variability of pollution or presence of new pollutants). Thus, they can help better supporting the selection of measures (operational monitoring), for assessing long-term trends in the overall status of the aquatic ecosystem (surveillance monitoring), or for establishing cause-effect relationships in water bodies that are considered as being at risk but for which causes of risk are unknown (investigative monitoring).

Based on the development of functions, the costs of using SMETs have been calculated and, where possible, compared to the costs of doing the same campaign with traditional methods. This cost calculation together with illustrations is described in more detail in 6.1.3. As a next step, the cost calculations have been included into the wider context of impact on decision making, taking into account the issue of uncertainties and the assessment of benefits and/or avoided costs that come along with better decisions making. This aspect is illustrated in 6.1.4.

## 6.1.3 THE COSTS OF USING SMETs

In order to estimate potential cost savings or additional costs that will arise from the application of SMETs there is a need to establish a common framework of costs to be considered:

Apart from investment costs that come along with buying a specific tool, variable costs for personnel and supplies come along with each use of the specific tool. Also, all relevant tasks – from taking a sample or deploying a specific tool to its analysis and data storage – within monitoring need to be considered.

Figure 6.1.1 illustrates the cost components that need to be considered within the different tasks within monitoring activities.

Generally, there are two major elements of costs which need to be considered when assessing the different steps of monitoring:

- *Investment costs (fixed costs):* costs of buying a monitoring device or of modernizing an existing monitoring device as well as costs for setting up and equipping a laboratory.

- *Variable costs:* all costs that occur while conducting the monitoring and all costs of maintaining the monitoring devices that come along with each analysis. Variable costs for monitoring water quality can be broken down into:
    1. Personnel cost (all cost for personnel including additional costs and administrative costs). Cost which occur for personal that is in charge of installing monitoring devices, taking samples and analysing the samples in laboratories. (Is a 'professional' needed for all procedures or is a semi-skilled worker sufficient?)

*The Costs of Using SMETs*

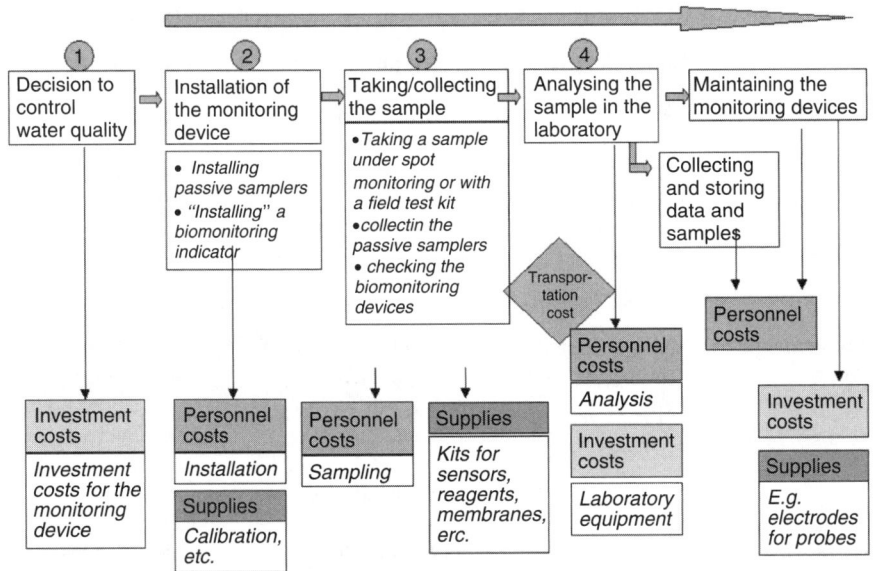

**Figure 6.1.1** Cost information broken down to the different steps of monitoring

2. Most SMETs require special supplies that are not reusable. This included solutions necessary for analysis, membranes or test kits.[†]

3. Other costs including transportation costs and energy costs.

As the specific cost assessment within SWIFT case studies might be outdated soon due to cost changes (e.g. a specific tool moves from prototype to commercial product) and partly depends on the national cost framework, the following description will focus on semi-quantitative assessment before giving some specific illustrations from case studies. The first sub-section of this chapter focuses on the cost assessment of sensors and portable instruments which might have a potential to replace traditional methods. The second subsection gives information on the cost aspects of passive samplers which, under the existing framework, need to be seen as SMETs that bring along additional monitoring costs to meet new requirements.

### 6.1.3.1 Costs of Sensors and Portable Instruments – Replacement of Traditional Methods

*General cost information on sensors and portable instruments*

For the group of sensors and portable instruments, the cost assessment is illustrated for an UV photometer, a colorimetric test kit as well as a multi-parameter probe

---

[†] In some cases it will be difficult to keep apart investment costs and material costs under recurring costs (e.g. it is difficult to allocate a special laboratory device into investment costs or recurring costs).

Table 6.1.1 Cost characteristics of sensors and portable instruments

|  | Equipment costs | Supplies | Personnel costs |
|---|---|---|---|
| **UV photometer** | UV photometer: ca. 6700 € | Glass bottles, filters | Preparation, measurement and sampling: 20 min. |
| **Colorimetric test kit** | Colorimeter: ca. 1200 € (depending on producer) | Test kits, kit reagent Bottles, filters | Preparation and sampling: 40–75 min. + digestion and cooling |
| **Multiparameter probe** | Multiparameter probe: ca. 6000 € | standard solution for conductivity and for different parameters; | Preparation: calibration: 60 min (+ calibration for conductivity: 30 min) |
|  |  | Electrodes for different parameters | Deployment: 10–20 min. |
|  |  | Oxygen membrane, batteries |  |

as they have differing cost characteristics. While the instrument itself is for some methods rather complex and thus cost-intensive, the major cost component of other sensors or portable instrument are the supplies, especially for the test kits. Also, the labour-intensity of the tools differs considerably. Table 6.1.1 gives a semi-quantitative assessment for the three SMETs discussed in this paragraph.

Table 6.1.1 makes clear that the three sensors have totally different cost implications. While the UV photometer is rather expensive to buy, it needs no expensive supplies and is rather quick to use (20 minutes). The colorimetric test kit, on the other hand involves considerably lower investment costs but brings along higher supply and personnel costs. For each analysis, a new kit needs to be used and preparation and sampling take more time. The multi-parameter probe brings along about the same investment costs than the UV photometer but needs more expensive supplies (especially standard solutions and electrodes) and due to necessary calibrations is more labour-intensive. However, SWIFT testing activities made clear that the calibration is not necessary for each analysis but that it is sufficient to do it every day or every other day. This means that the variable costs can be lower in optimized campaigns.

However, a final statement on the advantageousness of any of the tools cannot be made with this general cost information only. To assess the cost impact of sensors and portable instruments, the size of potential campaigns and the number of its uses per year need to be considered. This shows, that the UV photometer will be more expensive than classical methods in rather small campaigns which is mostly due to its high investment costs. However, within medium to large campaigns, this cost factor is spread over a high number of analysis and the cost per analysis of the UV photometer move below

*The Costs of Using SMETs*

the costs of classical methods (see illustration from case studies). For the colorimetric test kit, the so-called break-even point where the sensor becomes advantageous over the classical method in terms of cost, can already be achieved in smaller campaigns, but the total cost saving potential of the tool is lower than the potential of the UV photometer as its variable costs are much higher. This makes clear, that the cost assessment or cost comparison in the case of sensors and portable instruments does not only require a comparison between the sensor/portable instrument and classical methods but also within the group of sensors/portable instruments if several different tools could be used.

## *Illustration for sensors and portable instruments from case studies*

The UV photometer has been proposed both in the German and Czech case studies to support the supplementation and validation of the risk assessment. In order to show the necessary number of uses at which the costs per analysis of UV photometer move below the costs of traditional methods, the cost function for the UV photometer has been calculated. The costs of using the UV photometer are rather high if it is used for few analyses but fall considerably with higher numbers of uses. This cost function is then compared with the linear cost function of spot sampling where the cost of laboratory analysis (list prices) have been considered.

Figure 6.1.2 makes clear that the break-even point where the cost function of the UV photometer moves beyond the costs of classical methods is reached much earlier in the German case then in the Czech case. The figure illustrates that the break-even point where the UV photometer becomes advantageous over spot sampling lies at 17.5 analysis per year in Germany.

In the Czech Republic, where personnel costs are only one fifth of personnel costs in Germany, the linear cost function of classical methods lies much further down (30 € per analysis). This implies that the cost function of the UV photometer crosses the cost function of classical methods further right, at about 80 analyses per year. Thus, while the UV photometer already has a cost saving potential within smaller campaigns in Germany, it needs to be used in a more extensive way in the Czech Republic in order to become advantageous over traditional methods. The absolute cost saving potential

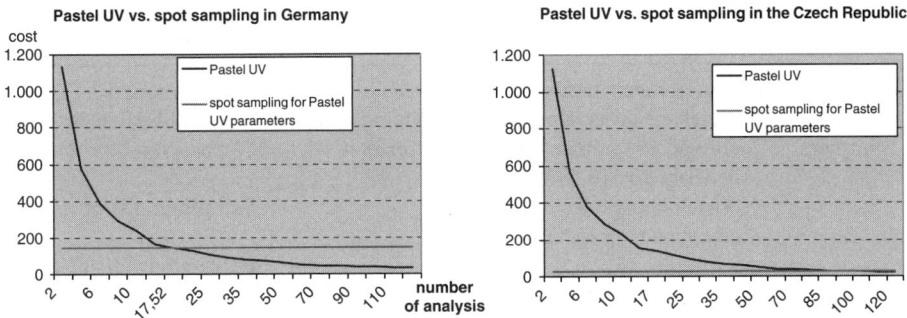

**Figure 6.1.2** Cost functions for UV photometer and traditional methods in Germany and the Czech Republic

also depends on the size of the specific campaign. When, for example, 70 analysis are done in the German case, the cost per analysis is 142 € and the cost of the UV photometer 45 €. Thus, the cost per analysis are nearly 100 € lower when using the UV photometer and the cost difference even increases with larger campaigns.

### 6.1.3.2 Passive Samplers – Obtaining Additional Information

*General information on costs of passive samplers*

Passive sampling devices deliver time-integrated information on water quality and thus a totally different quality of information than spot sampling. Similar to sensors and portable instruments, passive samplers cannot capture all relevant parameters but different passive samplers need to be deployed for different groups of parameters. Based on their characteristics, passive samplers can be used together with traditional spot sampling to obtain complementary information on the status of water quality or to check a water body for (long-term) trends. If an existing monitoring network builds on monthly spot sampling, a passive sampler campaign can easily be integrated into this approach. The collection of a spot sample and the deployment of a passive sampler can be organized within the same field trip and thus don't need any additional transport costs and only few additional cost for the field technician. Only the additional supply costs and cost for analysing the passive sampler need to be considered. Table 6.1.2 gives an overview of costs that come along with using passive samplers.

Table 6.1.2 makes it clear that passive samplers are in general a rather labour-intensive tool as preparation and extraction can take up to 2½ hours. In addition, the cost of buying the samplers and supplies need to be considered. When comparing to traditional spot sampling, all these cost factors are additional monitoring costs. Obtaining one result from passive sampling is clearly more expensive than obtaining one result from spot sampling.

However, these costs cannot be directly compared. Passive samplers are, for example, deployed for two weeks or one month and deliver a time-integrated information for this period. It is obvious, that one spot sample in the same period

**Table 6.1.2** Cost components of passive samplers

|  | Equipment | Supplies | Personnel |
|---|---|---|---|
| Costs of passive samplers | Passive sampler **body** (not reusable): costs depend on type of sampler: between 16–18 € for organic/inorganic compounds or metals; <br><br> Cold box, cage for passive samplers, buoy, fishing line | Various fluids for calibration and extraction | **Preparation:** 10–30 min. (+ calibration for some sensors) <br><br> **Sampling:** 15–60 min. <br><br> **Extraction:** 20–120 min. <br><br> Measurement/analysis: same than spot sampling |

*The Costs of Using SMETs*

does not deliver the same amount of information on water quality. Especially in water bodies with high temporal variability, it is difficult to capture the real status and trends within water quality, the current practice of taking spot samples, for example, on a monthly basis seems not adequate. Especially, for obtaining information on long-term trends which is necessary for planning the program of measures, much more spot samples would be necessary to get to the same level of information than passive samplers but clearly at much higher costs.

## *Illustration for costs of passive samplers from case studies*

Within the German and Czech case studies, passive samplers have been suggested as a monitoring method that can obtain more representative information on long-term trends in water quality. In both case studies, the same campaign has been considered, namely deploying a passive sampler for one month four time a year to cover the different seasonal characteristics. In the Czech case study, the Chemcatcher was proposed for monitoring Atrazin and alchlor while in Germany Chemcatcher (Atrazin and isoproturon) and DGT for metals have been considered. Within both case studies, passive samplers are deployed in duplicate in order to guarantee representativeness of results. The total cost of the passive sampling campaigns is depicted in Table 6.1.3.

Table 6.1.3 shows, that similar to the illustration for sensors and portable instrument, the passive sampler campaign (Chemcatcher) is less expensive in the Czech Republic than in Germany. As investment costs and supply costs are the same in both cases, this cost difference is due to the difference in personnel costs and the difference in costs for lab analysis (which can be traced back to the difference in labour costs).

In terms of cost comparison, it is interesting to take the costs of the different passive sampler campaigns and to check how many spot samples with classical methods could be conducted for this amount: With the same budget than using a passive sampler 4 times a year for one month, only a limited number of spot samples can be conducted. In the Czech case, the choice for monitoring the relevant pesticides is between time-integrated information for four whole months or one spot sample a month. In the German case, only 8 spot samples of the relevant pesticides could be conducted per year with the budget of using the passive sampler in the 4 times, 1 month campaign.

**Table 6.1.3** Cost of passive samplers and spot sampling in the Czech and German case studies

|  | Cost of passive samplers for 4 times 1 month | Cost of one spot sample with classical methods | Number of spot samples possible |
|---|---|---|---|
| **Chemcatcher in the Czech case** (2 pesticide parameters) | 997,3 € | 78 | 12,8 |
| **Chemcatcher in the German case** (2 pesticide parameters) | 1768,5 € | 227,2 | 7,8 |
| **DGT in the German case** (10 metal parameters) | 2840,5 € | 274 | 10,4 |

Information on long-term trends of metal pollution can in the German case be obtained through the passive sampler campaign or through 10 spot samples per year.

This cost comparison shows that passive samplers have a clear advantage for monitoring of long-term trends. The time-integrated information derived through passive samplers clearly gives a much better information on trends in water quality than one (or even fewer) spot sample per month. Especially if the temporal variability in water quality is high, there is a great risk that one spot sample per month does not adequately capture the real trends in water quality which can potentially lead to wrong decisions concerning the program of measures.

## 6.1.4 SUMMARY AND CONCLUSION

Assessing and comparing costs of screening methods and emerging tools (SMETs) clearly is a complex task. In a first step, the costs of using the specific tools need to be analysed: apart from investment and variable costs of using the tool, all relevant steps of monitoring need to be considered. Especially in the cases where SMETs have a potential to replace traditional methods this in-depth cost assessment is necessary to ensure that future monitoring networks are designed efficiently. The illustration for sensor and portable instruments makes clear that the advantageousness of sensors over traditional methods depends on the size of the campaign in which the sensor will be used. For small campaigns with only few analysis, sensors with high investment costs will bring along higher monitoring costs while other tools, especially the different test kits can bring along a cost-saving even in small campaigns.

Thus, when considering the use of sensors or portable instruments, not only the costs between sensors and traditional methods need to be compared but also the costs within the group of sensors need to be considered. The cost assessments of the SWIFT case studies makes clear that the cost-saving potential of sensors and portable instruments is considerable, especially if costs for laboratory analysis are high. In the German case, the UV photometer becomes cost-advantageous over traditional spot sampling at 17 analysis and then, with each number of analysis, the cost difference increases. In an aggregated view, this can sum up to cost savings of several thousand Euros for larger campaigns.

In other uses, SMETs will be used complementary to traditional methods and will thus bring along additional costs. These additional costs are, however, due to the WFD which poses new requirements for monitoring activities. Some of these new and additional requirements can be fulfilled with traditional methods while others require the use of SMETs. For those functions where no alternative to using SMETs is available, the cost assessment is less crucial but can however help for the internal budget planning. For other additional functions, there is an alternative between SMETs and traditional methods and the cost comparison is more important. For the case of monitoring long-term trends the SWIFT case studies have however made clear that passive samplers are clearly advantageous over traditional methods. To obtain an equally representative information on water quality, traditional methods would bring along much higher costs than passive samplers as a high number of analysis would be necessary.

In addition, to the direct cost comparison between SMETs and traditional methods the benefits of obtaining additional information on water quality need to be considered.

# REFERENCES

European Commission, 2003. *Monitoring under the Water Framework Directive*, CIS Guidance Document N°7, European Commission, Brussels.

# 6.2
# Acceptance of Screening Methods by Actors Involved in Water Monitoring

Didier Taverne

6.2.1 Introduction
6.2.2 Utility of Innovative Object
    6.2.2.1 Utility as Subjective Perception
    6.2.2.2 Perception of Problems and Solutions
    6.2.2.3 Risk and Acceptance
6.2.3 Usability
    6.2.3.1 Ergonomics
    6.2.3.2 Complexity of Use
6.2.4 Culture and Organization
    6.2.4.1 Cultural Aspects
    6.2.4.2 Organizational Aspects
6.2.5 Conclusion
References

## 6.2.1 INTRODUCTION

Screening methods, as all innovative products or technologies, necessarily meet obstacles. The issue of acceptance of innovative product or technology constitutes a specific field of sociology. It has been opened at the end of the seventies when computers and automatic machines arrived in the enterprises and organizations. Sometimes, people refused these machines. That fact led psychologists and sociologists to wonder about these behaviours. Some sociologists (Crozier, 1977) have developed the notion

---

*Rapid Chemical and Biological Techniques for Water Monitoring*   Edited by Catherine Gonzalez, Philippe Quevauviller and Richard Greenwood
© 2009 John Wiley & Sons, Ltd

of 'resistance to change' to explain these difficulties. It is not the way chosen in this paper, looking this resistance as rational behaviour. In fact, people 'resist to change' because every change modifies individual and social positions and so generates subjective risks.

The risk analysis has renewed the general approach of acceptance. Since a few years, the industrial society is described as high-risk society (Beck, 1986). Technologies don't help us to bring the world under control but generate a high uncertainty level. People don't accept to bear risks and ask for precautionary measures.

In spite of these works, giving a clear definition of social acceptance is still difficult. Only case studies are available and each one refers to singular factors. In fact, 'social acceptance' is not a concept but a very practical – but imprecise – notion. Surrounding this notion needs a large overview of works, in sociological and political fields and also in marketing issues.

Following this overview, it appears that three main dimensions must be taken into account: utility perceived by users, usability of an innovative product or method, and cultural and organizational changes introduced by that invention.

For each one, it is possible to decline some elements that can explain the reasons why actors involved in water monitoring may be refuse this kind of innovation. To conclude, as a counterpoint, it is possible to define a few ways, which could favour the diffusion of new screening methods.

## 6.2.2 UTILITY OF INNOVATIVE OBJECT

Three main dimensions of utility will be analyzed: (1) utility as a subjective perception; (2) linked with problems faced and other available solutions; (3) and thus with risks introduced by change. Nevertheless, we must remind that all these elements form a system. Concretely, it can be difficult to separate the different levels contributing to acceptance.

### 6.2.2.1 Utility as Subjective Perception

The economists consider that utility is always subjective. The inventor can't decide himself on the qualities of his product. Other persons, who will use it, will decide on these qualities. The first inventor's mistake often consists in thinking something like *'my project is so interesting that everybody will accept it without resistance'*. Latour (1997) speaks about 'ballistic temptation' to describe this error. Potential users should immediately recognize intrinsic qualities of the new thing. Its utility and technological superiority should be sufficient. In facts, utility and qualities are always relative and subjective.

The concept of human bounded rationality (Simon, 1957) invites us to consider that the human mind does not try to find the best solution, but the satisfactory solution. A new technology can effectively be better (on a technical point of view) but refused by users who can prefer to carry on with an 'old' technology. Two arguments can explain this behaviour. On one hand, the actual technology, tool or methodology can

*Utility of Innovative Object*

answer properly the main problems met by end-users even if it has less technical qualities. They are accustomed to use this technology that has given them acceptable results over a long time. So they have no reasons to change. It's a sort of loyalty phenomenon (Hirschman, 2006). On the other hand, change is always costly, and not only in monetary terms. Of course, it's impossible to deny the importance of economic considerations. Buying new devices requires for example an investment capacity. The monetary cost of a new thing can reduce its success. Also, investment patterns and cycles for substitution have to be taken into account: if for example a drinking water supplier has just invested in a new monitoring station he probably won't invest in another one in the short term, even if there is a new and even more innovative product. If a laboratory already has a wide range of classical tools, it will not be encouraged to buy new equipment. However, it may be important to remember that symbolic and also cultural elements also contribute to the cost of change. Adopting a new thing, involves the user in changing his custom. Before he adopts the new thing, he can't really be sure that the advantages will outweigh any disadvantages. Taking action is not going to happen where, above all, preferences are not really clear. It is the case with certain screening tools or methods (passive samplers for example) providing new kind of data about water quality that can't be compared with 'classical' data.

### 6.2.2.2 Perception of Problems and Solutions

The perceived utility of a new thing is always linked with an objective, a task to be fulfilled, a problem to be solved and the availability of other solutions. From that point of view, 'setting problem' and 'solving problem' must be distinguished. People can agree on the nature of a problem but disagree about acceptable solutions, and vice versa. For a chemical expert, the main objective may be to design a very sensitive tool giving very precise results. For a technician managing drinking water network, the objective may be mainly to make use of an alert indicator. He wants very simple information like: 'the light is green, all is OK; the light is red, I have to stop pumping from the river'. The nature of problems is different and the tool proposed by the expert is not convenient for the technician. The acceptance will be low whatever the technological or environmental qualities of the proposed change.

Even if protagonists agree on the problem, they can consider different solutions. For example, a general agreement exists on the $CO_2$ reduction, but, some people want to develop new sources of energy whereas other people want to reduce the consumption of energy. For the first approach, windmills constitute a convenient solution, for the other solution they do not. That means that the designer of new technology or tool must at first consider the problems faced, the objectives to be fulfilled, the concrete tasks to be performed by end-users in order to create a common language. The problems he must consider are those of the user, linked with available solutions he perceives as convenient. For a single situation, a lot of interpretations are possible and nobody can decide which the pertinent one is. That is why particular procedures such as the use of 'focus groups' or 'consensus conferences' have been developed, in order to build a common sense of the nature of problems and situations.

## 6.2.2.3 Risk and Acceptance

Furthermore, new practices always introduce risks. The human being faced with a risk can be analysed from two ways. Firstly, by their nature human beings do not like risk. So they resist change (Herzberg, 1959). They are not rational, but immature. Secondly human beings are always – whatever the type of rationality – rational. A change modifies the situation, advantages and contributions the participant gains and bears from its participation. To accept a change, the potential user must be able to assess the new advantages and new contributions resulting from the change. If he has good reasons to think that the result of the balance between advantages and contributions will be less interesting for him, then he has an excellent reason to refuse the change.

Conversely, using a classical technology which has been in use over a long period, and is familiar, reduces the risks of error. The process is well controlled. Changing to a new technology can cause the user to commit errors in the interpretation of either the instructions and/or of the results. A user's competence could be brought into question. Before changing and adopting a new thing, the user must be able to estimate the potential difficulties and risks linked to the change in order to be able to adapt to the new situation resulting from the change. People will still compare new and classical tools, old and new situations, even in the absence of sufficient information. A new thing is compared with an old but well-known one even if it is not a pertinent comparison, even if assessment criteria are not adjusted. Potential users have a preference for tools that are closely near to the tools they usually use. The acceptance is higher for tools which operate according to principles that are the same for the classical tools. Emerging tools that work on the same physico-chemical principles as laboratory devices are more easily adopted. Equipment or methods such as immuno-assays or passive samplers where the principles are different may be rejected even if they could be interesting.

This situation characterizes the paradox associated with innovative products. It is subject to a 'reality test'. To be adopted, a new thing should have already proved its worth. The wait and see policy seems to be the best because potential users don't have criteria for assessing the new situation. This approach shows that the available information plays a very important part: without information, an actor cannot assess the contributions and advantages linked to the change. But the relationship with risks is not only an economic calculation. Risk and perception of risk are psychological and social constructions (Covello, 1989). The propensities for risk are different from one person to another, and from one social group to another. The propensity varies according to personal history, and to the intellectual, financial and social resources that the person or group can mobilize. This explains why the diffusion of innovative products often depends on imitative behaviours. The wait and see policy is a standard practice, until someone else adopts the product. Then, other potential users can assess the advantages/disadvantages or costs/opportunities linked with the use of that product. They are now able to assess the risks of adopting that new thing.

## 6.2.3 USABILITY

The usability of a tool can be defined as the capacity for the user to use concretely that tool. This definition refers to three main issues.

*Usability* 401

### 6.2.3.1 Ergonomics

If twelve fingers are needed to operate a machine, the potential user will be very disappointed! However, several examples show that bad ergonomic qualities can be accompanied by commercial success. Nevertheless, ergonomic considerations cannot be discounted. There are very significant differences between laboratory and ground conditions. For instance, inside the laboratory the tool is put down on a table, doesn't move. In practical working conditions of use, user will carry the tool in his car; and may have to carry it manually over distances of thousands of metres. An insignificant excess of weight inside the laboratory becomes totally unacceptable in real conditions of use, particularly if given equipment only measure a limited number of parameters.

Inside the laboratory, the environment is stable. There are no variations in temperature, no change in electric power levels, and the humidity is under control. In contrast in the field, all these parameters can become important variables. The tool can be sensitive to the atmospheric conditions; results can be distorted or uncertain and thus unacceptable. Some of the emerging tools are not really suitable for particular conditions because they are too sensitive to low temperature for instance.

### 6.2.3.2 Complexity of Use

Sometimes, usability tests are not realized before launching a new product in the market. There is a high risk that the technology will suffer a setback because of disjunction between manufacturers and users. This disjunction can appear at two levels. The first level is connected with the way in which the new technology or tool is used. The knowledge and experience of producers and users are not of the same nature. An expert or specialist developing a new thing has an intimate knowledge of his product. This makes it easy for him to believe that the functioning is easy and so to forget to give clear and complete information about how the new thing works. The expert can forget that the user will never have the same relationship with this product. That's why the information given to users is very important. If the user's guidance is not efficient, then the acceptance will be low because of perceived risk of errors. The second level is related to the interpretation of results. From the user's point of view, controls can be unclear and the interpretation of results can be too complex. That complexity also introduces risks of error. Nobody likes to take additional risks. Particularly when other available and well-known products exist or when the potential consequences of the error are very important.

For example, there are certain biological constraints on some species of South American electric fish. The conditions for fish's survival fall within a very narrow range: the temperature of water has to be strictly stable. Here is a first constraint, a first risk of error. The fish's death can be due to a variation of temperature as well as pollutant. Another example is provided when the result of the measurement takes form of a list of numbers, and in order to interpret this, it is necessary to refer to a book. This does not permit a rapid interpretation of the results. The technician in charge of water pumping cannot react quickly. Where other less constraining techniques exist, he will not be interested by that tool. This example shows that designers of new tools or methodologies must take into account the user's constraints. Finally a convenient measurement tool is a 'black box' giving result without any additional cognitive operation in order to provide the necessary information for immediate decision. This is why it is necessary for the transfer of knowledge to be as clear, rapid and easy as possible. For example: the guidance has to be written in user's terms, considering the user's faculties and problems. Introducing a new tool must go with training and demonstrations in order to demonstrate and concretely prove the qualities associated with the change in a way that perceived risks are reduced and the acceptance is higher.

## 6.2.4 CULTURE AND ORGANIZATION

Various forms of rationality exist. An organization is crossed by axiological rationality, by values that can be incompatible with the impacts of the new technology. To be acceptable, a new product or technology has to create the conditions for a compromise between these different rationalities and cultures. In the case of screening methods, difficulties are increased because of a large number of organizations concerned.

### 6.2.4.1 Cultural Aspects

Administrative culture can favour the checking of progress. Taking preventive action is not yet usual. Wherein order to carry out their functions administrations need to use normalized and standardized tools and methods with a high level of accuracy, then they will recommend classical tools that have been validated over a long period of time, and upon which there is a general agreement. Then, laboratories favour these tools. Most laboratories interviewed have a simple position about screening tools: 'we just use the classical tools that are validated and advised by authorities'. This position is quite general. The lack of quality standards is mentioned as the main reason for refusing to use new screening methods. In addition, policy makers don't like to get a wide choice. In conclusion the acceptance of new tools by policy makers should be a priority because it's one of the main barriers to the development of these tools.

For laboratories, the general preference for classical tools and laboratory methods is not only an issue of administrative culture and legal requirements. They also have a 'culture of accuracy'. Their views are always reducing the value of emerging tools because of detection limits and lack of accuracy. They are sure that emerging tools are not comparable with established laboratory equipment. For them, these tools look like gadgets. People are used to having a high level of accuracy, and so the data obtained using emerging tools seems to fall too short of their needs.

In addition, economical evolutions could reinforce that cultural preference. The new market's rules are increasing the competition between laboratories. They are engaged in a development race. In that culture and legal organization, development is synonymous with the use of new equipment. In order to catch large markets, they need to have the equipment that is perceived as important.

Distributors and manufacturers have an economic culture: they do not offer for sale equipment for which there is no demand. Distributors follow the market demand. Thus, if there is no demand for new tools they do not supply them to the market. If manufacturers or distributors think the market for new tools is not promising, then they don not develop these tools, particularly if the manufacturers are not specialized in the area of water measurement. They just offer or develop classical tools that are dominating the market and are able to answer the main problems/demands.

### 6.2.4.2 Organizational Aspects

Organizational aspects can also be analysed at three levels: industries, laboratories and public institutions. Most of the industrial companies subjected to self-regulation do not conduct sampling and analysis themselves. They commission a laboratory to do it. They regularly send a report to the Public Administration but they don't feel concerned about water monitoring or water policy. Fortunately, some enterprises feel concerned by these issues especially when risk of accidental pollution with important environmental impacts exists or when they want to be careful about their image. In addition, cost of accidental pollution and polluter-pays principle should encourage manufacturers to prevent pollution and use new tools and methods to this end. However, at present, industrial firms are concentrating their core activities and on their main jobs. Others are externalized. Currently, they do not undertake the sampling and analysis themselves, and nobody inside companies could implement new tools and interpret results.

Laboratories sometimes suggest that their reluctance to change is based on a lack of necessary competences. For instance if they are organized around very specific (e.g. chemical) competences, then changing and adopting biological tools for example, would require the laboratories to employ biologists as a necessary precondition to enable them to carry out biological analysis.

A new thing can modify the organization or the roles and status of employees within the organization. The change can modify the competences needed, and those with competences necessary for the replaced technology can be moved to a lower status or be made redundant. The actor's system will be changed and associated with this position in the organization. Persons can fear becoming unemployed.

The institutional organization has also to be taken into account. France can be seen as typical of these problems. A large number of public institutions are playing a part in water policy. Each institution has its own regulations and territory in which other institutions are not welcome. Real barriers still exist between institutions. A lot of coordination problems exist because of the number of actors involved in the water policy area. Territories and actors are superimposed on one other. Normally, Water Agencies should coordinate all the actors. In fact, in real life, powers are very different from one institution to another. Few institutions have a regulatory power (EEA, 2005). Departments of Agriculture and Industry have the upper hand on water policy. Administrations not involved in the environmental issues have a very low power. As result, only a few institutions have a real environmental approach, and are developing innovative or specific methods. Interrogating the Fishing Council about screening tools, the answer was very clear: 'of course we are interested, but who has the regulatory power?'

In absence of prescriptions, subjects are free to choose their own methods. Classical tools can appear as convenient if the user just needs standardized information in order to meet legal requirements. This can be reinforced where public institutions favour the classical methods, and the problems/demands can be resolved with classical equipment.

## 6.2.5 CONCLUSION

Our experience leads us to emphasize the difficulties that the screening methods have to face. An innovative product has no 'natural friend'. Its diffusion is closely linked with its acceptance in social, industrial, and political networks. To paraphrase Bruno Latour, we could write that an innovative product is not used because it is technically good: it is good, because and when it is used. Nevertheless, this approach also allows the recognition of three main aspects that could help in the development of the innovative screening methods.

Manufacturers and providers could improve the information available on their products. User's guidance could be clearer, field demonstrations and training could be developed in order to familiarize end-users with these kind of tools. Information reduces the perceived risks and plays and important part in the process of acceptance.

This information should also take form of social and political justification. Why should laboratories, industries, administrations need these tools? What kind of information are they able to deliver, and what for? It could be efficient to emphasize the specific and new functions these tools and methods can fulfil rather to insist on technical results that generate a comparison and then a competition with classical methods. This also means that the emergence of a new culture should be favoured, linked with an innovative approach to risks and to protecting the environment. Currently it is uncertain whether the requirements of the Water Framework Directive are well known. A larger dissemination of these requirements may help the development of screening methods through a renewed understanding of monitoring.

## REFERENCES

Beck U., 1986. *Risikogesellschaft*, Suhrkamp verlag, Frankfurt am Main.
Covello T.V., 1989. Informing people about risks from chemicals, radiation and other toxic substances: a review of obstacles to public understanding and effective risk communication. In: *Prospects and Problems in Risk Communication*, University of Waterloo Press.
Crozier M., 1977. L'acteur et le système. Paris, Le Seuil.
EEA, 2005. *Effectivness of urban wastewater treatment policies in selected countries: an EEA pilot study*, EEA Report N° 2/2005, October 2005.
Herzberg F., 1959. *The Motivation to Work.* New York, John Wiley & Sons, Inc.
Hirschman A.O., 2006. *Exit, Voice and Loyalty: Responses to Decline in Firms, Organizations and States*, Harvard University Press.
Latour B., Protee U.E., 1997. Contract N°ST-97-SC.2093.
Simon H.A., 1957. *Models of Man.* New York, John Wiley & Sons, Inc.

# Index

AA *see* annual average
AAS *see* atomic absorption spectroscopy
Ab-Ag *see* antibody-antigen
acceptance by actors   397–404
  complexity of use   401–2
  costs   399
  cultural aspects   402–3
  ergonomics   401
  organizational aspects   403
  perception of problems and solutions   399
  risk   398, 400
  subjectivity   398–9
  usability   400–2
  utility of innovative object   398–400
accidental pollution events
  passive sampling   250–2
  screen-printed electrodes   270–1
  UV spectrophotometry   104
accuracy   265, 267
acetylcholine esterase   136–7, 141
acoustic transducers   146
AFNOR *see* Association Française de Normalization
AhR *see* aryl hydrocarbon receptor
algal BEWS   202, 213
alkylphenols   375–8
alternative test procedures (ATP)   22, 23–5
Ames test   131
ammonium   81–2, 85, 87, 98
analysis of variance (ANOVA)
  environmental variability   320–5

microbial assay for risk assessment   112, 121
reference materials   342, 346–9
uncertainty   309, 313–14, 316
analytical uncertainty   305–6, 310–11
androgen receptors (AR)   182–3, 184, 379
annual average (AA) concentration   54–5, 244–5, 258
anodic stripping voltammetry (ASV)   264–5
ANOVA *see* analysis of variance
anthracene   277, 281–4
antibody-antigen (Ab-Ag) equilibria   157–9
antimony   322, 324–7
AQC data   113–15
Aqua-Tox-Control fish monitor   210–11
aquaculture   207–8
AR *see* androgen receptor
arsenic sensing protein (ArsR)   179, 184
aryl hydrocarbon receptor (AhR)   180, 184, 189
Association Française de Normalization (AFNOR)   80
ASTM Standard   209
ASV *see* anodic stripping voltammetry
atomic absorption spectroscopy (AAS)   244
ATP *see* alternative test procedures
atrazine   167–70, 357, 363–5
automated immunochemical methods   166–7
AWACSS assays   146–7, 151, 167, 171

*Rapid Chemical and Biological Techniques for Water Monitoring*   Edited by Catherine Gonzalez, Philippe Quevauviller and Richard Greenwood
© 2009 John Wiley & Sons, Ltd

BaA *see* benzo[a]anthracene
background review documents (BRDs) 379
bacterial BEWS 202
bacterial ecotoxicity tests 110
bacterial luciferase (*luxCDABE* gene) 185, 187
BaP *see* benzo[a]pyrene
Beckman's Operating Manual 131
benzo[a]anthracene (BaA) 277, 281–5
benzo[a]pyrene (BaP) 277, 281–4
benzo[k]fluoranthene (BkF) 277, 282–4
BEWS *see* biological early warning systems
bias 265, 267, 306
bioaccumulation 58
bioassays 125–34
  chemical methods 126
  estrogenicity 127, 131–4
  genotoxicity 127, 131
  hierarchy of complexity of living systems 126
  legal requirements 128–34
  passive sampling 65
  screening methods 125–6, 297–8
  toxicity 128–31, 133
  Water Framework Directive 46–7
bioavailability 55–6, 243–5, 258
biochemical oxygen demand (BOD) 80–1, 84–6, 89
  *see also* biological oxygen demand
biofouling 208, 294–5
biological early warning systems (BEWS) 197–219
  algae 202, 213
  applications 205–8
  aquaculture 207–8
  bacteria 202
  basic requirements 199–201
  biofouling 208
  biomolecular recognition systems 176
  bivalve molluscs 203–4, 206–7, 211–14
  case studies 210–15
  commonly used systems 201–4
  continuous monitoring of waters 197–219
  drinking water 205–6
  effluents 207

emerging water quality monitoring tools 47, 288
  fish 204, 209–11
  groundwater 207
  invertebrates 202–3, 213–15
  organism selection 200
  quality assurance 208–10
  response parameters 199–200
  spot sampling 197–8
  surface waters 206
  tuning chlorination of cooling water 208
biological methods
  bacterial tests 110
  bioassays 125–34, 297–8
  biological early warning systems 176, 197–219, 288
  biomarkers 47–8, 221–39
  biomolecular recognition systems 175–95
  biosensors 125, 127–8, 134–52
  ecotoxicity testing 109–10, 116–17
  Environmental Protection Agency 32–3
  future developments 149–52
  hierarchy of complexity of living systems 126
  immunoassays 125, 127
  immunochemical methods 157–73
  legal requirements 128–34
  microbial assay for risk assessment 109–24
  pollutant effects 45–8
  screening methods 125–6
  Water Framework Directive 40–1, 45–8
biological oxygen demand (BOD)
  bioassays 127
  biosensors 148
  rapid mapping 290, 292
  *see also* biochemical oxygen demand
bioluminescence 130–1, 135
biomarkers 47–8, 221–39
  case studies 225–35
  classification 221–5
  effect 48, 221–2, 224–5, 236
  exposure 47–8, 221–2, 224–5, 236
  fluorescence of PAH metabolites 224, 232–6
  imposex/intersex 224, 225–32

# Index

integrated approach  236
susceptibility  48, 221–3
tributyltin  225–32, 236
biomolecular recognition systems (BRS)  175–95
  applications  175–8
  aryl hydrocarbon receptor  180, 184
  bioreporter organisms  185–8
  bioreporter technology  188–91
  emission phase  178–9
  estrogen receptor  179–82, 184, 189–91
  promoter interaction  183–4
  receptor proteins  178–84, 189–91
  recognition phase  177, 178–84
  reporter systems  185–8
  response characteristics  176–7
  transduction phase  178–9, 185–91
  *see also* biological early warning systems
bioreporter organisms  185–8
bioreporter technology  188–91
biosensors  134–52
  DNA biosensors  147
  electrical sensors  139
  electrochemical transducers  138–9, 145–6
  environmental analysis/monitoring  136–7, 142–4, 150
  enzyme biosensors  141, 145
  future developments  149–52
  immunosensors  145–7
  magnetic devices  140
  mass sensitive sensors  139–40, 146
  microbial biosensors  148
  MIP-based sensors  148–9
  optical transducers  135, 138
  principles/classification  125, 127–8, 135–40
  recognition element classification  140–9
  thermometric devices  140
bisphenol A  375–7
bivalve mollusc BEWS  203–4, 206–7, 211–14
BkF *see* benzo[k]fluoranthene
BOD *see* biochemical oxygen demand; biological oxygen demand
boron  314–15, 320–3, 324–7

boron-doped silicon nanowires (SiNWs)  151–2
box and whisker plots  315, 317
BRDs *see* background review documents
British Standards Institute (BSI)  254–5
BRS *see* biomolecular recognition systems
BSI *see* British Standards Institute
butylcholine esterase  136–7, 141

cadmium resistance operon (cadC)  178, 184
calibrants  342–4
CALM *see* Consolidated Assessment and Listing Methodology
carbamate pesticides  358
CCDs *see* charge-coupled devices
cell proliferation assays  372
CEN *see* European Standardization Organization
certified reference materials (CRMs)
  metrology  10–11
  proficiency testing  336
  uncertainty  305–6, 310–11
CFA *see* continuous flow analysis
CFR *see* Code of Federal Regulations
charge-coupled devices (CCDs)  151
Chemcatcher®  57–63, 66, 248–59, 293–5
CHEMFET analysers  139
chemical methods
  bioassays  126
  chemical parameters  80–2
  continuous monitoring of waters  198
  detection methods  85–8, 91–106
  early warning systems  104
  Environmental Protection Agency  33–4
  estrogenicity  377–8
  existing methods  79–90
  heavy metals  243–61, 263–73
  laboratory-based methods  84–8
  monitoring and prevention  105
  nonpolar organic compounds  63–6, 71–4
  organometalic compounds  63–6
  parameters  80–3
  polar organic compounds  58–63, 71–3, 74–6
  priority substances  43–5

chemical methods  (continued)
  reference materials  335–50
  sample preparation  85
  sampling design  83–5
  screen-printed electrodes  263–73
  spatial/temporal variability  96–100
  study design  73
  trophic state assessment of lakes  100–5
  yeast-based assays  377–8
  see also passive sampling; physico-chemical parameters
Chemical Monitoring Activity (CMA)  5
chemical oxygen demand (COD)
  rapid mapping  290, 292
  water quality monitoring  81, 83, 85–6, 89
chlorination of cooling water  208
chlorophenols  131
chlorpyrifos  357, 362–3
chrysene  277, 281–4
CIS see Common Implementation Strategy
CL see confidence limits
Clean Water Act (CWA)  15–22
  see also Environmental Protection Agency
CMA see Chemical Monitoring Activity
coating antigen format  160–2
COD see chemical oxygen demand
Code of Federal Regulations (CFR)  16
coefficient of variation (CV)  111–14
colloidal material  245
colorimetric test kits  85–7, 389–90
comitology  8
Common Implementation Strategy (CIS)  5, 304, 386
competitive immunoassays  161–2
complexity of use  401–2
conductivity  88
confidence limits (CL)  111–14, 123
confounding variables  116, 118
Consolidated Assessment and Listing Methodology (CALM)  19
contamination of samples  309
Contamination Warning System (CWS)  26

continuous flow analysis (CFA)  82–3
continuous monitoring of waters see bioassays; biological early warning systems; passive sampling
cooling water  208
cooperative networks  26–7
Cooperative State Research, Education and Extension Service (CSREES)  26–8
CRMs see certified reference materials
CSREES see Cooperative State Research, Education and Extension Service
cultural aspects  402–3
CV see coefficient of variation
CWS see Contamination Warning System
cyanotoxins  34
cytochrome P450  47

Daphnia Toximeter  214–15
Daphtoxkit  358, 366
DBS see dodecyl benzene sulfonate
DDT see dichlorodiphenyltrichloroethane
DGT see diffusion gradient in thin films
diazinon  136, 141
dichlorodiphenyltrichloroethane (DDT)  152, 183, 189
3,5-dichlorophenol  113–16, 120–3
dichlorvos  117–18, 136, 141
diffusion gradient in thin films (DGT)  49, 57–9, 247–59, 293–5
DIN standards  80–2
dioxins
  biomarkers  224
  biomolecular recognition systems  180, 191
  biosensors  146
  immunochemical methods  159
  Water Framework Directive  47
direct fluorimetric analysis  275–86
direct toxicity assessment (DTA)  110
dissolved organic material (DOM)  252–3, 256, 260
dissolved oxygen content (DOC)  5, 88
diuron  167–70, 213, 357, 363–6
DNA biosensors  147
DOC see dissolved oxygen content
dodecyl benzene sulfonate (DBS)  290, 292
DOM see dissolved organic material
drinking water

# Index

biological early warning systems 205–6
  uncertainty 312–13
Drinking Water Directive 293
DTA *see* direct toxicity assessment
duplicate samples 331
dynamic Daphnia test 202–3

E-SCREEN assays 134, 374–6
EAQC-WISE *see* European Analytical Quality Control in support of Water Information System for Europe
early warning systems 104
  *see also* biological early warning systems
$EC_{50}$ values 113–19
ecotoxicity testing 109–10, 116–17, 358–9
EDCs *see* endocrine-disrupting compounds
EEF *see* estradiol equivalency factors
effect biomarkers 48, 221–2, 224–5, 236
effluents 120–3, 207
EIA *see* enzyme immunoassay
electrical sensors 139
electrochemical methods 82
electrochemical transducers 138–9, 145–6
electrostatic self-assembly (ESA) 146
ELISA *see* enzyme-linked immunosorbent assays
ELRA assays 134
EMAP *see* Environmental Monitoring and Assessment Program
emerging pollutants 295–7
endocrine-disrupting compounds (EDCs)
  bioassays 131–4
  biomolecular recognition systems 178
  biosensors 143
  yeast-based assays 372–7
endosulphan 205
Environmental Monitoring and Assessment Program (EMAP) 19
Environmental Protection Agency (USEPA) 15–37
  alternative test procedures 22, 23–5
  applications of screening methods 32–4
  approved methods 22–3

  Clean Water Act 15–22
  data management 20, 22, 30–1
  field kits 28
  homeland security issues 25–6
  monitoring and assessing water quality 18–22
  new method approval 23–5
  Quality Assurance Project Plan 31
  river and lake parameters 29–30
  screen-printed electrodes 272
  state water and monitoring programs 19–20
  Volunteers Monitoring Programs 26–31
  water quality indicators 21
environmental quality standards (EQS)
  bioassays 297
  exposure biomarkers 221
  heavy metals 244–6, 258, 260
  passive sampling 54–6, 65
  rapid mapping 290
  screen-printed electrodes 272
  uncertainty 311–13
environmental risk assessment (ERA) 221–3, 236
  *see also* microbial assay for risk assessment
Environmental Technology Verification (ETV) Program 272
environmental variability 303–4, 316–29
  assessment procedure 316–18
  major components 324, 327
  spatial variability 96–100, 268–70, 290–3, 318–25, 328
  surveys 318–24
  temporal variability 96–100, 270–1, 304, 318–28, 394
enzyme biosensors 141, 145
enzyme immunoassay (EIA) 158, 160–1
enzyme-linked immunosorbent assays (ELISA) 158–9
  formats 161–3
  performance/limitations 169–71
  proficiency testing 357, 362–6
  screening methods 34
  Water Framework Directive 167–9
enzyme-tracer format 160–2, 168–9

EPA *see* Environmental Protection Agency
EQS *see* environmental quality standards
ER *see* estrogen receptor
ER-CALUX assay  134, 373, 374–6
ERA *see* environmental risk assessment
ERE *see* estrogen-responsive elements
ergonomics  401
ESA *see* electrostatic self-assembly
estradiol equivalency factors (EEF)  373
estrogen receptor (ER)
 binding assays  373
 biomolecular recognition systems 179–82, 184, 189–91
 competitive binding assays  372
estrogen-responsive elements (ERE) 181, 184, 191, 373
estrogenicity
 bioassays  127, 131–4
 biosensors  143
 chemical methods  377–8
 future developments  378–9
 interlaboratory exercises  374–7
 non-yeast-based assays  374
 relative estrogenic potency  373–5
 screening methods  377
 yeast-based assays  371–81
estrogens
 biomarkers  224
 biomolecular recognition systems 178, 190
 immunochemical methods  165
 passive sampling  65
ETV *see* Environmental Technology Verification
European Analytical Quality Control in support of Water Information System for Europe (EAQC-WISE)  12
European Standardization Organization (CEN)  7–8, 80
eutrophication *see* trophic state index
exposure biomarkers  47–8, 221–2, 224–5, 236
external calibrations  9

Federal Water Pollution Control Act (FWPCA) *see* Clean Water Act
FES *see* flame emission spectroscopy
FF *see* fixed fluorescent
FIA *see* flow injection analysis; fluoro immunoassay

field kits  28
field studies
 chemical methods  83–4
 emerging water quality monitoring tools  287–301
 fluorimetric analysis  275–86
 passive sampling  73–5, 253–7
 proficiency testing  339–40
 rapid mapping  289–93
 reference materials  339–40
 screen-printed electrodes  270–1
 uncertainty  310–16
 Water Framework Directive  42–3
firefly luciferase (*lucFF* gene)  185–7
fish BEWS  204, 209–11
Fisher's least significant difference (LSD) 314–15
fixed costs  387, 388
fixed fluorescent (FF) technique  233–4
flame emission spectroscopy (FES) 343–4
flow injection analysis (FIA)  82–3
fluorimetric analysis  275–86
 analytical procedure  280–1
 front-face measurements  276, 280, 284
 materials and methodology  277–80
 performance/limitations  275–7, 284–5
 phase preparation and operation 278–80
 polyaromatic hydrocarbons  275–86
 results and discussion  281–4
 spiking solutions  277–8
 tank studies  277
fluoro immunoassay (FIA)  158
front-face fluorimetric analysis  276, 280, 284
FWPCA *see* Clean Water Act

gas chromatography (GC)  171, 189
gas chromatography–mass spectrometry (GC-MS)  5, 234, 249, 296, 346
genetically engineered microorganisms (GEMs)  148, 185
genetically modified organisms (GMO) 140
genistein  374–7
genotoxicity bioassays  127, 131
genotoxins  224

# Index

GFP *see* green fluorescent protein
global uncertainty   309
GLP *see* good laboratory practice
GMO *see* genetically modified organisms
good laboratory practice (GLP)   167
green fluorescent protein (GFP)   185, 187–8, 373
groundwaters
   biological early warning systems   207
   rapid mapping   289–93
   Water Framework Directive   4, 5
Grubbs test   306, 311
GUM standard   305, 307

HACH Kit Colo   360–2
heat shock protein (HSP)   47
heavy metals
   bioassays   297–8
   bioavailability   55–6, 243–5, 258
   biomarkers   224
   field studies   253–7
   monitoring   244–6, 257–9
   passive sampling   243–61, 293–5
   performance criteria   263–73
   screen-printed electrodes   263–73
   screening methods   243–61
   screening methods and emerging tools   355–7
   spatial variability   268–70
   surface waters   249–57
   tank study   250–3
   uncertainty   305–16
hER *see* human estrogen receptor
herbicides
   biological early warning systems   213
   biomolecular recognition systems   178, 182
   *see also* individual herbicides
hierarchy of complexity of living systems   126
high performance liquid chromatography (HPLC)   171, 175, 189, 232–4
Homeland Protection Act   206
homeland security issues   25–6
horizontal surveys   318
hormones   165, 295
   *see also* estrogenicity; estrogens
HPLC *see* high performance liquid chromatography
HS *see* humic substances

HSP *see* heat shock protein
human estrogen receptor (hER)   373, 376
humic substances (HS)   279–80
hydromorphological monitoring   41

ICCVAM   379
ICP-AES   343–4
ICP-MS *see* inductively coupled plasma mass spectrometry
immunoassays   125, 127, 157–9, 161–2
immunochemical methods   157–73
   antibody development   159–61
   antibody-analyte equilibria   157–9
   automation   166–7
   formats of competitive immunoassays   161–2
   future developments   171
   immunoassays   157–9, 161–2
   multi-analyte systems   166–7
   performance/limitations   169–71
   quality control   162–4
   test-kits   164–6
immunosensors   145–7
immunotoxins   224
imposex   224, 225–32
inductively coupled plasma atomic absorption spectroscopy (ICP-AES)   343–4
inductively coupled plasma mass spectrometry (ICP-MS)
   passive sampling   244–5, 249
   reference materials   346
   screen-printed electrodes   268–72
   uncertainty   305, 308–11
industrial chemicals   165
industrial effluents   120–3, 207
insecticides   182
intake water   205–6
intercalating pollutants   147, 178
interlaboratory exercises   374–7
internal calibrations   9
International Standardization Organization (ISO)
   approved methods   22
   biological early warning systems   209
   chemical methods   80–3
   metrology   7–8, 12
   reference materials   337
   uncertainty   306–8, 329

intersex 224, 225–32
invertebrate BEWS 202–3, 213–15
investment costs 387, 388
ISO *see* International Standardization Organization
isoproturon 167–70, 357, 363–5

kinetic samplers 247

lab on a chip 148
laboratory accreditation 12
laboratory reference materials 10
LC-MS *see* liquid chromatography–mass spectrometry
LC-MS/MS *see* liquid chromatography-tandem mass spectrometry
$LC_{50}$ tests 129
LDPE *see* low-density polyethylene
least significant difference (LSD) 314–15
limits of detection (LOD) 246
   ecotoxicity tests 359–60, 368
   fluorimetric analysis 285
   yeast-based assays 378
limits of quantification (LOQ)
   passive sampling 245–6, 258
   proficiency testing 360
   screen-printed electrodes 265, 267
linearity tests 265–6
liquid chromatography–mass spectrometry (LC-MS) 5
liquid chromatography-tandem mass spectrometry (LC-MS/MS) 169, 175
LOD *see* limits of detection
LOQ *see* limits of quantification
low-density polyethylene (LDPE) 58–9, 73–4
LRMs *see* quality control materials
LSD *see* least significant difference
*lucFF* gene 185–7
luminescence 130–1, 135
*luxCDABE* gene 185, 187

MAC *see* maximum allowable concentration
magnetic devices 140
major components
   environmental variability 324, 327

screening methods and emerging tools 353–4
MARA *see* microbial assay for risk assessment
Marine Protection, Research and Sanctuaries Act 15
mass sensitive sensors 139–40, 146
maximum allowable concentration (MAC) 54–5, 244, 312–13
MCF-7 cell-based proliferation assays 134
membrane-enclosed sorptive coating (MESCO) 57, 59, 64
mercury 257–9
MESCO *see* membrane-enclosed sorptive coating
Metal Detector 297–8
metal oxide semiconductor (MOS) sensors 139
metals *see* heavy metals; major components; trace elements
methoxychlor 376–7
metrology 4, 6–13
   certified reference materials 10–11
   laboratory accreditation 12
   proficiency testing schemes 11–12
   reference methods 9–12
   standardization 7–8
   Système International units 6–7, 9, 11
   traceability 6–7, 9, 11, 13
MFB *see* Multispecies Freshwater Biomonitor®
microbial assay for risk assessment (MARA) 109–24
   bacterial tests 110
   ecotoxicity testing 109–10, 116–17
   inter-laboratory trial 119–23
   intra-laboratory trial 111–19
   repeatability 111–13
   reproducibility 113–16
   sensitivity 116–19
microbial biosensors 148
microbial toxic concentration (MTC) 111–14, 116–19, 121–3
Microtox® 115, 118–19, 131, 297–8, 358–9, 366
molecularly imprinted polymers (MIPs) 148–50, 152

*Index*

monitoring strategies
  chemical methods  81, 83, 85–6, 89, 105, 198
  emerging tools  42, 288, 385–95
  screening methods  18–22, 26–31, 287–301, 385–95, 397–404
  *see also* bioassays; biological early warning systems; passive sampling
MOS *see* metal oxide semiconductor
MTC *see* microbial toxic concentration
multi-analyte systems  166–7
multifactor ANOVA  309
multiparameter probes  389–90
Multispecies Freshwater Biomonitor® (MFB)  203, 210
Musselmonitor®  203–4, 206–7, 211–14

N-butyl benzene sulfonamide (N-BBSA)  211–12
National Environmental Methods Index (NEMI)  28
National Stream Accounting Network (NASQAN)  18
National Water Quality Assessment Program (NAWQA)  18, 19
National Water Quality Monitoring Council (NWQMC)  27
NEMI *see* National Environmental Methods Index
nephelometry  82, 88
new method approval  23–5
nitrates  289–93, 357–8, 360–2
nitrification inhibition test  114
nitro aromatics  157
nonpolar organic compounds  63–6, 71–4
nonspecific promoters  183
non-yeast-based assays  374
normal distribution  306, 314–16, 329
NWQMC *see* National Water Quality Monitoring Council

OECD *see* Organization for Economic Cooperation and Development
oestrogen *see* estrogen
on-site methods, rapid mapping  289–93
open pore (OP) samplers  247, 252, 254–7

optical brighteners  33
optical transducers  135, 138
Organization for Economic Cooperation and Development (OECD)  198–9, 208, 378
organizational aspects  403
organometalic compounds  63–6
organophosphate pesticides  358
organotins  224, 225–32, 236, 249, 257–9
orthophosphate  85, 87
OSPAR guidelines  226–30
outliers  306, 360–1, 364

PAHs *see* polyaromatic hydrocarbons
PALL Versaflow™ filters  339
PalmSens  265–72
passive sampling  53–69, 71–7
  applications  48–9, 60–6
  bioavailability  243–5, 258
  calibration methods  58–60, 62
  costs  392–4
  devices  57–60
  emerging pollutants  295–7
  field studies  253–7
  heavy metals  243–61, 293–5
  instrumentation  247–9
  monitoring  244–6, 257–9
  nonpolar organic compounds  63–6, 71–4
  organometalic compounds  63–6
  performance/limitations  259–60, 294–5
  polar organic compounds  58–9, 60–3, 71–3, 74–6
  screening methods  243–61, 293–7, 299
  study design  73
  surface waters  249–57
  tank study  250–3
  techniques  56–60
  *see also* diffusion gradient in thin films; polar organic chemical integrative sampler; semi-permeable membrane devices
Pastel UV  93–100, 104, 289–93, 360–1, 368, 391–2
PCBs *see* polychlorinated biphenyls
PCDD/PCDF  180–1
PCNs *see* polychlorinated naphthalenes

PCR *see* polymerase chain reaction
performance reference compounds
  (PRCs)  60, 63–4, 74
pesticides
  biomarkers  224
  biosensors  136–7, 142–3, 145, 152
  immunochemical methods  157,
    165–6
  passive sampling  60–1
  reference materials  337–8, 345–9
  screening methods and emerging tools
    353–4, 357–9, 362–6
  Water Framework Directive  43, 45,
    47
  yeast-based assays  372
pH  81–2, 87
pharmaceuticals  60–1
  biomolecular recognition systems  182
  passive sampling  75, 295–7
phenols
  bioassays  131
  biosensors  142–3
  reference materials (RM)  340
phthalates  179, 182
physico-chemical methods
  chemical methods  82–3, 88, 98–100,
    102–4
  continuous monitoring of waters  198
  environmental variability  318–20
  passive sampling  58–9
  uncertainty  308
  Water Framework Directive  42, 43
phytotoxicity testing  115–16
phytotoxins  178, 182
piezoelectric sensors  139–40, 146
polar organic chemical integrative
  sampler (POCIS)  71–3
  applications  60–3, 65
  calibration  58–9
  screening methods  296–7
polar organic compounds  58–63, 71–3,
  74–6
pollution events
  passive sampling  250–2
  screen-printed electrodes  270–1
  UV spectrophotometry  104
polyaromatic hydrocarbons (PAHs)
  bioassays  131
  biomarkers  224, 232–6

  biomolecular recognition systems
    179–82
  biosensors  143, 146, 149
  fluorimetric analysis  275–86
  immunochemical methods  157, 159
  passive sampling  63–4, 72, 74
  reference materials  337–8, 344–9
  screening methods and emerging tools
    353–4, 358–9, 366
  Water Framework Directive  43, 45,
    47
polychlorinated biphenyls (PCBs)
  bioassays  131
  biomolecular recognition systems
    179–82, 184
  biosensors  147
  immunochemical methods  159
  passive sampling  64, 72
  Water Framework Directive  42–3,
    45, 47
polychlorinated naphthalenes (PCNs)
  180
polymerase chain reaction (PCR)  49
portable systems *see* field studies
potassium dichromate  111–13,
  120–3
potentiometric sensors  138–9, 145
PRCs *see* performance reference
  compounds
pregnane X receptor (PXR)  182–3
priority substances  43–5
process uncertainty  308
proficiency testing  329
  calibrants  342–4
  data evaluation  359–67
  design of QC tools  336–9
  ecotoxicity tests  358–9
  field studies  339–40
  immunochemical methods  167, 169
  inorganic analysis  341–4
  metrology  11–12
  organic analysis  344–6
  reference materials  11–12, 335–50
  screening methods and emerging tools
    351–70
  spiking solutions  342–4, 345–6
  stability testing  346–9
  standard solutions  345–6
  variance homogeneity  346–9
promoter interaction  183–4

# Index

PXR *see* pregnane X receptor
pyrene 277, 281–4

QA *see* quality assurance
QAPP *see* Quality Assurance Project Plan
QC *see* quality control
QCM *see* quartz crystal microbalances
quality assessment 31
Quality Assurance Project Plan (QAPP) 31
quality assurance (QA)
  biological early warning systems 208–10
  passive sampling 53, 56–8, 60, 66
  proficiency testing 336
  screening methods 300
  screening methods and emerging tools 352, 367–9
  uncertainty 329
  Volunteers Monitoring Programs 30–1
quality control materials 10
quality control (QC)
  alternative test procedures 23–5
  immunochemical methods 162–4
  passive sampling 53, 56–8, 60, 66
  proficiency testing 335–50
  reference materials 335–50
  screening methods 300
  screening methods and emerging tools 352, 367–9
  uncertainty 329
  Volunteers Monitoring Programs 30–1
quartz crystal microbalances (QCM) 146, 191

radio immunoassay (RIA) 158
rapid mapping 289–93
reality tests 400
reference materials (RM) 305–6, 310–11
  calibrants 342–4
  design of QC tools 336–9
  field studies 339–40
  inorganic analysis 341–4
  metrology 9–12
  organic analysis 344–6
  proficiency testing 11–12, 335–50
  screening methods and emerging tools 352–65

  spiking solutions 342–4, 345–6
  stability testing 346–9
  standard solutions 345–6
  variance homogeneity 346–9
  *see also* certified reference materials
regulatory proteins 178–9
relative estrogenic potency 373–5
relative standard deviation (RSD) 266–7
remote sensing 49–50
repeatability
  microbial assay for risk assessment 111–13
  screen-printed electrodes 265–7
  screening methods and emerging tools 368
replicate samples 331
reporter gene assays 373
reporter systems 185–8
representative samples 330
reproducibility
  fluorimetric analysis 285
  microbial assay for risk assessment 113–16
  screen-printed electrodes 265–7
  screening methods and emerging tools 368
resistance to change 398
respiration inhibition (RI) test 115
restricted pore (RP) samplers 247, 252, 254–7
retinoid receptor (RR) 182, 184
RI *see* respiration inhibition
RIA *see* radio immunoassay
RIANA immunosensors 146–7
ribotoxins 178
risk assessment *see* environmental risk assessment; microbial assay for risk assessment
River Basement Management Plans 4
RM *see* reference materials
RP *see* restricted pore
RR *see* retinoid receptor
RSD *see* relative standard deviation

Safe Drinking Water Act 15
sample preparation 308, 309, 318, 330–1
sampling uncertainty *see* uncertainty
SAMs *see* self-assembled monolayers

sandwich format   162
scandium   315–17
scatter plots   320–1, 326
screen-printed electrodes (SPEs)
    263–73
  accuracy/bias   265, 267
  applications   268–71
  characteristics   264–8
  field studies   270–1
  heavy metals   263–73
  linearity tests   265–6
  performance criteria   263–73
  pollution events   270–1
  repeatability   265–7
  reproducibility   265–7
  sensitivity   266
  spatial variability   268–70
  specificity   265–6
  surface waters   268–70
  temporal variability   270–1
screening methods
  acceptance by actors   397–404
  assessment   385–95
  bioassays   125–6, 297–8
  bioavailability   243–5, 258
  costs and decision making   385–95
  emerging pollutants   295–7
  emerging water quality monitoring
    tools   287–301
  environmental variability   303–4,
    316–29
  estrogenicity   377
  fluorimetric analysis   275–86
  heavy metals   243–61, 293–5,
    297–8, 305–16
  nitrates   289–93
  passive sampling   243–61, 293–7,
    299
  performance/limitations   259–60,
    263–73, 287–301
  polyaromatic hydrocarbons   275–86
  rapid mapping   289–93
  screen-printed electrodes   263–73
  trace elements   305–16, 320–7
  uncertainty   303–16, 327–9
  US water regulation   15–37
  water monitoring strategies   385–95,
    397–404
  yeast-based assays   377
screening methods and emerging tools
    (SMETs)   351–70, 385–95
  analytical methods   355–8
  applications   355–9
  case studies   391–4
  costs and decision making   386–95
  data evaluation   359–67
  ecotoxicity tests   358–9
  instrumentation   389–94
  participation in SWIFT-WFD PT
    schemes   353–5
  performance/limitations   356, 367–9
  principles   352–3
  qualitative analysis   358–9, 366–7
  water monitoring strategies   385–95
  *see also* passive sampling
selectivity   368
self-assembled monolayers (SAMs)   146
semi-permeable membrane devices
    (SPMDs)
  calibration   58–9
  heavy metals   258
  nonpolar organic compounds   63–5
  organometalics   63–5
  principles   49, 57–8, 71–4
sensitivity
  fluorimetric analysis   285
  microbial assay for risk assessment
    116–19
  screen-printed electrodes   266
sewage treatment plants (STP)   292–3
SFS *see* synchronous fluorescent
    spectroscopy
SI *see* Système International
*Silent Spring*   371
SiNWs *see* boron-doped silicon
    nanowires
SMETs *see* screening methods and
    emerging tools
soil leachate   120–3
solid-phase extraction disks   66
SOPs *see* standard operating procedures
spatial variability   318–25, 328
  chemical methods   96–100
  rapid mapping   290–3
  screen-printed electrodes   268–70
specific promoters   184
specificity   265–6, 368
SPEs *see* screen-printed electrodes
spiking solutions   342–4, 345–6

*Index* 417

SPMDs *see* semi-permeable membrane devices
spot sampling 197–8, 391, 393
SPR *see* surface plasmon resonance
square wave anodic stripping voltammetry (SWASV) 264–5, 355–7, 360
SRF *see* State revolving funds
stability testing 346–9
STAC UV analysers 93–5
standard operating procedures (SOPs) 31, 167–8
standard solutions 345–6
State revolving funds (SRF) 17
Stern–Volmer plots 279
STP *see* sewage treatment plants
surface acoustic wave sensors 140
surface plasmon resonance (SPR) 49
   biomolecular recognition systems 188–9
   biosensors 138, 146, 151–2
surface waters
   bioassays 297–8
   biological early warning systems 206
   environmental variability 303–4, 316–27
   fluorimetric analysis 275–86
   heavy metals 249–57
   passive sampling 249–57, 293
   proficiency testing 335–50
   rapid mapping 290
   reference materials 335–50
   screen-printed electrodes 268–70
   uncertainty 303–16
   Water Framework Directive 4, 5
surfactants
   immunochemical methods 165
   rapid mapping 290, 292
susceptibility biomarkers 48, 221–3
SWASV *see* square wave anodic stripping voltammetry
SWIFT-WFD
   chemical methods 80, 83, 100, 104
   emerging water quality monitoring tools 42, 288
   environmental variability 304
   fluorimetric analysis 277
   immunochemical methods 167
   passive sampling 56, 250, 259
   proficiency testing 336–9, 349, 351–70

screen-printed electrodes 266
screening methods and emerging tools 351–70, 389–90, 394
uncertainty 304
synchronous fluorescent spectroscopy (SFS) 233
Système International (SI) units 6–7, 9, 11

TA *see* transcriptional activation
tamoxifen 375–7
tank studies 250–3, 277
TBT *see* tributyltin
temporal variability 304, 318–28
   chemical methods 96–100
   passive sampling 394
   screen-printed electrodes 270–1
   *see also* pollution events
test-kits 164–6, 272
tetrazolium red (TZR) 110
Thamnotoxkit 358
thermometric devices 140
thyroid receptor (TR) 182, 184
TIE *see* toxicity identification evaluation
time-weighted average (TWA) *see* passive sampling
TIRF *see* total internal reflection fluorescence
TMDLs *see* total maximum daily loads
TN *see* total nitrogen
TOC *see* total organic content
total internal reflection fluorescence (TIRF) 138, 146
total maximum daily loads (TMDLs) 17
total nitrogen (TN) 81–2, 85–6, 96–100
total organic content (TOC)
   chemical methods 81, 85–6, 100–1
   rapid mapping 290, 292
total phosphorus (TP) 81–2, 85, 87
total suspended solids (TSS)
   chemical methods 32, 82–3, 97–8
   rapid mapping 290, 292
toxicity bioassays 128–31, 133
toxicity fingerprints 121
toxicity identification evaluation (TIE) 72
ToxScreen 297–8
TP *see* total phosphorus
TR *see* thyroid receptor

trace elements
  biomarkers  224
  reference materials  337–8, 347, 349
  screening methods  305–16, 320–7
  SMETs  353–4, 360
traceability  6–7, 9, 11, 13
transcriptional activation (TA) assays  379
transparency  32
tributyltin (TBT)  225–32, 236, 258
trophic state index (TSI)  33–4, 100–5
trueness tests  306, 330
TSI *see* trophic state index
TSS *see* total suspended solids
tuning chlorination of cooling water  208
turbidity  32, 88
turbulence  58
TWA (time-weighted average) *see* passive sampling
TZR *see* tetrazolium red

UK Technical Advisory Group (UKTAG)  226
ultrafiltration  245
ultraviolet (UV) spectrophotometry  33–4, 91–106
  accidental pollution events  104
  analysers  93–6
  applications  96–105
  chemical methods  89, 91–106
  costs  389–92, 394
  deconvolution methods  92–3
  early warning systems  104
  monitoring and prevention  105
  online analysers  95
  portable analysers  94, 96–100, 104
  software  95–6
  spatial/temporal variability  96–100
  trophic state assessment of lakes  100–5
uncertainty  303–16, 327–9
  analytical  305–6, 310–11
  assessment procedure  305–9
  estimation  307–8
  field studies  310–16
  sampling  306–8, 311–16, 330
  test design  308–9
United States Geological Survey (USGS)  18, 19
usability  400–2

USEPA *see* Environmental Protection Agency
USGS *see* United States Geological Survey
utility of innovative object  398–400
UV *see* ultraviolet

variable costs  388–9
variance homogeneity  306, 313, 325, 330, 346–9
Vas Deferens Sequence Index (VDSI)  225–7
vertical surveys  318–24
visible spectrophotometry  89
Visual MINTEQ  250, 254–6, 294
volatile organic compounds (VOCs)  83
voltammetric sensors  138
Volunteers Monitoring Programs  26–31

wastewater effluents  207, 295
Water Framework Directive (WFD)  3–13, 39–50
  acceptance by actors  404
  binding rules  6
  biological early warning systems  197–8, 199, 208
  biological methods  40–1, 45–8
  biomarkers  226–7, 231
  biosensors  127–8
  certified reference materials  10–11
  chemical methods  41, 43–5, 54–6, 72, 79–80, 91–2
  emerging techniques  42–8
  emerging water quality monitoring tools  287–8, 299–300
  field studies  42–3
  heavy metals  244, 263–73
  hydromorphological monitoring  41
  immunochemical methods  159, 167–9
  laboratory accreditation  12
  legal requirements  4
  metrology  4, 6–13
  new trends  48–50
  non-legally binding recommendations  5
  passive sampling  48–9, 54–6, 72, 244
  performance criteria  263–73
  physico-chemical methods  42, 43

pollutant effects   45–8
principles   3–4
priority substances   43–5
proficiency testing   335–9, 351–70
proficiency testing schemes   11–12
reference methods   9–12
screen-printed electrodes   263–73
screening methods and emerging tools   351–70, 385–7
standardization   7–8
Système International units   6–7, 9, 11
traceability   6–7, 9, 11, 13
uncertainty   304, 327
*see also* SWIFT-WFD
water monitoring *see* monitoring strategies
water quality indicators   21
Water Quality Standards (WQS)   16–17, 20
WaterSentinel (WS) system   26

WFD *see* Water Framework Directive
World Health Organization (WHO)   34
WQS *see* Water Quality Standards
WS *see* WaterSentinel

xenoestrogens   179

yeast reporter gene assays   134
yeast-based assays   371–81
　chemical methods   377–8
　classification   372–3
　future developments   378–9
　interlaboratory exercises   374–7
　performance/limitations   378
　relative estrogenic potency   373–5
　screening methods   377
yEGFP assay   373
YES assays   134, 373, 376

z-scores   120–3
zinc   315–17, 322–3, 324–7